SEDIMENTOLOGY AND GEOCHEMISTRY OF MODERN AND ANCIENT SALINE LAKES

Based on a Symposium
Sponsored by SEPM (Society for Sedimentary Geology)

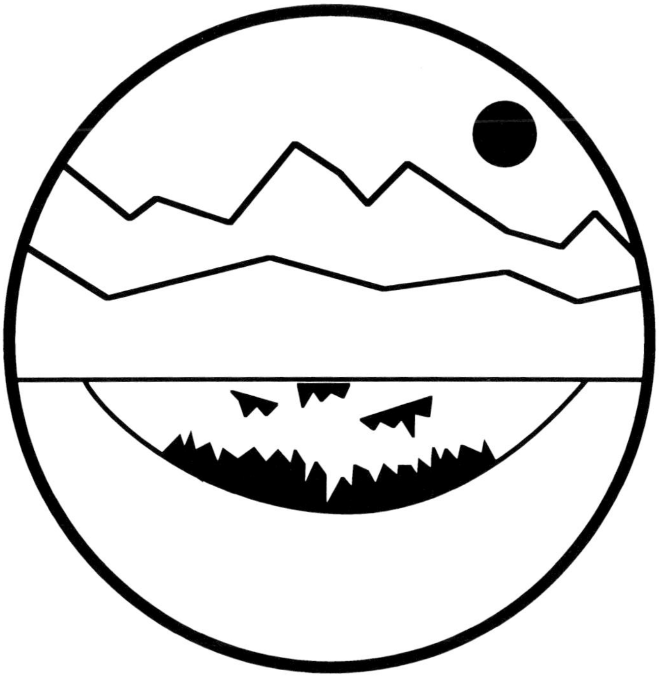

Edited by

Robin W. Renaut, University of Saskatchewan, Saskatoon, Canada
and
William M. Last, University of Manitoba, Winnipeg, Canada

Peter A. Scholle, Editor of Special Publications
Special Publication No. 50

Tulsa, Oklahoma, U.S.A. *September, 1994*

A Publication of

SEPM (Society for Sedimentary Geology)

ISBN 1-56576-014-X

Printed in the United States of America

CONTENTS

IV. ANCIENT SALT LAKE/MARINE SYSTEMS

SALT LAKE SEDIMENTOLOGY AND GEOCHEMISTRY: INTRODUCTION AND HISTORICAL PERSPECTIVE

ROBIN W. RENAUT
Department of Geological Sciences, University of Saskatchewan, Saskatoon, Canada S7N 0W0
AND
WILLIAM M. LAST
Department of Geological Sciences, University of Manitoba, Winnipeg, Canada R3T 2N2

INTRODUCTION

This volume of papers grew out of a four-day symposium entitled "Sedimentary and Paleolimnological Records of Saline Lakes" held at Saskatoon, Canada in August, 1991. The meeting was held as a Canadian contribution to Projects 219 and 324 of the International Geological Correlation Programme. This international conference, co-sponsored by the Canadian National Committee of the International Geological Correlation Programme (CNC-IGCP) and the National Hydrology Research Institute (Environment Canada), attracted nearly 100 participants from a dozen countries and covered literally all aspects of salt lake sedimentology, geochemistry, and paleolimnology.

As with most conferences and meetings dealing with lacustrine environments and paleolimnology, the Saskatoon conference was broad in scope. Participants had a remarkable diversity of backgrounds and scientific specialities, including sedimentologists, geochemists, hydrologists, geographers, paleontologists, ecologists, and biologists from universities, governments, industry, and private organizations. Papers presented at this conference that dealt primarily with biological or biochemical paleolimnology have been collected and published separately as a special theme issue of *Journal of Paleolimnology* (Evans, 1993). The aim of *SEPM Special Publication Number 50* is to bring together selected papers from this conference that deal specifically with the sedimentological, inorganic geochemical, and hydrological aspects of salt lakes and their stratigraphic records. We have not tried to compile a comprehensive review of saline lake sedimentology, but rather reflect the scope of current research. An excellent review of much of the saline lake sedimentological and geochemical literature has already recently been presented by Smoot and Lowenstein (1991).

HISTORICAL PERSPECTIVE

The timing of the Saskatoon salt lake symposium and resulting publications was ideal. For many years, saline lakes have ranked among the least understood and most poorly studied environments in the entire spectrum of sedimentary geology. Despite the fact that some of the earliest scientific research done on lakes was centered on *saline lakes* and that closed basin continental environments have long been recognized as very important sources of paleoenvironmental information, geolimnological investigations of saline lakes in most of the world have lagged behind those of freshwater settings. Indeed, salt lakes have received little more than passing reference in such preeminent monographs and edited volumes as *Chemical Processes in Lakes* (Stumm, 1985), *Lake Sediments and Environmental History* (Haworth and Lund, 1984), *Principles of Lake Sedimentology* (Håkanson and Jansson, 1983), and *Limnology* (Wetzel, 1983). This past lack of attention is somewhat surprising considering the widespread occurrence of modern saline lakes and the fact that the volume of inland salt water is approximately equal to that of freshwater lakes and rivers (Williams, 1986; Hammer, 1986).

This neglect is currently changing. Approximately 1000 papers, reports, and postgraduate theses on the general topic of saline lake geoscience have been done during the past decade. As shown in Figure 1, the early part of the 1980's was a period of rapid increase in saline lake publications. Since about 1985, however, the rate of production of geoscience publications dealing with salt lake topics has remained relatively constant at ~70 per year. Although maintaining a steady output per annum, Figure 1 also shows that there has been a significant increase in the *proportion* of publications dealing with saline lakes versus freshwater lacustrine sediments and environments, such that over the past several years about 30 percent of all lacustrine geoscience publications have dealt with salt lakes.

Most of the salt lake papers from the past decade have been published as journal articles, with a much smaller proportion appearing as chapters or articles in monographs, published government reports, or unpublished masters and doctorate degree theses (Fig. 2). An encouraging trend is the recent increase in salt lake research by graduate students, implying not only that young geoscientists have taken an increasing interest in salt lakes, but also suggesting that the number of journal publications dealing with salt lakes should increase at least over the next several years. In total for the past decade, slightly over 90 percent of the journal articles have been published in just nine journals, with *Palaeogeography, Palaeoclimatology, Palaeoecology, Chemical Geology,* and *American Association of Petroleum Geologists Bulletin* comprising the source for well over half the saline lake references (Fig. 3). Saline lake research is flourishing. Further evidence of the burgeoning interest came in 1992 with the first issue of the new interdisciplinary *International Journal of Salt Lake Research,* edited by W. D. Williams (Adelaide).

It should be no surprise that most of the geoscience material published on salt lakes during the past 10 years has dealt with Cenozoic sediments (Fig. 4), and most has been on lakes located in United States (Fig. 5). Researchers have

Sedimentology and Geochemistry of Modern and Ancient Saline Lakes, SEPM Special Publication No. 50

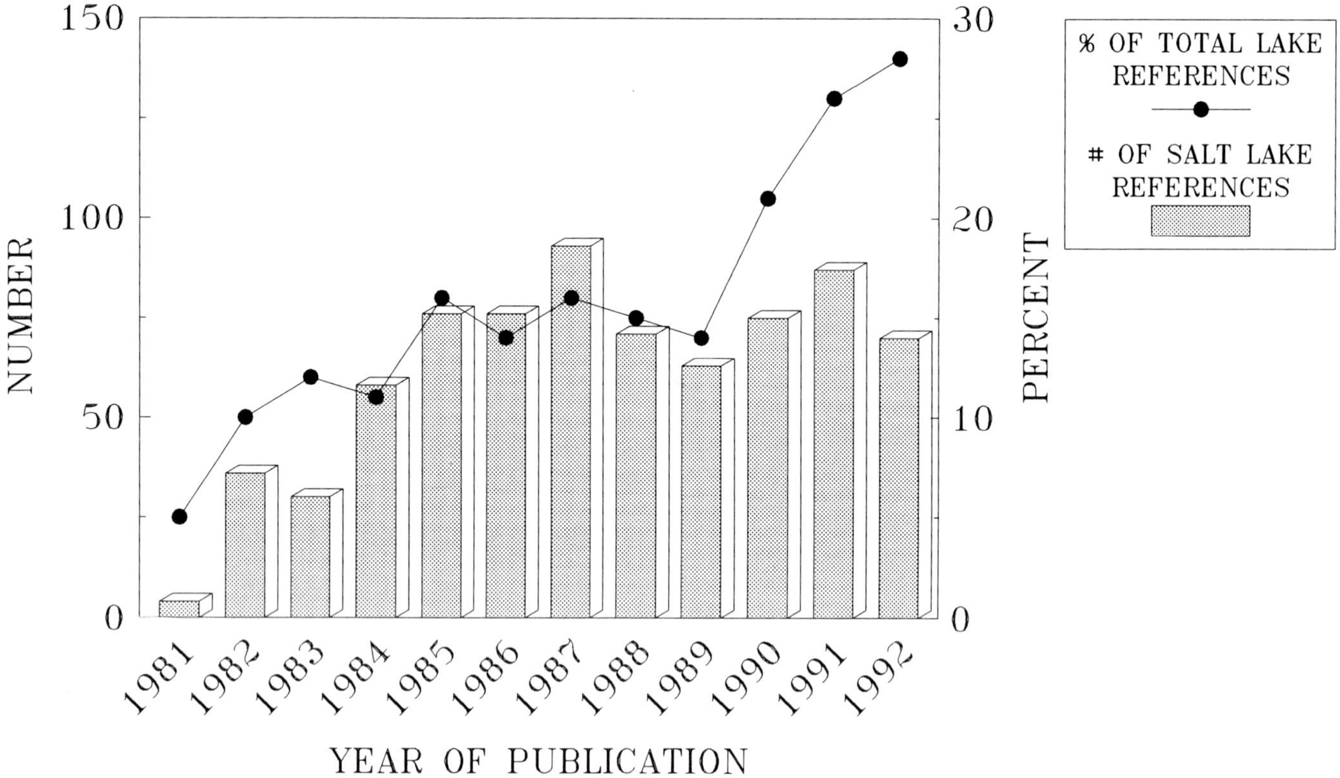

FIG. 1.—Bar-line diagram showing the number of salt lake geoscience references per year, 1981–1992 (bars), and the percentage of salt lake references for each year relative to the total number of geoscience references dealing with all types of lakes (line). Data acquired using AGI GeoRef Index and ISI Science Citation Index. Included in this data base are non-English references, government reports, and unpublished graduate theses, in addition to published journal articles and monograph papers/chapters.

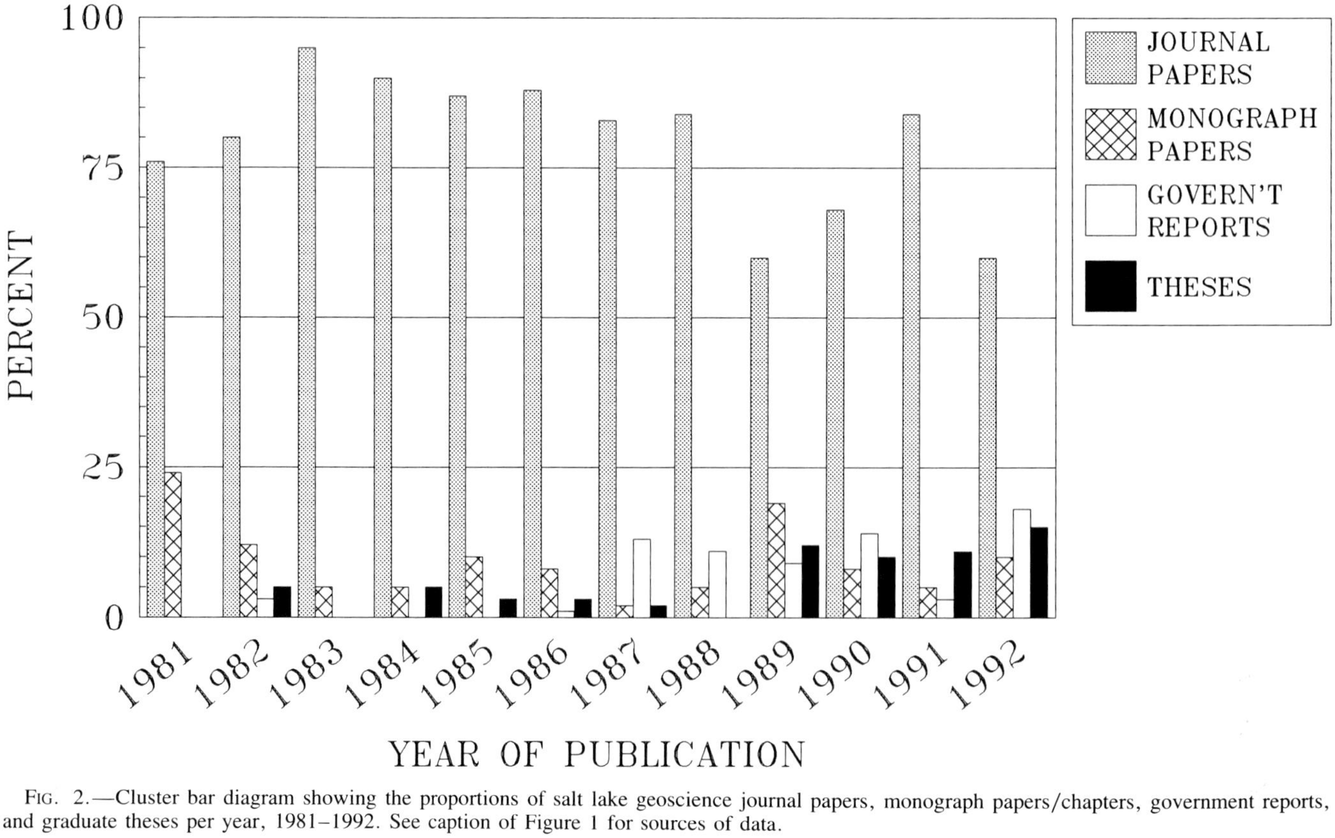

FIG. 2.—Cluster bar diagram showing the proportions of salt lake geoscience journal papers, monograph papers/chapters, government reports, and graduate theses per year, 1981–1992. See caption of Figure 1 for sources of data.

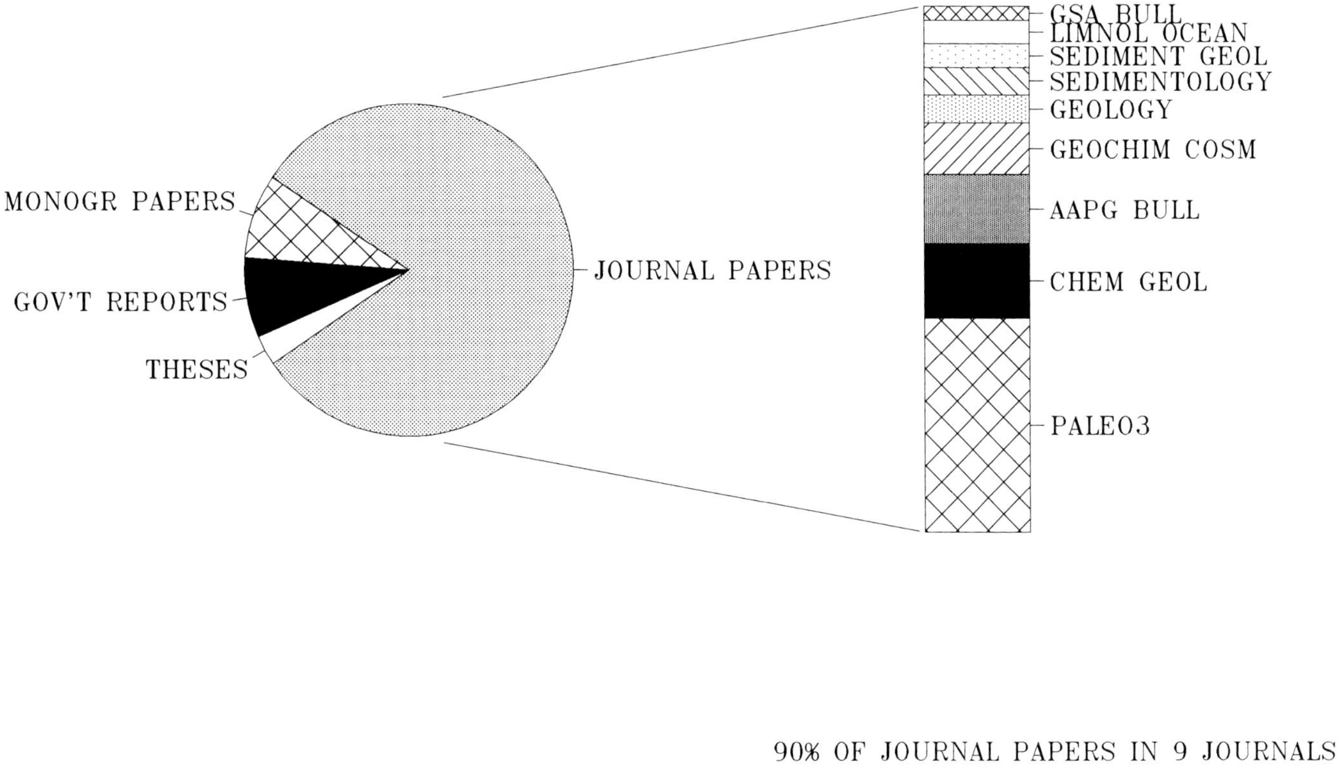

FIG. 3.—Pie-bar diagram showing the overall proportion of the various types of salt lake geoscience references, 1981–1992, and the source journals for approximately 90% of the journal references. See caption of Figure 1 for sources of data.

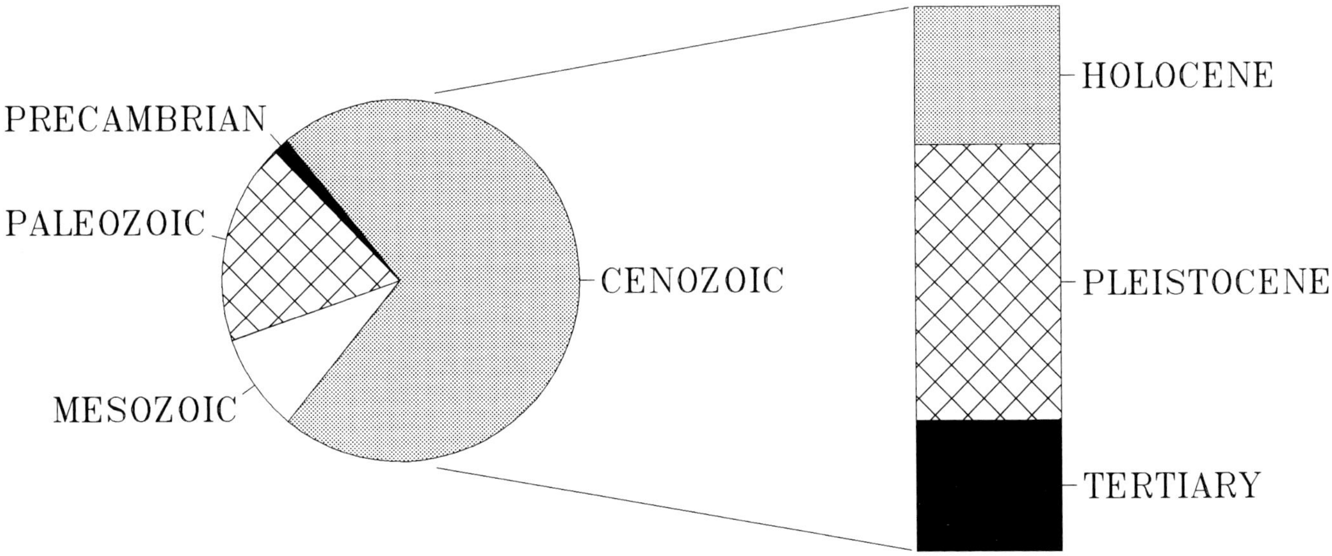

FIG. 4.—Pie-bar diagram showing the overall proportion of salt lake geoscience references with respect to age of the deposit, 1981–1992. See caption of Figure 1 for sources of data.

placed considerable emphasis on the study of modern process sedimentology and geochemistry of saline lakes (Fig. 6); hopefully, this advance in our understanding of the physical and geochemical processes that control sedimentation and facies distribution in modern salt lakes will provide paleolimnologists with better tools to help decipher the stratigraphic records preserved in this important, but often neglected, environment. The emphasis on modern and Quaternary saline lake systems is also reflected in the contents of this book.

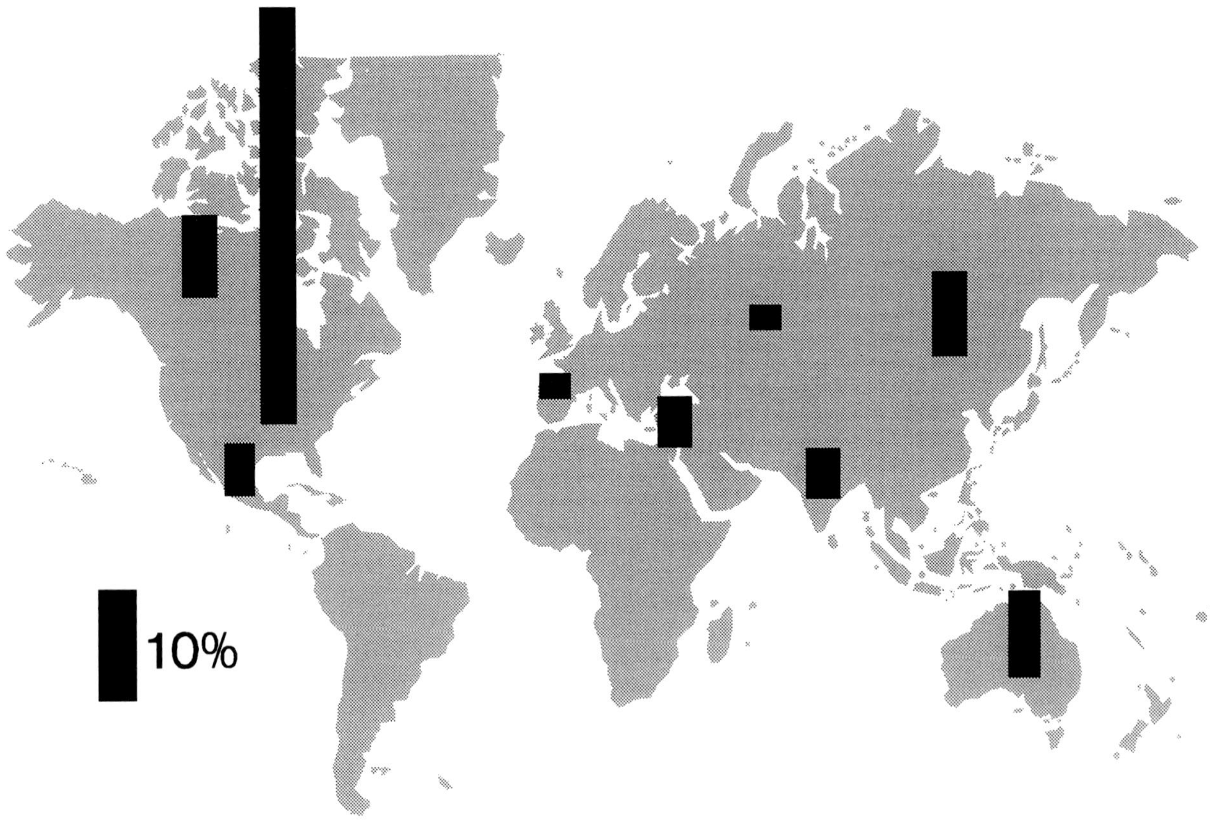

FIG. 5.—Sketch map showing the relative proportion of the total number of salt lake geoscience references with respect to geographic location of the deposit. Nine countries account for more than 90% of all references. See caption of Figure 1 for sources of data.

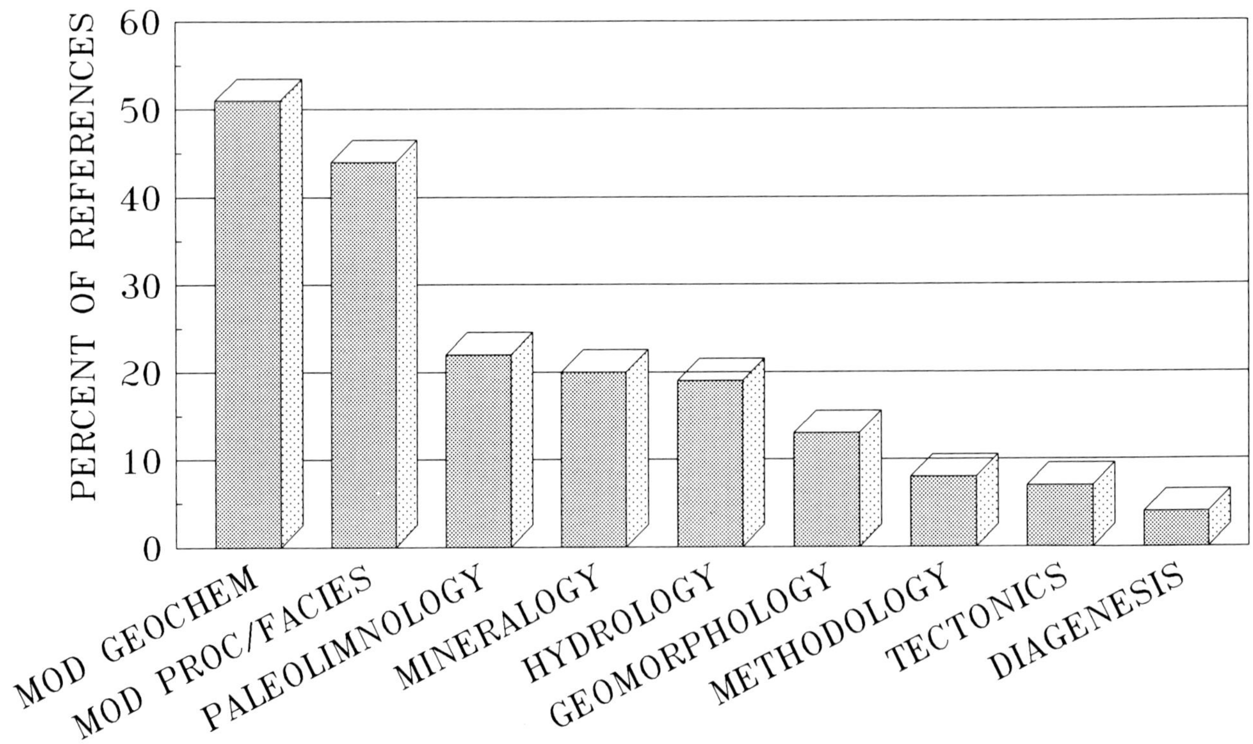

FIG. 6.—Bar diagram showing the major subject matter of salt lake geoscience references, 1981–1992. Subject matter identified according to key words and title words. See caption of Figure 1 for sources of data.

SEPM SPECIAL PUBLICATION 50

Following the format of the 1991 Saskatoon Symposium, this volume is divided into four sections. The first and largest section contains ten papers that deal with modern saline lakes. Brine geochemistry and geochemical processes that are important in controlling water composition in salt lakes are the focus of most of these papers. ***Bryant, Drake, Millington, and Sellwood*** document the dramatic compositional changes and short-term evolution that occurred in a large saline playa in Tunisia after a major flooding event. ***Lyons, Hines, Last, and Lent*** address the problem of organic carbon preservation in lakes of varying composition and different levels of sulfate reduction. Similarly, ***Komor*** provides new data on sulfate reduction, organic matter degradation, and the influence of these processes on brine geochemistry of the well-studied Devils Lake in North Dakota. In an unusually detailed investigation of solute flux using a mass balance approach, ***Donovan*** supplies the basis for quantitatively evaluating salt sources and budgets of the many saline lakes of the northern Great Plains of western North America. Also within the Great Plains region, ***Last*** documents unusually high sedimentation rates in deep-water hypersaline lakes. The La Mancha region of central Spain contains many salt lakes and saline playas. ***Ordóñez, Sánchez Moral, García del Cura, and Rodriguez Badiola*** review the major salt–mineral geochemistry of these lakes and make use of an artificial evaporating pond to establish a predictive model for mineral precipitation and brine geochemistry. In one of two invited contributions, ***Smoot and Castens-Seidell*** provide a particularly well illustrated synopsis of sedimentary structures produced by efflorescent salt crusts in the saline lakes of eastern California. One reviewer commented that this paper is one of the most important references on the generation and preservation of structures in playa lakes. In the other invited paper in this volume, also dealing with efflorescent crusts in playa lakes, ***Arakel and Hongjun*** combine field observations, experimental precipitation results, and thermodynamic considerations to decipher and interpret the modern evaporite facies in a series of groundwater discharge lakes in central Australia. ***Renaut and Tiercelin*** provide an overview of sedimentation in saline, alkaline Lake Bogoria, one of the most intriguing and beautiful East African Rift basin lakes. In the final paper of this section, ***Drake, Bryant, Millington, and Townshend*** show how satellite imagery can help in mapping and monitoring the sedimentology and hydrology of modern playa lakes in remote regions.

The second section of this volume contains five papers dealing with sedimentation and diagenesis of late Quaternary salt lakes. ***Mees and Richardson*** use bedding characteristics, mineralogy, and magnetic susceptibility to interpret periods of meromixis alternating with episodes of low water or dry conditions, and ultimately holomictic conditions in the Holocene record of a crater lake in Sudan. ***Newton*** provides the first account of sediments recovered from the floor of Mono Lake, a well-known saline, alkaline lake in California. He then uses the evidence from cores to reconstruct the lake's Holocene history. The mineralogy of the clay fraction of saline lake sediments, an often neglected topic, is used by ***Webster and Jones*** to help determine paleoenvironmental conditions in Quaternary playa sediments of the southern High Plains region, Texas. ***Coshell and Rosen*** examine the 6,000 year history of a hypersaline meromictic marginal-marine lake containing abundant benthic microbial mats in western Australia. In the final paper of this section, ***Salama*** reviews the long and complex history of lacustrine sedimentation in the Yilgarn River region of Western Australia, using a combination of sedimentological and geomorphological evidence.

The third and fourth sections of this volume both contain papers devoted to ancient (pre-Quaternary) lacustrine sequences. The Tertiary saline lake sediments of central and eastern Spain are among the most spectacular and best studied ancient lacustrine deposits in the world, and have been the subject of intensive petrographic and stratigraphic examination by a very active group of Spanish researchers. ***Salvany and Orti*** present results of a detailed petrographic examination of glauberite-rich lacustrine rocks of the Miocene age Ebro Basin. The Miocene lake sediments of the large Madrid Basin to the west are the subject of two other papers. ***Ordóñez and García del Cura*** discuss the petrology, diagenesis and depositional environments of a sequence of glauberitic deposits, while ***Sanz, Rodriguez-Aranda, Calvo, and Ordóñez*** provide new criteria for recognition of detrital lacustrine gypsum in the geological record. ***Buchheim*** uses the lithostratigraphy, mineralogy, stable isotope ratios, and paleontology to help describe the conditions of deposition and paleosalinity of the Eocene Green River Formation in southwestern Wyoming. ***Gierlowski-Kordesch and Rust*** provide a much needed critical examination of past ideas on the Jurassic East Berlin Formation in northeastern North America and a reinterpretation of this important graben-fill sequence. In the final paper of the third section, ***Valero Garces and Gisbert Aguilar*** provide a comprehensive examination and interpretation of the late Paleozoic lacustrine carbonates in the Aragón-Béarn Basin of northern Spain.

Differentiation between lacustrine sediments and marine deposits is often difficult, particularly in hypersaline settings. In the first paper of the final section, ***Williams-Stroud*** discusses the transition from marine to non-marine sedimentation in the Pennsylvanian Paradox Salt deposit. ***Chipley and Kyser*** provide important new stable isotopic and fluid inclusion data to show that some types of chevron halite can form in the subsurface, and cannot always be assumed to indicate sedimentation from shallow subaerial brines. In the final paper, ***Benalioulhaj, Schreiber, and Philp***, in a novel approach, show how organic geochemistry can be combined with more conventional methods to determine the marine to non-marine transition in the Lorca Basin, Spain.

The contents of a proceedings volume of this type are a reflection not only of the researchers who contributed their scientific findings but also of the reviewers who helped to evaluate the data and interpretations presented. We express our sincere gratitude to the following reviewers for their efforts: R. Anderson, J. Bada, A. Behrensmeyer, R. Burns, P. De Deckker, R. Forester, J. I. Drever, H. Geldsetzer, L. A. Hardie, R. L. Hay, P. Hare, R. Hecky, M. Hickman,

I. Hutcheon, N. P. James, B. F. Jones, K. Kelts, H. Kraus, T. K. Kyser, L. S. Land, T. K. Lowenstein, G. MacDonald, E. Mazzullo, K. McNamara, D. Naldrett, R. B. Owen, J. Parnell, B. Pratt, J. Risberg, E. Roedder, L. Sack, C. Schenk, L. Snowdon, S. Stein, M. Talbot, J. T. Teller, G. van der Kamp, J. Warren, M. A. J. Williams, S. Williams-Stroud, W. D. Williams, B. H. Wilkinson, W. Wood, J. A. T. Young, and R. Yuretich.

We are especially grateful to SEPM Special Publication Editors, Peter Scholle, Dana Ulmer-Scholle, and Barbara Lidz, for their exceptional tolerance and guidance in producing this volume.

REFERENCES

EVANS, M., ed., 1993, Special Issue: Salt Lake Paleolimnology: Journal of Paleolimnology, v. 8, 129 p.

HAWORTH, E. Y., AND LUND, J. W. G., eds., 1984, Lake Sediments and Environmental History: Minneapolis, University of Minnesota Press, 411 p.

HÅKANSON, L., AND JANNSON, M., 1983, Principles of Lake Sedimentology: New York, Springer-Verlag, 316 p.

HAMMER, U. T., 1986, Saline Lake Ecosystems of the World: Dordrecht, Dr. W. Junk Publishers, 616 p.

SMOOT, J. P., AND LOWENSTEIN, T. K., 1991, Depositional environments of non- marine evaporites, *in* Melvin, J. L., ed., Evaporites, Petroleum and Mineral Resources: Amsterdam, Elsevier, p. 189–347.

STUMM, W., ed., 1985, Chemical Processes in Lakes: New York, John Wiley & Sons, 265 p.

WETZEL, R. G., 1983, Limnology: Toronto, W. B. Saunders Company, 743 p.

WILLIAMS, W. D., 1986, Limnology, the study of inland waters: a comment on perceptions of studies of salt lakes, past and present, *in* De Deckker, P., and Williams, W. D., eds., Limnology in Australia: Melbourne, CSIRO Australia, p. 471–486.

PART I
MODERN SALINE LAKES

Sedimentology and Geochemistry of Modern and Ancient Saline Lakes, SEPM Special Publication No. 50
 ISBN 1-56576-014-X

THE CHEMICAL EVOLUTION OF THE BRINES OF CHOTT EL DJERID, SOUTHERN TUNISIA, AFTER AN EXCEPTIONAL RAINFALL EVENT IN JANUARY 1990

ROBERT G. BRYANT

Department of Environmental Science, University of Stirling, Stirling, FK9 4LA, Scotland

NICK A. DRAKE

Department of Geography, King's College, University of London, The Strand, London, WC2R 2LS, United Kingdom

ANDREW C. MILLINGTON

Department of Geography, The University of Reading, Whitenights, Reading RG6 1DL, Berkshire, United Kingdom

AND

BRUCE W. SELLWOOD

The Postgraduate Research Institute for Sedimentology, The University of Reading, Whitenights, Reading RG6 2AB, Berkshire, United Kingdom

ABSTRACT: In January 1990, an exceptional rainfall event in southern Tunisia caused the Chott el Djerid, an ephemeral salt playa, to fill with water. Under an arid climate, the ephemeral lake on Chott el Djerid evaporated to dryness in ten months. During March, May and September 1990 we sampled the lake brines. Chemical analysis of the major solutes showed that the dilute waters that flow into the Chott el Djerid basin (groundwater, wadis and aquifer waters) have a consistent chemistry, generally saturated with respect to gypsum. This may result from the uniform basin geology, which is made up of Cretaceous, Mio-Pliocene, and Quaternary sediments, dominated by marine evaporites. Potassium was conserved throughout the evaporation sequence, suggesting the saturation of sorption surfaces within the playa. With increasing evaporation, the precipitation of gypsum and halite are predicted and observed, the final, most-concentrated, brines being saturated with respect to sylvite. XRD analyses of salt crusts from the Chott el Djerid reveal a mineral assemblage of gypsum, halite and carnallitite (carnallite with halite). The overall nature of both the predicted and observed salt phases suggests that the main control on the geochemistry of the playa is the recycling of ancient marine evaporites.

INTRODUCTION

In many arid regions, playas form very significant hydrological, sedimentological and biological domains (Teller and others, 1982; Teller and Last, 1990; Last, 1989; Millington and others, 1987). There have been many studies of the geochemical evolution of closed-basin continental brines (Garrels and Mackenzie, 1967; Hardie and Eugster, 1970; Eugster and Jones, 1979; Eugster, 1980; Drever, 1982), but they have generally concentrated on a limited number of basins. More recently, there has been increasing interest in the sedimentological and biological processes that occur in playas. However, the remote and harsh nature of playa environments and their generally poor accessibility, make the sedimentological, chemical and geomorphological monitoring of playas a difficult task. As a result, the amount of geochemical and sedimentological data that exist for salt playas in most continents remains sparse. Indeed, Eugster (1980) stated that the number of closed basins for which sufficient water chemistry data were available to document solute behavior during evaporative concentration numbered no more than ten. Consequently, the geochemical models that have been developed for the behavior of major solutes in closed basins rely on a restricted data base.

The opportunity to extend the data base of extensively studied closed basins was provided in 1990 through research on Chott el Djerid, a large, accessible, continental salt playa in southern Tunisia. In January 1990, the Chott el Djerid basin received an exceptional amount of rainfall (Ouezdou and others, 1990). As a result, much of the playa filled with water, creating an ephemeral saline lake. We sampled water from the ephemeral lake (groundwater, spring water and water from feeder channels) in March, May and September 1990 while the lake evaporated and contracted in area. The purpose of this paper is to follow the chemical evolution of the Chott el Djerid brines as they experienced evaporative concentration between March and September 1990. We then examined the brine evolution in the context of current models.

REGIONAL SETTING

Geographical Background

Chott el Djerid is an ephemeral salt playa in southern Tunisia situated in the center of a closed, arid basin that has an areal extent of 10,500 km^2 (Gueddari, 1984). Chott el Djerid itself has a surface area of ~5,360 km^2 (Millington and others, 1989) and has an elongated northern arm that stretches eastward toward the coastal city of Gabes. This arm, the Chott el Fedjadj, has an area of 770 km^2. The Chott el Djerid basin is situated at a latitude of around 34°N (Fig. 1). The mean annual rainfall for this area is between 80 mm and 140 mm. The mean annual temperature is 21°C, and evaporation, which is at a maximum between the May and September, has a mean annual value of 1,500 mm. From the 21 to 23 January 1990, a weak anticyclonic depression was situated over southern Tunisia as a result of an influx of polar air from eastern Europe (Ouezdou and others, 1990). Precipitation over southern Tunisia in these three days varied from 8 to 50 times the mean monthly average and 0.5 to 4 times the annual mean. Ouezdou and others (1990) state that run-off within southern Tunisia over this period, which they calculated to be 2,200 $\times$ 10^6 m^2, was six times the annual mean.

Geology and Hydrology

To the north of the Chott el Djerid basin lies the Gafsa-Medinine fault trend which is bounded by Djebel Bou Ramli and Djebel Ben Younis. East-west trending anticlines, generally plunging westward, are the dominant exposed structures north and east of the Chott el Djerid basin, separating the Gafsa-Medinine fault trend and the Pre-Saharan platform to the south (Coque, 1962). The Chott el Djerid basin

Sedimentology and Geochemistry of Modern and Ancient Saline Lakes, SEPM Special Publication No. 50

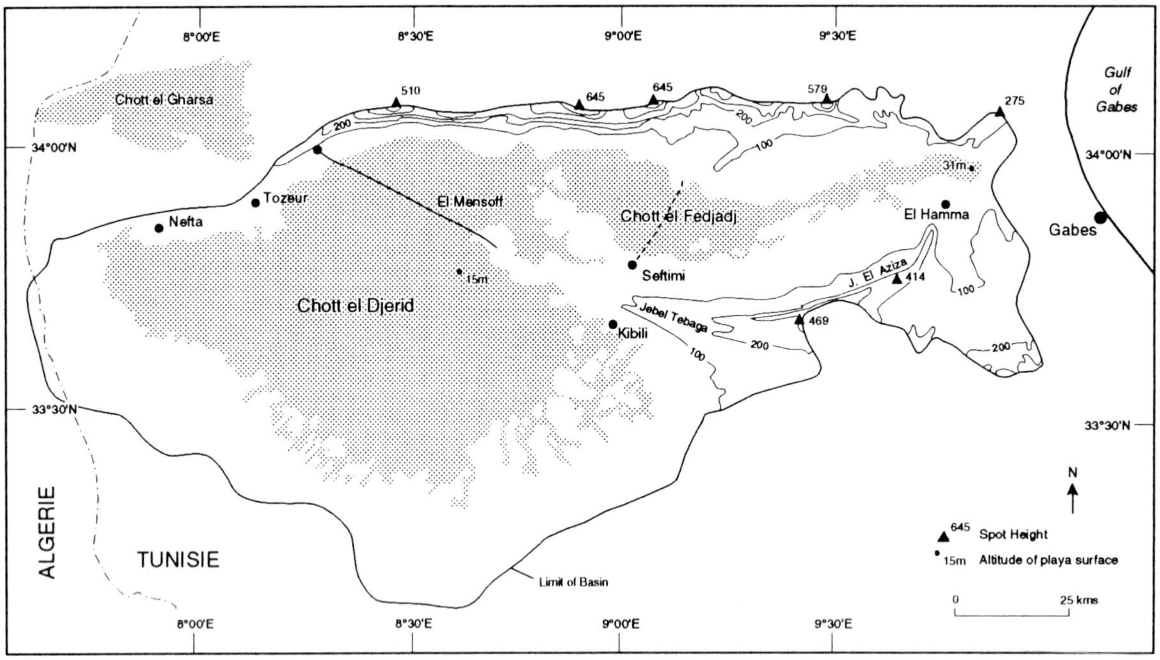

FIG. 1.—Location map of Chott El Djerid, Southern Tunisia.

is bordered by an assemblage of Cretaceous, Mio-Pliocene and Quaternary sequences made up of gypsiferous deposits, marls, limestones and dolomitic carbonates (Gueddari and others, 1983; Coque and Jausien, 1967). The dominant formations exposed within this basin are of Cretaceous age and are characterized by a basal unit that is largely made up of "Wealden-like" continental sediments (Coque, 1967). The sandstones, sands and variegated clays of this basal unit can be seen on the northern flanks of the Chott el Djerid depression. Cuestas of Cenomanian and Turonian dolomitic limestones form the Djebel Tebaga range, which overlooks the southern shore of Chott el Fedjadj. They form part of the southern limb of an east-west oriented, collapsed anticline that plunges toward the Chott el Djerid. To the south of Chott el Djerid lie the dune fields of the Grand Erg Oriental.

The Chott el Djerid basin is fed by two aquifers: the Complexe Terminale and the Continental Intercalaire (Gueddari, 1984; Mamou, 1976; Roberts and Mitchell, 1987). The Complexe Terminale is made up of Senonian limestones and gypsum marls. The Continental Intercalaire is composed of sandy clays, sandstones and gypsiferous marls. In Chott el Djerid, the main aquicludes to the Complexe Terminal are the Mio-Pliocene sands and gypsiferous clays. As a result of aquifer resurgence from the Complexe Terminal within the Chott el Djerid basin, the depth to groundwater, although seasonally variable, is generally shallow, never exceeding more than 1.5 m in the dry summer season (Coque, 1962; Millington and others, 1987). In the winter months, small bodies of water are commonly observed on the playa surface; these are attributable to the seasonal rise in groundwater and an accompanying, marginal increase in rainfall (Meckelein, 1977).

METHODS

Field Methods

The Chott el Djerid is traversed in the north by an elevated causeway, 30 km in length (Fig. 2). The causeway was built in the early 1980's and facilitates access to the northern playa. After the initial, exceptional rainfall event of January 1990, the playa filled up with water forming two lakes that were separated by the road. The causeway was not covered by water due to its elevation above the playa surface (almost 1.7 m), allowing easy access to the lake and playa for sampling. We sampled lake waters and groundwaters at 2 km intervals, adjacent to the kilometer posts, on both sides of the causeway in March, May, and September 1990 (Fig. 2). In March, samples were also taken at different depths in the lake for one location. These samples were augmented by further samples of aquifer waters (from wells and pumping stations) and waters in channels from the catchment area. By September 1990, the playa had almost totally desiccated and extensive areas of salt crust covered the surface. We sampled these crusts at the same localities where brine samples had been taken throughout the year.

The brine samples were collected in 200 ml bottles that had been pre-washed in double distilled water before sampling. We placed the bottles in sealed plastic bags before transportation to the laboratory. We sampled salt crusts by taking the top 3 cm of the sediment surface with a clean trowel; the material was then placed in sealed plastic bags that were double-sealed in a larger plastic bag. We obtained groundwater samples from holes dug into the sediment to a depth below the predicted water-table (commonly 1 m). These holes were left for 24 hours before sampling.

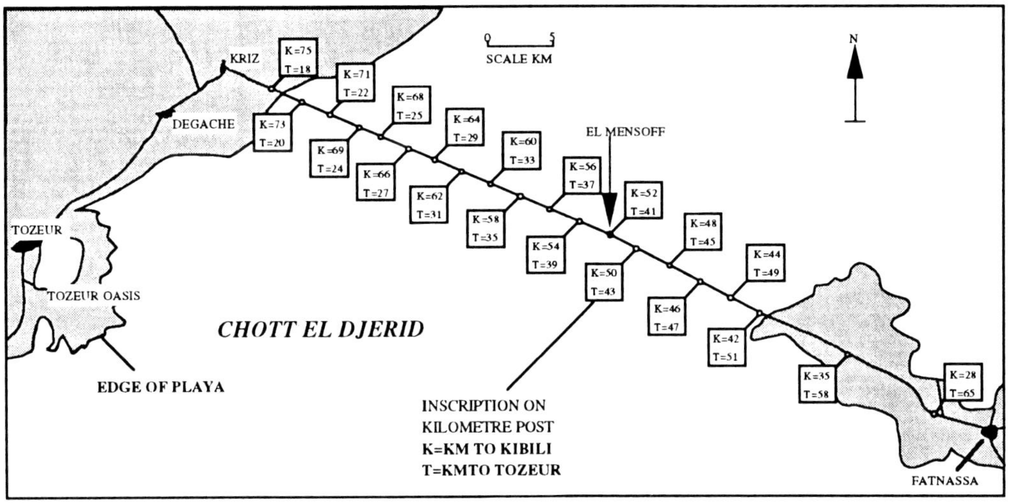

FIG. 2.—Location of kilometer markers on the road that crosses Chott el Djerid. Samples were taken at these points on either side of the road.

Laboratory Methods

The brine samples were stored at constant temperature (0°C) on return to the laboratory. The samples were diluted by 100 to 1,000 times, depending on the tolerance of the analytical methods and instruments used. Ca^{2+}, SiO_2, Mg^{2+}, and Na^+ were determined using an ARL 35000 C sequential ICPMS. K^+ was analyzed on a Perkin-Elmer Atomic Absorption Spectrometer. Cl^-, SO_4^{2-}, and NO_3^- were analyzed on an automated Dionex Ion Chromatograph running Dionex AI-450 software with AS4A columns and Na_2CO_3 eluent. The pH was measured using a standard meter, and HCO_3–CO_3 (alkalinity) was determined by titration (Jackson, 1958). The mineralogy of salt crusts was determined by running whole-samples on a Philips X-Ray diffractometer, using standard methods.

RESULTS AND DISCUSSION

To observe and identify the changes that occurred within the brines of Chott el Djerid as they underwent evaporation, we used various standard methods. These included construction of Piper diagrams (Piper, 1944), solute concentration plots (Eugster, 1980), and output from thermodynamic computer programs.

Brine Fractionation Mechanisms: The Hardie-Eugster Model

Closed-basin brine evolution has been studied in great detail both theoretically and by observation (Garrels and MacKenzie, 1967; Hardie and Eugster, 1970; Al-Droubie, 1976; Eugster, 1980; Eugster and Jones, 1979; Smith and Drever, 1976; Drever, 1982; Herczeg and Lyons, 1991). Hydrologically-closed basins provide a unique situation in which dilute inflow waters can be compared with concentrated brines. Eugster (1980) stressed the importance of the catchment geology as the prime source of dissolved solutes and stated that "two salt lakes on either side of a watershed may have a completely different mineral assemblage associated with them." Essentially, these workers have shown that solute fractionation of a brine can occur by five different methods: (1) mineral precipitation, (2) dissolution of efflorescent crusts and sediment coatings, (3) exchange and sorption reactions on active surfaces, (4) degassing, and (5) redox reactions.

Solute fractionation by mineral precipitation has an extreme effect on the composition of a brine as its concentration increases (Eugster and Jones, 1979; Garrels and MacKenzie, 1967). Figure 3 shows the general evolutionary trend of a closed basin brine with respect to the possible mineral phases that may precipitate as evaporative concentration increases, after an inflow of dilute waters into the basin (Eugster, 1980). Smith and Drever (1976) have shown that solute fractionation can also occur within the shallow sediment if the water table is close to the surface. In this case, brine fractionation is driven by evaporative pumping of the brine to the sediment surface.

As efflorescent crusts are present in all arid basins that have sufficient inflow (Eugster and Jones, 1979), the periodic introduction of inflow waters into the basin must cause dissolution of these crusts and soil coatings. Thus, fractionation will be observed as the most-soluble salts dissolve first and the least-soluble salts normally remain behind. Selective dissolution and removal of the more-soluble salt crusts leads to the formation of more-soluble efflorescences in a lower part of the basin (Eugster and Jones, 1979; Eugster, 1980). Such selective enrichment of certain solutes must affect solute evolution as evaporative concentration progresses. In this case, the composition of the resultant solution is controlled by the kinetics of dissolution of the original precipitated phases and not strictly by their solubilities (Drever, 1982).

Differential exchange and sorption on active surfaces, such as those provided by some clays, can cause the loss of solutes from solution (Eugster, 1980). Eugster and Jones (1979) document the potential loss of K^+ by this process in several situations. Degassing, as a result of equilibrium with the

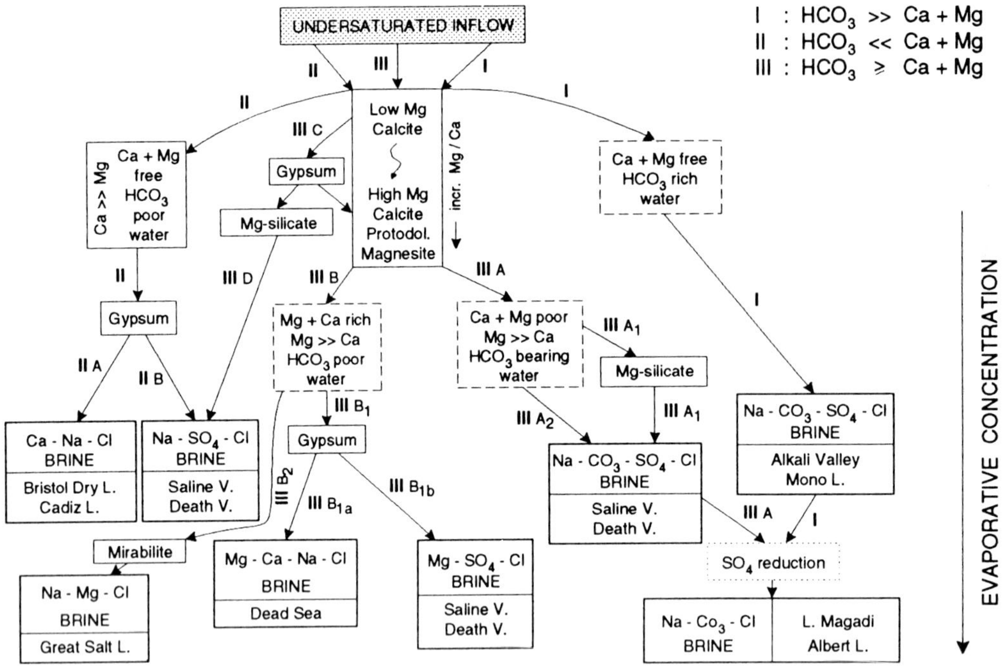

FIG. 3.—Brine evolution flow diagram (from Eugster, 1980) showing critical precipitates (solid rectangles) and resulting brines, together with examples of salt lakes (brine classification).

atmosphere, an increase in temperature, a decrease in solubility with salinity, or photosynthesis, causes precipitation of Ca-Mg carbonate in some situations (Eugster and Jones, 1979). Redox reactions in closed basins generally involve the removal of SO_4^{2-} from solution by microbial sulfate reduction under anaerobic conditions, thereby limiting the amount of sulfate mineral phases present in the closed basin. It is important, therefore, to view the chemical evolution of closed-basin brines with close reference to the various factors that may affect major solute concentration and/or mineral precipitation.

The Brine Chemistry of the Chott el Djerid Basin

The chemistry of the dilute inflow waters.—

The dilute inflow waters sampled in the Chott el Djerid basin were as follows: (1) samples from springs and wells (pumping stations) within the basin, taken in March 1990, (2) samples of surface waters from small pools on the edge of the basin in March 1990, and (3) samples from ephemeral streams in March 1990.

The Piper diagram (Fig. 4) shows the dominant cations and anions in the dilute waters. All samples plot within the same general area. Thus, for dilute waters, the dominant cations were of the calcium and mixed type, and the dominant anions were of the sulfate and chloride type. These plots suggest that the dilute waters flowing into the Chott el Djerid basin were relatively enriched with respect to calcium and sulfate.

These results generally concur with those of Mamou (1976) and Roberts and Mitchell (1987) who analyzed waters from the Continental Intercalaire and the Complexe Terminale. Gueddari (1984), who also analyzed the groundwaters from the basin, found that the dilute waters had a similar chemistry. Thus, the bulk chemistry of the dilute waters from the wadis, the surface waters, the two aquifers, and some groundwaters is generally the same. This suggests that all the dilute waters will undergo the same general chemical evolution. Such compositional uniformity is caused by the nature of the basin geology. Essentially, all waters entering the basin must flow through, or over, extensive sequences of marine evaporites (which are generally gypsum-rich) and carbonates (mainly dolomites). The evolution of the Chott el Djerid brines should, therefore, reflect the nature of the marine evaporites and sediments that are being recycled within the basin.

The chemistry of the concentrated waters.—

The concentrated waters within a closed basin are initially made up of a combination of the different types of dilute waters flowing into the basin. All dilute waters have been shown to have a similar chemistry. Once they have entered the closed basin they are subject to the chemical processes outlined by Eugster (1980). The relatively concentrated waters sampled in the Chott el Djerid basin were: (1) surface water samples taken between March and September 1990, and (2) groundwater samples taken on the edge of the ephemeral lake between March and September 1990.

For concentrated waters, the dominant cations were sodium and potassium and the dominant anion was chloride (Fig. 4). Figure 5 shows the variation in concentration of major solutes (sodium and chloride) with depth in the lake. The lake had drowned a halite salt crust that was approx-

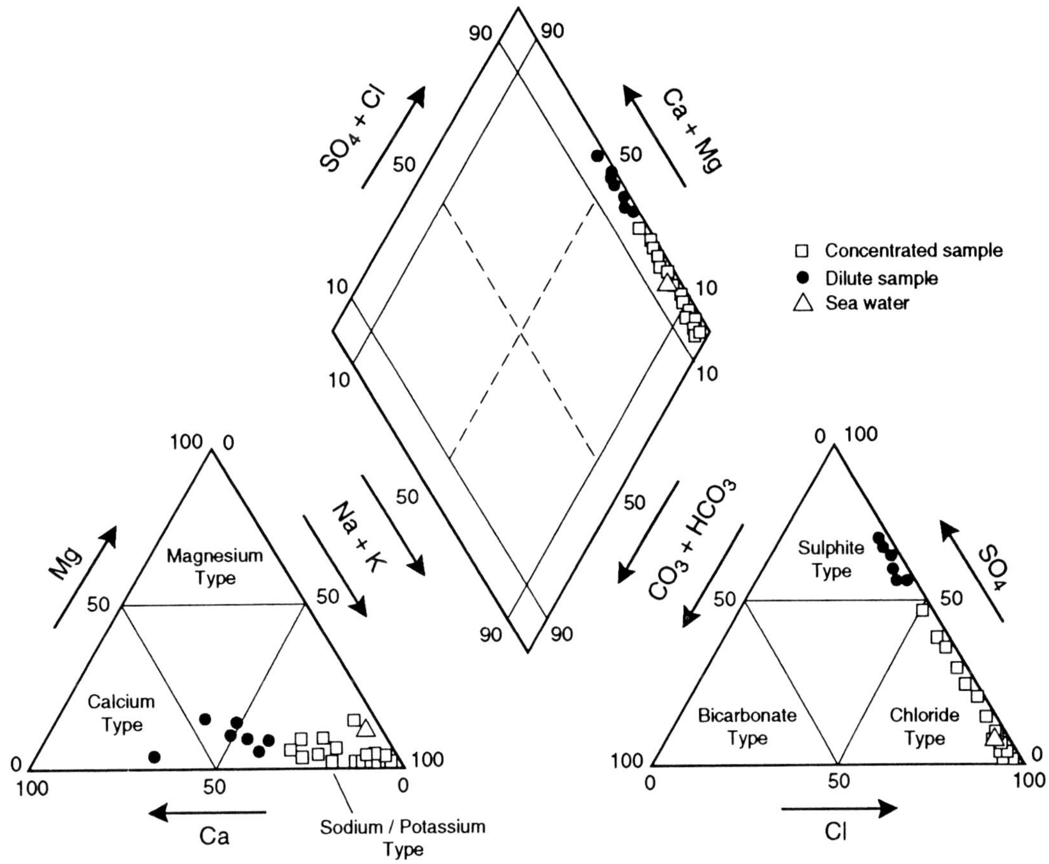

FIG. 4.—Piper diagram representing the chemical composition of brines from the Chott el Djerid.

imately 10 cm thick. Sodium and chloride concentrations were slightly higher, both at the surface of the lake and in the immediate vicinity of the halite crust, causing a dual concentration gradient within the water body. From Figure 5, it is reasonable to suggest that the concentrations of sodium and chloride for this location increased as a result both of evaporative concentration and dissolution of the halite crust. Both processes, therefore, may have contributed to the dominance of sodium and chloride in the concentrated waters.

The Behavior of Major Solutes during Evaporation: The Precipitation of Mineral Phases

The use of potassium as a conservative element.—

Conservative elements are used to provide a reliable measure of the degree of evaporative concentration of a brine. In natural waters, chloride is generally conserved over the widest concentration range (Eugster and Jones, 1979). Indeed, in their study of Lake Magadi, Kenya, Jones and others (1977) used chloride as their major conservative element. However, the possible onset of halite saturation in waters leads to the net loss of chloride from solution. In the Chott el Djerid, halite is a very common salt on the playa surface (Coque, 1962; Millington and others, 1989). Consequently, halite saturation should commonly be reached in the ground- and surface waters of the playa.

Figure 6A gives a sensitive test of sodium and chloride behavior (Eugster and Jones, 1979) and shows the Na:Cl ratio plotted against chloride concentration. As can be seen, chloride was not a conservative element during the period of the sampling of the brines. At low chloride concentrations, the Na:Cl ratio remained constant. After further concentration of sodium and chloride by evaporation, a sharp drop in the ratio Na:Cl is observed. Therefore, an alternative conservative element was needed to monitor solute behavior. Figure 6B shows the behavior of the Mg:K ratio with increasing potassium concentration. This graph shows that the Mg:K ratio is generally constant over a wide range of potassium concentration, suggesting that neither element is lost as a result of mineral precipitation. Therefore, we have used potassium as the conservative element in this study. Normally, potassium is not lost from solution until potassium salts (e.g., sylvite and carnallite) are precipitated or if sorption processes occur. Figure 6B suggests, therefore, that most sorption surfaces within the playa sediments were saturated during this study.

All observations of evaporative concentration presented have been normalized against relative potassium concentration. The graphs (Figs. 7 to 11) show trends of major solute concentrations for all brine types with increasing

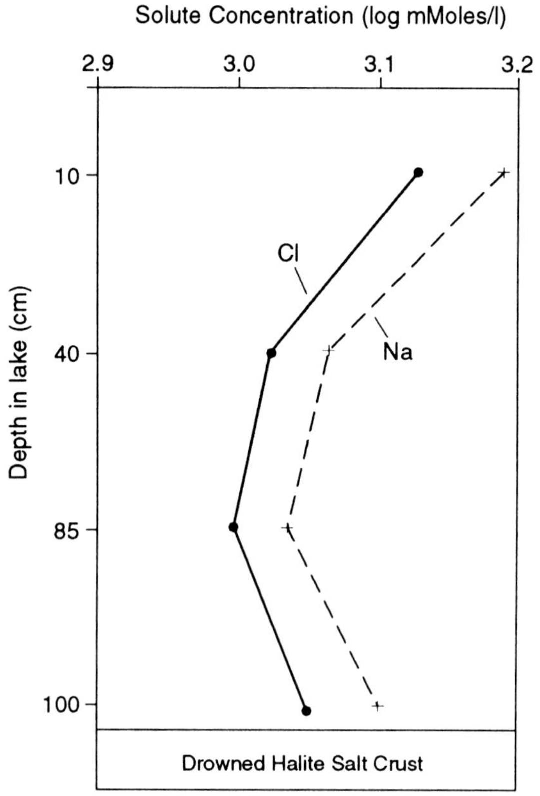

FIG. 5.—The variation of major solutes with depth in the ephemeral lake, March 1990.

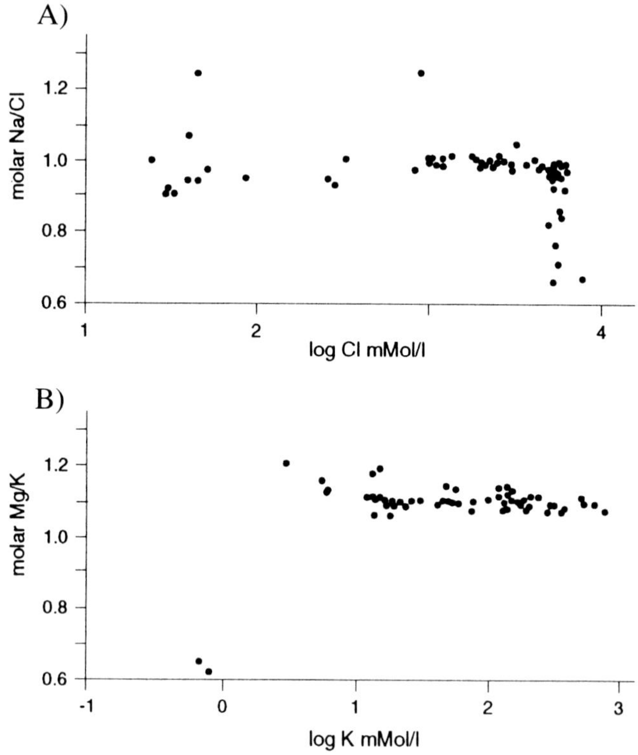

FIG. 6.—(A) The non-conservative nature of chloride relative to sodium with increasing evaporative concentration. (B) The conservative nature of potassium with increasing evaporative concentration.

evaporation. The behavior of silica (SiO_2) has been ignored because its concentration rarely exceeded 10 ppm, suggesting that silica is relatively unimportant in this environment.

The behavior of alkalinity and pH.—

Alkalinity was very low in the Chott el Djerid (Fig. 7A), rarely exceeding 6 meq/l. Although low, the behavior of alkalinity with increasing evaporation within the basin was not constant, a marked increase being seen at higher evaporative concentration. Low alkalinity levels within the Chott el Djerid are fairly consistent with the process of recycling of ancient marine evaporites within the basin. Figure 7B shows the general relationship between pH and evaporative concentration. Between the concentrations 0 and 2.5 log K mMoles/l the general trend of pH is a gradual decline from 7.5 to about 6.0, Krumgaltz (1981) suggests that at increasingly high solute concentrations in natural waters, hydrogen ion activity may be suppressed, effectively reducing the pH of the solution. Alternatively, the reduction in pH may have been a result of hydroxyl and chloride ions complexing with magnesium at higher concentrations (B. F. Jones, pers. commun.).

The behavior of calcium and sulfate.—

Figure 7A shows the behavior of calcium and sulfate during the evaporative concentration of the brines in the Chott el Djerid basin. In the relatively dilute brines, the ratio of $Ca:SO_4$ was approximately 1:2.5. At concentrations of between 0 and 1.5 log K mMoles/l, both calcium and sulfate followed the same general trend, increasing in their relative concentrations at the same rate. Above 1.5 log K mMoles/l, there was a change in this trend and, with increased evaporative concentration, the relative concentration of calcium within the brine decreased rapidly; sulfate showed the opposite behavior. The divergent trend of these two solutes at this point was caused by a chemical divide. Hardie and Eugster (1970), following the work of Garrels and MacKenzie (1967), define a chemical divide within a closed basin brine as occurring when a binary salt is precipitated.

In this case, the chemical divide of calcium and sulfate seen in Figure 7A can be attributed to gypsum ($CaSO_4 \cdot 2H_2O$) precipitation. The simple relationship between calcium and sulfate within these brines was further characterized by observing the variation of gypsum saturation with increasing evaporation, using the program SOLMINEQ. With increasing evaporative concentration, a positive solubility index (SI) for gypsum was observed between 0 and 1.5 log K mMoles/l, with a stronger positive gypsum SI for brine samples after this point (Fig. 8B). A positive SI indicates saturation of the water with respect to the mineral phase. Within the brine, calcium and sulfate concentrations were observed to increase in concentration, even though saturation with respect to gypsum had been reached. Indeed, the chemical divide between calcium and sulfate is signified by an increase to supersaturation within the brine. Cody (1991) observed that sub-stoichiometric levels of organic

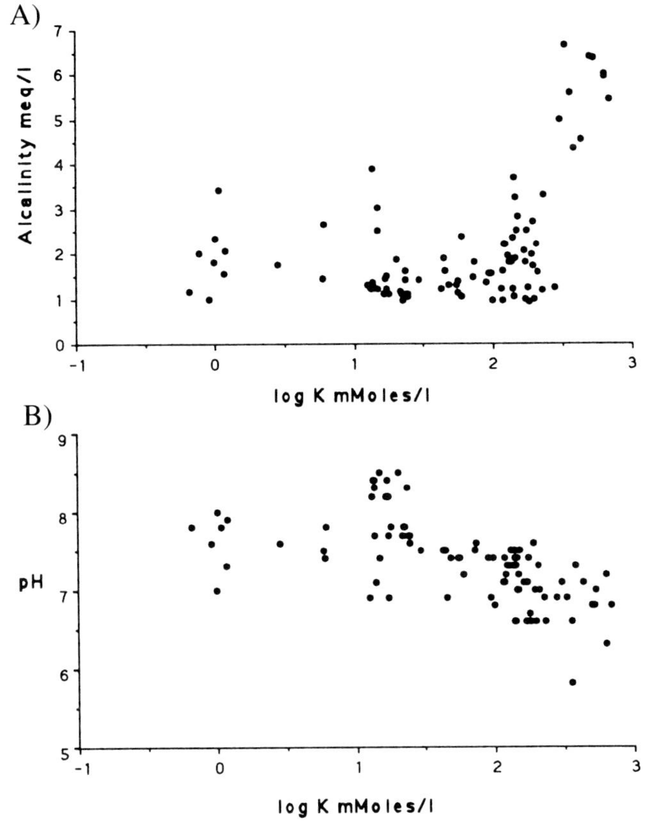

FIG. 7.—(A) The behavior of alkalinity with increasing evaporative concentration. (B) The behavior of pH with increasing evaporative concentration.

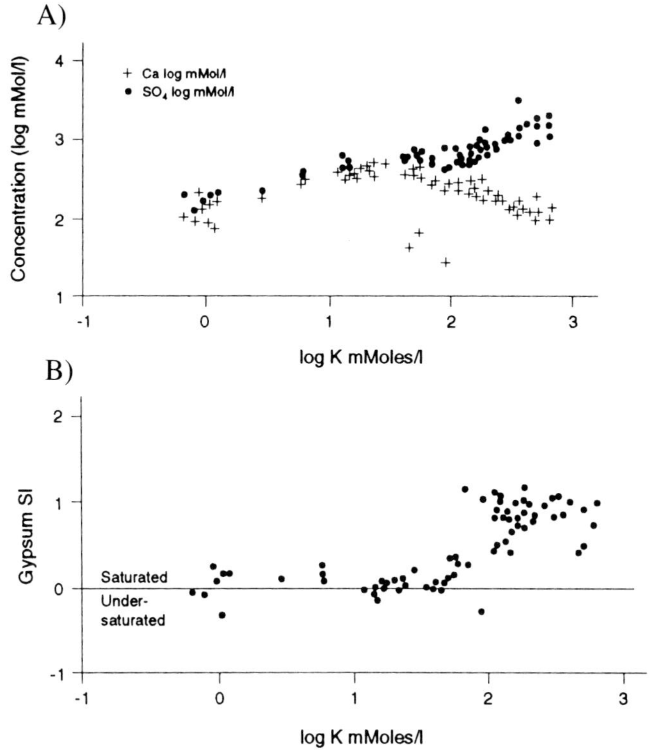

FIG. 8.—(A) The chemical evolution of calcium and sulfate with increasing evaporative concentration. (B) The variation of gypsum saturation in the brine with increasing evaporation.

molecules can delay gypsum precipitation to the point of supersaturation within a brine.

Overall, however, the precipitation of gypsum in the Chott el Djerid seems to have followed the same general pattern (i.e., a chemical divide caused by saturation with respect to, and precipitation of, solid phases) presented by Hardie and Eugster (1970), Garrels and Mackenzie (1967), Drever (1982), and Rosen and Warren (1990). If the brine at this point, is compared to the paths of chemical evolution suggested by Hardie and Eugster (1970), Eugster (1980) and Drever (1982), then the major solutes present after gypsum precipitation would be Na, SO_4, and Cl (with Mg and K also being of importance). Assuming that $HCO_3 << Ca + Mg$, this suggests that the Chott el Djerid brine would compare either to those of Dead Sea and marine-type (Drever, 1982) or to the Na-Cl-SO_4 type (Eugster, 1980; see Fig. 3). The relatively high concentrations of magnesium and potassium within the Chott el Djerid brines, therefore, reflect the input of these solutes from the dissolution of marine evaporites within the basin.

The behavior of sodium and chloride.—

Figure 9A shows the behavior of the solutes, sodium and chloride, during evaporative concentration. Figure 9B shows the behavior of the halite SI with increasing evaporation. Essentially, the trends shown in Figure 9a can be divided

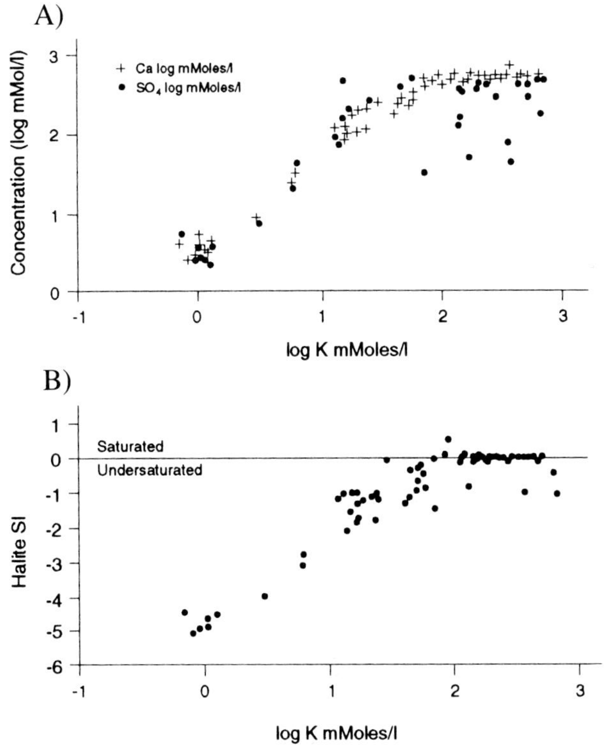

FIG. 9.—(A) The chemical evolution of sodium and chloride with increasing evaporative concentration. (B) The variation of halite saturation in the brine with respect to increasing evaporation.

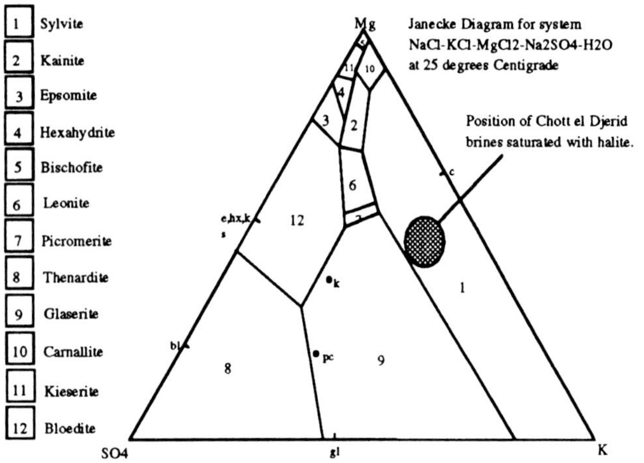

FIG. 10.—Jänecke diagram for the system NaCl-KCl-$MgCl_2$-Na_2SO_4-H_2O (at a constant 25°C) showing the trend towards the precipitation of sylvite in halite saturated brines (after Krauskopf, 1985).

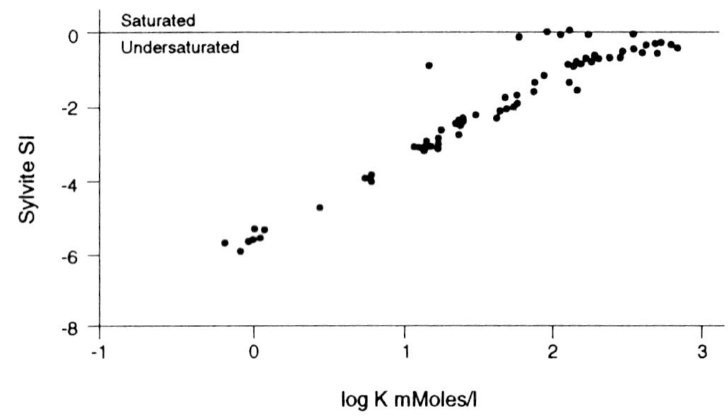

FIG. 11.—The variation of sylvite saturation with respect to evaporative concentration.

into two parts: (1) concentration in solution and (2) saturation and precipitation of a solid mineral phase. Between the concentrations 0 and 1.8 log K mMoles/l, there was a constant increase in the log concentration of the two solutes. Both solutes followed the same general trend at this stage. The Na:Cl ratio was approximately 1:1, and the halite SI for brines in this range is strongly negative. The Na:Cl ratio of 1:1 suggests that Na and Cl are probably derived from the dissolution of halite. As previously stated, such dissolution may have occurred both within the basin (dissolution of ancient evaporites) or on the surface of playa (dissolution of saline pan crusts). Above a concentration of 1.8 log K mMoles/l, the concentration of the two solutes did not increase. The halite SI for the brine at this point was positive (Fig. 9B). This stage, therefore, can be interpreted as the result of saturation with respect to halite. After a concentration of 1.8 log K mMoles/l, Na concentrations were seen to decrease. The loss of Na may have resulted from the precipitation of halite from brines with a Na:Cl ratio of less than 1:1.

Potassium salt phases.—

The literature describing modern potash deposits is very sparse (Lowenstein and Spencer, 1990). It has been demonstrated that potassium was a conservative element in this study. For brine samples in which halite saturation had been reached, major solutes were plotted on a Jänecke diagram for the system NaCl-KCl-$MgCl_2$-Na_2SO_4-H_2O at a constant 25°C (Fig. 10) (Krauskopf, 1985). Most points fall within the sylvite (KCl) stability field, suggesting that this mineral should be the next precipitate after, or with, halite. This is also the case when the sylvite SI is calculated for the same samples (taking into account the variation in temperature of the sampled brines). Nearly all the halite-saturated samples have an SI index close to zero (Fig. 11), suggesting the onset of saturation with respect to potash phases in these brines. Gueddari (1984) suggests that, with further evaporation, kainite, carnallite, kieserite and bischofite may also precipitate from Chott el Djerid brines. Sonnenfeld (1984), however, suggests that sylvite, kainite, kieserite and bischofite are extremely rare primary minerals in contemporary arid basins, carnallite having been found on a few occasions. The evaporative concentration of a brine to the point of potash precipitation is very rare in modern evaporitic environments. Smoot and Lowenstein (1991), however, suggest that such concentrations may be reached if repeated dissolution and reprecipitation of the saline pan occur over time. This may be the case in the Chott el Djerid, because extensive dissolution of the pre-existing salt-pan was observed after the January flood (see also Fig. 5).

The chemistry of the salt crusts on the Chott el Djerid (September 1990).—

The surface salts that we collected and analyzed in September 1990 are of four main types: (1) mixed halite, gypsum and clastic material (quartz and calcite); (2) mixed halite and gypsum; (3) halite; and (4) carnallite and halite (carnallitite). These samples were collected at the same sample localities as the brines.

Carnallitite was found in a small pool of yellow brine in the north-central playa. It was distinguished by a marked yellow coloring caused by a thin hematite coating on the euhedral grains. The associated brine sample was saturated with respect to potassium salt phases, suggesting a primary origin for the potash mineral. Carnallite ($KMgCl_3 \cdot 6H_2O$) is a very rare primary mineral phase in modern evaporitic basins. Indeed, Hardie (1990) reports only two previous cases in which contemporary carnallite may have been observed: the Qaidam Basin, China (Lowenstein and Spencer, 1990), and the Danakil Depression, Ethiopia (Holwerda and Hutchinson, 1968). To precipitate carnallite from a brine, a surface temperature of 41–47°C and a relative humidity of less than 46 percent are needed (Sonnenfeld 1984). In arid basins like Chott el Djerid, these values are not uncommon. The brine in which the carnallitite was sampled had a temperature of 36°C, and the relative humidity was about 40 percent at the time of sampling.

The presence of gypsum and halite as precipitated salt phases supports the results of the water analyses discussed above. The main chemical processes to affect the major solutes in the brines were precipitation of gypsum and halite. The precipitation of carnallite, however, was not predicted because the most concentrated brines were theoretically saturated with respect to sylvite. Nevertheless, carnallite is commonly associated with sylvite in many ancient, and some modern, saline pans (Sonnenfeld, 1984). If the Chott el Djerid brines are directly influenced by the recycling of ancient marine evaporites, then the observed mineral assemblage should reflect this.

Harvie and Weare (1980) define the ideal mineral assemblage derived from the evaporation of seawater as: gypsum, halite, sylvite, kainite, carnallite, kieserite and bischofite, precipitated sequentially. Gueddari (1984) suggests, from thermodynamic analysis of groundwaters, that the evaporation sequence of minerals within the Chott el Djerid Basin should be: gypsum, halite, sylvite, kainite, carnallite, kieserite and bischofite. This sequence is, therefore, fairly indicative of a marine evaporite mineral assemblage enriched with potassium relative to magnesium. It could be the case that the suite of minerals observed on the Chott el Djerid (gypsum, halite, carnallitite) were an abbreviated form of either of the two predicted sequences. This may be due to (1) changes in local climate that affected the stability of mineral phases, (2) the incomplete evaporation of the brines, or (3) incomplete sampling of the mineral phases present. Field evidence suggests that (1) and (2) occur, and that (3) is a possibility. Smoot and Lowenstein (1991) suggested that repeated dissolution and reprecipitation of the saline pan by flooding will result both in increased solute concentration within shallow groundwaters (possibly up to the point of potash precipitation) and in the eventual precipitation of a largely monomineralic saline pan crust. Both processes occur on the Chott el Djerid, suggesting that they may be a controlling factor on the nature and distribution of primary evaporite phases within the playa.

Hardie (1990) suggested that ancient potash evaporites fall into two groups: (1) an $MgSO_4$-rich group of minerals that are generated from the evaporitic concentration of seawater, and (2) an $MgSO_4$-poor group, characterized by a combination of halite, sylvite, carnallite and minor amounts of tachyhydrite and bischofite. The origin of this second group is unclear. Hardie (1990) hypothesizes a hydrothermal influence in closed, rifted basins, whereas Braitsch and Krinesen (1978) suggest that it is a modified marine sequence. If complete sampling of the mineral phases of Chott el Djerid is assumed, then, purely on mineralogical grounds, it appears that this mineral assemblage falls into the $MgSO_4$-poor group. However, according to the Hardie-Eugster model, the brines the Chott el Djerid exhibited a normal marine chemistry. Furthermore, the repeated dissolution of the saline pan by successive flooding events may fundamentally affect the terminal mineral assemblage present on the playa. Therefore, the results of this study suggest that potash evaporite assemblages may form in closed, continental basins as a result of the simple recycling of ancient marine evaporites within the basin.

CONCLUSIONS

1. Following a rare rainfall event of January 1990 in the Chott el Djerid basin, all dilute inflow waters had a uniform chemistry. All the waters were relatively enriched with respect to calcium, sodium, chloride and sulfate, and, to a lesser extent, with potassium and magnesium.
2. The chemistry of these dilute waters is a function of the basin geology. The natural waters that flow into the basin recycle ancient marine evaporite deposits.
3. As the dilute brines have a uniform chemistry, the waters within the Chott el Djerid must have undergone the same chemical (major solute) evolution with increasing evaporative concentration.
4. Concentration of major solutes within the ephemeral lake occurred both by evaporation and dissolution of pre-existing saline pan evaporites.
5. Potassium was conserved throughout the evaporation sequence, becoming an important solute in the brine after the point of halite saturation. The conservation of potassium suggests that all active sorption surfaces were saturated. Consequently, potassium was used as a conservative element to indicate degree of evaporative concentration.
6. The evolution of major solutes with evaporation indicates that both gypsum and halite were precipitated from the lake water by simple brine fractionation. Concentrated brines were saturated with respect to potash phases.
7. The suite of precipitated minerals found on the Chott el Djerid basin was gypsum, halite and carnallite. Periodic dissolution and reprecipitation of pre-existing saline pan evaporites by flooding control the terminal mineral assemblage.
8. Potash evaporites can form from a brine with a marine-like chemistry in a continental playa setting, by simple recycling of ancient marine evaporites.

ACKNOWLEDGMENTS

The authors would like to thank S. A. Malik, D. Gara, J. Watkins, S. Dance and H. Browning for technical assistance. Prof. Max Coleman, Dr. Joy Rae and Dr. Paul Wright are thanked for helpful discussions. Drs. B. F. Jones and J. I. Drever are thanked for their constructive reviews of the text. R. G. Bryant undertook the work as part of N. E. R. C. (U.K.) research training award GT4/89/GS/101. B. P. and N. E. R. C. are also thanked for supporting the travel costs of R. G. B to the conference in Saskatoon. For R. G. B. and B. W. S., this work represents P. R. I. S. Contribution #227.

REFERENCES

Al-Droubie, A., 1976, Géochimie des sols et des solutes concentrées par évaporation: Modele thermodynamique de simulation. Application aux sols salées du Tchad: Sciences Géologie Mémoires, v. 46, 177 p.

Braitsch, O., and Krinsen, D. J. J., 1978, Evaporites—physiochemical conditions of origin, *in* Fairbridge, R. W., and Bourgeois, J., eds., The Encyclopedia of Sedimentology: New York, Hutchinson and Ross, p. 901–902.

Cody, R. D., 1991, Organo-crystalline interactions in evaporite systems: the effects of crystallisation inhibition: Journal of Sedimentary Petrology, v. 61, p. 704–718.

Coque, R., 1962, La Tunisie Presaharienne: Paris, Armand Colin, 476 p.

Coque, R., and Jausien, A., 1967, The geomorphology and Quaternary geology of Tunisia, *in* Martin, L., ed., Guidebook to the Geology and History of Tunisia: Tripoli, The Petroleum Exploration Society of Libya, Handbook, v. 3, p, 227–257.

DREVER, J. I., 1982, The Geochemistry of Natural Waters: Englewood Cliffs, Prentice-Hall, 388 p.

EUGSTER, H. P., 1970, Chemistry and origin of the brines of Lake Magadi, Kenya: Washington, Mineralogical Society of America Special Publication, v. 3, p. 215–235.

EUGSTER, H. P., 1980, Geochemistry of evaporitic lacustrine deposits: Annual Reviews of Earth and Planetary Sciences, v. 8, p. 35–63.

EUGSTER, H. P., AND JONES, B. F., 1979, Behavior of major solutes during closed-basin brine evolution: American Journal of Science, v. 279, p. 609–631.

GARRELS, R. M., AND MACKENZIE, F. T., 1967, Origin of the chemical composition of some springs and lakes, *in* Equilibrium Concepts in Natural Water Systems: Advances in Chemistry, American Chemical Society, v. 67, p. 222–242.

GUEDDARI, M., 1984, Géochimie et thermodynamique des évaporites continentales: étude du Lac Natron (Tanzanie) et du Chott Jerid (Tunisie): Unpublished Ph.D Dissertation, Université Louis Pasteur, Strasbourg, 129 p.

GUEDDARI, M., MONNIN, C., PERRET, D., FRITZ, B., AND TARDY, Y., 1983, The geochemistry of the brines of Chott el Djerid in southern Tunisia: applications of Pitzer's equations: Chemical Geology, v. 39, p. 165–178.

HARDIE, L. A., 1990, The roles of rifting and hydrothermal $CaCl_2$ brines in the origin of potash evaporites: an hypothesis: American Journal of Science, v. 290, p. 43–106.

HARDIE, L. A., AND EUGSTER, H. P., 1970, The evolution of closed basin brines: Mineralogical Society of America Special Publication, v. 3, p. 273–290.

HARVIE, C. E, AND WEARE, J. H., 1980, The prediction of mineral solubilities in natural waters: the Na-K-Mg-Ca-Cl-SO_4-H_2O system from zero to high concentration at 25°C: Geochimica et Cosmochimica Acta, v. 44, p. 981–997.

HERCZEG, A. L., AND LYONS, W. B., 1991, A chemical model for the evolution of Australian sodium chloride lake brines: Palæogeography, Palæoclimatology, Palæoecology, v. 84, p. 43–53.

HOLWERDA, J. G., AND HUTCHINSON, R. W., 1968, Potash-bearing evaporites in the Danakil area, Ethiopia: Economic Geology, v. 63, p. 124–150.

JONES, B. F., EUGSTER, H. P., AND RETTIG, S. L., 1977, Hydrochemistry of the Lake Magadi Basin, Kenya: Geochimica et Cosmochimica Acta, v. 41, p. 53–72.

JACKSON, J., 1958, Soil Chemical Analysis: London, Prentice-Hall, 465 p.

KRAUSKOPF, K. B., 1985, Introduction to Geochemistry (3rd ed.): London, McGraw-Hill, 617 p.

KRUMGALTZ, B., 1981, Salt effect on the pH of hypersaline solutions, *in* Nissenbaum, A., ed., Hypersaline Brines and Evaporitic Environments: Amsterdam, Elsevier, p. 73–83.

LAST, W. M., 1989, Continental brines and evaporites of the northern Great Plains of Canada: Sedimentary Geology, v. 64, p. 207–221.

LOWENSTEIN, T. K., AND SPENCER, R. J., 1990, Syndepositional origin of potash evaporites: petrogenic and fluid-inclusion evidence: American Journal of Science, v. 290, p. 1–42.

MAMOU, A., 1976, Contribution à l'étude hydrogéologique de la presqu'ile de Kibili: Unpublished Ph.D. Dissertation, Université Pierre et Marie Curie, Paris, 265 p.

MECKELEIN, W., 1977, Zür geomorphologie des Chott Djerid: Stuttgarter Geographische Studien, v. 91, p. 247–301.

MILLINGTON, A. C., JONES, A. R., QUARMBY, N., TOWNSHEND, J. R. G., 1987, Remote sensing of sediment transfer processes in playa basins, *in* Frostick, L. E. and Reid, I., eds., Desert Sediments: Ancient and Modern: London, Geological Society of London Special Publication 35, p. 369–383.

MILLINGTON, A. C., DRAKE, N. A., TOWNSHEND, J. R. G., QUARMBY, N. A., SETTLE, J. J., AND READING, A. J., 1987, Monitoring of salt playa dynamics using Thematic Mapper data: IEEE Transactions on Geoscience and Remote Sensing, v. 27, p. 754–761.

OUEZDOU, H. B., KALLEL, R., AND MAMOU, A., 1990, Les pluies exceptionalles de janvier 1990 en Tunisie: les conséquences hydrologiques et hydrogéologiques: Paris, Comptes Rendues Académie des Sciences, v. 311, p. 1375–1382.

PIPER, A. M., 1944, A graphic procedure in the geochemical interpretation of water analyses: Transactions of the American Geophysical Union, v. 25, p. 914–928.

ROBERTS, C. R., AND MITCHELL, C. W., 1987, Spring mounds in Southern Tunisia, *in* Frostick, L. E., and Reid, I., eds., Desert Sediments: Ancient and Modern: London, Geological Society of London Special Publication 35, p. 321–336.

ROSEN, M. R., AND WARREN, J. K., 1990, The origin and significance of groundwater seepage gypsum from Bristol Dry Lake, California, USA: Sedimentology, v. 37, p. 983–996.

SMITH, C. L., AND DREVER, J. I., 1976, Controls on the chemistry of springs at Teels Marsh, Mineral County, Nevada: Geochimica et Cosmochimica Acta, v. 40, p. 1081–1093.

SMOOT, J. P. AND LOWENSTEIN, T. K., 1991, Depositional environments of non-marine evaporites, *in* Melvin J. L., ed., Evaporites, Petroleum and Mineral Resources (Developments in Sedimentology 50): Amsterdam, Elsevier, p. 189–347.

SONNENFELD, P., 1984, Brines and Evaporites: New York, Academic Press, 613 p.

TELLER, J. T., BOWLER, J. M., AND MACUMBER, P. G., 1982, Modern sedimentation and hydrology in Lake Tyrell, Victoria: Journal of the Geological Society of Australia, v. 29, p. 159–175.

TELLER, J. T., AND LAST, W. M., 1990, Palæohydrological indicators in playas and salt lakes, with examples from Canada, Australia and Africa: Palæogeography, Palæoclimatology, Palæoecology, v. 76, p. 215–240.

SULFATE REDUCTION RATES IN MICROBIAL MAT SEDIMENTS OF DIFFERING CHEMISTRIES: IMPLICATIONS FOR ORGANIC CARBON PRESERVATION IN SALINE LAKES*

W. BERRY LYONS

Hydrology/Hydrogeology Program, Mackay School of Mines, University of Nevada, Reno, Reno, Nevada 89557–0138 (present address: Department of Geology, University of Alabama, Tuscaloosa, Alabama 35487-0338)

MARK E. HINES

Biogeochemical Systems Center, Institute for the Study of Earth, Oceans and Space, University of New Hampshire, Durham, New Hampshire 03824

WILLIAM M. LAST

Department of Geological Sciences, University of Manitoba, Winnipeg, Manitoba, Canada R3T 2N2

AND

ROBERT M. LENT

Institute for the Study of Earth, Oceans and Space, University of New Hampshire, Durham, New Hampshire 03824 (present address: United States Geological Survey, Water Resources Division, Bismarck, North Dakota 58501)

ABSTRACT: Kelts (1988) was the first to point out that the ionic composition of lacustrine waters could exert an important effect on the preservation of organic matter in the sediments of these systems. In lakes with high sedimentation rates, the sediments become devoid of molecular oxygen at or very near the sediment-water interface. The rates of anoxic (anaerobic) decomposition of organic matter vary with the amount and "type" of organic matter being buried as well as with the electron acceptor being reduced. The amount of energy produced per organic molecule being oxidized is more for microbial sulfate reduction than for methane production. Therefore, where sulfate is abundant, organic carbon should be more readily oxidized than where other anions such as Cl^- and HCO_3^- dominate the chemistry.

We have tested this hypothesis by determining *in-situ* sulfate reduction via radiolabel techniques in sediments from Freefight Lake, a Na^+-Mg^{2+}-SO_4^{2-} saline lake in southwestern Saskatchewan. These data are then compared to sulfate reduction rate measurements from other sedimentary environments. Of particular interest is our previous work on algal flat sediments from Na^+-Mg^{2+}-Cl^- salt lakes and pans in Bonaire, N. A. The maximum sulfate reduction rate for the nearshore, algal flat sediments of Freefight Lake was 400 nmol/ml/d. This is 2 to 2.5 times lower than rates from similar settings in Bonaire, even though the sulfate concentrations were 10–20 times higher at Freefight Lake. In addition, the S:OC ratios of the Freefight Lake sediments were extremely low, indicating the lack of sulfate reduction in these sediments. Although the total reduced sulfur concentration increased with depth at Freefight Lake, an extremely high percentage of this total S^{2-} is in the form of acid-volatile sulfur. This would not have been predicted from the low sulfate reduction rates.

A number of important conclusions can be drawn from this work. They include the following: (1) sulfate reduction rates in salt lake sediments are not solely dependent on SO_4^{2-} concentrations nor on the ionic composition of the lake waters; (2) even though the organic carbon concentrations of these sediments are high, sulfate reduction is not rapid; and (3) the low sedimentation rate (0.03 cm/yr) may lead to the utilization of the most metabolizable of the carbon present in the oxic zone by aerobic bacteria, thereby slowing sulfate reduction rates.

INTRODUCTION

The northern Great Plains of North America are characterized by flat to gently rolling topography and a cold semi-arid climate. This region contains an estimated 3.5×10^6 salt lakes and saline wetlands. The composition of these waters varies dramatically, although Na^+-Mg^{2+}-SO_4^{2-} waters dominate (Last, 1992). Because of the chemical diversity of the waters, these lakes can serve as important natural "beakers" in which to conduct various experiments. For example, Marino and others (1990) have utilized the sulfate-rich waters of northern Great Plains' lakes to establish what factors control nitrogen-fixation in cyanobacteria. These experiments were undertaken in these lakes solely because of their unique water compositions (Cole and others, 1986).

It has been argued that the ionic composition of the lacustrine waters should exert an important effect on the preservation of organic matter in these systems (Kelts, 1988). In lakes with high sedimentation and/or high primary production rates, depletion of molecular oxygen can occur at or above the sediment-water interface (Wetzel, 1975; Kelts, 1988). In aquatic systems, the rates of anoxic decomposition of organic matter vary with the amount and type of organic matter being deposited and buried in the sediment (Berner, 1978), as well as with the electron acceptor being reduced (Froelich and others, 1979). The amount of energy produced per organic molecule being oxidized is more for microbial sulfate reduction than for other anoxic biochemical pathways.

$$2CH_2O + SO_4^{2-} \rightarrow H_2S + 2HCO_3^-$$

Therefore, where the sulfate ion is abundant, organic carbon should be more readily oxidized and, hence, not preserved in the sediments where other anions such as Cl^- and $HCO_3^- + CO_3^{2-}$ dominate the solution chemistry (Kelts, 1988). This hypothesis is of great consequence in understanding why organic matter is preserved in some lacustrine environments and not in others.

IMPORTANCE OF LAKE STUDIES TO PETROLEUM GEOLOGY

From the perspective of petroleum geology, one of the major tasks challenging earth scientists is to gain a more complete understanding of the mechanisms and conditions which promote the production and preservation of organic matter in lacustrine (and marine) environments. As detailed in Kelts (1988), in order to better understand the source rock potential of ancient lake sediments, details of the car-

*Palliser Triangle Global Change Project Contribution Number 11

Sedimentology and Geochemistry of Modern and Ancient Saline Lakes, SEPM Special Publication No. 50

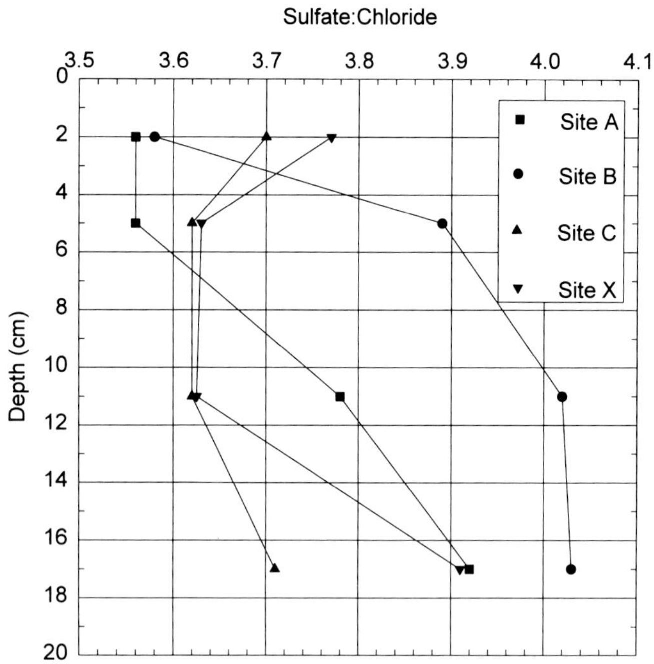

FIG. 3.—Sulfate:chloride porewater ratios for the four Freefight Lake sites.

regarding these data: (i) sulfate:chloride ratios are greater than the overlying water value, (ii) A_T:Cl are greater than the overlying water value, and (iii) the Ca:Cl ratios are less than the overlying water value (Table 3).

The increase in the SO_4:Cl ratio in the pore waters is undoubtedly due to the dissolution of authigenic and endogenic SO_4^{2-} minerals that have been deposited within the mat. These minerals include mainly gypsum, bloedite, epsomite, and mirabilite (Last, 1993a). We have observed a similar phenomenon in evaporitic-mat sediments in Bonaire (Spader and Lyons, 1985) where gypsum dissolution dominates SO_4^{2-} input to the pore fluids. Similar observations have been made in Solar Lake. By observing the pore water profile alone, one would conclude that sulfate reduction is not occurring because in many cases SO_4^{2-} increases with depth. On the other hand, the increase in A_T indicates that sulfate reduction is occurring because HCO_3^- is a major product of sulfate reduction (Berner, 1980). In addition, the decrease in Ca^{2+} relative to Cl^- suggests that authigenic $CaCO_3$ is precipitating in the mat. This is similar to that documented in both Solar Lake (Aharon and others, 1977; Lyons and others, 1984b) and Bonaire (Spader and Lyons, 1985) mat sediments. Analyses of the Freefight Lake sediments indicate that the following carbonate minerals are present: aragonite, magnesian calcite, normal (low-Mg) calcite, stoichiometric and well-ordered dolomite, Ca-rich "protodolomite" (poorly ordered, nonstoichiometric dolomite), magnesite, hydromagnesite, and huntite (Last, 1993a). Stable carbon isotopic measurements of the aragonite in the upper meter of sediment in the algal flat area indicate a gradual decrease (i.e., increase in ^{12}C) from +3.2‰ to −0.7‰ with depth in the core (Last, 1990). These data support the notion that *in-situ* precipitation of $CaCO_3$ is occurring in this environment. The increase in $CaCO_3$ enriched in ^{12}C with depth suggests that this CO_3^{2-} is derived from the decomposition of organic matter ($\delta^{13}C = -18.6$ to -26.3‰) within the sediment.

Organic Carbon Preservation

The most surprising observation of this study is the relatively low SRR seen at Freefight Lake. Although the dissolved sulfate concentrations are high and the sedimentation rate is low, a large amount of organic carbon is buried rather than oxidized by sulfate reduction or any other microbial pathway. Compared to other hypersaline algal mat environments like Solar Lake and Bonaire (Table 4), this is an unusual finding. In these latter two settings, organic matter is rapidly converted to inorganic carbon via sulfate reduction.

Of course, the obvious question is why Freefight Lake's organic matter is preserved even while concentrations of high dissolved sulfate exists. There is no doubt that in coastal marine environments, sulfate reduction is the major process in organic matter mineralization (Jørgensen, 1982). This prompted the notion that sulfate-rich waters are not conducive to organic matter preservation in lacustrine sediments (Kelts, 1988). It is obvious from our work, that the preservation of organic matter in lacustrine environments is controlled, in part, by factors other than the anionic composition of water and pore water. Recent work on the origin of marine, organic-rich rocks has suggested that high primary production rather than anoxia is the most important control on the preservation of organic matter (Pederson and Calvert, 1990). Using our integrated SRR data and assuming that for every mole of sulfate reduced, two moles of organic carbon are consumed, we have calculated the rate of OC consumption via SR at our four sites at Freefight Lake (Table 5). The organic carbon burial rate at Freefight Lake is ~500 nmol $cm^{-2}d^{-1}$. Therefore, the gross OC primary production rate in the algal mat in this location at Freefight Lake varies from ~3000 nmol $cm^{-2}d^{-1}$ to 1160 nmol $cm^{-2}d^{-1}$. These should be considered minima as loss by other modes of decomposition have not been accounted for. The photosynthetic rate at Solar Lake (assuming a 12 hour period of light and the hourly rates of Cohen and others, 1977) is 50 μmol $cm^{-2}d^{-1}$. Therefore, the photosynthetic rate at Solar Lake is about sixteen times greater than at Freefight Lake. This value must be considered a minimum. These calculations suggest that rates of primary production are not the only factors controlling preservation of organic carbon in Freefight and Solar Lakes. The rate of primary production in Solar Lake is 16 times greater than

TABLE 5.—ESTIMATED ORGANIC CARBON CONSUMPTION RATE AT FREEFIGHT LAKE

Site	Rate (nmol $cm^{-2}d^{-1}$)
A	960
B	2514
C	1070
X	660

in Freefight Lake but far more organic matter is preserved in Freefight Lake sediments.

Degens and Paluska (1979) suggested that organic matter is preserved and petroleum formed in hypersaline environments due to a "pickling" effect. Although they offer no detailed information on how this should occur, it is an intriguing notion. Obviously it does not occur everywhere as organic matter is rapidly decomposed in some hypersaline, anoxic settings such as Solar Lake. Perhaps higher pH in the mixolimnion of the Freefight Lake waters aids in organic matter preservation. However, the mystery of why organic matter accumulates here and not in similar marine hypersaline settings remains. There is no doubt that these high accumulations of organic matter at Freefight Lake could make it a potential oil-shale environment.

CONCLUSIONS

Our work indicates that sulfate reduction rates in salt lakes are not directly related to the anionic composition of the lake water. As in marine sediments, the rate of decomposition and, hence preservation, of organic matter in salt lake sediments is related to the quality and quantity of the organic matter being buried as well as the sedimentation rate.

ACKNOWLEDGMENTS

We are extremely grateful to Dr. Deb Meese for the use of her ion chromatographic facilities. We thank Ms. Joyce Tugel for her laboratory assistance. We acknowledge Ms. Julie Kotraba and Monica Tolliver for typing the manuscript. We appreciate Steve and Nora Palanuik of the Cypress Hills Motor Inn for the use of their "laboratory" facilities. Acknowledgement is made to the Donors of the Petroleum Research Fund, administered by the American Chemical Society for the support of this research. We appreciate the reviews of the original manuscript by two anonymous referees who greatly helped the paper. Thanks also to Robin Renaut for his editorial help and patience.

REFERENCES

AHARON, P., KOLODNY, Y., AND SASS, E., 1977, Recent hot brine dolomitization in the Solar Lake, Gulf of Elat, isotopic, chemical and mineralogical study: Journal of Geology, v. 85, p. 25–48.

ALLER, R. C., AND YINGST, J. Y., 1980, Relationships between microbial distributions and the anaerobic decomposition of organic matter in surface sediments of Long Island Sound, USA: Marine Biology, v. 56, p. 29–42.

BEIN, A., ALMOGI-LAVIN, A., AND SASS, E., 1990, Sulfur sinks and organic carbon relationships in Cretaceous organic-rich carbonates: Implications of evaluation of oxygen-poor depositional environmentals: American Journal of Science, v. 290, p. 882–911.

BERNER, R. A., 1970, Sedimentary pyrite formation: American Journal of Science, v. 268, p. 1–23.

BERNER, R. A., 1978, Sulfate reduction and the rate of deposition of marine sediments: Earth and Planetary Science Letters, v. 37, p. 492–498.

BERNER, R. A., 1980, Early diagenesis: A Theoretical Approach: Princeton, Princeton University Press, 241 p.

BERNER, R. A., 1985, Sulphate reduction, organic matter decomposition and pyrite formation: Philosophical Transactions of Royal Society of London A, v. 315, p. 25–38.

BERNER, R. A., AND RAISWELL, R., 1983, Burial of organic carbon and pyrite sulfur in sediments over Phanerozoic time: a new theory: Geochimica et Cosmochimica Acta, v. 47, p. 855–862.

BERNER, R. A., AND RAISWELL, R., 1984, C/S method for distinguishing freshwater from marine sedimentary rocks: Geology, v. 12, p. 365–368.

BOYER, B. W., 1982, Green River laminants: Does the playa lake model really invalidate the stratified-lake model?: Geology, v. 10, p. 124–160.

CALVERT, S. E., AND KARLIN, R. E., 1991, Relationships between sulfur, organic carbon, and iron in the modern sediments of the Black Sea: Geochimica et Cosmochimica Acta, v. 55, p. 2483–2490.

COHEN, Y., CRUMBEIN, W. E., AND SHILO, M., 1977, Solar Lake (Sinai) 2. Distribution of photosynthetic microorganisms and primary production: Limnology and Oceanography, v. 22, p. 609–620.

COLE, J. J., HOWARTH, R. W., NOLAN, S. S., AND MARINO, R., 1986, Sulfate inhibition of molybdate assimilation by planktonic algae and bacteria: some implications for the aquatic nitrogen cycle: Biogeochemistry, v. 2, p. 179–196.

DEGANS, E. T., AND PALUSKA, A., 1979, Hypersaline solutions interact with organic detritus to produce oil: Nature, v. 281, p. 666–668.

DESBOROUGH, G. A., 1978, A biogenic-chemical stratified lake model for the origin of oil shale of the Green River Formation: An alternative to the playa-lake model: Geological Society of America Bulletin, v. 89, p. 961–971.

EUGSTER, H. P., AND HARDIE, L. A., 1975, Sedimentation in an ancient playa-lake complex: The Wilkins Peak Member of the Green River Formation of Wyoming: Geological Society of America Bulletin, v. 86, p. 319–334.

FROELICH, P.N., KLINKHAMMER, G.P., BENDER, M.L., LUEDTKE, N.A., HEATH, G.R., CULLEN, D., DAUPHIN, P., HAMMOND, D., HARTMAN, B., AND MAYNARD, V., 1979, Early oxidation of organic matter in pelagic sediments of the eastern equatorial Atlantic: suboxic diagenesis: Geochimica et Cosmochimica Acta, v. 43, p. 1075–1090.

GOLDHABER, M. B., AND KAPLAN, I. R., 1975, Controls and consequences of sulfate reduction rates in recent marine sediments: Soil Science, v. 119, p. 42–55.

HINES, M. E., BAZYLINSKI, D. A., TUGEL, J. B., AND LYONS, W. B., 1991, Anaerobic microbial biogeochemistry in sediments from the two basins in the Gulf of Maine: Evidence from iron and manganese reduction: Estuarine Coastal and Shelf Science, v. 32, p. 313–324.

HINES, M. E., AND JONES, G. E., 1985, Microbial biogeochemistry in the sediments of Great Bay, New Hampshire: Estuarine Coastal and Shelf Science, v. 20, p. 729–742.

HINES, M. E., KNOLLMEYER, S. L., AND TUGEL, J. B., 1989, Sulfate reduction and other sedimentary biogeochemistry in a northern New England salt marsh: Limnology and Oceanography, v. 34, p. 578–590.

HINES, M. E., LYONS, W. B., LENT, R. M., AND LONG, D. T., 1992, Sedimentary biogeochemistry of an acidic, saline groundwater discharge zone in Lake Tyrrell, Victoria, Australia: Chemical Geology, v. 96, p. 53–65.

JOHNSON, R. C., 1981, Stratigraphic evidence for a deep Eocene Lake Uinta, Piceance Creek Basin, Colorado: Geology, v. 9, p. 55–62.

JØRGENSEN, B. B., 1982, Mineralization of organic matter in the sea bed—The role of sulphate reduction: Nature, v. 296, p. 643–645.

JØRGENSEN, B. B., AND COHEN, Y., 1977, Solar Lake (Sinai) 5. The sulfur cycle of the benthic cyanobacterial mats: Limnology and Oceanography, v. 22, p. 657–666.

KELTS, K., 1988, Environments of deposition of lacustrine petroleum source rocks: an introduction, *in* Fleet, A. J., Kelts, K. and Talbot, M. R., eds., Lacustrine Petroleum Source Rocks: Oxford, Geological Society Special Publication No. 40, p. 3–26.

KIRKLAND, D. W., AND EVANS, R., 1981, Source rock potential of evaporitic environments: Association of Petroleum Geologists Bulletin, v. 65, p. 181–190.

LAST, W. M., 1990, Sedimentology and Geochemistry of Freefight Lake, Saskatchewan: Winnipeg, Unpublished Final Report to University of New Hampshire, Contract Number 89–09, 250 p.

LAST, W. M., 1992, Chemical composition of saline and subsaline lakes of the northern Great Plains, western Canada: International Journal of Salt Lake Research, v. 1, p. 47–76.

LAST, W. M., 1993a, Geolimnology of Freefight Lake: an unusual hypersaline lake in the northern Great Plains of western Canada: Sedimentology, v. 40, p. 431–448.

LAST, W. M., 1993b, Rates of sediment deposition in a hypersaline lake in the northern Great Plains, western Canada: International Journal of Salt Lake Research, v. 2, p. 47–58.

LAST, W. M., AND SLEZAK, L. A., 1987, Geolimnology of an unusual saline lake in the Great Plains of western Canada, *in* Chivas, A. R. and De Deckker, P., eds., SLEADS Workshop 87: Canberra, Australian National University, p. 9–11.

LORD, C. J., AND CHURCH, T. M., 1983, The geochemistry of salt marshes: Sedimentary ion diffusion, sulfate reduction, and pyritization: Geochimica et Cosmochimica Acta, v. 47, p. 1381–1391.

LYONS, W. B., HINES, M. E., AND GAUDETTE, H. E., 1984a, Major and Minor element pore water geochemistry of modern marine sabkhas, *in* Microbial Mats: Stromatolites: New York, A.R. Liss, Inc., p. 411–423.

LYONS, W. B., LONG, D. T., HINES, M. E., GAUDETTE, H. E., AND ARMSTRONG, P. B., 1984b, The calcification of cyanobacterial mats in Solar Lake, Sinai: Geology, v. 12, p. 623–626.

MARINO, R., HOWARTH, R. W., SHAMESS, J., AND PREPAS, E., 1990, Molybdenum and sulfate as controls on the abundance of nitrogen-fixing cyanobacteria in saline lakes in Alberta: Limnology and Oceanography, v. 35, p. 245–259.

MINSTER, T., NATHAN Y., AND RAVEH, A. 1992, Carbon and sulfur relationships in margin Senonian organic-rich, iron-poor sediments from Israel—A case study: Chemical Geology, v. 97, p. 145–161.

MORSE, J. W., AND CORNWELL, J. C., 1987, Analysis and distribution of iron sulfide minerals in recent anoxic marine sediments: Marine Chemistry, v. 22, p. 55–69.

PEDERSEN, T. F., AND CALVERT, S. E., 1990, Anoxia vs. productivity: What controls the formation of organic-carbon-rich sediments and sedimentary rocks?: American Association of Petroleum Geologists Bulletin, v. 74, p. 454–466.

SLEZAK, L. A., 1989, Modern sedimentary environments and early sediment diagenesis of Freefight Lake in southwestern Saskatchewan: Unpublished M.S. Thesis, University of Manitoba, Winnipeg, 177 p.

SMOOT, J. P., 1983, Depositional subenvironments in an arid closed basin: The Wilkins Peak Member of the Green River Formation (Eocene), Wyoming, USA: Sedimentology, v. 30, p. 801–827.

SPADER, D., AND LYONS, W. B., 1985, Early diagenesis of carbonate sediments in an arid, supratidal environment: Bonaire, N. A.: Stygologia, v. 1, p. 160–185.

SULLIVAN, R., 1985, Origin of lacustrine rocks of Wilkins Peak Member, Wyoming: American Association of Petroleum Geologists Bulletin, v. 69, p. 913–922.

SURDAM, R. C., AND STANLEY, K. O., 1979, Lacustrine sedimentation during the culminating phase of Eocene Lake Gosiute, Wyoming (Green River Formation): Geological Society of America Bulletin, v. 90, p. 93–110.

SWEENEY, R. E., AND KAPLAN, I. R., 1973, Pyrite framboid formation: laboratory synthesis and marine sediments: Economic Geology, v. 68, p. 618–634.

TISSOT, B., DEROO, G., AND HOOD, H., 1978, Geochemical study of the Uinta Basin: formation of petroleum from the Green River Formation: Geochimica et Cosmochimica Acta, v. 42, p. 1469–1485.

WARREN, J. K., 1986, Shallow water evaporitic environments and their source rock potential: Journal of Sedimentary Petrology, v. 56, p. 442–454.

WESTICH, J. T., AND BERNER, R. A., 1984, The role of sedimentary organic matter in bacterial sulfate reduction: The G model tested: Limnology and Oceanography, v. 29, p. 236–249.

WETZEL, R. G., 1975, Limnology: Philadelphia, W. B. Saunders Co., 743 p.

BOTTOM-SEDIMENT CHEMISTRY IN DEVILS LAKE, NORTHEAST NORTH DAKOTA

STEPHEN C. KOMOR

United States Geological Survey, 2280 Woodale Drive, Mounds View, Minnesota 55112

ABSTRACT: Devils Lake is a 200-km^2 terminal lake that contains sodium sulfate type water. Dissolved-solids concentrations range from about 3,500 mg/L to 10,000 mg/L depending on location. To investigate geochemical processes in the bottom sediments of Devils Lake, sediment cores were collected at two sites in the western half of the lake during a period of bottom-water oxygen depletion. The upper 10 cm of the sediments consist of about 60 weight percent silicates (quartz, feldspar, and clays), 35 weight percent carbonates, and 5 weight percent organic material. At depths between 1 and 3 cm in the sediments, bacterial sulfate reduction and associated degradation of organic material cause minima in sulfate concentrations and $\delta^{13}C$ values of dissolved inorganic carbon and maxima in alkalinity, ammonia, phosphate, and sulfide concentrations and $\delta^{34}S$ values of dissolved sulfate. Downward increases of sodium, magnesium, potassium, and calcium concentrations result from upward diffusion of ions from saline pore water and dissolving sulfate minerals below 30-cm depth in the sediments.

High-magnesium calcite (8 mole percent $MgCO_3$) is the most abundant carbonate at the sediment surface. With increasing depth, abundances of high-magnesium carbonate decrease and abundances of low-magnesium calcite, aragonite, and dolomite increase. Carbon-isotope compositions of bulk carbonates range from $\delta^{13}C = -0.7$ to $+0.5$‰. These values are close to equilibrium with dissolved inorganic carbon in lake water ($\delta^{13}C = -2.0$‰), but far from equilibrium with dissolved inorganic carbon in pore water ($\delta^{13}C = -16.3$ to -10‰). Disequilibrium between pore water and carbonates suggests that the carbonates did not recrystallize substantially in the presence of pore water. Therefore, the change of carbonate-mineral proportions with depth in the sediments is due mainly to temporal changes in the proportions of endogenic, detrital, and biologic carbonates that were deposited on the lake bottom rather than postdepositional carbonate diagenesis.

INTRODUCTION

The chemistry of salt lakes can change dramatically in response to changing hydrologic conditions. Salinity increases when evaporation exceeds inflow or when inflow is more saline than lake water (Eugster and Hardie, 1978). Salinity increases can lead to precipitation of evaporite minerals in the water column and on the lake bottom. Conversely, the salinity of lake water decreases and previously deposited evaporite minerals may dissolve when inflow exceeds evaporation or when inflow is less saline than lake water. A record of chemical variations in lake water may be preserved by the compositions of pore water and evaporite minerals in the bottom sediments.

This study was undertaken to document the compositions of pore water and sediment on the bottom of Devils Lake, a saline lake in northeast North Dakota (Fig. 1). Chemical and isotopic analyses were obtained for the top 10 cm of bed material in Creel Bay and Main Bay in Devils Lake. Sediments in this 10-cm interval accumulated over approximately 25 years between 1965 and 1990 (Jacobsen and Engstrom, 1989). This comparatively thin sediment interval contains evidence for sulfate reduction, upward fluxes of dissolved ions from buried evaporites, and temporal changes in the compositions of carbonates deposited on the lake bottom. Information on these and other geochemical features in the bottom sediments provides insight into the chemical evolution of Devils Lake and other closed-basin lakes on the northern prairies.

LAKE-LEVEL AND SALINITY VARIATIONS

Devils Lake is a 200-km^2 lake that drains a closed basin of 9,870 km^2. Devils Lake is not integrated with regional drainage systems and its water level is highly dependent on local climatic conditions. In the 1940's, the lake was only about 0.9 m deep and dissolved-solids concentrations exceeded 25,000 mg/L (Callender, 1968). Figure 2A summarizes salinities and lake-level variations during the 25 years when the top 10 cm of sediment accumulated. Between 1965 and 1987, the lake level rose by approximately 5.4 m and the average salinity of the western half of the lake decreased by about 9,000 mg/L. Drought conditions prevailed during 1987–90 and the lake level fell by about 1.3 m while salinities increased by 1,000 mg/L. The present-day (1991) average lake depth is about 3.7 m and dissolved-solids concentrations range from about 3,500 to 10,000 mg/L, depending on location in the lake. Sodium and sulfate are the dominant dissolved ions in Devils Lake (Fig. 2B). Significant increases in the concentrations of these ions occurred between August 15, 1989, and May 9, 1990, a period when the lake level fell by about 25 cm.

Calculated water balances indicate that ground-water inflow into Devils Lake is minimal (Wiche, 1991). Therefore, lake-water chemistry is not significantly affected by upward advection of ground-water through the bottom sediments.

SAMPLING AND ANALYTICAL METHODS

Sediment cores (6.75 cm in diameter and about 50 cm in length) were collected in Devils Lake on July 23 and 25, 1990, when the bottom water was depleted in oxygen. Water depths were 5.58 m at site T7 in Creel Bay and 7.44 m at site T2 in Main Bay. Cores were collected from an anchored platform that drifted a small amount between successive core recoveries, but at each site the sampled area of lake bottom probably was not larger than about 20 m^2. Pore water was extracted from the cores within 24 hours of collection.

Sediments and pore water were processed under a nitrogen atmosphere to preserve ambient redox conditions from the lake bottom. Samples of lake bottom water were collected from about 1 cm above the sediment surface in each core tube and treated as described for pore water. Sediments were extruded from the core tubes, cut into slices 1

Sedimentology and Geochemistry of Modern and Ancient Saline Lakes, SEPM Special Publication No. 50

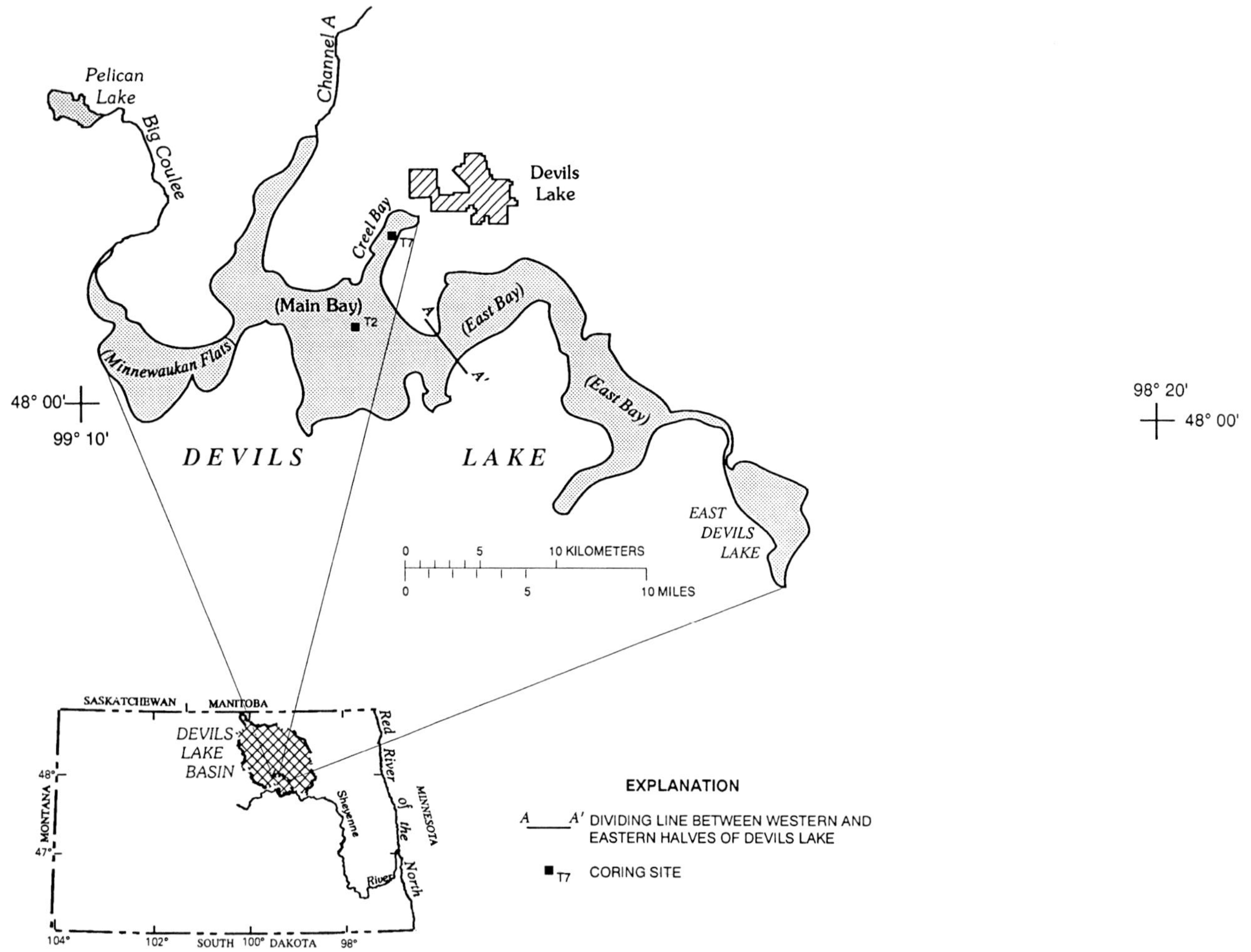

FIG. 1.—Generalized map of Devils Lake in northeast North Dakota. Sediment cores were recovered at sites T2 in Main Bay and T7 in Creel Bay. Big Coulee and channel A are the two major sources of inflow to Devils Lake. Line A-A′ divides the lake into western and eastern halves, as discussed in the text.

or 2 cm long, and centrifuged for 10 minutes at 10,000 r/min. Supernatant liquid was decanted and filtered through polycarbonate filters with 0.2-μm pore size. Pore water from each depth interval in the cores was combined to produce an average pore water for each interval at each site. Pore water was combined from five cores for Creel Bay samples and from 10 cores for Main Bay samples. Filtered pore water was treated as follows: 20 mL was treated with 0.1 mL of concentrated HNO_3 and analyzed for cations; a similar volume was chilled and analyzed for anions and nutrients; 1 mL was treated with N,N-dimethyl-p-phenylenediamine oxalate, diluted to 25 mL with distilled water, and analyzed for sulfide concentrations; between 2 and 10 mL was treated with 1,10 phenathroline indicator, diluted to 50 mL with distilled water, and analyzed for ferrous-iron concentrations. Samples from the Main Bay site were prepared for stable-isotope analyses. Twenty-five mL of filtered pore water was treated with 1 mL of amnoniacal strontium chloride to precipitate strontium carbonate (Bishop, 1990) and analyzed for carbon-isotope ratios of dissolved inorganic carbon (DIC). Two 15-mL volumes were treated with 0.1 mL of mercuric chloride; one was analyzed for oxygen- and hydrogen-isotope ratios and the other was analyzed for sulfur-isotope ratios in dissolved sulfate. Sediments from which pore water was extracted were analyzed for mineralogy, carbon- and oxygen-isotope ratios in solid carbonate, and sulfur-isotope ratios in solid sulfide. Two samples of gases from between 4- and 6-cm depth in the sediments were collected by the method of Oremland and Des Marais (1983).

Analytical methods and errors, instrumental detection limits, and dilution factors for water analyses are summarized in Table 1. Pore-water samples were diluted to provide sufficient volumes for analysis; instrumental detection limits must be multiplied by the dilution factors to determine the actual detection limits for these samples. Analytical errors were determined from replicate analyses of standards. Concentrations of major ions (sodium, magnesium, potassium, calcium, sulfate, and chloride), trace constituents (arsenic, boron, fluoride, iron, lead, lithium, manganese, mercury, molybdenum, selenium and strontium), and nutrients (ammonia, nitrate plus nitrite, total phosphate, and silicon) were measured at the North Dakota State

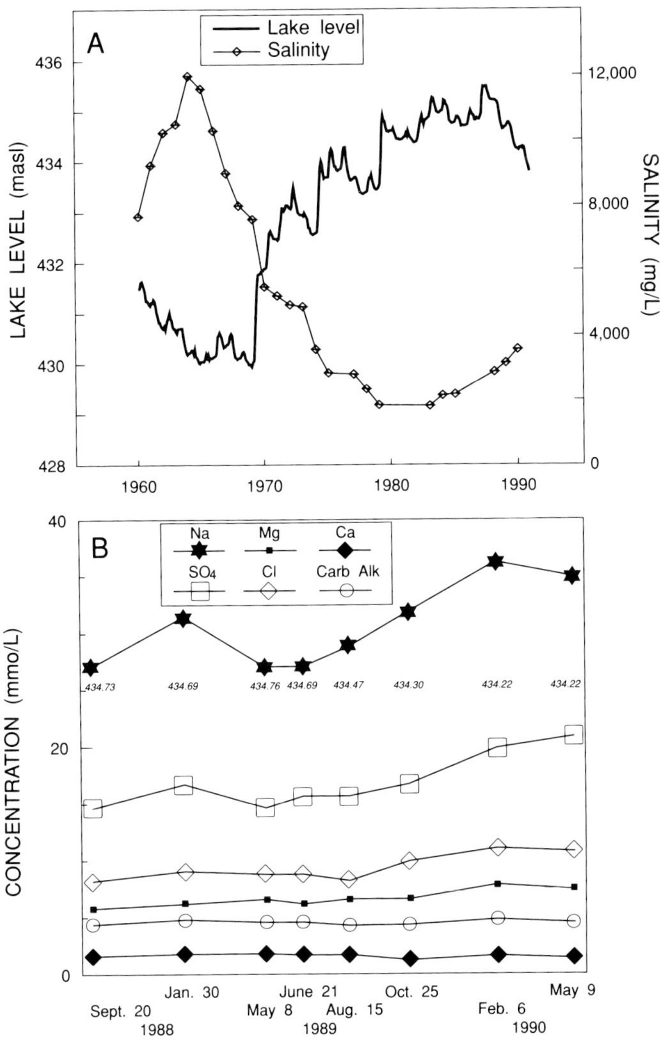

FIG. 2.—(A) Changes in lake level and average salinities in the western half of Devils Lake (west of A-A′ in Fig. 1) between 1960 and 1990. Lake-level data are from U. S. Geological Survey gaging stations. Salinity data are from Steve Sando, U. S. Geological Survey (written commun., 1991). Sediments analyzed in this study accumulated between about 1955 and 1990, a period dominated by rising lake levels and decreasing salinity. (B) Concentrations of major cations and anions in near-surface lake water at site T2 in Main Bay between September 20, 1988, and May 9, 1990. Carb-alk is carbonate alkalinlty. Numbers are lake levels in meters above sea level. Concentrations of sodium, magnesium, sulfate, and chloride are inversely correlated with lake levels.

TABLE 1.–ANALYTICAL METHODS, DETECTION LIMITS, ERRORS, AND DILUTIONS FOR WATER ANALYSIS

Constituent	Analytical method[1]	Detection limit (mg/L)	Analytical error (percent)[2]	Dilution
Major ions				
Calcium (Ca)	ICP	0.012	6	10
Magnesium (Mg)	ICP	.090	5	10
Potassium (K)	ICP	.10	5	10
Sodium (Na)	ICP	.48	7	10
Sulfate (SO_4)	FIA	15.0	7	10
Chloride (Cl)	FIA	5.0	8	10
Alkalinity	TITR	5.0	5	2 to 2.5
Trace metals				
Arsenic (As)	GFAA	0.001	7	10
Boron (B)	ICP	.006	5	10
Fluoride (F)	SIE	.10	33	10
Iron (Fe)	ICP	.012	6	10
Lead (Pb)	GFAA	.001	8	10
Lithium (Li)	GFAA	.001	7	10
Manganese (Mn)	ICP	.010	5	10
Mercury (Hg)[3]	MCV	.0002	—	100
Molybdenum (Mb)[3]	ICP	.057	—	10
Selenium (Se)	GFAA	.001	8	10
Strontium (Sr)	ICP	.004	5	10
Reduced species				
Sulfide (S^{2-})	PM	0.001	5	25 to 250
Ferrous iron (Fe^{2+})	MBM	.001	5	5 to 25
Nutrients				
Nitrogen, total	FIA	0.010	5	10
Phosphorous, total	FIA	0.10	6	10
Silicon (Si)	ICP	.025	7	10

[1]Analytical method: ICP, inductively-coupled plasma spectroscopy; FIA, flow-injection analysis; SIE, specific-ion electrode; TITR, titration with sulfuric acid; GFAA, graphite-furnace atomic absorption; MCV, manual cold-vapor atomic absorption; PM, 1,10 phenathroline method, measured in glove-box; MBM, methylene-blue method, measured in glove-box.

[2]Analytical error: Values represent two standard deviations of measured values, expressed as percentages of measured values; —indicates no data.

[3]Concentrations of mercury and molybdenum were below detection limits in all samples.

Department of Health and Consolidated Laboratories in Bismarck (Table 2). Concentrations of dissolved sulfide and ferrous iron were measured with a Hach DR/2000 spectrophotometer. Alkalinity was determined by titration with 0.16N H_2SO_4 to a pH of 4.3.

Sediment porosity was determined by weighing known volumes of sediment before and after drying. Weight-loss data were equated to water volumes by assuming a constant water density of 1.0 g/cm^3 and divided by the total sample volumes to yield sediment porosities. Porosities average 94 percent in the top 10 cm of sediment in both Creel Bay and Main Bay. The assumption that pore-water density is 1.0 g/cm^3 does not cause significant errors in the porosity estimates because the pore water contains <5,000 mg/L dissolved solids. For comparison, seawater with about 35,000 mg/L dissolved solids has a density of 1.024 g/cm^3 (Henderson, 1982).

Sediment mineralogy (Table 3) was determined by X-ray diffraction with Cu-K_a radiation at a scan rate of 1.8° 2θ per minute. The modal mineralogy in weight percent was determined by comparing diffraction-peak heights in the samples to those in standards with known mineralogical proportions. The modal values have relative errors of ± 10 percent.

Concentrations of methane in pore water were analyzed by gas chromatography. Methane concentrations were below the detection limit of 0.5 volume percent in the samples, which equates to a detection limit of about 5,300 ppm in the pore water.

Stable-isotope ratios (Table 4) were measured by mass spectrometry in U. S. Geological Survey laboratories. The CO_2-equilibration method (Epstein and Mayeda, 1953) was used for measurements of oxygen-isotope ratios in water. Hydrogen was extracted quantitatively from water by passage over hot zinc (Kendall and Coplen, 1985). Carbonates were reacted with 100 percent phosphoric acid under vacuum to liberate CO_2 for isotopic measurements (McCrea,

TABLE 2.—BOTTOM-WATER AND PORE-WATER CHEMISTRY FOR SITE T7 IN CREEL BAY AND SITE T2 IN MAIN BAY

Depth interval (cm)[1]	pH		Ca (mg/L)		Mg (mg/L)		K (mg/L)		Na (mg/L)		SO_4 (mg/L)		Cl (mg/L)		Alkalinity (mg/L as $CaCO_3$)	
	CB[2]	MB[3]	CB	MB	CB	MB	CB	MB	CB	MB	CB	MB	CB	MB	CB	MB
B-W	8.76	8.27	62	64	190	195	95	102	762	796	1,640	1,560	361	389	475	432
0-1	7.46	8.59	67	62	198	184	105	98	772	751	1,320	1,400	405	495	520	608
1-2	7.75	8.08	69	70	202	199	105	103	777	817	1,040	1,120	1,040	519	800	836
2-3	8.47	8.98	72	67	210	196	112	107	802	812	926	1,180	461	519	1,260	1,180
3-4	8.38	8.43	73	74	207	216	110	112	783	888	989	1,320	486	503	1,240	1,100
4-6	7.51	8.69	80	75	212	225	112	116	807	945	1,170	1,450	749	510	724	1,030
6-8	8.44	8.27	91	81	227	254	118	126	858	1,050	1,390	1,750	458	533	1,090	1,060
8-10	7.83	8.19	99	84	232	271	119	130	865	1,100	1,480	2,060	551	553	975	1,040

Depth interval (cm)	As (µg/L)		B (µg/L)		F (µg/L)		Fe_{total} (mg/L)		Pb (µg/L)		Li (µg/L)		Mn (mg/L)		Se (µg/L)	
	CB	MB	CB	MB	CB	MB	CB	MB	CB	MB	CB	MB	CB	MB	CB	MB
B-W	29	31	1,560	1,340	440	200	0.18	0.13	45	25	291	297	0.09	0.07	43	25
0-1	44	24	1,960	1,390	ND[4]	210	.45	.16	39	31	302	292	.76	.49	24	3
1-2	114	72	1,600	1,450	230	190	.24	.33	207	57	313	308	.61	.65	32	65
2-3	100	78	1,210	1,640	210	180	.24	.15	47	49	319	310	.52	.64	26	49
3-4	153	130	1,720	1,480	190	180	.34	.19	32	31	310	326	.59	.76	32	37
4-6	112	77	1,560	1,370	220	180	.16	.18	25	26	322	336	.64	.83	20	27
6-8	14	71	2,020	1,380	190	180	.29	.24	27	30	337	377	.77	.69	27	28
8-10	60	69	2,080	1,530	190	190	.29	.21	24	26	330	387	.73	.55	18	26

Depth interval (cm)	Sr (µg/L)		S^{2-} (mg/L)		Fe^{2+} (mg/L)		NH_4 as N (mg/L)		$NO_2 + NO_3$ (mg/L)		P_{total} (mg/L)		Si (mg/L)	
	CB	MB	CB	MB	CB	MB	CB	MB	CB	MB	CB	MB	CB	MB
B-W	426	415	BDL[5]	BDL	BDL	BDL	0.2	0.0	0.30	0.10	0.21	0.23	6.3	5.7
0-1	417	446	14.8	4.3	0.5	BDL	7.5	24.7	.40	.20	1.52	.85	21.6	13.3
1-2	423	488	35.5	30.3	BDL	BDL	13.8	35.2	.10	.10	2.44	1.93	27.8	24.1
2-3	455	449	71.0	24.9	.1	0.3	14.3	57.4	.20	.10	2.42	2.20	34.5	28.8
3-4	461	454	71.8	41.0	BDL	BDL	13.4	32.2	.10	.04	2.20	2.17	34.3	29.1
4-6	479	460	49.0	43.5	BDL	.1	12.3	30.2	.20	.20	2.20	2.02	30.9	26.6
6-8	525	522	51.5	58.0	BDL	BDL	10.9	25.1	.10	.20	2.30	2.52	30.5	26.8
8-10	500	811	22.5	48.9	BDL	BDL	11.7	46.7	.20	.30	2.68	3.06	28.6	27.9

[1]Depth interval: Depths of sampled intervals, in cm. Sample B-W is water from about 1 cm above the sediment surface. Data points in figures are plotted at the midpoints of depth intervals (e.g., 0.5 cm for the interval 0.1 cm).
[2]CB, site T7 in Creel Bay.
[3]MB, site T2 in Main Bay.
[4]ND, not determined.
[5]BDL, below detection limits (see table 1).

TABLE 3.—MINERALOGY OF SEDIMENTS FROM SITE T2 IN MAIN BAY (WEIGHT PERCENT) AND $MgCO_3$ CONTENT OF CALCITE (MOLE PERCENT)

Depth interval (cm)	Weight percent									
	Quartz	Plagioclase	Potassium feldspar	Calcite	Dolomite	Aragonite	Kaolinite	Montmorillonite	Muscovite	Total
0-1	32	8	5	34	9	BDL[1]	2	2	5	97
1-2	35	7	6	29	11	BDL	2	2	4	96
2-3	35	5	6	21	11	9	2	2	5	96
3-4	40	8	7	15	15	BDL	3	2	5	95
4-6	44	5	BDL	12	15	12	2	2	4	96
6-8	40	8	BDL	10	14	15	2	2	5	96
8-10	40	6	5	9	11	14	2	2	5	94

Depth interval (cm)	Other minerals	$MgCO_3$ in calcite (mole percent)
0-1	1-2 weight percent hematite, 2 weight percent other carbonates	8.3
1-2	1-2 weight percent hematite	6.1
2-3	1-2 weight percent hematite	5.5
3-4	2-3 weight percent hematite, 2-3 weight percent other carbonates	3.0
4-6	2-3 weight percent hematite	3.5
6-8	2-3 weight percent hematite	1.6
8-10	1-2 weight percent hematite	3.0

[1]BDL, below detection limit (approximately 2 weight percent).

TABLE 4.—VALUES OF δD (V-SMOW), $\delta^{18}O$ (V-SMOW), $\delta^{13}C$ (PDB), AND $\delta^{34}S$ (CDT) IN WATER AND SEDIMENTS AT SITE T2 IN MAIN BAY, DEVILS LAKE

Depth interval (cm)	Water				Carbonate		Sulfide
	δD (‰)	$\delta^{18}O$ (‰)	$\delta^{13}C$ (‰)	$\delta^{34}S$ (‰)	$\delta^{18}O$ (‰)	$\delta^{13}C$ (‰)	$\delta^{34}S$ (‰)
B-W[1]	−47.0	−4.0	−2.0	1.5			
0.1	−49.0	−4.3	−10.0	4.2	25.8	0.0	−27.5
1-2	−47.5	−4.1	−14.4	9.8	26.2	−0.2	−30.4
2-3	−48.0	−4.0	−16.3	7.9	27.0	−0.3	−29.9
3-4	−46.0	−3.9	−14.6	6.2	27.0	−0.7	−30.9
4-6	−46.5	−4.2	−13.5	12.0	28.0	−0.1	−35.7
6-8	−48.5	−4.3	−13.3	10.2	27.4	0.5	−40.1
8-10	−49.5	−4.4	−12.7	10.9	27.8	0.5	−39.6

[1]Sample B-W is water from about 1 cm above the sediment surface.

1950). Strontium carbonate precipitates of DIC in pore water were reacted at 25°C for 16 hours; bulk sediment samples containing mixed carbonates were reacted at 50°C for 72 hours to ensure that dolomite dissolved. Barium chloride was added to water samples to precipitate dissolved sulfate as $BaSO_4$. Dried $BaSO_4$ precipitate and solid sulfides in the sediments were converted to SO_2 for isotopic measurements.

Isotopic data (Table 4) are presented in the standard δ notation:

$$\delta_x = [(R_x/R_{std}) - 1] \times 10^3 \qquad (1)$$

where x is ^{18}O, deuterium (D), ^{13}C, or ^{34}S; R_x is the ratio of isotopes ($^{18}O/^{16}O$, D/H, $^{13}C/^{12}C$, or $^{34}S/^{32}S$) in a sample; and R_{std} is the corresponding ratio in a standard (Vienna Standard Mean Ocean Water (X-SMOW) for oxygen and hydrogen, the Pee-Dee belemnite (PDB) for carbon, and the Canyon Diablo troilite (CDT) for sulfur). Analytical standard deviations are 1‰ for δD and 0.1‰ for $\delta^{18}O$, $\delta^{13}C$, and $\delta^{34}S$.

CORE DESCRIPTION

The top 1–2 cm of sediment in Creel Bay and Main Bay is soft gray mud. Green algal debris is common at the sediment/water interface and decreases in abundance downwards. Between 2- and 6-cm depth, the sediments are light gray to brown, unlaminated, and have a fine, loose appearance. Below 6-cm depth, the sediments are gray to dark brown or black and take on a more massive, dense appearance. Numerous irregular black patches that probably are rich in metal sulfides are common below 3-cm depth in the sediments. Between depths of 15 and 50 cm, the sediments are dark brown to black and unlaminated. A few small red chironomid larvae were found at the sediment/water interface, but no larvae burrows were found below the interface.

SEDIMENT MINERALOGY

Sillcate minerals compose about 55 to 60 weight percent of the sampled sediments (Table 3). Quartz and feldspar are the most abundant silicates, followed by muscovite. Clay minerals include kaolinite and montmorillonite. Small concentrations of hematite (1–3 weight percent) were detected in all samples. Much of this hematite may represent fine-grained iron sulfide that became oxidized during sample drying.

Total carbonate concentrations decrease from 43 weight percent at 0-1 cm depth to 34 weight percent at 8–10 cm depth. Calcite concentrations decrease with depth from 34 to 9 weight percent, whereas dolomite concentrations increase from 9 to 15 weight percent and aragonite concentrations increase from 0 to 15 weight percent (Fig. 3). The fraction of $MgCO_3$ in calcite, as determined from the position of the calcite *d*(104) diffraction peak (Goldsmith and Graf, 1958), decreases with depth from about 8 to 2 mole percent.

The upper meter of sediments in Devils Lake contains an average of about 5 weight percent organic carbon (Callender, 1968). Organic-carbon concentrations may be considerably larger at the sediment/water interface.

PORE-WATER CHEMISTRY

Dissolved Constituents

Devils Lake pore water is alkaline, with pH values varying between about 7.5 and 9.0 in the sampled sections. Concentrations of major cations (sodium, magnesium, potassium, and calcium) in pore water increase almost linearly with depth in the sampled sections (Fig. 4). The downward increases in concentrations indicate that cations are diffusing toward the sediment/water interface from below 10-cm depth in the sediments. Sulfate concentration profiles have pronounced minima at 2–3 cm depth (Fig. 5A). At about the same depth, concentrations of alkalinity, ammonia, silica, and phosphate reach local maxima (Figs. 5B-D). In Creel Bay, sulfide concentrations reach a maximum at depths of 2–3 cm (Fig. 5E). In Main Bay, sulfide concentrations reach a local maximum at 1–2 cm depth and are largest between 6- and 8-cm depth. Pore-water chloride

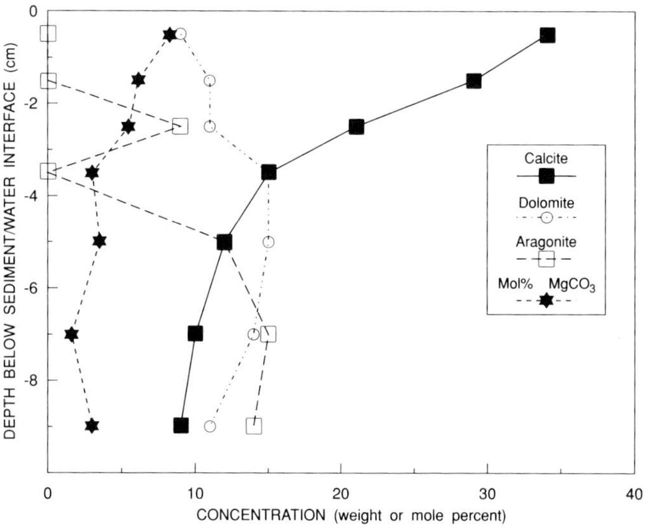

FIG. 3.—Weight percentages of calcite, dolomite, and aragonite and mole percent of $MgCO_3$ at site T2 in Main Bay. Modal calcite and mole percentages of $MgCO_3$ in calcite decrease with depth, whereas modal dolomite and aragonite increase with depth.

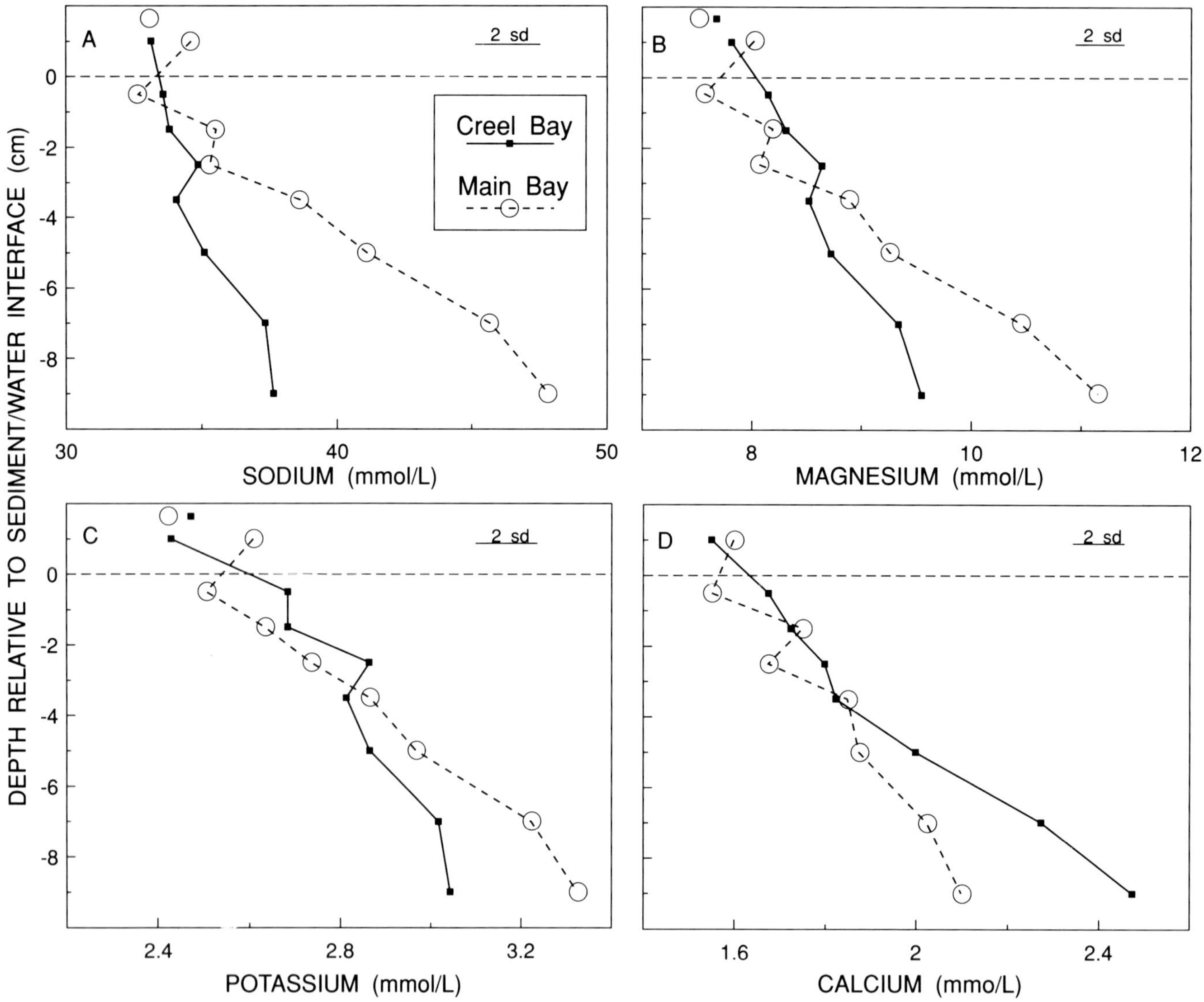

FIG. 4.—Depth variations in bottom water and pore water of (A) sodium, (B) magnesium, (C) potassium, and (D) calcium. Symbols not connected by lines at the tops of graphs (A), (B), and (C) show the compositions of near-surface lake water at the sampling sites on August 7, 1990. Calcium concentrations on this date were 1.22 mmol/L in Creel Bay and 1.34 mmol/L in Main Bay. Bars labeled 2 sd in this and following figures show two analytical standard deviations. Horizontal dashed lines in this and following depth profiles mark the sediment/water interface. Concentrations of major cations increase with depth in the sediments along roughly linear trends. These trends indicate that the cations diffuse upward in the upper 10 cm of sediment.

concentrations in Main Bay increase gradually with depth (Fig. 5F). Chloride concentrations in Creel Bay follow the same general trend, with the exceptions of concentration peaks at depths of 1–2 cm and 4–6 cm.

Concentration profiles of the trace constituents are characterized by pronounced peaks (Fig. 6). The peaks for iron occur at 0-1 cm depth in Creel Bay and 1–2 cm depth in Main Bay; for manganese at 0-1 cm depth in Creel Bay and 4–5 cm depth in Main Bay; for lead at 1–2 cm depth in both bays; for selenium at 1–2 cm depth in Main Bay; and for arsenic at 3–4 cm depth in both bays. Below depths of 6 cm in the sampled sections, concentrations of iron, lead, selenium, and arsenic vary within comparatively restricted ranges. These ranges are 3.8 to 5.2 μmol/L for iron, 0.12 to 0.14 μmol/L for lead, 0.23 to 0.35 μmol/L for selenium, and 0.19 to 0.95 μmol/L for arsenic.

Stable Isotopes

Values of δD and $\delta^{18}O$ in water covary within restricted ranges in the sampled sections (Fig. 7). Values of $\delta^{13}C$ in dissolved inorganic carbon ($\delta^{13}C_{DIC}$) decrease from −2‰ above the sediment/water interface to a minimum of −16.3‰ at 2–3 cm depth (Fig. 8A, squares). Sedimentary carbonate has $\delta^{13}C$ values between −0.7 and +0.5‰ (Fig. 8A, circles) and $\delta^{18}O$ values that increase from 25.8‰ below the sediment water interface to a maximum of 28‰ at 4–6 cm depth (Fig. 8B, circles). Values of $\delta^{34}S$ in dis-

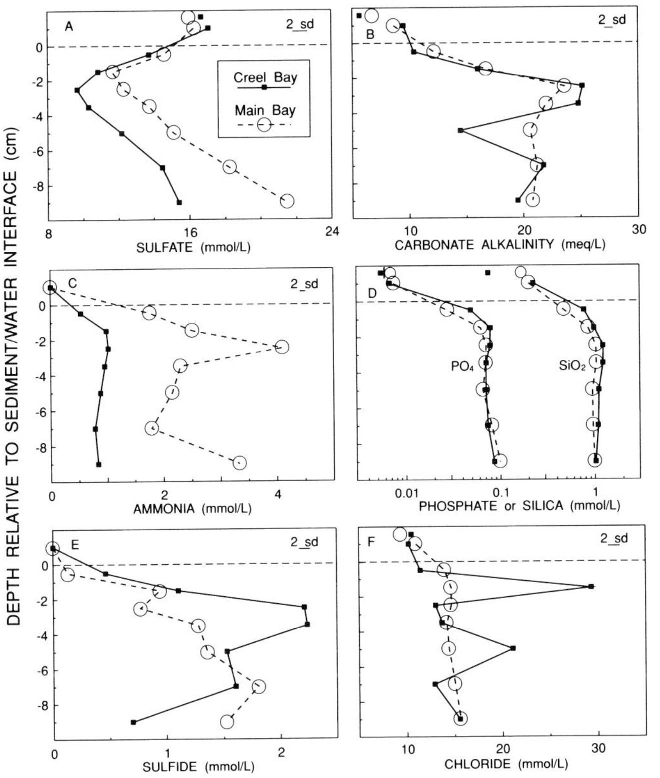

FIG. 5.—Depth variations in bottom water and pore water of (A) sulfate, (B) carbonate alkalinity, (C) dissolved ammonia, (D) phosphate and silica, (E) dissolved sulflde, and (F) chloride. Symbols not connected by lines at the top of graphs (A), (B), and (D) show the compositions of near-surface lake water at the sampling sites on August 7, 1990. Concentrations of dissolved ammonia on this date were 0.001 mmol/L in both bays. Between 1- and 3-cm depth, there are minima in sulfate concentrations and maxima in alkalinity, ammonia, phosphate, and, in Creel Bay, sulfide concentrations. These abrupt changes in chemical gradients suggest that sulfate-reduction rates peak between 1- and 3-cm depth.

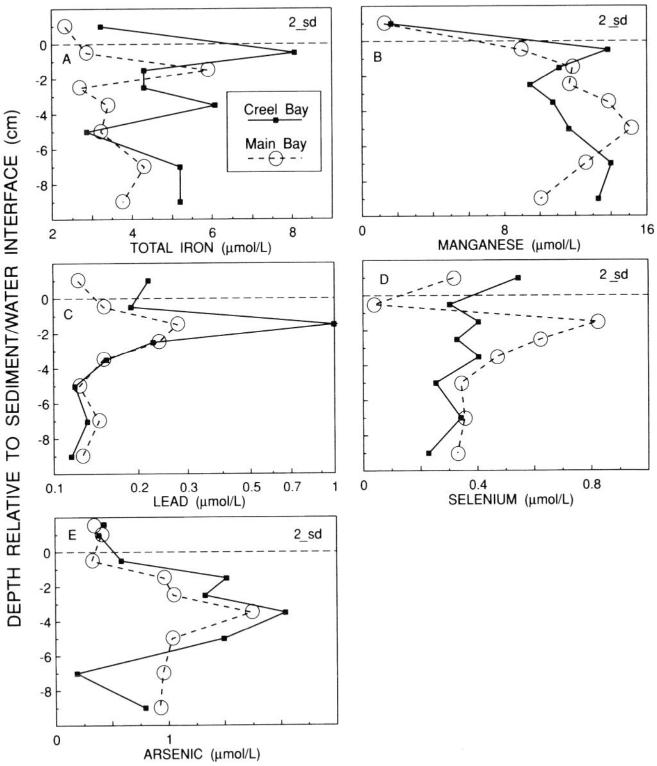

FIG. 6.—Depth variations in bottom water and pore water of (A) total iron, (B) manganese, (C) lead, (D) selenium, and (E) arsenic. Symbols not connected by lines at the top of graph (E) show the concentrations of arsenic in near-surface lake water on August 7, 1990. Near-surface lake-water concentrations on August 7, 1990, for graphs (A), (B), (C), and (D) were as follows: iron, 0.54 μmol/L in both bays; manganese, <0.2 μmol/L; lead, <0.01 μmol/L; and selenium, <0.01 μmol/L.

solved sulfate ($\delta^{34}S_{sulfate}$) increase with depth and reach a local maximum of 9.8‰ at 1–2 cm depth (Fig. 9A). Values of $\delta^{34}S$ in sedimentary sulfide ($\delta^{34}S_{sulfide}$) decrease downward from −27.5‰ just below the sediment/water interface to about −40‰ below 6-cm depth (Fig. 9B).

INTERPRETATION OF CHEMICAL PROFILES

Dissolved Constituents

Between 1- and 3-cm depth in the sampled sections, significant changes occur in the concentrations of sulfate, alkalinity, ammonia, phosphate, and sulfide, and in the values of $\delta^{13}C_{DIC}$ and $\delta^{34}S_{sulfate}$. These changes, and the strong smell of hydrogen sulfide in the sediments, are consistent with sulfate reduction:

$$(CH_2O)_x(NH_3)_y(H_3PO_4)_z + xSO_4^{2-} \rightarrow 2xHCO_3^- + xH_2S + 2yNH_3 + 2zH_3PO_4 \quad (2)$$

This reaction (Aller, 1980) describes the oxidation of organic matter that contains hydrocarbon, ammonia, and phosphate components. Sulfate is consumed while alkalinity, dissolved ammonia and phosphate, and hydrogen-sulfide gas are produced. Production of isotopically light alkalinity is characteristic of bacterially mediated sulfate reduction and accounts for the minimum in pore-water $\delta^{13}C$ values (Anderson and Arthur, 1983). Similarly, sulfate-reducing bacteria preferentially reduce ^{32}S, which leaves residual dissolved sulfate enriched in ^{34}S (Kaplan, 1983). This process explains the peak in $\delta^{34}S_{sulfate}$ values and the local minimum in $\delta^{34}S_{sulfide}$ values at 1–2 cm depth.

Chemical evidence of sulfate reduction is strongest between 1- and 3-cm depth and sulfate-reduction rates probably peak in this interval. Above this interval, sulfate reduction rates probably are smaller because sulfate-reducing bacteria are outcompeted for organic matter by bacteria that mediate thermodynamically more efficient reactions. Such reactions include oxygen reduction, denitrification and reduction of manganese and iron oxides (Stumm and Morgan, 1981). Sulfate reduction may occur at decreased rates below 1- to 3-cm depth because of depletion of labile organic matter. The lack of measurable methane in pore water suggests that competition of sulfate-reducing bacteria with methanogenic bacteria is not an important cause of decreased sulfate-reduction rates below 1- to 3-cm depth.

Sulfate concentrations increase both upward and downward from 1- to 3-cm depth, indicating that sulfate reduc-

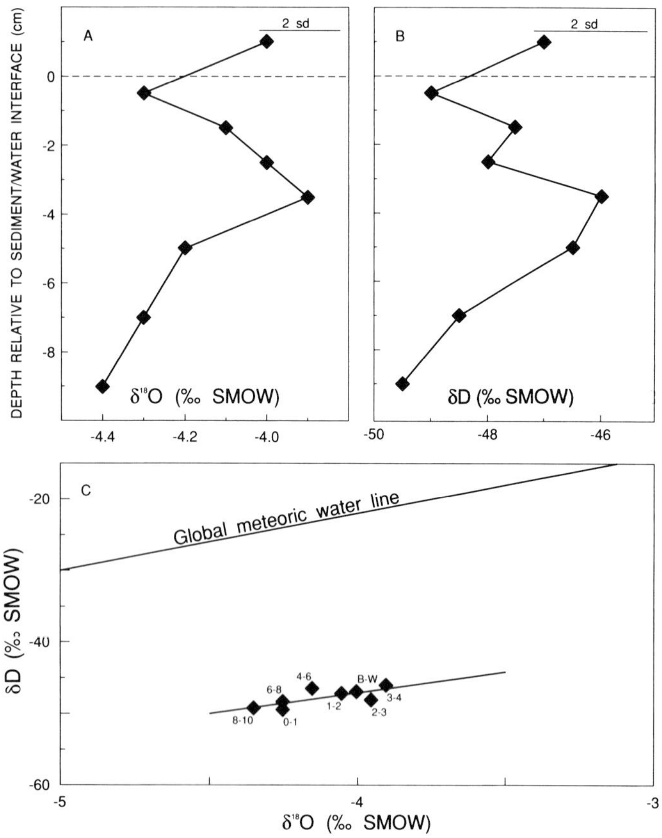

FIG. 7.—Depth variations of (A) $\delta^{18}O$ and (B) dD in pore water, and (C) δD versus $\delta^{18}O$. The global meteoric water line in (C) is from Craig (1961). Pore-water compositions define a regression line (r = 0.77; dashed) with a shallower slope ($\delta D/\delta^{18}O = 5.9$) than the global meteoric water line ($\delta D/\delta^{18}O = 8$). The shallow slope of the regression line indicates that pore water has been affected by evaporation.

tion is a sink for both water-column sulfate and for sulfate diffusing upward from deeper in the sediments (Komor, 1992). The source of the deeper sulfate probably is saline pore water at depths of 30–40 cm in the sediments where sulfate concentrations reach 238 mmol/L (Callender, 1968). This saline water accumulated in the sediments about 500 years ago when parts of Devils Lake were very shallow or completely desiccated (Callender, 1968). Sulfate minerals, also deposited during previous low stands of Devils Lake, are additional sources of dissolved sulfate and cations in the sediments (Callender, 1968). Sulfate minerals that commonly form in present-day desiccating water bodies in eastern North Dakota include epsomite ($MgSO_4$), gypsum ($CaSO_4 \cdot 2H_2O$), hexahydrite ($MgSO_4 \cdot 6H_2O$), konyaite ($Na_2Mg(SO_4)_2 \cdot 5H_2O$), mirabilite ($Na_2SO_4 \cdot 10H_2O$), and thenardite ($Na_2SO_4$; Keller and others, 1986). It is likely that these same minerals precipitated when Devils Lake was desiccated.

Possible evidence for reduction of iron and manganese oxyhydroxides in the upper 2 cm of sediment is provided by iron concentration peaks at both sampling sites (Fig. 6A) and by a manganese concentration peak at the Creel Bay site (Fig. 5B). Lead and selenium concentrations (Figs. 6C, D) peak at about the same depth as iron and manganese. It may be that lead and selenium were adsorbed on iron or manganese oxyhydroxides and released to pore water when the oxyhydroxides dissolved (Fotstner, 1982). Alternatively, lead and selenium may have resided in organic matter that degraded during reduction of the oxyhydroxides (Morel and Hudson, 1985). These explanations can account for the similar depths of peak concentrations for iron, manganese, lead, and selenium. Similar reduction or degradation processes may account for the peaks in arsenic concentrations at 3- to 4-cm depth (Fig. 6E). Thermodynamic calculations (WATEQ4F; Ball and others, 1987) indicate that the pore water is saturated with amorphous iron-sulfide precipitate (FeS), orpiment (As_2S_3), galena (PbS), and a selenium analog of pyrite ($FeSe_2$). These and other sulfide phases may buffer pore-water concentrations of trace constituents. Such buffering may account for the tendency of iron, lead, selenium, and arsenic to vary within only restricted ranges below 6-cm depth in the sediments.

Interpretation of Stable-Isotope Profiles

Hydrogen and oxygen isotopes in water.—

The isotopic composition of water generally does not change during early diagenesis in hydrologic systems. Therefore, δD and $\delta^{18}O$ values in Devils Lake pore water should represent the original compositions of lake water that were buried in the sediments. This interpretation of pore water as unaltered lake water is strengthened by the strong correlation of δD and $\delta^{18}O$ in pore water. This correlation would not be expected if hydrogen or oxygen were fractionated by diagenetic reactions, or if the δD and $\delta^{18}O$ profiles were significantly affected by diffusion of water molecules. The regression line defined by δD and $\delta^{18}O$ values has a shallower slope ($\delta D/\delta^{18}O = 5.9$) than the global meteoric-water line ($\delta D/\delta^{18}O = 8$; Fig. 7C). This shallow slope indicates that the pore water has been evaporated (Gat and Gonfiantini, 1981), and supports the idea that pore water is buried, evaporated lake water. The regression line defined by pore-water compositions intersects the global meteoric water line at $\delta^{18}O = -16‰$, $\delta D = -115‰$. These values are very similar to the average isotopic composition of groundwater in eastern and central North Dakota (Rehm and others, 1982; Robert B. Shaver, North Dakota State Water Commission, unpublished data, 1991) and probably approximate the average catchment composition of Devils Lake.

Carbon and oxygen isotopes and carbonate mineralogy.—

The $\delta^{13}C_{DIC}$ value in Main Bay bottom water is −2‰, which is lower than would result from equilibrium with atmospheric CO_2. If the DIC in Main Bay bottom water was in equilibrium with atmospheric CO_2 at 20°C, $\delta^{13}C_{DIC}$ would be approximately 1.3‰ (Deines and others, 1974). Low $\delta^{13}C_{DIC}$ values in bottom water must be partly due to an upward flux of isotopically light alkalinity from the bottom sediments.

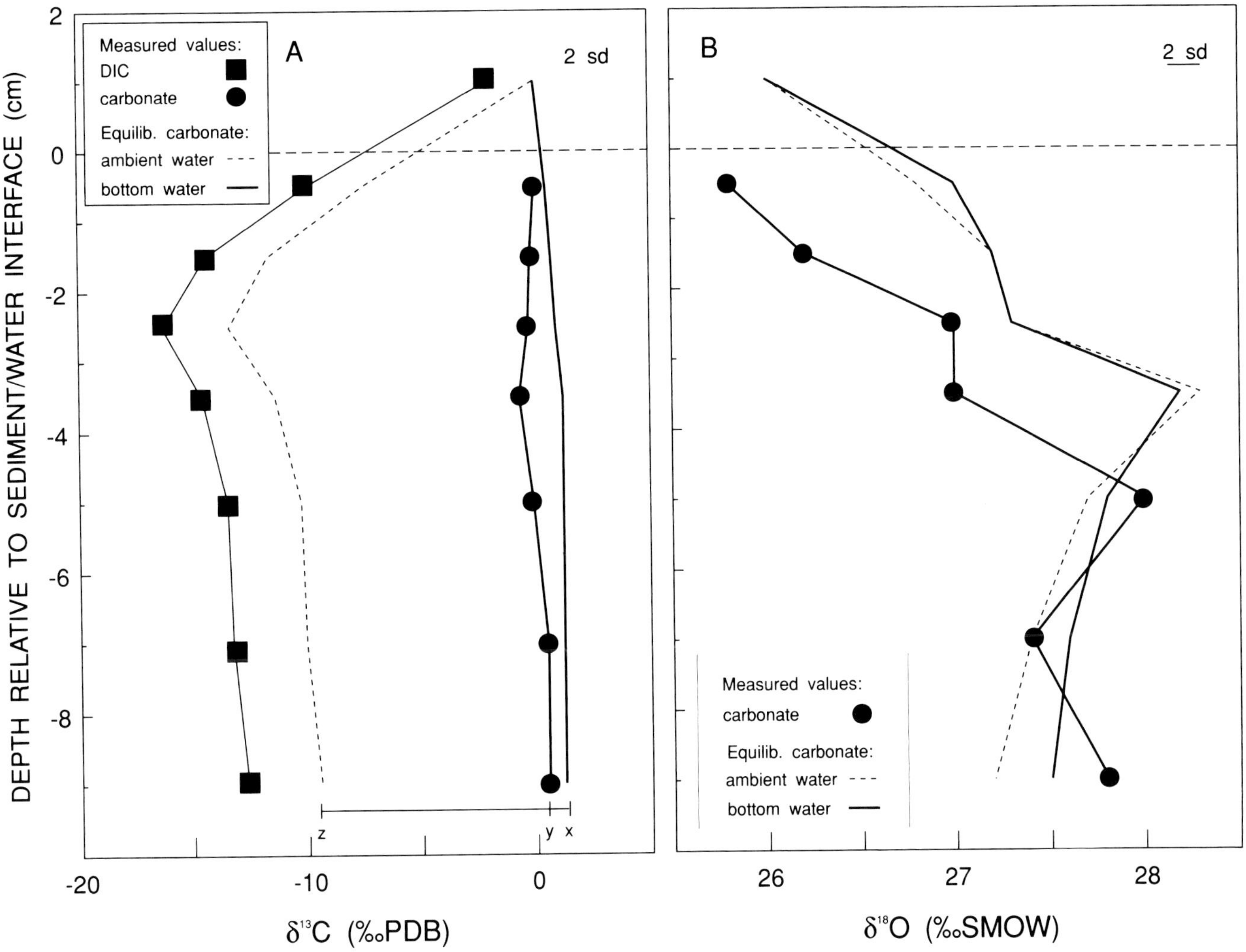

FIG. 8.—(A) Values of $\delta^{13}C$ in dissolved inorganic carbon (DIC, squares) and solid carbonate (circles). (B) Values of $\delta^{18}O$ in solid carbonate (filled circles). Dotted lines show carbonate compositions in equilibrium with ambient water (ambient water is pore water below the sediment/water interface and lake-bottom water above the sediment/water interface). Dashed lines show carbonate compositions in equilibrium with bottom water. In (A), measured $\delta^{18}C$ values in carbonates are close to equilibrium with bottom water, but far from equilibrium with pore water. The ratio of the line-segment lengths xy/xz indicates that no more than about 15 percent of carbonate originally in equilibrium with lake water may have been replaced by carbonate in equilibrium with pore water. In (B), the increase of carbonate $\delta^{18}O$ values in the top 4 to 6 cm of sediments may be due to increases in the $\delta^{18}O$ value of the water column, limited amounts of carbonate recrystallization, and to varying proportions of diagenetic and biological carbonate.

Carbonates in Devils Lake bottom sediments are a combination of endogenic carbonate that precipitated in the water column, detrital carbonate from glacial sediments surrounding Devils Lake, carbonate formed by diagenetic reactions in the bottom sediments, and biological carbonate. According to Callender (1968), at least 60 percent of the carbonate in Devils Lake bottom sediments originated as inorganic precipitates in the water column (i.e., endogenic carbonate). Callender (1968) also noted that glacial till surrounding Devils Lake contains calcite and dolomite but no magnesian calcite or aragonite. Therefore, magnesian calcite and aragonite in the bottom sediments cannot be detrital in origin and must have precipitated from the water column or formed by postdepositional diagenetic reactions. Biological carbonate in the sediments includes ostracode shells composed of low-magnesium carbonate (Callender, 1968).

Sediments in the upper 1 cm of bottom sediment in Main Bay contain 34 weight percent magnesian calcite (8.3 mole percent $MgCO_3$) and 9 weight percent dolomite (Table 3). With increasing depth, abundances and magnesium contents of calcite decrease while abundances of aragonite and dolomite increase. The changes in carbonate mineralogy with depth in the bottom sediments have been attributed to post–depositional diagenetic processes (Callender, 1968). The diagenesis explanation holds that high-magnesium calcite is the primary inorganic carbonate precipitate, and that it begins to recrystallize to dolomite, aragonite, and low-mag-

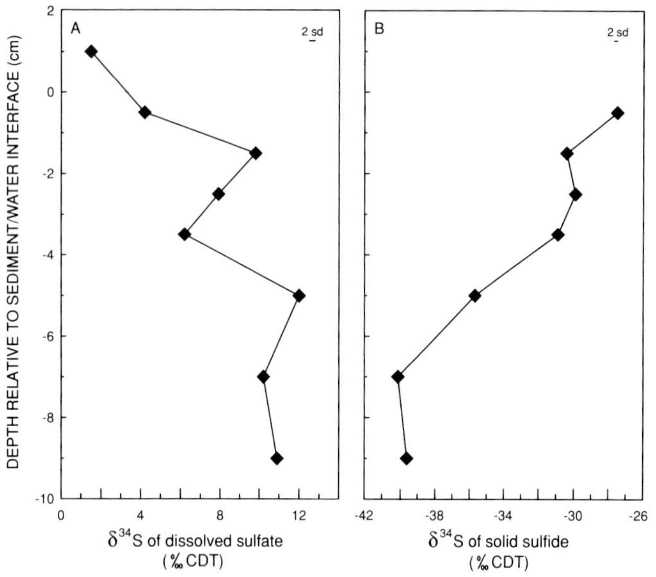

FIG. 9.—(A) Values of $\delta^{34}S$ in dissolved sulfate. (B) Values of $\delta^{34}S$ in sedimentary sulfide.

nesium calcite soon after incorporation into the sediments. The reactions that describe this process are:

dissolution of high-magnesium calcite:

$$Ca_{(1-x)}Mg_xCO_3(s) = xMg^{2+} + (1-x)Ca^{2+} + CO_3^{2-} \quad (3)$$

crystallization of low-magnesium calcite or aragonite:

$$Ca^{2+} + CO_3^{2-} = CaCO_3 \quad (4)$$

dolomitization of $CaCO_3$:

$$2CaCO_3 + Mg^{2+} = CaMg(CO_3)_2 + Ca^{2+} \quad (5)$$

These reactions convert Mg and Ca released by dissolution of high-magnesium calcite to dolomite and low-magnesium carbonate. Variations of mineral proportions and concentrations of Mg^{2+} and Ca^{2+} in pore water generally are permissive of this diagenetic process in Devils Lake. However, it is difficult to separate the effects of diagenesis on mineral proportions and pore-water chemistry from the influence of other processes. Carbonate-mineral proportions may vary with depth not only because of diagenesis, but also because of temporal changes in the proportions of biological, detrital, and endogenic carbonate deposited on the lake bottom. Similarly, changes in Mg^{2+} and Ca^{2+} concentrations may be influenced not only by carbonate diagenesis, but also by upward fluxes from saline pore water and dissolving evaporites deeper in the sediments.

Stable-isotope data can be used to evaluate the importance of carbonate diagenesis in Devils Lake bottom sediments. Endogenic inorganic carbonates generally precipitate in isotopic equilibrium with the water column (McKenzie, 1985), whereas diagenetic carbonate should be in isotopic equilibrium with pore water (Mullins and others, 1985; Middelburg and others, 1990). Therefore, post depositional modification of endogenic carbonates, which compose the bulk of the carbonate assemblage, should be recognizable in the stable-isotope compositions. The difference between endogenic and diagenetic carbonate should be particularly evident in the carbon-isotope compositions because the $\delta^{13}C_{DIC}$ values of lake water and pore water are substantially different.

To evaluate whether carbonates in the bottom sediments reflect isotopic equilibrium with lake water or pore water, equilibrium compositions at 20°C were calculated from measured $\delta^{13}C_{DIC}$ and $\delta^{18}O$ values in lake bottom water and pore water. The calculations used fractionation factors for carbon exchange among bicarbonate, calcite (Deines and others, 1974), aragonite (Rubinson and Clayton, 1969), and dolomite (Sheppard and Schwarz, 1970); and fractionation factors for oxygen exchange among water, calcite (O'Neil and others, 1969), aragonite, magnesian calcite (Tarutani and others, 1969), and dolomite (Sheppard and Schwarz, 1970). Fractionation factors were weighted according to the molar percentages of carbon and oxygen contributed by each phase in the total carbonate assemblage (Table 3). The carbonate that precipitated from water just above the sediment/water interface is assumed to have been calcite with 8.5 mole percent $MgCO_3$. Dotted lines in Figure 8 show carbonate compositions in equilibrium with ambient water at each depth. Ambient water is lake bottom water (sample MB, B-W) above the sediment/water interface and pore water below the interface. Dashed lines in Figure 8 show carbonate compositions in equilibrium with bottom water. If carbonates in the sediments are in equilibrium with coexisting pore water, their compositions should fall along the dotted line; if carbonates at all depths in the sediments are in equilibrium with present-day bottom water, their compositions should fall along the dashed line.

Carbonate $\delta^{13}C$ values in equilibrium with bottom water (dashed line) range from 0.5 to 1.2‰ below the sediment/water interface and are similar to measured carbonate compositions (circles). In contrast, carbonate $\delta^{13}C$ values in equilibrium with pore water (dotted line) range from −7.5 to −13.4‰ (per mil) below the sediment/water interface, and are much lower than measured compositions. These relationships are consistent with the interpretation that much of the carbonate precipitated in near-equlibrium with lake water that had a $\delta^{13}C_{DIC}$ composition similar to that of present-day bottom water. The results also show that carbonates are not in carbon-isotope equilibrium with pore water.

Although carbonates in the sediments are *close* to equilibrium with bottom water, carbonate $\delta^{13}C$ values *are* consistently as much as 2‰ lower than bottom-water equilibrium values. These comparatively low $\delta^{13}C$ values may be due to the presence of ostracode shells in the carbonate assemblage; such biological carbonates commonly are depleted in ^{13}C and ^{18}O compared to equilibrium compositions (McConnaughey, 1989). Another explanation for the measured values that are as much as 2‰ lower than bottom-water equilibrium values is that a limited amount of carbon-isotope exchange with pore water occurred during recrystallization of magnesian calcite to low-magnesium calcite, aragonite, and dolomite. This explanation is supported by the occurrence of the largest difference between measured and bottom-water equilibrium $\delta^{13}C$ at 3- to 4-cm depth, near the level where pore-water $\delta^{13}C_{DIC}$ values reach a minimum. By applying the lever rule to calculated and measured carbonate compositions (i.e., the ratio $(xy)/(xz)$ in Fig. 8A), it is estimated that 10–15 mole percent of the carbonate in equilibrium with lake water may have been replaced by carbonate in equilibrium with pore water.

Unlike $\delta^{13}C_{DIC}$ in pore water, $\delta^{18}O$ values do not vary widely and are not much different than the $\delta^{18}O$ value of bottom water (Fig. 7A). Therefore, there are not large differences between the $\delta^{18}O$ compositions of equilibrium carbonates calculated from pore water (Fig. 8B, dotted line) and bottom water (Fig. 8B, dashed line). Both calculated curves roughly reproduce the trend of the measured carbonate $\delta^{18}O$ values; i.e., a decrease of 2.2‰ from the sediment/water interface to between 4- and 6-cm depth and roughly constant values below that depth.

The decrease of carbonate $\delta^{18}O$ values from 4- to 6-cm depth to the sediment/water interface may indicate that the $\delta^{18}O$ value of the water column decreased during the period when the carbonates precipitated. Some evidence of such a decrease is preserved in pore water, which represents buried lake water and which decreases in $\delta^{18}O$ by 0.4‰ from 3 to 4-cm depth to the sediment/water interface. Part of the change in carbonate $\delta^{18}O$ values in the upper 6 cm also may be due to limited replacement of magnesian calcite by aragonite and dolomite, both of which concentrate $\delta^{18}O$ to a greater extent than magnesian calcite. For example, if 15 percent of the magnesian calcite was replaced by dolomite, the $\delta^{18}O$ composition of the bulk carbonate assemblage would increase by 0.6‰. Thus, changes in water-column composition and carbonate recrystallization may account for a total of about 1‰ of the change in $\delta^{18}O$ values between the sediment/water interface and 4- to 6-cm depth, about half of the measured variation. The remaining 1.2‰ may represent variations in the $\delta^{18}O$ values of detrital carbonate and varying proportions of biological carbonate in the sediments.

CONCLUSIONS

Two major processes in pore water in the upper 10 cm of sediments are: (1) sulfate reduction and (2) upward diffusion of dissolved ions from buried saline pore water and evaporite minerals. These processes are related because sulfate reduction affects sulfate that diffuses upward from the buried sources as well as sulfate that diffuses downward from the water column.

Carbon-isotope data suggest that a maximum of about 15 mole percent of carbonates in the sediments are in equilibrium with bicarbonate in the pore water. Because carbonate proportions change by much more than 15 mole percent in the sampled interval, most of the depth variation in carbonate mineralogy must represent temporal variation in the proportions and compositions of endogenic, detrital, and biological carbonate that was deposited on the lake bottom. Such changes in carbonate sedimentation could have occurred in response to changes in the level and salinity of Devils Lake. Previous suggestions that dolomite in Devils Lake bottom sediments is diagenetic in origin (Callender, 1968) are at odds with experimental evidence that dolomite cannot form in sulfate-rich water (Baker and Kastner, 1981). The finding in this study that most carbonate, including dolomite, in Devils Lake bottom sediments is *not* diagenetic in origin removes this seeming contradiction from our understanding of dolomite formation.

A caveat is that the carbon-isotope composition of pore water may change seasonally. Pore water for this study was collected during the summer when sulfate-reduction rates and production of isotopically light alkalinity probably are greatest. Rates of sulfate reduction and alkalinity production may be smaller when the bottom water is well oxygenated or when organic material is depleted in the sediments. To use rigorously carbon-isotope compositions of pore water to assess the degree to which sedimentary carbonates are in equilibrium with pore water, one would need to use a time-averaged pore-water $\delta^{13}C$ composition. Nevertheless, the available carbon-isotope data indicate that sedimentary carbonate did not substantially recrystallize in the presence of pore water.

Because the carbonates have not recrystallized substantially, their stable-isotope compositions can be used to infer changes in the isotopic composition of the water column. One such inference is that the $\delta^{18}O$ composition of lake water decreased between the times when carbonates at 4 to 6-cm depth precipitated and those at 0 to 1-cm depth precipitated. This inference generally is supported by the upward decrease of pore-water $\delta^{18}O$ values in the top 4 cm of sediment. It is possible that a long-term record of $\delta^{18}O$ variations in Devils Lake could be obtained through stable-isotope analyses of pore water and carbonates in longer cores.

ACKNOWLEDGMENTS

Fieldwork was accomplished with the assistance of Brad Sether, Jim Wald, and Gary Burkart. Gregg Wiche provided general information about the hydrology of Devils Lake and shared his geochemical and hydrologic data. Steve Sando supplied the salinity data in Figure 2A. Field equipment used in this project was graciously provided by Ted Callender. The manuscript was constructively reviewed by Ted Callender, Kathryn Kuivila, Bill Horak, Jim LaBaugh, Bob Lent, Ted McConnaughey, and Marjorie Pokladnik. The project was funded by the North Dakota State Department of Health and Consolidated Laboratories and the U. S. Geological Survey.

REFERENCES

Aller, R. C., 1980, Diagenetic processes near the sediment-water interface of Long Island Sound: I. Decomposition and nutrient element geochemistry (S, N, P), *in* Saltzman, B., ed., Advances in Geophysics, v. 22: New York, Academic Press, p. 238–350.

Anderson, T. F., and Arthur, M. A., 1983, Stable isotopes of oxygen and carbon their application to sedimentologic and paleoenvironmental problems, *in* Arthur, M. A., Stable Isotopes in Sedimentary Geology: Tulsa, Society of Economic Paleontologists and Mineralogists Short Course No. 10, p. 1–1 to 1–151.

Baker, P. A., and Kastner, M., 1981, Constraints on the formation of sedimentary dolomite: Science, v. 213, p. 214–216.

Ball, J. W., Nordstrom, D. K., and Zachman, D. W., 1987, WATQ4F—A personal computer translation of the geochemical model WATEQ2 with revised data base: Menlo Park, U. S. Geological Survey Open-File Report 87–50, 108 p.

Bishop, P. K., 1990, Precipitation of dissolved carbonate species from natural waters for $\delta^{13}C$ analysis—A critical appraisal: Chemical Geology, v. 80, p. 251–259.

CALLENDER, E., 1968, The post-glacial sedimentology of Devils Lake, North Dakota: Unpublished Ph.D. Dissertation, University of North Dakota, Grand Forks, 312 p.

CRAIG, H., 1961, Isotopic variations in meteoric waters: Science, v. 133, p. 1702–1703.

DEINES, P., LANGMUIR, D., AND HARMON, R. S., 1974, Stable carbon isotopic ratios and the existence of a gas phase in the evolution of carbonate ground waters: Geochimica et Cosmochimica Acta, v. 38, p. 1147–1164.

EPSTEIN, S., AND MAYEDA, T. K., 1953, Variation of O^{18} contents of waters from natural sources: Geochimica et Cosmochimica Acta, v. 4, p. 213–224.

EUGSTER, H. P., AND HARDIE, L. A., 1978, Saline Lakes, *in* Lerman, A., ed., Lakes: Chemistry, Geology, Physics: New York, Springer, p. 237–289.

FORSTNER, U., 1982, Accumulative phases for heavy metals in limnic sediments, *in* Sly, P. G., ed., Sediment/Freshwater Interaction: The Hague, Junk, p. 269–284.

GAT, J. R., AND GONFIANTINI, R., 1981, Stable isotope hydrology, deuterium and oxygen-18 in the water cycle: Vienna, International Atomic Energy Agency, 339 p.

GOLDSMITH, J. R., AND GRAF, D. L., 1958, Relation between lattice constants and composition of the Ca-Mg carbonates: American Mineralogist, v. 43, p. 84–101.

HENDERSON, P. H., 1982, Inorganic Geochemistry: London, Pergamon Press, 353 p.

JACOBSON, H. A., AND ENGSTROM, D. R., 1989, Resolving the chronology of recent lake sediments: An example from Devils Lakc, North Dakota: Journal of Paleolimnology, v. 2, p. 81–97.

KAPLAN, I. R., 1983, Stable isotopes of sulfur, nitrogen, and deuterium in recent marine environments, *in* Arthur, M. A., Stable Isotopes in Sedimentary Geology: Tulsa, Society of Economic Paleontologists and Mineralogists Short Course No. 10, p. 2–1 to 2–108.

KELLER, L. P., MCCARTHY, G. J., AND RICHARDSON, J. L., 1986, Mineralogy and stability of soil evaporites in North Dakota: Soil Science Society of America Journal, v. 50, p. 1059–1071.

KENDALL, C., AND COPLEN, T. B., 1985, Multisample conversion of water to hydrogen by zinc for stable isotope determination: Analytical Chemistry, v. 57, p. 1437–1440.

KOMOR, S. C., 1992, Bidirectional sulfate diffusion in saline-lake sediments: Evidence from Devils Lake, northeast North Dakota: Geology, v. 20, p. 319–322.

MCCONNAUGHEY, T., 1989, ^{13}C and ^{18}O isotopic disequilibrium in biological carbonates: II. *In vitro* simulation of kinetic isotope effects: Geochimica et Cosmochimica Acta, v. 53, p. 163–171.

MCCREA, J. M., 1950, On the isotopic chemistry of carbonates and a paleotemperature scale: Journal of Chemical Physics, v. 18, p. 849–857.

MCKENZIE, J., 1985, Carbon isotopes and productivity in the lacustrine and marine environment, *in* Stumn, W., ed., Chemical Processes in Lakes: New York, Wiley, p. 99–118.

MIDDELBURG, J. J., deLANGE, G. J., AND KRVELEN, R., 1990, Dolomite formation in anoxic sediments of Kan Bay, Indonesia: Geology, v. 18, p. 399–402.

MOREL, F. M. M., AND HUDSON, R. J. M., 1985, The geobiological cycle of trace elements in aquatic systems: Redfield revisited, *in* Stumn, W., ed., Chemical Processes in Lakes: New York, Wiley, p. 251–282.

MULLINS, H. T., WISE, S. W., JR., LAND, L. S., SIEGEL, D. I., MASTERS, P. M., HINCKEY, E. J., AND PRICE, K. R., 1985, Authigenic dolomite in Bahamian peri-platform slope sediment: Geology, v. 13, p. 292–295.

O'NEIL, J. R., CLAYTON, R. N., AND MAYEDA, T. K., 1969, Oxygen isotope fractionation in divalent metal carbonates: Journal of Chemical Physics, v. 51, p. 5547–5558.

OREMLAND, R. S., AND DES MARAIS, D. J., 1983, Distribution, abundance, and carbon isotopic composition of gaseous hydrocarbons in Big Soda Lake, Nevada: An alkaline meromictic lake: Geochimica et Cosmochimica Acta, v. 47, p. 2107–2114.

REHM, B. W., MORAN, S. R., AND GROENEWOLD, G. H., 1982. Natural ground water recharge in an upland area of central North Dakota: Journal of Hydrology, v. 59, p. 293–314.

RUBINSON, M., AND CLAYTON, R. N., 1969, Carbon-13 fractionation between aragonite and calcite: Geochimica et Cosmochimica Acta, v. 33, p. 997–1002.

SHEPPARD, S. M. F., AND SCHWARZ, H. P., 1970, Fractionation of carbon and oxygen isotopes and magnesium between coexisting metamorphic calcite and dolomite: Contributions to Mineralogy and Petrology, v. 26, p. 162–198.

STUMM, W., AND MORGAN, J. J., 1981, Aquatic Chemistry: New York, Wiley, 780 p.

TARUTANI, T., CLAYTON, R. N., AND MAYEDA, T. K., 1969, The effect of polymorphism and magnesium substitution on oxygen isotope fractionation between calcium carbonate and water: Geochimica et Cosmochimica Acta, v. 33, p. 987–996.

WICHE, G. J., 1991, Evaporation computed by energy-budget and mass-transfer methods and water-balance estimates for Devils Lake, North Dakota, 1986–88: Bismarck, North Dakota State Water Commission Water-Resources Investigation 11, 200 p.

ON THE MEASUREMENT OF REACTIVE MASS FLUXES IN EVAPORATIVE GROUNDWATER-SOURCE LAKES

JOSEPH J. DONOVAN

Department of Geology and Geography, University of West Virginia, Morgantown, West Virginia 26506-6300

ABSTRACT: The character and sequence of reaction products formed within evaporative groundwater-dominated lakes are controlled by source water chemistry, intensity of evaporation, residence time of lake water, and hydraulic flux ratio between groundwater inflow to the lake and lake water outflow to groundwater. Conventional hydrologic data (aqueous chemical concentrations, stage fluctuations, *in-situ* or other estimates of evaporation, aquifer characteristics) may be employed to calculate mass budgets for fluid and solutes (both conservative and reactive) in lake systems; by this approach, details of saline lake geochemical reactions may be resolved, supplementing inferences based on calculations of mineral equilibria. For lakes of known age, comparison of calculated mineral precipitation fluxes, based on aqueous chemistry, to sediment volume estimates allows projection of brine out-seepage and rates of eolian salt erosion, slow processes by which chemical mass is removed from storage. In an example case from the Clear Lake outwash aquifer in eastern Montana, mass budgets for two alkaline lakes in series—one a carbonate-producing marl-tufa lake, the other a sodium-sulfate-producing playa—suggest a major component (about 80 percent) of solute inflow to the playa basin is lost to brine outflow and wind erosion.

INTRODUCTION

As for lakes in semi-arid climates worldwide, many mineralized lakes of the Northern Great Plains are groundwater-dominated, fed almost wholly by groundwater and incident precipitation. Regional evaporation is well in excess of precipitation, forcing lakes to salinize; however, the climate is sufficiently humid to provide intermittent recharge and drive mass transport by groundwater flow. Spatial variation in the chemistry of such lakes and their sediments results from complex interaction between climate, hydrogeology, topography, and limnological processes. Understanding this interaction is a key in applying observations from these lakes to study of Holocene climate and hydrology.

While not new, the awareness of the link between saline lakes and groundwater transport has been heightened in recent research. A traditional view of saline lakes is as "closed" basins in which cyclic variations in climate cause changes in water volume, thereby inducing chemical variations (Langbein, 1961). Such closed lakes were considered to have essentially no fluid outflow. For example, Spencer and others (1985) developed a chemical mass-balance model to relate stage-induced chemical variations in Great Salt Lake over the last 30,000 years to chemical variations in source water inputs. In a contrasting approach, Wood and Sanford (1990) applied fluid mass balance techniques in conjunction with mineral equilibrium calculations to relate evaporite basin salt mineralogy and fluid chemistry to hydraulic parameters and groundwater leakage rates from lakes. They hypothesized that lake-groundwater hydraulics in conjunction with equilibrium mineral reactions could account for a wide range of aqueous chemical variations without resort to variations in source inputs. These two approaches to the geochemistry of saline lakes attain fundamentally different conclusions by distinct approaches; the former emphasizes source control of chemical mass balance in an essentially "closed" system, while the latter emphasizes the role of groundwater hydraulics in an "open" system.

Clearly both chemical and fluid mass balances are important controls on the geochemistry of saline lake systems, and both are influenced by groundwater flow phenomena. If one treats the closed (or nearly closed) basin setting as a "special" case of an open lake, then a variety of new techniques integrating both approaches (chemical and fluid mass balances) may be developed to give quantitative insights into products and rates of geochemical processes. When applied to systems amenable to approximate closure, chemical mass balance models augment the application of mineral equilibria models and allow consideration of biological and kinetically-controlled reactions important in many lakes.

However, application of mass balance models is beset by some practical difficulties. Relevant reactions must first be identified, and then the net fluxes involved in these reactions must in some fashion be quantified according to field relationships. This calibration process is commonly subjective and prone to errors related to data availability. The quality of conclusions drawn from such models is closely related to how effectively calibration may be performed.

The purpose of this paper is to formulate a modeling approach for quantitative examination of coupled fluid-chemical mass balance in groundwater-fed saline lake systems. This approach should assist interpolation of approximate rates of geochemical processes active at points along the reaction path of evaporating groundwater. The basis for demonstration of the technique is a set of field observations of lake sediments and hydrology collected in conjunction with aqueous geochemical measurements. The technique integrates chemical mass balance (similar to Spencer and others, 1985) with closed-form fluid mass balance expressions (similar to Wood and Sanford, 1990) to describe groundwater-lake interaction. The desired results are estimates of transport rates and the fate of solutes in groundwater basins. Particularly useful would be an approach to estimate the slow transport rates associated with wind and brine escape in saline lake basins. Such rates are sufficiently slow to be difficult to measure accurately in the field and may induce large errors in mass-balance investigations.

Field Context

As an example for application of the technique, a field setting relatively amenable to hydraulic closure is employed. This area is located among the glacial-formed lakes of the Montana-North Dakota Great Plains (Fig. 1), which vary widely in size, chemistry, and physical configuration. These lakes are supplied with water primarily from shallow

Sedimentology and Geochemistry of Modern and Ancient Saline Lakes, SEPM Special Publication No. 50

vations of low O/C ratio (Fig. 9) are consistent with (but do not in themselves prove) bio-induced water-column aragonite precipitation; this mechanism for precipitation has been outlined by Wetzel (1975), Kelts and Hsu (1978), and Dean (1981). The role of microorganisms is significant for this study as it affects interpretation of solute mass balance.

While investigation of nutrient flux in Brush Lake has not been performed, it appears oligotrophic (dissolved NO_3^- < 0.5 mg/L as N in summer 1990). No data are available for phosphorus. This low nitrate does not preclude cyanobacterial abundance, many species of which are able to develop heterocysts capable of a nitrifying function (Wetzel, 1975).

White Lake

Sediments and brine.—

White Lake is an alkaline ephemeral playa, inundated intermittently during periods of rain or snowmelt and underlain by a permanent brine (pH 9.5–9.6, TDS >200,000 mg/L; Table 1, Rows 15–16) and a single sodium sulfate salt bed. Binyon (1952) reports results of a U. S. Bureau of Mines (USBM) drilling investigation to quantify reserves of salt beneath White Lake. Based on four test holes (Fig. 2), average salt bed thickness is 2.0 m; reported dry salt weight is 1.58×10^6 short tons (1.44×10^9 kg). According to the reported chemical analyses of bulk sample (as salt plus brine; Table 3) the bed is 77 percent $Na_2SO_4 \cdot 10H_2O$, with 12.8 percent brine and 7.0 percent insoluble residue. Other chemical species (Ca^{2+}, Mg^{2+}, Cl^-, dissolved inorganic carbon) reported in the coring investigation occur in relatively low concentration relative to Na^+ and SO_4^{2-} and in similar concentration to those observed in chemical analyses of the brine itself (Table 1, Rows 15-16). The reported quantity of salt is therefore likely primarily mirabilite with minor thenardite, its anhydrous form. However, the mineralogy and degree of zonation of the evaporite were not determined from these core samples.

Hydrology.—

Stage elevations of surrounding lakes and water well measurements suggest apparent hydraulic closure around White Lake (Fig. 4). Freshwater seeps and fresh- or brackish-water-tolerant vegetation (*Scirpus balticus*, "chair-maker's bulrush") are found in several locations on the north playa margin, marking groundwater entry. One of these seepage zones is found at the north end of the playa, separated from Brush Lake by a ridge of sediment 100 m wide (hydraulic gradient >0.01). This south-flowing seepage from Brush Lake is sampled by shallow piezometer 90–3 (Table 1, Row 17); this water has dissolved solids of 4600 mg/L, only slightly less saline than Brush Lake and suggesting possible recharge in the upper portion of the water table. If geologic conditions may be projected from the cross-section of Figure 3 east of White Lake, the floor of White Lake is underlain by approximately 70 m of lacustrine clay and till, the aquitard overlying the Missouri channel gravels. While permeable near-surface outwash allows water to both enter and leave Brush Lake, seepage beneath White Lake evidently may escape only slowly.

Similar to brine lakes described by Eugster and Hardie (1978) and Teller and Last (1990), this playa may in detail be mineralogically zoned, with mixing between laterally-inflowing relatively fresh water and a central hypersaline brine. The brackish-water and salt-tolerant vegetation fringing playas like White Lake are surface evidence for lateral mixing. Brine-fluid mixing is also influenced by periodic incident precipitation causing ponding of very fresh water at the surface, similar to periodic flooding in Saskatchewan playas described by Last (1989). No salt or brine samples from the center of the lake are available.

TABLE 3.—BULK CHEMISTRY OF THE WHITE LAKE SALT DEPOSIT (BINYON, 1952).

Geologic Description:	Crust, Mud, Crystals	Permanent Salt Bed	Mud, Crystals
Thickness (m)	0.12	1.72	0.19
% NaCl	0.66	0.31	0.94
% Na_2CO_3	1.45	1.0	1.9
% $NaHCO_3$	0.27	0.08	0.77
% $CaSO_4$	0.06	0.12	0.02
% $KHCO_3$	0.27	0.15	0.46
% $MgSO_4$	0.84	1.09	0.26
% Insoluble	35.93	7.01	49.7
% Total H_2O	40.52	56.14	39.88
% Free H_2O	28.63	12.83	32.18
% Na_2SO_4	9.36	34.1	6.06
% Mirabilite	21.26	77.43	13.77

Lake and Aquifer Stage Variations

Periodic stage variations occur both in the aquifer and the lake which can be used to estimate lake evaporation. In wells at some distance from the lake (for example, well 84–20, upper hydrographs, Fig. 10), the two dominant frequencies of water level variation are annual (amplitude of fluctuation about 30 cm) and about 3–8 days (amplitude 2–10 cm). These frequencies correspond in period to regional-scale recharge processes from snowmelt and barometric fluctuations, respectively.

In well 89-1 installed in beach sands 10 m from the northeast shore (1990 stage position) of Brush Lake, fluctuations at these frequencies are observed, but with damped barometric response. Superimposed on this "background" aquifer water level signal are characteristic-shaped diurnal fluctuations between mid-March and early December. The height of these daily "spikes" apparently reflect lake-stage fluctuations, interpreted to be caused primarily by free-water evaporation. The downward limb of these spikes—typically spanning the time period between about 9:00 AM and early evening—may be measured directly and used as an estimate of evaporation, following correction for low-frequency "background" water level decline of the summer potentiometric surface. Such background correction is accomplished by taking derivatives of the "background" trend (filtered for diurnal spikes) at daily intervals (noon) and subtracting half this trend (e.g., for the approximate 12 hours that evaporation occurs each day) from each evaporative spike. This correction approach is similar to the graphical technique employed by Meyboom (1967), but applied to

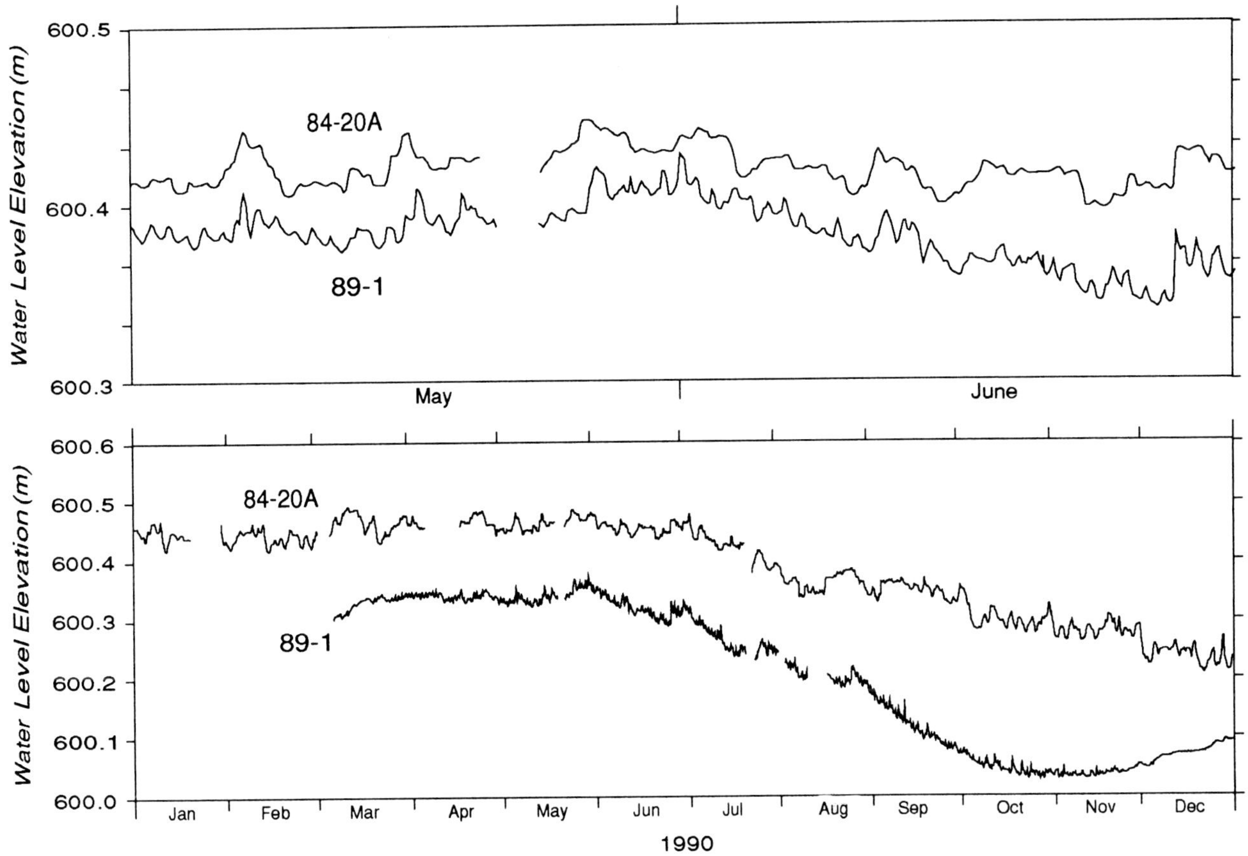

FIG. 10.—Stage fluctuations (A) in May-June 1990 and (B) over all of 1990 for Grenora aquifer [well 84–20A] and well 89–1. Diurnal spikes in 89–1 are interpreted to represent primarily lake evaporation. Elevation of 89–1 stage is scaled 4 m higher than 84–20A stage to appear on similar scale. Vertical scale in meters above sea level.

lake stage rather than that of a water table aquifer. The short distance of the well from the lake damps high-frequency wind- and wave-induced noise in lake stage but does not significantly alter diurnal peak amplitude, as confirmed by concurrent observation at both well 89-1 and an additional well closer to the lake (one meter from the water line). By concurrent observations at these two wells, the possibility that observed diurnal well fluctuations are driven by phreatophytic evapotranspiration has been ruled out.

Evaporation rates in 1990 at Brush Lake (Table 4), expressed as daily averages in mm/day, are similar to both the 1990 and long-term average (1980–90) monthly data from the nearest Class A pan station near Froid, 43 km southwest. Measurement error related to determining daily lake stage-change height is estimated at about ±10 percent. Additional errors relate to data gaps and to short-term hydrograph disturbances during and for several days after substantial rainfall when lake-aquifer flow reversals occur (e.g., Winter, 1983; Sacks and others, 1992); these preclude accurate estimates for these periods. The well hydrograph data show that lake evaporation commences in mid-March as soon as melt-off begins and continues until freeze-up in late fall. Annual totals for the pan station are about 25 percent lower than for the lake, but the lake evaporation record is longer than observations at the pan station (February-December vs. mid-April to September). The indicated "pan

TABLE 4.—COMPARISON OF EVAPORATION ESTIMATES FOR BRUSH LAKE (CALCULATED AT WELL 89–1 BY HYDROGRAPH ANALYSIS AS DESCRIBED IN TEXT) AND FROID, MT, USDA-ARS CLASS A PAN (LOCATION 43 KM SOUTHWEST OF BRUSH LAKE). NUMBER OF DAYS OF MISSING DATA IN PARENTHESES (SOURCE D. HARRIS, UNPUBLISHED FILE DATA, U. S. DEPARTMENT OF AGRICULTURE, AGRICULTURAL RESEARCH SERVICE, SIDNEY, MT. 1992).

	Brush Lake 1990 (mm/day)	Froid ARS 1990 (mm/day)	Froid ARS 1980–89 Mean (mm/day)
January	0.0	—	—
February	0.0	—	—
March	3.88	—	—
April	5.44	5.34	4.61 (6)
May	7.43	6.50	7.11
June	9.95	6.49	8.51
July	8.72 (2)	10.14	9.59
August	7.01 (7)	9.43	8.68
September	8.09	9.61	6.44
October	4.28	—	—
November	3.38	—	—
December	1.26	—	—
Annual Total	1816	1451	1374
Total, April–Sept.	1555	1451	1374

factor" for extrapolating Froid pan data to Brush Lake is on the order of 1.0; similarly high factors have been calculated by Bowler (1986) and Shih (1980). Although caution is required in correlating between the pan station and the lake due to the distance between the two and other factors, both are in a similar topographic and wind-aspect setting, and the observed similarity in rates between the two is not unrealistic. Estimated 1990 Brush Lake evaporation from the hydrograph was 1816 mm. Effective evaporation (evaporation minus incident rainfall) was 1511 mm. While these values are somewhat higher than "average" regional rate estimates (Kohler and others, 1959), the year 1990 was drier than average (+6 percent at the Froid evaporation pan station, with ten years of record; +13 percent at the Williston, North Dakota station, with 34 years of record). The arid conditions at Brush Lake in 1990 resulted in a net lake level decline of 200 mm.

The evaporation estimation method used here is less precise than pan measurements, but are consistent with a conclusion that the lake undergoes roughly comparable evaporation to that observed at the Froid pan (e.g., an approximate pan factor of about 1.0) (Veihmeyer, 1964; Andersen and Jobson, 1982; Morton, 1983). The data for 1990 are employed in mass balance calculations, although they are probably somewhat higher than modern "long-term" evaporation.

MASS BALANCE CALCULATIONS

Brush Lake

Fluid mass balance is based on the 1990 effective evaporation rate (E_l) for Brush Lake (Table 4). Within a 1 percent range, lake chemistry in July 1990 was both spatially vertically uniform: a single sample from 12 m water depth (Table 1, Row 11) is chosen as representative of this date. Ascertaining mean source chemistry is more difficult: both groundwater flow rates and chemistry vary spatially west and east of the lake. Field data indicate that aquifer transmissivity is higher west of the lake, so more flux is expected from this direction, but how much more flux? Although an infinite number of possible calibrations exist to obtain a mixing ratio for the source, one may be arbitrarily selected based on two simplifying assumptions: that both flow and chemistry are different (but uniform) on the two sides of the lake, and that average chemistry may be estimated using the means of bed seepage samples from either side (Table 1, Rows 1–3 for the east side; Rows 4–6 for the west side).

Under these assumptions, the "optimum" mixing fraction Γ (fraction of total fluid inflow derived from the Clear Lake aquifer) is 0.72 ± 0.2 ($\alpha_l = 8.77$), obtained where π_i's of the other halides, F^- and Br^-, intersect π_{Cl} at near unity (boxed area, Fig. 11). The binary mixing curves for the various concentration factors are parabolic with respect to mixing ratio. Also plotted is the concentration factor a_l (solid line; left axis scale) versus mixing ratio. An estimated value of O/C ratio may be calculated under the assumption that all [$r_{Ca} + r_{Sr}$] react with inorganic carbon (r_C) to form aragonite and that residual r_C is utilized for organic carbon synthesis. At the "optimum" Γ (determined using halide

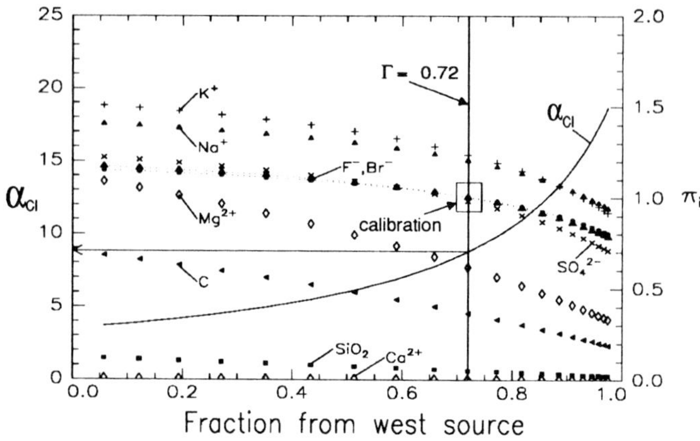

FIG. 11.—Calculated values of π_i (symbols; right scale) as a function of mixing fraction Γ of water from the Clear Lake aquifer. Also shown: evaporative concentration factor for chloride α_{Cl} (curve, left scale). Vertical line is "optimum" Γ value, based on intersection at $\pi_i = 1.0$ for F- and Br- (in boxed area).

convergence), SO_4^{2-} also exhibits nearly conservative behavior ($\pi_{SO4} = 0.98$). Table 5 gives mass fluxes for E_l = 1.51 m/year, the observed 1990 value.

Deviations from conservative behavior for reactive species reflect processes active in the lake. Ca^{2+} (and Sr^{2+}, not plotted) is removed nearly completely from the water column by carbonate production (e.g., positive reaction fluxes); lesser fluxes of Mg^{2+} and SiO_2 are also lost to reaction. Na^+ and K^+ influx show negative reaction fluxes (Table 5, Column 6), possibly suggesting some allogenic transport of K^+ (small flux) and Na^+ (very large flux) into the lake. Column 8 (Table 5) shows the residual reaction fluxes following removal of strontium-containing aragonite utilizing available r_{Ca} and r_{Sr}, in 1:2 stoichiometry with r_C (one mole of aragonite per mole of organic carbon; see equations (9)–(10)). Strontium is incorporated into the mass balance calculations based on its observed decline from groundwater concentrations (1–3 mg/L) to less than 0.3 mg/L in the lake, combined with the commonly described tendency for sedimentary aragonite to accommodate up to 3 percent Sr^{2+}, as well as Pb^{2+} and Zn^{2+}, by isomorphous substitution in its lattice structure (Heinrich, 1965; Muller and others, 1972). Column 9 shows residual fluxes following calculated removal of iron and manganese as metal hydroxides, products observed in tufa precipitates. At this point, no further mass subtraction is performed as the nature of further reactions involving other components is ambiguous. For example, some flux of silica is likely incorporated into biological production, but no estimate of this production is available from sediments. Carbon residuals following removal of carbonate and an equimolar quantity of organic carbon are small and could easily be ascribed to "excess" biological production, to CO_2 outgassing, or to formation of Mg-carbonates. Br^-, F^-, NO_3^-, and Li^+ all exhibit negligibly small fluxes. This leaves Mg^{2+} and SO_4^{2-} (flux into sediments) and Na^+ and K^{2+} (flux from sediments) as problematic to explain.

Dolomite in the marl is likely detrital, but in tufa it is authigenic (Donovan, 1992). Authigenic hydromagnesite is

TABLE 5.—CALCULATED MASS BALANCE FOR BRUSH LAKE, USING $\Gamma = 0.72$; $\alpha_1 = 8.77$; AND $E_1 = 1.51$ M/YR (0.04977 M^3/S).

	[1] Aquifer mg/L*	[2] Brush Lake mg/L**	[3] α_i	[4] π_i	[5] inflow flux q_i, kMol/yr	[6] outflow flux s_i, kMol/yr	[7] flux r_i, kMol/yr	[8] reaction r_i, aragonite removed	[9] r_i, $Fe(OH)_3$ removed
Ca	89.9	4.2	0.04	0.005	3974	21	3953	0	0
Mg	49.9	269	5.39	0.62	3634	2234	1399	1399	1399
Na	125.8	1328	10.56	1.20	9692	11670	−1977	−1977	−1977
K	8.4	91	10.85	1.24	380	470	−90	−90	−90
Fe	1.95	0.06	0.03	0.003	62	0	62	62	0
Mn	0.33	0.01	0.04	0.005	11	0	11	11	0
Si	17.6	7.9	0.45	0.05	1111	57	1054	1054	1054
C	7.73	24.66	3.19	0.36	13695	4982	8713	731	731
Cl	8.75	76.8	8.77	1.00	437	438	0	0	0
SO_4	305.2	2625	8.60	0.98	5628	5521	108	108	108
NO_3(N)	0.09	0.43	4.54	0.52	11	6	5	5	5
F	0.30	2.7	8.80	1.00	28	29	−1	−1	−1
Br	0.12	1.1	8.80	1.00	3	3	0	0	0
Sr	1.93	0.19	0.10	0.01	39	0	39	0	0
Li	0.10	0.58	5.26	0.60	26	17	9	9	9

* calculated by weighted average at $\Gamma = 0.72$ of analyses [1]–[6], Table 1
** analysis [11], Table 1

also found in tufa. The estimated mass of Mg-carbonates in tufa (about 10^8 kg) corresponds to a flux of about 120 kMol/yr as hydromagnesite, assuming a constant rate of formation over the Holocene (as discussed, a simplified and likely erroneous assumption). This is far less than the calculated reaction flux of Mg^{2+} for lake sediments (1399 kmol/year), even allowing a ±20 percent uncertainty in source Mg^{2+} averaging. Even if magnesium carbonate phases could account for the total Mg^{2+} flux, the large residual excess of Na^+ would still be unexplained. Additional possibilities include wind transport of Na-sulfate salts into the lake; Mg^{2+} assimilation into marl carbonate phases, in low concentration but high volume; and exchange between aqueous Mg^{2+} and Na^+ on detrital clays. The estimated flux of detrital clay to Brush Lake in the marl (on the order of 20,000 kg/year, based on modern sedimentation rates and clay content) is insufficient to account by ion exchange for the estimated Na^+ flux, although alteration of mixed-layer Na-smectite minerals to form stevensite interstratifications (Jones, 1986) has been observed in several high-Mg alkaline lakes. Relatively minor adjustments in the mixing algorithm, however, could substantially influence calculated reaction fluxes of Mg^{2+} and Na^+. For Na^+ in particular, there is moderate discrepancy between aquifer concentrations on either side of the lake. The calculated mass-balance fluxes for these ions therefore may not be subject to uncertainty in spatial integration of source chemistry.

The relatively small residual flux of SO_4^{2-} is within the range of accuracy in estimating source SO_4^{2-} concentration, but could reflect anaerobic sulfate reduction below the sediment-water interface. Some anaerobic decomposition may be inferred from apparent reduction of organic carbon with depth in diagenetic profiles of the upper 30 cm of marl.

The calculated mass flux rate of carbonate is 3.9 x 10^5 kg/year; Holocene production at this rate over a 10,000 year period of similar influent chemistry and hydraulics would be 3.9 x 10^9 kg. The sediment volume estimate based on modern sedimentation rate is 3.0 x 10^9 kg. The two values are on the same order, although the sediment volume estimate is crude. The components of the mass balance and the assumptions on which they were based are not inconsistent.

White Lake

Table 6 shows the White Lake mass balance based on calculated Brush Lake fluid outflow ($Q_1 = 0.0064\ m^3/s$) and chemistry. This fluid mass balance is only partial, as the Brush Lake outflow is a fraction (perhaps 30–50 percent) of total fluid inflow to the playa; however, because the chemistry of Brush Lake is substantially more saline than groundwater, this component may represent the largest portion of mass flux to White Lake. Therefore, the results of the mass balance calculations are conservative in terms of chemical mass flux to White Lake. Table 7 shows the fluid mass balance summary for Brush and White Lakes.

Average Cl^- concentration for brine from piezometer samples at the north end of White Lake (Table 1) yields a value of $\alpha_2 = 108$, at which all components are present in conservative or lower quantities ($\pi_1 < 1$) except for Br^- (1.14). At such a high concentration factor, any flux less than about 10 kMol/yr is likely insignificant. The largest reaction fluxes are of Na^+ and SO_4^{2-}, in nearly molar ratio to mirabilite stoichiometry. With mirabilite precipitation removed from the mass balance (Table 6, Column 7), the largest residual fluxes are Mg^{2+} and carbonate; a minor calculated reaction flux of SO_4^{2-} may be ascribed to sulfate reduction in playa-flat cyanobacterial mats.

As for Brush Lake, Mg^{2+} is problematic for White Lake mass balance. Feasible reactions may include dolomitization of subsurface carbonate phases (little excess Ca^{2+} is available for formation of primary dolomite); formation of primary Mg-carbonates (hydromagnesite), Mg-sulfate (bloedite, epsomite), or Mg-silicate phases. Unlike Brush Lake, ion exchange displacing Na^+ on clays with Mg^{2+} is not a likely process, due to the extreme Na/Mg ratios in this brines. All the aforementioned magnesian salts have been observed in saline lakes of the region (Teller and Last, 1990). Despite the salinity, gypsum activity products are below saturation because of depletion of Ca^{2+} earlier in the reaction path by marl precipitation.

TABLE 6.—CALCULATED MASS BALANCE FOR WHITE LAKE, USING Q_1 = 0.0064 m³/S AND Q_2 = 0.000059 m³/s.

Component	[1] Brush Lake mg/L*	[2] White Lake mg/L**	[3] α_i	[4] π_i	[5] inflow flux q_i, kMol/yr	[6] outflow flux s_i, kMol/yr	[7] reaction flux r_i, kMol/yr	[8] r_i Mirabilite removed
Ca	4.2	42.0	10.00	0.09	21	2	19	19
Mg	268.8	192	0.71	0.007	2234	15	2219	2219
Na	1328	68000	51.20	0.47	11670	5529	6141	0
K	91.0	3100	34.07	0.32	470	148	322	322
Fe	0.06	0.38	6.23	0.06	0	0	0	0
Mn	0.01	0.11	8.46	0.08	0	0	0	0
Si	7.9	39.7	5.03	0.05	57	3	54	54
C	24.66	652.0	26.44	0.25	4982	1219	3763	3763
Cl	76.8	8300	108.07	1.00	438	438	0	0
SO_4	2625	97200	37.03	0.34	5521	1892	3629	559
NO_3(N)	0.43	14.0	32.48	0.30	6	2	4	4
F	2.70	77.0	28.52	0.26	29	8	21	21
Br	1.10	150	136.36	1.26	3	4	−1	−1
Sr	0.19	1.8	9.47	0.09	0	0	0	0
Li	0.58	3.2	5.57	0.05	17	1	16	16

* analysis [11], Table 3.1
** average of analyses [15]–[16], Table 3.1

TABLE 7.—FLUID MASS BALANCE SUMMARY, BRUSH LAKE CHAIN.

Brush Lake	White Lake
(area = 1.03 × 10⁶ m²)	(area = 1.42 × 10⁶ m²)
Q_0 = 0.0562 m³/s	Q_1 = 0.0064 m³/s
Q_1 = 0.0064 m³/s	Q_2 = 0.000059 m³/s
E_1 = 0.04977 m3/s	E_2 = 0.00632 m3/s
= 1.51 m/year	= 0.14 m/year
α_1 = 8.77	α_2 = 108
$Q_{H2O-mirab}$ = 0.000017 m³/s	

Assuming that all the reactive mass flux r_{Na} forms mirabilite, the calculated Holocene mass of anhydrous mirabilite in White Lake is 4.4 x 10^9 kg by mass balance, much greater than the observed USBM reserve estimate (1.4 x 10^9 kg), considered relatively accurate. Addition of the omitted groundwater flux to White Lake would only exacerbate this discrepancy. The reaction flux does not include brine outflow, but does include loss to mineral and ion-exchange reactions as well as wind erosion of salts from the playa surface, a visibly active process. If no sodic phases are formed in the lake besides mirabilite and thenardite, than the estimated Holocene wind transport of salt out of the basin is the difference between the reaction flux and the USBM reserve estimate, or 3 x 10^9 kg. Chemical mass flux leaving the basin in brine outflow, on the other hand, depends not on the reaction flux but on the seepage rate out of the playa (Q_2 in Table 7) and on brine chemistry. The concentration of Na^+ and SO_4^{2-} in brine is about 165 g/L, for a Holocene brine flux of 3 x 10^9 kg, similar to the wind erosion flux.

Most uncertainty in this analysis relates to the suspected zoned nature of White Lake brine chemistry (unlike the chemically well-mixed Brush Lake) and to sampling brine chemistry. Brine samples from closer to the center of the playa may yield a different, possibly higher estimate of α_2, resulting in decreased brine seepage mass flux and increased wind erosion of salt. The net mass inflow of salt, on the other hand, depends not on α_2 but on inflow hydraulics and chemistry.

UNCERTAINTY ANALYSIS

Like most mass balance calculations, these are subject to various uncertainties, including those in:

1. The evaporation estimate E_1 (influencing fluid fluxes Q_0, Q_1, Q_2),
2. The sample estimate of mean groundwater chemistry and flux-weighting of groundwater inflow,
3. The sample estimate of mean lake chemistry,
4. Assumptions inherent in the serial-lake model, and
5. Long-term representativeness of mean estimates in both flow and chemistry.

As hydrograph estimates of annual evaporation were made over the period of only one year, no statistical estimate of either error or temporal variance may be drawn. The measured rate is almost certainly higher than modern average evaporation by perhaps as much as 15 percent due to the relatively dry conditions in 1990. The 1990 rate seems unrealistically (perhaps 10–40 percent) high in comparison to the evaporation pan station, particularly for one month (June); this observation may indicate some unaccounted source of systematic error in the hydrograph data. However, the seasonal pattern of diurnal fluxes indicate that evaporation occurs from this lake even in cool fall and spring months, periods during which pan observations are normally not made. Evaporation at this lake appears grossly similar to that observed at the topographically-similar pan station at Froid, which exhibits the highest average pan evaporation in the region. Long-term rates may well be substantially lower than the 1990 measured estimate, but they are nonetheless high, likely one meter/year or greater. In any case, there is of course little reason to believe this short-term rate estimate is representative of Holocene conditions.

The sample estimate of mean groundwater source chemistry to Brush Lake includes error from two sources: error in estimating flux weights for groundwater flow into the lake and error resulting from natural spatial heterogeneity in groundwater chemistry. The flux-weighting method employed is a highly simplified approach that might be improved were local-scale flow information available; however, no local determinations of hydraulic conductivity or of Darcy flux are available, and direct seepage measurements indicate a high level of local heterogeneity in flow. It is clear from results of aquifer tests on pumping wells around the lake that transmissivity is highest west of Brush Lake; the empirical flux-weighting based strictly on the assumption of difference from west to east in Darcy flux and on halide mass-balance is consistent with this observation.

Observed sample (spatial) variability in groundwater seepage chemistry is relevant to the statistical significance of the mean estimates for individual ions, upon which the mass-flux calculations are based. Table 8 gives error statistics relating to sample variation in geochemistry for the six calculated lake samples (Table 1, Rows [1]–[6]). Shown in Table 8 are weighted standard errors of individual ion concentrations and 95 percent confidence limits on the mean estimate (t_{crit} = 2.23). Also shown are the *t*-distribution confidence levels that the flux-weighted mean groundwater concentrations are significantly different from the "conservative" concentrations calculated for lake water divided by α_{Cl}; this latter value would be the expected source chemistry were there no reactions adding or removing that ion. The null hypothesis is that there is no reaction flux involving each ion in the lake. The confidence level for Cl^- is, by default, 0 percent, as α is calculated using dissolved chloride in the lake. The confidence levels are close to zero for the other halides; allowing rejection of the null hypothesis at a high probability for these ions. Similarly, there is a low level of statistical certainly that reaction flux of SO_4 occurs in Brush Lake. For the other ions except Na^+ and K^+, the significance that there is some reaction flux is quite high, >95 percent for all but NO_3^-. For Na^+ and K^+, the significance of rejecting the null hypothesis that there is no reaction flux is lower but still sizable, on the order of 50 percent. It may be concluded that the halide ions and sulfate appear nearly conservative in Brush Lake, while other ions with the possible exceptions of Na^+ and K^+ are clearly involved in chemical reactions within the lake.

These error estimates in variability of groundwater chemistry require caution in interpretation. The magnitude of sample variation on the west side of the lake is much lower than on the east side for several ions, particularly Ca^{2+}, Mg^{2+}, Na^+, and SO_4^{2-}. The larger variance on the east side derives primarily from a single rather uncharacteristic sample at GE, a site for which the water analysis has been reproduced by sampling on successive years. Water geochemistry at this location is sufficiently dissimilar both from other lake-seepage samples and from nearby wells to cast some doubt on how free from lake-water contamination (presumably as back-infiltration) these samples were. Without the influence of this analysis, the standard error values in Table 8 for the above-mentioned ions are reduced by about 40 percent. The uncharacteristic analysis was nonetheless included in all mass-balance and error calculations and tends to bias the Na^+ concentration in the source to the high side; however, omission of this sample from the mass balance would tend to exacerbate, rather than reduce, the already difficult-to-explain negative reaction flux of Na^+ calculated for Brush Lake (Table 5).

Regarding the other sources of error (assumptions inherent in the serial-lake model, and the long-term representativeness of 1990 mean estimates for flow and chemistry), little conclusive may be said. The sampling of groundwater chemistry of groundwater sources flowing into Brush Lake was done comprehensively, while, on the other hand, the outflow from Brush Lake to White Lake is only a fraction of the fluid flux to the playa. Because of the concentration of dissolved solids in Brush Lake, however, the Brush Lake outflow is thought to be a substantial fraction of the total mass flux to White Lake. Any additional mass flux not considered would only tend to increase the calculated discrepancy between solid salt volume in the basin and the time-integrated inflow mass flux. Thus although not all sources of groundwater to the lower lake were considered, additional inflow beyond those noted here would only further support the stated arguments for salt transport out of the basin by brine seepage and wind erosion.

Despite the not-inconsequential uncertainties, two conclusions may be drawn. First, the predicted mass balance and reaction fluxes are consistent with the sequence and relative abundances of all preserved sediments except for mirabilite salt, which is substantially less abundant than predicted by mass balance. This discrepancy is outside the

TABLE 8.—ERROR ESTIMATES FOR SAMPLING VARIATION IN LAKE-BED SEEPAGE CHEMISTRY (TABLE 1, ROWS [1]–[6])

Species	Source Mean	lake mean/α_{Cl}	east total variation	west total variation	weighted std. deviation	weighted std. error	std. error (%)	95% confidence limits		H_o rejection (%)
								lower	upper	
Ca	89.9	0.4	11676	2719	38.49	11.76	13.1	63.6	116.1	>99
Mg	49.8	30.6	3185	454	19.35	5.91	11.9	36.6	63	>99
Na	125.7	151.4	112675	29133	120.82	36.91	29.3	43.4	208	50
K	8.38	10.37	426	169	7.82	2.39	28.5	3	13.7	57
Fe	1.95	0.007	19.1	8.06	1.67	0.51	26.1	0.81	3.09	>99
Mn	0.33	0.001	0.09	0.3	0.2	0.06	18.2	0.2	0.47	>99
SiO_2	17.6	0.9	3.1	102.1	3.29	1	5.7	15.3	19.8	>99
C	7.73	2.81	19.4	7.3	1.66	0.5	6.6	6.6	8.86	>99
Cl	8.75	8.75	1300	204	12.44	3.8	43.4	0.28	17.23	0
SO_4	305.1	299.1	708478	109330	290.1	88.6	29	107.4	502.8	5
NO_3	0.095	0.049	0.019	0.06	0.09	0.027	28.9	0.03	0.15	88
F	0.307	0.308	0.675	0.168	0.295	0.09	29.4	0.1	0.5	<1
Br	0.125	0.125	0.152	0.024	0.134	0.041	32.8	0.03	0.21	0
Sr	1.939	0.022	0.962	5.689	0.827	0.253	13	1.37	2.5	>99
Li	0.109	0.066	0.03	0.009	0.064	0.019	17.8	0.06	0.15	95

range of all the known uncertainties except the assumption of similar rates throughout Holocene time, and is interpreted as support for the hydraulic and geochemical models. Second, depending on the influence of these uncertainties, as much as 80 percent of the influent mass of dissolved sodium sulfate to White Lake may escape the playa, both as brine seepage through its base and as wind-blown salt, and in roughly equal proportion.

Is such a flux of brine escape hydrogeologically reasonable? Simplistically, it may be assumed that brine seeps from beneath the full area of the playa and that the underlying glaciolacustrine aquitard is homogeneous with 30 percent effective porosity. Neglecting dimensional and hydraulic requirements for free-convection, the hypothesis that vertical brine movement is possible through the aquitard may be tested solely on mass-balance grounds. The Darcy (flux) velocity required for mass balance (and, under unit head gradient, the hydraulic conductivity at brine density of 1.15 g/cm^3) is 4 x 10^{-11} m/s, for a seepage (spatial) velocity of 4 mm/year (40 m/10,000 years). The top of the underlying alluvial aquifer is about 70 m below the base of the playa. Under the calculated mass balance and assumed porosity, then vertical brine seepage may not yet have reached the deeper aquifer and might still be in the process of establishing a new steady-state flow regime. The required very small intrinsic vertical permeability for the aquitard (corresponding to a hydraulic conductivity of 10^{-11} to 10^{-10} m/s) is low but hydrogeologically reasonable, given the nature of these deposits.

On the other hand, one might ask why, given the numerous manifestations of heterogeneity in aquifer materials, the aquitard should be expected to be hydraulically homogeneous. Effectively reducing the connected porosity, "leaky" flow paths along more permeable layers in the lakebed deposits could accelerate the seepage (but not the Darcy) velocity by an order of magnitude or more. The occurrence of fracture-flow through otherwise tight clays would also increase transport velocity. It is not inconceivable that, during the Holocene, saline water could have migrated up to 100 vertical meters or more under any of a number of conceivable flow configurations, under a high enough rate of flow to reach deeper more transmissive deposits. Future work will focus on definition of the range of feasible flow-field configurations under which such free convection might operate.

SUMMARY

The mass-balance model has been applied to quantify reaction fluxes for a well-mixed tufa lake, within the limits imposed by uncertainty in mean source chemistry. The estimate of mass balance from this lake to a lower, less homogeneous brine is less well-constrained, but indicates reactions consistent with the salt products found in the playa. Mg-carbonate formation under kinetic control (dolomite is highly supersaturated) may be one possible reaction to account for observed Mg-flux in Brush Lake, where sedimentary evidence for authigenic dolomite formation has been found (Donovan, 1992).

The lumped-parameter approach utilizes system mass balance based on a free-water evaporation estimate and chemical ratios between lake and source, both attainable field data objectives. The reactive-flux calculations infer rates of formation of mineral phases. The serial-lake hydraulic model projects mass flux from lakes of low dissolved solids, for which evaporation can be measured with some degree of confidence, to ephemeral playa lakes, where such evaporation measurements are much more difficult. While the fluxes measured in this study are based on a very short period of hydrologic observation that may be quite different from long-term mean values, the estimate they provide of flux rates are consistent with observations of Holocene sediment volumes and a starting point for study of transient variations in these fluxes induced by climate and recorded in sediments.

The mass balance developed here is likely only approximately accurate; at best, it serves to confirm interpretations of process based on observations of apparently authigenic mineral phases in the lakes. On the other hand, the evaporation rate is simply a scaling factor for the mass balance; proportions of mass and reaction fluxes are determined by the ion ratios of lake to groundwater chemistry, and by assumptions inherent in the chain-lake model.

Brush Lake and White Lake form a serial "cascade" that chemically differentiates groundwater into carbonate (Brush Lake) and sulfate (White Lake) reaction products. Controlling this fractionation is the early, near-equilibrium removal of calcium carbonate from the evolving water, as would be predicted by the chemical divide model (Hardie and Eugster, 1970). Calculated evaporative concentration factors are 8.7x for the marl lake and 108x (net 950x source groundwater) for the playa, although brine and salt distribution may in fact be concentrically zoned within the playa (Jones, 1965), thus influencing the latter concentration estimate. The marl lake is chemically well-mixed and amenable to the lumped-parameter approach; the playa may be chemically heterogeneous, at least for species undersaturated with respect to mineral phases (not Na^+ and SO_4^{2-}, whose activities are controlled by saturation with mirabilite).

An unaccountably high negative estimate of reaction flux for Na^+, if accurate, is best explained by eolian influx of salts and sodic clays to the lake from surrounding glacial till soils. Mg^{2+} appears to be lost to lake sediments via several processes, including carbonate precipitation and ion exchange. Modern Brush Lake may be analogous in some ways to the "dry" period of mid-Holocene (6000–7200 years BP) Elk Lake, Minnesota (Dean and others, 1984), during which aragonite marl was precipitated. Under its modern temperate climate, it precipitates only magnesian calcite.

The most abundant salt in the playa is mirabilite; mass balance indicates that less abundant Mg-sulfate and/or carbonate salts may also be present, but these were not encountered at the four available drillhole locations of Binyon (1952). A large inorganic carbon flux to the playa apparently supports microbial growth and is consumed by sulfate reduction. Calculated Holocene mass flux of sodium-sulfate to the playa from Brush Lake is greater than the actual volume of salt estimated by Binyon (1952). This discrepancy is attributed to brine escape and to wind erosion of salts in some proportion, which together may remove approximately 80 percent of the total mass inflow to the playa.

Brine out-seepage from the playa may have reached deeper aquifers during the Holocene; if so, such brine has not to date been detected in nearby water wells, or alternately has dispersed sufficiently as not to be noted. The fact that additional more dilute groundwater flows into White Lake along its other margins increases rather than reduces this sodium-sulfate discrepancy. Consideration of this additional groundwater inflow also would effectively raise the brine-lake evaporation estimate (E_2). This cannot be done with available information.

Interpreted fluxes associated with organic carbon synthesis, ion exchange, and Mg-carbonate formation in the marl lake would have been difficult to predict by application of the chemical divide model, based on mineral equilibria alone. The identity and extent of such reactions early in the reaction path of brine may influence subsequent reaction products, as noted by Eugster and Jones (1979). Biological activity, kinetically-slow Mg-carbonate formation, and ion-exchange reactions involving clays in moderately fresh carbonate-producing lakes, widespread in the semi-arid Great Plains, may influence the ultimate geochemistry of water and sediments in saline lakes in the region.

ACKNOWLEDGEMENTS

Financial support was provided by the Geological Society of America, Hydrogeology Division (Grant No. 4410-90) and by the Krynine Fund, Department of Geosciences, Penn State University. John Sonderegger of Montana Tech College, Jon Reiten of the Montana Bureau of Mines and Geology, and Virginia Carter of the U. S. Geological Survey loaned field gear. Jim McCallum of the Medicine Lake Wildlife Refuge (U. S. Fish and Wildlife Service) provided a research boat. Rom Hedges and Ken Hanson were dive buddies. Dave Harris of the U. S. Agricultural Research Service, Sidney, Montana, provided unpublished Class A pan measurements of evaporation rate from their Froid, Montana, research farm. Helpful reviews of the manuscript were provided by Art Rose, Susan Brantley, Garth Van der Kamp, Blair Jones, Ward Sanford, Warren Wood, and Herb Wright. While help was obtained from many, sole responsibility for results of this research lies with the author.

REFERENCES

ANDERSEN, M. E., AND JOBSON, H. E., 1982, Techniques for estimating annual lake evaporation using climatological data: Water Resources Research, v. 18, p. 630–636.

BINYON, E. O., 1952, North Dakota sodium sulfate deposits: Washington, United States Bureau of Mines Report of Investigations 4880, 41 p.

BOWLER, J. M., 1986, Spatial variability and hydrologic evolution of Australian lake basins: analogue for Pleistocene hydrologic change and evaporite formation: Paleogeography, Paleoclimatology, Paleoecology, v. 54, p. 21–41.

CHRISTIANSEN, E. A., AND PARIZEK, R. R., 1961, A summary of studies completed to date of the ground-water geology and hydrology of the buried Missouri and Yellowstone valleys near Estevan: Regina, Saskatchewan Research Council, Circular 1, 30 p.

CLAYTON, L., AND MORAN, S. R., 1982, Chronology of late Wisconsinan glaciation in middle North America: Quaternary Science Reviews, v. 1, p. 55–82.

CROWE, A. S., AND SCHWARTZ, F. W., 1981, Simulation of lake-watershed systems I. Description and sensitivity analysis of the model: Journal of Hydrology, v. 52, p. 71–105.

DEAN, W., 1981, Carbonate minerals and organic matter in sediments of modern north temperate hard-water lakes, *in* Ethridge, F. G., and Forest, R. M., eds., Recent and Ancient Nonmarine Depositional Environments: Models for Exploration: Tulsa, Society of Economic Paleontologists and Mineralogists Special Publication 31, p. 213–231.

DEAN, W. E., BRADBURY, J. P., ANDERSON, R. Y., AND BARNOSKY, C. W., 1984, Variability of Holocene climatic change: evidence from varved lake sediments: Science, v. 266, p. 1191–1194.

DEAN, W. E., AND GORHAM, E., 1976, Major chemical and mineral components of profundal surface sediments in Minnesota lakes: Limnology and Oceanography, v. 21, p. 259–284.

DONOVAN, J. J., 1988, Groundwater geology and high-yield aquifers of northeastern Montana: Butte, Montana Bureau of Mines and Geology Open-File Report MBMG-209, 116 p.

DONOVAN, J. J., 1992, Geochemical and hydrologic dynamics in evaporative groundwater-dominated lakes of glaciated Montana and North Dakota: Unpublished Ph.D. Dissertation, Penn State University, University Park, 252 p.

EUGSTER, H. P., 1970, Chemistry and origin of the brines of Lake Magadi, Kenya: Washington D. C., Mineralogical Society of America Special Paper No. 3, p. 215–235.

EUGSTER, H. P., AND HARDIE, L. A., 1978, Saline lakes, *in* Lerman, A., ed., Lakes: Chemistry, Geology, Physics: New York, Springer-Verlag, p. 237–293.

EUGSTER, H. P., AND JONES, B. F., 1979, Behavior of major solutes during closed-basin brine evolution: American Journal of Science, v. 279, p. 609–631.

GROSSMAN, I. G., 1968. Origin of the sodium sulfate deposits of the northern Great Plains of Canada and the United States: United States Geological Survey Professional Paper 600-B, p. B104-B109.

HANSEN, D. E., 1967, Geology and ground-water resources, Divide County, North Dakota: Part I, Geology: North Dakota State Water Commission Bulletin 45, 90 p.

HARDIE, L. A., AND EUGSTER, H. P., 1970, The evolution of closed-basin brines: Washington, Mineralogical Society of America Special Paper 3, p. 273–290.

HARRIS, D., 1992, Unpublished Class A evaporation pan data for Culbertson, Montana, Research Farm, Water Year 1990: United States Department of Agriculture, Agricultural Research Service, Sidney, Montana.

HEINRICH, E. W., 1965, Microscopic identification of minerals: New York, McGraw-Hill, 406 p.

JONES, B. F., 1965, The hydrology and mineralogy of Deep Springs Lake, Inyo County, California: United States Geological Survey Professional Paper 502-A, 56 p.

JONES, B. F., 1986, Clay mineral diagenesis in lacustrine sediments: United States Geological Survey Bulletin 1578, p. 291–300.

JONES, B. F., EUGSTER, H. P., AND RETTIG, S. L., 1977, Hydrochemistry of the Lake Magadi basin, Kenya: Geochimica et Cosmochimica Acta, v. 41, p. 53–72.

KELTS, K., AND HSU, K. J., 1978, Freshwater carbonate sedimentation *in* Lerman, A., ed., Lakes: Chemistry, Geology, Physics: New York, Springer-Verlag, p. 295–323.

KOHLER, M. A., NORDENSEN, T. J., AND BAKER, D. R., 1959, Evaporation maps for the United States: United States Weather Bureau Technical Paper No. 37.

LANGBEIN, W. B., 1961, Salinity and hydrology of closed lakes: Washington, D.C., United States Geological Survey, Professional Paper 412, 20 p.

LAST, W. M., 1989, Continental brines and evaporites of the northern Great Plains of Canada: Sedimentary Geology, v. 64, p. 207–221.

LEE, D. R., 1977, A device for measuring seepage in lakes and estuaries: Limnology and Oceanography, v. 22, p. 140–147.

LEVINGS, G. W., 1986, Preliminary appraisal of ground water in and near the ancestral Missouri River valley, northeastern Montana: Helena, United States Geological Survey Water Resources Investigations 86-4138, 44 p.

MCCONNAUGHEY, T. A., AND FALK, R. H., 1991, Calcium-proton exchange during algal calcification: Biological Bulletin, v. 180, p. 185–195.

MEYBOOM, P., 1967, Mass-transfer studies to determine the groundwater regime of permanent lakes in hummocky moraine of western Canada: Journal of Hydrology, v. 5, p. 117–142.

MORTON, F. I., 1983, Comment on "Techniques for estimating annual lake evaporation using climatological data": Water Resources Research, v. 19, p. 1347–1357.

MULLER, G., IRION, G., AND FOERSTNER, U., 1972, Formation and diagenesis of Mg-carbonates in the lacustrine environment: Naturwissenschaften, v. 59, p. 158–164.

O'CONNOR, D. J., 1989, Seasonal and long-term variations of dissolved solids in lakes and reservoirs: Journal of Environmental Engineering, v. 115, p. 1213–1234.

SACKS, L. A., HERMAN, J. S., KONIKOW, L. F., AND VEDA, A. L., 1992, Seasonal dynamics of groundwater-lake interactions at Donana National Park, Spain: Journal of Hydrology, v. 136, p. 123–154.

SCHULTZ, L. G., 1964, Quantitative interpretation of mineralogical composition from x-ray and chemical data for the Pierre shale: Washington, D. C., United States Geological Survey Professional Paper 391-C, 31 p.

SHIH, S. F., 1980, Water budget computation for a shallow lake: Lake Okeechobee, Florida: Water Resources Bulletin, v. 16, p. 724–728.

SPENCER, R. J., EUGSTER, H. J., JONES, B. F., AND RETTIG, S. L., 1985, Geochemistry of Great Salt Lake, Utah. I: Hydrochemistry since 1850: Geochimica et Cosmochimica Acta, v. 49, p. 727–737.

STABEL, H. H., 1986, Calcite precipitation in Lake Constance: chemical equilibrium, sedimentation, and nucleation by algae: Limnology and Oceanography, v. 31, p. 1081–1093.

TELLER, J. T., AND LAST, W. M., 1990, Paleohydrological indicators in playas and salt lakes, with examples from Canada, Australia, and Africa: Palaeogeography, Palaeoclimatology, Palaeoecology, v. 76, p. 215–240.

THOMPSON, J. B., AND FERRIS, F. G., 1990, Cyanobacterial precipitation of gypsum, calcite, and magnesite from natural alkaline lake water: Geology, v. 18, p. 995–998.

VEIHMEYER, F. J., 1964, Evapotranspiration, *in* Chow, V. T., ed., Handbook of Applied Hydrology: New York, McGraw-Hill, p. II-1–38.

WANEK, A., 1985, Skjermo Lake Aquifer, Yearly Update, April 1985: Bismarck, North Dakota State Water Commission, unpublished report, 10 p.

WETZEL, R. G., 1975, Limnology (2nd ed.): Philadelphia, Saunders College Publishing, 412 p.

WINTER, T. C., 1983, The interaction of lakes with variably-saturated porous media: Water Resources Research, v. 19, p. 1203–1218.

WINTER, T. C., LABAUGH, J. W., AND ROSENBERRY, D. O., 1988, The design and use of a hydraulic potentiomanometer for direct measurements of differences in hydraulic head between groundwater and surface water: Limnology and Oceanography, v. 33, p. 1209–1213.

WITKIND, I. J., 1959, Quaternary geology of the Smoke Creek-Medicine Lake-Grenora area, Montana and North Dakota: United States Geological Survey Bulletin 1073, 80 p.

WOOD, W. W., AND SANFORD, W., 1990, Ground water control of evaporite deposition: Economic Geology, v. 85, p. 1226–1335.

DEEP-WATER EVAPORITE MINERAL FORMATION IN LAKES OF WESTERN CANADA*

WILLIAM M. LAST
Department of Geological Sciences, University of Manitoba, Winnipeg, Canada R3T 2N2

ABSTRACT: Despite more than a century of multidisciplinary study, our understanding of the genesis of evaporite minerals in deep-water sedimentary environments remains poor. This is most likely due to a paucity of modern settings in which deep-water evaporites are forming. Several of the present-day deep, saline to hypersaline lakes in the northern Great Plains of western Canada provide excellent opportunities to examine subaqueous salt mineral formation, and to document the processes by which these salts are deposited, preserved, and altered.

Deadmoose, Little Manitou, and Freefight lakes are lacustrine basins in which soluble and sparingly soluble evaporite minerals are accumulating today in the relatively deep offshore areas of the lakes. Both Deadmoose and Freefight lakes are hypersaline and meromictic, with strongly anoxic bottom waters. Little Manitou is also hypersaline but seasonal density and chemical stratification usually breaks down by early to midsummer. The salts in these basins, which are comprised of a complex assemblage of hydrated sodium and magnesium sulfates and carbonates, reflect, in general terms, the high alkalinity of the brines and the overall dominance of Na^{+}, Mg^{2+}, and SO_4^{2-} ions. Modern sediment–trap data show that salt precipitation and sedimentation are extremely rapid, with the precipitates being derived from a variety of positions in the water columns. Salt accumulation is most rapid in the deepest parts of the lakes. The evaporites are generally coarsely crystalline, massive to thickly bedded, and interfinger laterally with laminated organic-rich clays and poorly sorted carbonate-clastic debris derived from nearshore areas. Even in the shallowest basin, the salts show no indication of dissolution or recycling. Indeed, preliminary augering and coring of the late Holocene sediment reveals that relatively thick sections of evaporites have accumulated and are being preserved in the basins.

INTRODUCTION

It has long been recognized that there are very few examples of modern sedimentary environments in which true deep-water evaporite mineral formation is occurring. Even the basinal sediments of the Dead Sea, cited by Warren (1989) as one of the best approximations to a modern deep-water evaporite forming environment, contain very little soluble evaporite mineral material. In contrast, many ancient evaporite deposits are interpreted to have been deposited in deep, subaqueous settings (see summaries in Warren, 1989; Dronkert, 1985; Till, 1978). The strong emphasis on theoretical considerations of deep-water evaporites without a rigorous understanding of modern deep-water analogues has led to considerable discussion and controversy in the literature over the interpretation of these ancient suspected deep subaqueous deposits. The northern Great Plains region of western Canada contains lacustrine basins in which deep-water, soluble and sparingly soluble endogenic precipitates are forming and being preserved in the Holocene stratigraphic record. The objective of this paper is to describe these deep-water lacustrine settings and discuss the formation of the soluble salts.

Terminology

As discussed by Sonnenfeld (1984), the terms "deep" and "shallow" are very poorly defined in most evaporite mineral depositional settings. Absolute values of the usual morphological measures of water depth of a basin, such as mean depth, maximum depth, and relative depth, have little meaning in saline/evaporitic basins because the water column is often chemically stratified. This brine stratification, which can develop even in relatively shallow basins (less than 3 m), effectively isolates the sediment-water interface from wave action, physical sorting, and immediate dissolution of the very soluble precipitates. "Deep-water" conditions are present in each of the lacustrine basins discussed in this paper by virtue of the presence of a relative stable chemical and/or density gradient within the water columns.

An evaporite mineral is generally taken to be a salt deposited by chemical precipitation from a natural brine due to the process of evaporation. Although a wide variety of sparingly soluble carbonate minerals occur in the deep basinal sediments of the salt lakes of the Great Plains (Last, 1990; Ghebre-Egziabhier and St. Arnaud, 1983a, b), this paper will concentrate only on the most soluble salts precipitated.

SETTING AND LOCATION

The northern Great Plains of western Canada stretch from the Precambrian Shield near Winnipeg, Manitoba, westward for nearly 1600 km to the foothills of the Rocky Mountains (Fig. 1). This is a vast region of flat to gently rolling terrain interrupted only by occasional steeply sided, deeply entrenched river valleys. Within this region there are millions of salt lakes, with estimates ranging as high as 5.5 million individual saline basins and wetlands. Indeed, nearly all of the surface water present in the Great Plains of western Canada is saline. Based on a morphological survey of over 500 of these lakes (Fig. 2), the vast majority of the basins are small (less than 100 km^2) and shallow (maximum depth less than 3 m). Although few in number, the permanent, deep-water lakes are important because their sediment records are usually undisturbed by either bioturbation, desiccation, or erosion and, therefore, often contain finely laminated sequences which form the basis of high resolution paleoenvironmental reconstructions (Last, 1992).

The climate of the northern Great Plains ranges from a cold, sub-humid continental regime in the eastern and northern parts of the region to semi-arid and arid conditions in the west. The high salinity of most of the surface water in the region is controlled by the high evaporation/precipitation ratios and the presence of large areas of closed drainage. Over 100,000 km^2 of southern Saskatchewan and eastern Alberta are characterized by internal drainage (Fig. 1). Although the origin and ultimate source of ions in the salt

*Palliser Triangle Global Change Contribution Number 9

Sedimentology and Geochemistry of Modern and Ancient Saline Lakes, SEPM Special Publication No. 50
 ISBN 1-56576-014-X

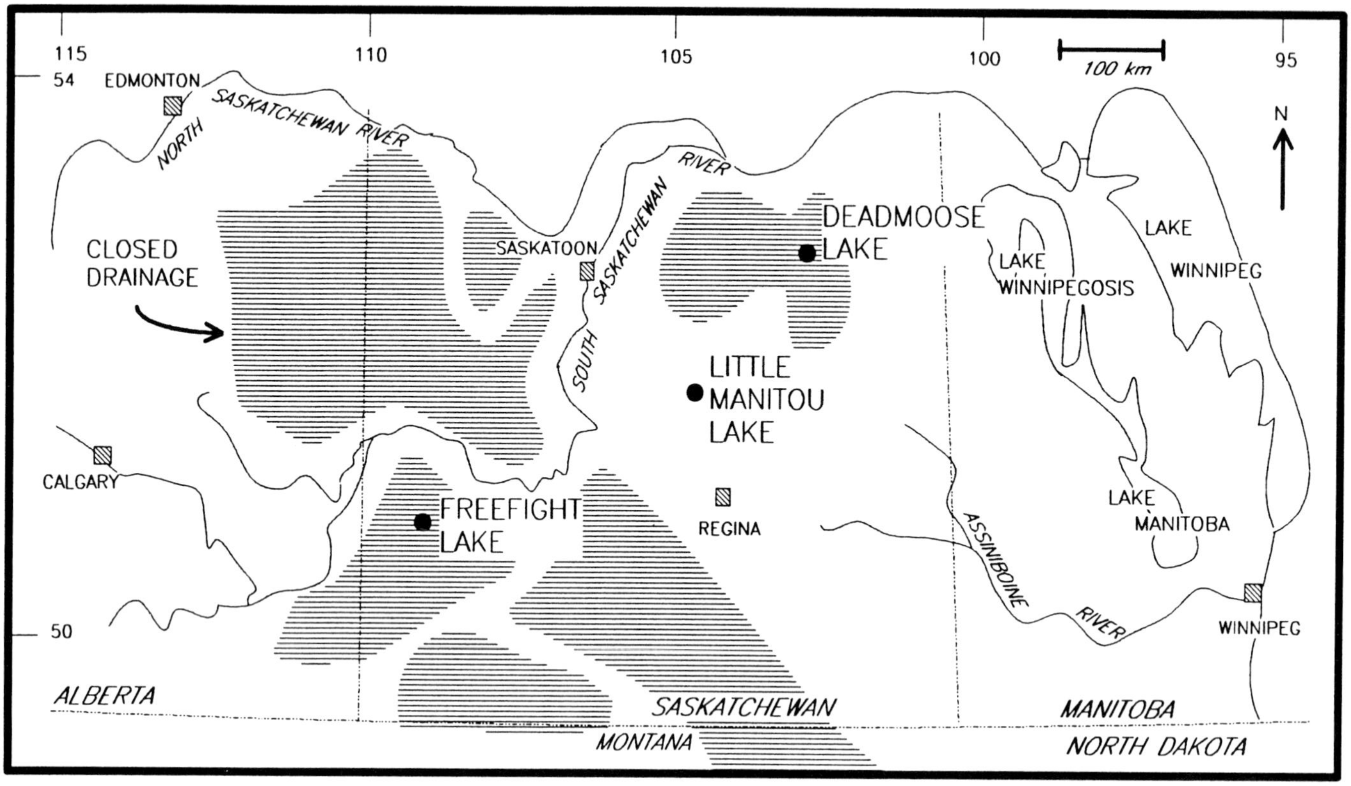

FIG. 1.—Map showing the location of Freefight, Deadmoose, and Little Manitou lakes in the northern Great Plains region. The shaded areas are areas of internal drainage.

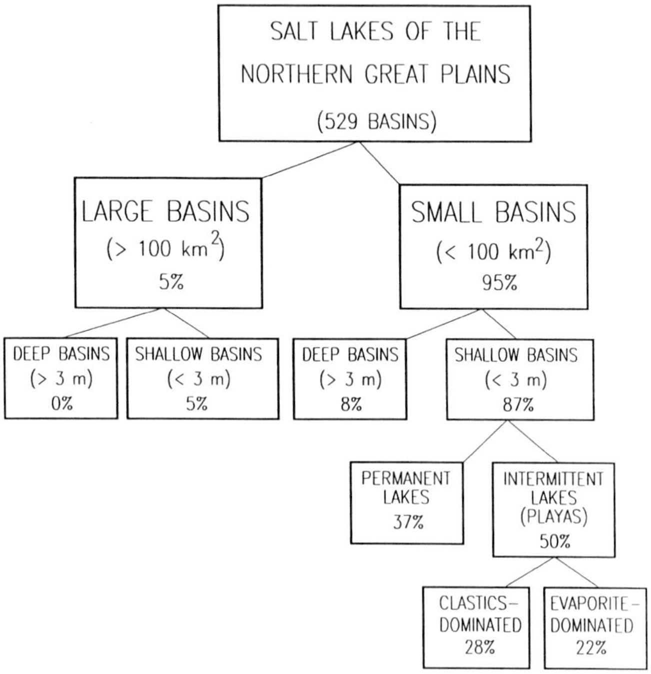

FIG. 2.—Morphological summary of salt lakes in the northern Great Plains.

lakes have been topics of considerable discussion in the scientific literature, most investigators suggest that weathering, leaching, and cation exchange reactions taking place in the thick, unconsolidated glacial drift of the region can provide an adequate source of dissolved components to the closed drainage basins.

The locations of the lakes discussed in this paper are shown in Figure 1. Freefight Lake is located in the Great Sand Hills region of southwestern Saskatchewan, the most arid portion of the northern Great Plains. Deadmoose Lake, situated about 100 km east of Saskatoon, is located in aspen parkland at the northern margin of the Great Plains. Little Manitou Lake is located about 100 km north of Regina.

PREVIOUS WORK

Of the millions of salt lakes in this region, the hydrochemical characteristics (total salinity, major ion concentrations) and morphology (area, depth) of about 500 basins are known (Last, 1988; Hammer, 1978; Hammer and Haynes, 1978; Rutherford, 1970). Because salt lakes are the only surface water present in large areas of the Great Plains, the lakes have been examined in some detail by biological limnologists and water resources managers (e.g., Hammer and others, 1983; Haynes and Hammer, 1978). Much of this work, including previously published information on Deadmoose Lake and Little Manitou Lake, is summarized by Hammer (1986). Data on the nature of the modern offshore sediments (mineralogy, texture, organic content) and the modern facies relationships within the basins are available for about 100 of the lakes (Last, 1989). Details of the Quaternary stratigraphy and post-glacial history of twenty of the basins in this large region have been published (see summaries in Last, 1992; Schweger and Hickman, 1989; Last and Slezak, 1988). Slezak (1989) and Boden (1985) provide details of the modern facies relationships and geochemistry in Freefight Lake and stable isotope stratigraphy of the endogenic carbonates in Deadmoose Lake, respectively.

METHODS

This report is based on information and data collected from the three modern saline lacustrine basins over a period of eight years. Bathymetry of the basins was determined during the summers of 1983–1989 using an echo sounder and weighted line. Modern facies mapping and grab samples of surficial sediment and shallow subsurface sediment were collected from the basins during the period 1983–1990. Water levels, brine chemistries, and various environmental parameters of the basins have been monitored since 1984. Water samples were collected using a 2-l Kemmerer sampling bottle. Bottom dredge samples were collected during various times of the year using a modified Ponar grab sampler. Cores were acquired during the winter seasons using various methods: a modified Livingstone piston corer, an Acker 5 cm split tube sampler, and a mechanized rotary drill.

Sediment traps have been installed in two of the basins: Freefight Lake and Deadmoose Lake. In each case, the trap configuration consisted of an array of solitary cylindrical PVC vessels of 9 cm diameter suspended at various depths by a surface buoy–subsurface anchor system. Following suggestions by Håkanson and Jansson (1983), the aspect ratio of the vessels (i.e., height/diameter) was approximately 10. A compensating fin and rotating swivels assured vertical orientation of each collection vessel. The traps were positioned such that accumulation data were acquired from: (a) immediately above the chemocline, (b) immediately below the chemocline, and (c) just above the sediment-water interface.

Sediment samples were analysed for bulk mineralogy and detailed evaporite mineralogy by X-ray diffraction. Estimates of the percentages of the various minerals were obtained from the diffractograms using a method modified after Schultz (1964). Standard analytical techniques were used to determine the chemistry of the water samples. Temperature, pH, Eh, O_2, H_2S, and conductivity were measured in the field; cation and anion concentrations were determined in the laboratory by atomic absorption, gravimetric, and turbidimetric methods.

RESULTS

Basin Morphology

Figure 3 and Table 1 summarize the bathymetry and various morphological parameters of the three lakes discussed in this paper. All of the lakes are deep and contain permanent water bodies. Freefight Lake is the deepest of the three and, with a mean depth of over 19 m, is the deepest saline lake in the Great Plains. It has very steeply sloping sides and a flat, featureless bottom. Little Manitou Lake is shallower and occupies a long, steep-walled, riverine channel-shaped basin. It similarly has steeply sloping sides and a flat bottom. In contrast to Freefight and Little Manitou basins, Deadmoose Lake has a very irregular basin morphology and bottom topography with numerous deep (greater than 35 m) troughs and holes.

Water Chemistry

All three lakes contain hypersaline brines. The water columns in both Freefight Lake and Deadmoose Lake were chemically stratified during the entire observation period. Little Manitou Lake was chemically stratified for much of the period between 1987–1989, but was isochemical during 1984–1986 and again in 1990. In general, the lakes' waters are highly alkaline, with pH values between 8 and 11. The ionic composition of the water in the three basins, as well as that of many of the other saline lakes of the region, is dominated by sodium, magnesium, and sulfate ions (Fig. 4).

The chemocline in Freefight Lake is located at approximately 6-m depth. The average salinity of the mixolimnion is 112 ppt TDS (total dissolved solids), whereas the monimolimnion has a salinity of about 180 ppt. Both water masses are dominated by Na^+, Mg^{2+}, and SO_4^{2-}. The water below the chemocline has a considerably higher proportion of HCO_3^- than that of the surface water (Fig. 4). The lower water mass is also very strongly anoxic and has H_2S contents as high as 1200 mg l^{-1}.

Little Manitou Lake water is likewise usually in excess of 100 ppt TDS but has different ionic ratios than Freefight Lake water; Little Manitou shows a strong dominance of Mg^{2+} and a higher proportion of Cl^-. During the episode of temporary meromixis (1987–1989), the ionic ratios of waters in both layers were similar. Because it is the shallowest of the three lakes, the salinity of Little Manitou Lake showed the greatest temporal variation, ranging from a low of 67 ppt (surface water during spring of 1984) to 192 ppt (bottom water in winter, 1989). The water composition of this lake is also quite variable on a spatial basis, presumably due to subaqueous spring discharge. The water below about 3 m depth in Little Manitou Lake was reducing even during isochemical periods.

Deadmoose Lake is the least saline of the lakes discussed here, with an average salinity of the mixolimnion of 34 ppt and a monimolimnion salinity of 85 ppt. The chemocline is located at approximately 8 m depth. The brines in both water masses have nearly identical ionic ratios: $Na > Mg >> K \simeq Ca$ and $SO_4 > Cl >> HCO_3$. Deadmoose Lake has the highest Na^+ and Cl^- proportions of the three basins. Like Freefight Lake, the monimolimnion of Deadmoose Lake is permanently strongly reducing (Eh values to −400 mv) and has high levels of H_2S (to 200 mg l^{-1}).

Modern Basinal Sediments

The modern, offshore (deep-water) sediment of each of the basins is dominated by very soluble salts. In general,

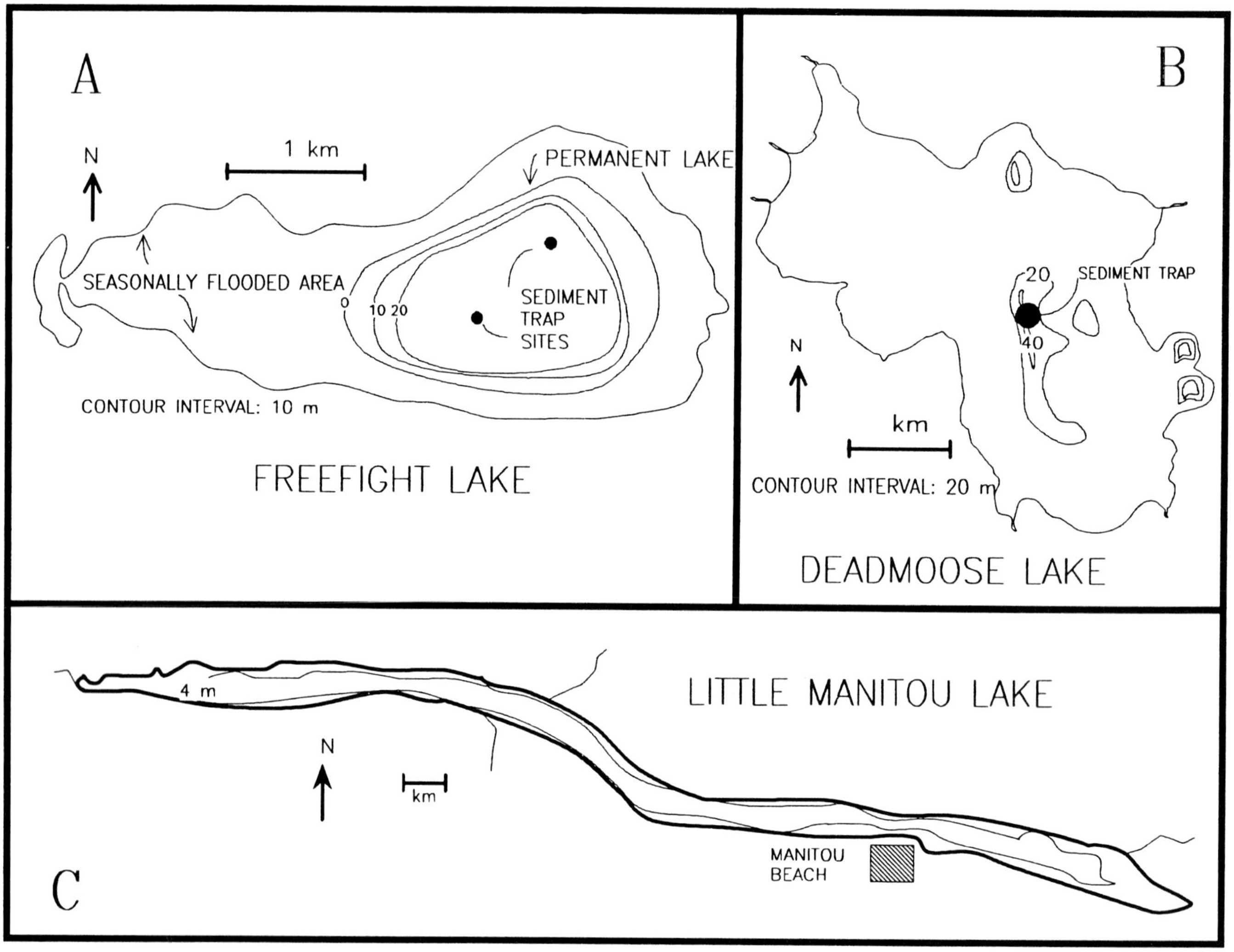

FIG. 3.—Bathymetry and basin morphology of (A) Freefight Lake, (B) Deadmoose Lake and (C) Little Manitou Lake.

TABLE 1.—SUMMARY OF MORPHOMETRIC DATA FOR FREEFIGHT, DEADMOOSE, AND LITTLE MANITOU LAKES.

	Freefight	Deadmoose	Little Manitou
Area (km^2)	2.59	10.50	13.9
Volume ($10^6 m^3$)	30	109	53
Drainage basin area (km^2)	55.20	70.19	169
Maximum length (km)	2.95	5.69	20.1
Maximum width (km)	1.25	3.23	1.1
Maximum depth (m)	24.92	48.10	5.9
Mean depth (m)	19.52	9.90	4.0

these salts are coarsely crystalline, massive to very poorly bedded, with only minor amounts of sparingly soluble and insoluble inorganic components. At the sediment-water interface, the salts range from loose, porous accumulations of single, euhedral crystals (Fig. 5A) to well-lithified crystalline rock (Fig. 5B). Organic matter content of these modern sediments is variable but often quite high (up to 38 percent). Over twenty non-detrital minerals have been identified, however most of the minerals making up the non-carbonate, soluble salts in the deposits are composed of variable proportions of the sodium and sodium + magnesium sulfates: bloedite [$Na_2Mg(SO_4)_2 \cdot 4H_2O$], epsomite ($MgSO_4 \cdot 7H_2O$), or mirabilite ($Na_2SO_4 \cdot 10H_2O$). Lesser amounts of gypsum and halite are also commonly present. The major endogenic/authigenic carbonate minerals identifed include: aragonite, both Ca-rich and Mg-rich dolomite, natron, and huntite. Pyrite is also found in relatively high proportions in the deep-water sediment of the lakes.

Deadmoose Lake contains the simplest salt mineral assemblage: mainly mirabilite with minor gypsum (Fig. 6A). The soluble basinal salts in Little Manitou Lake are composed of a mixture of epsomite and bloedite with secondary abundances of mirabilite, gypsum, and halite (Fig. 6B). Freefight Lake sediments contain the most complex assemblage of evaporite minerals: bloedite and mirabilite are usually dominant but significant proportions of epsomite and gypsum also occur (Fig. 6C). Glauberite [$Na_2Ca(SO_4)_2$], halite, hexahydrite [$MgSO_4 \cdot 6H_2O$], kieserite [$MgSO_4 \cdot H_2O$], konyaite [$Na_2Mg(SO_4)_2 \cdot 5H_2O$], loeweite [$Na_{12}Mg_7(SO_4)_{13} \cdot 5H_2O$], and tychite

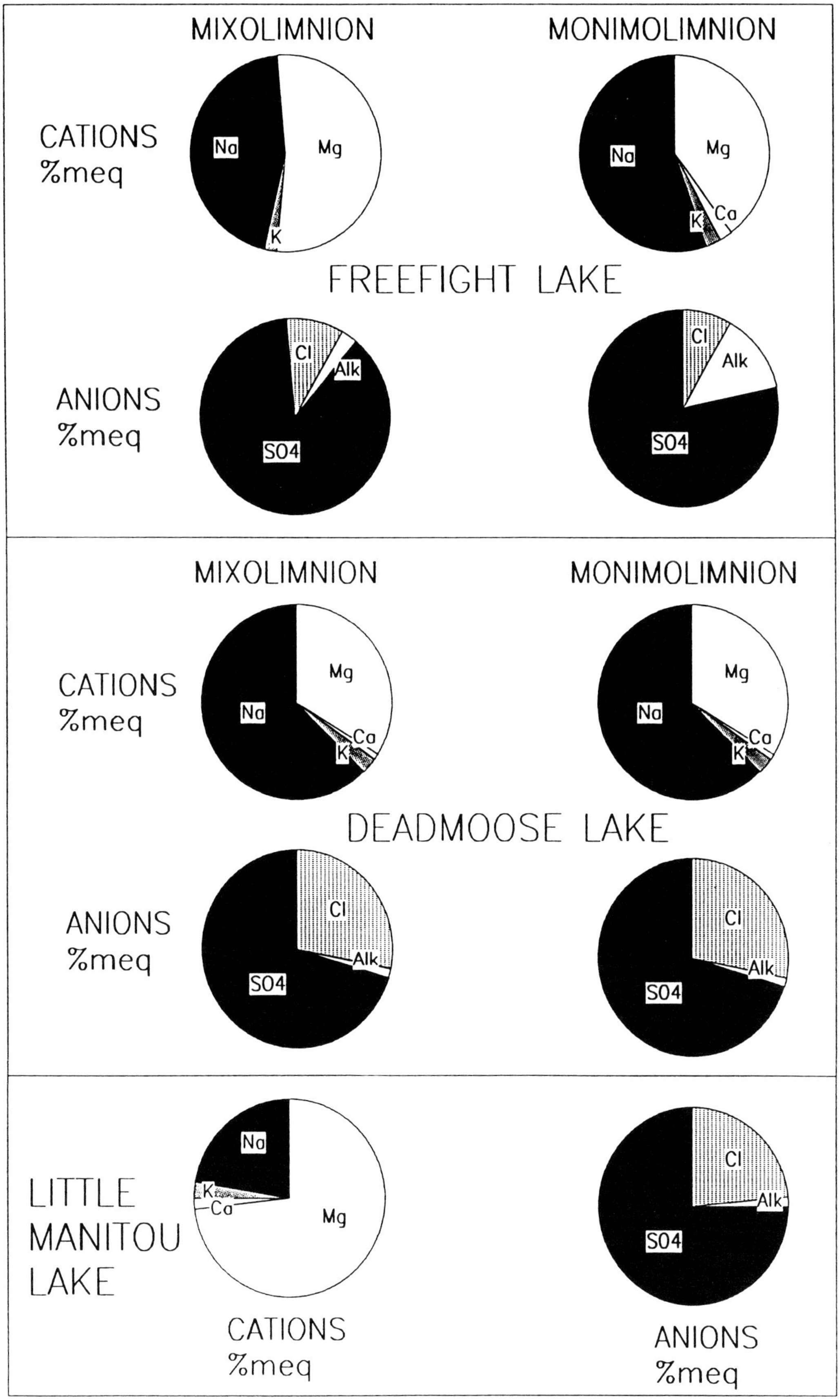

FIG. 4.—Pie charts showing the cation and anion ratios (in percent milliequivalents) of water in Freefight, Deadmoose, and Little Manitou lakes.

TABLE 2.—AVERAGE VALUES OF WATER COMPOSITION OF FREEFIGHT, DEADMOOSE, AND LITTLE MANITOU LAKES.

	Na	K	Ca	Mg (mmol l^{-1})	Cl	HCO_3	SO_4
Freefight Lake							
mixolimnion	708	59	4	450	197	51	665
monimolimnion	1313	64	11	504	201	263	956
Deadmoose Lake							
mixolimnion	208	7	4	56	144	7	174
monimolimnion	389	14	11	104	270	16	285
Little Manitou Lake	480	20	22	703	455	18	741

FIG. 5.—(A) Example of loose, unconsolidated sediment composed of a mixture of mirabilite and bloedite crystals from the modern, deep-water surficial sediment of Freefight Lake. Long dimension of sediment dredge is 35 cm. (B) Example of lithified aggregates of euhedral mirabilite crystals from the modern, deep-water surficial sediment of Deadmoose Lake. Diameter of coin is 2 cm.

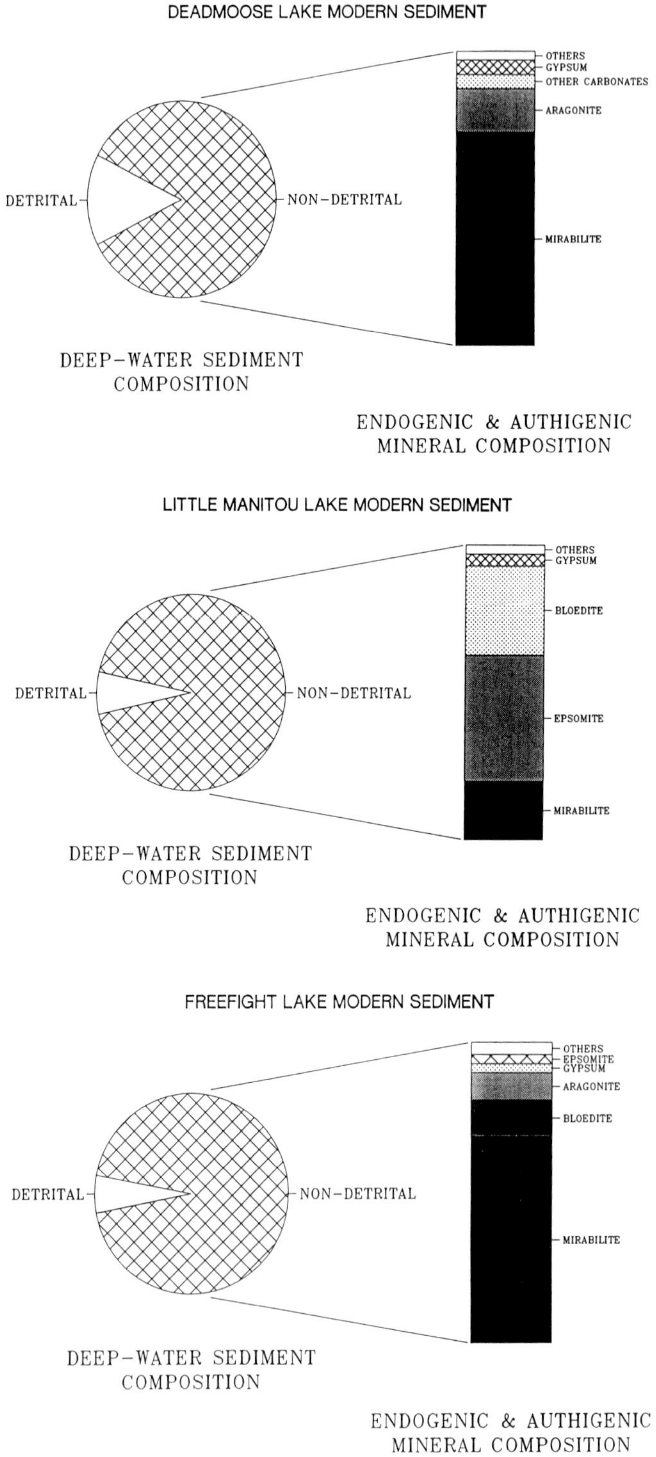

FIG. 6.—Modern basinal sediment composition in (A) Deadmoose Lake, (B) Little Manitou Lake, and (C) Freefight Lake.

$[Na_6Mg_2(CO_3)_4SO_4]$ have all been identified in small quantities.

The spatial distribution of these salts within the basins varies. In Freefight Lake, the entire lake bottom below water depths about 7.5 m is uniformly covered with salt. Total maxiumum thickness of the salt in this lake is still unknown because coring has not yet penetrated a complete section in the basin center. In water depths greater than 23 m, the salt is at least 1 m thick, whereas near the edges of the lake the precipitates thin to several centimeters and interfinger shoreward with nearshore shallow-water, evaporitic and organic carbonates. Little Manitou Lake shows a similar widespread spatial distribution of the basinal salts in water depths greater than about 4 m. Maximum thickness of salt penetrated by augering and coring in this basin is 1.25 m. Shoreward in Little Manitou Lake the salts interfinger with finely laminated aragonite-clay couplets and nonbedded shoreline clastics. In contrast to the even, uniform spatial distribution of salts in Freefight and Little Manitou lakes, the distribution of deep-water evaporites in Deadmoose Lake is much less regular and is confined just to the deep troughs and holes (greater than 25 m) where at least 1 m of salt is present.

Sediment Trap Data

Annual and seasonal sediment trap data have been collected from two sites in the Freefight basin for a total of four years between 1984 and 1990. Deadmoose trap data span only two years during this same period and is limited to one site near the center of the basin. Figure 7 summarizes the composition and quantities of material caught in the sediment-water interface traps.

Freefight Lake is currently experiencing very high sedimention rates. The average accumulation rate at the sediment-water interface in depths greater than 22 m is approximately 12 kg m^{-2} yr^{-1}, with rates reaching as high as 55 kg m^{-2} yr^{-1}. Nearly all of this material is soluble salt and is composed mainly of sodium and sodium + magnesium sulfate minerals. About a quarter of the sediment trap material is composed of sparingly soluble endogenic precipitates (mainly finely crystalline aragonite and pyrite), clastic and biogenic carbonates derived from nearshore and mudflat areas, and detrital silicates. Less than 10 percent of the sediment trap material is organic matter. Striking differences in the amount of material in traps positioned above versus below the chemocline indicate that most of the sediment is originating from within the monimolimnion.

The two years of trap data from Deadmoose Lake suggest that this basin is undergoing more moderate sedimentation rates in the deep basinal areas. The average accumulation rate is about half that of Freefight Lake. In addition, the trap sediment in Deadmoose Lake has a higher proportion of sparingly soluble carbonates, gypsum, and detrital silicates and a correspondingly lower proportion of soluble salts relative to Freefight Lake. Finally, the above versus below chemocline traps show that a much larger amount of material is originating from the mixolimnion in Deadmoose than is the case in Freefight.

CONCLUSIONS AND FUTURE WORK

The genesis of deep-water evaporites is one of the most problematic topics in all of sedimentology. There is abundant evidence from ancient rocks that deep-water evaporite mineral formation did occur in a variety of marine and nonmarine settings (Schreiber, 1986; Schmalz, 1969), however there are very few modern environments where such sedimentation has been identifed. The primary focus of this paper is to document the occurrence of deep-water evaporite mineral formation in several modern saline lakes of the

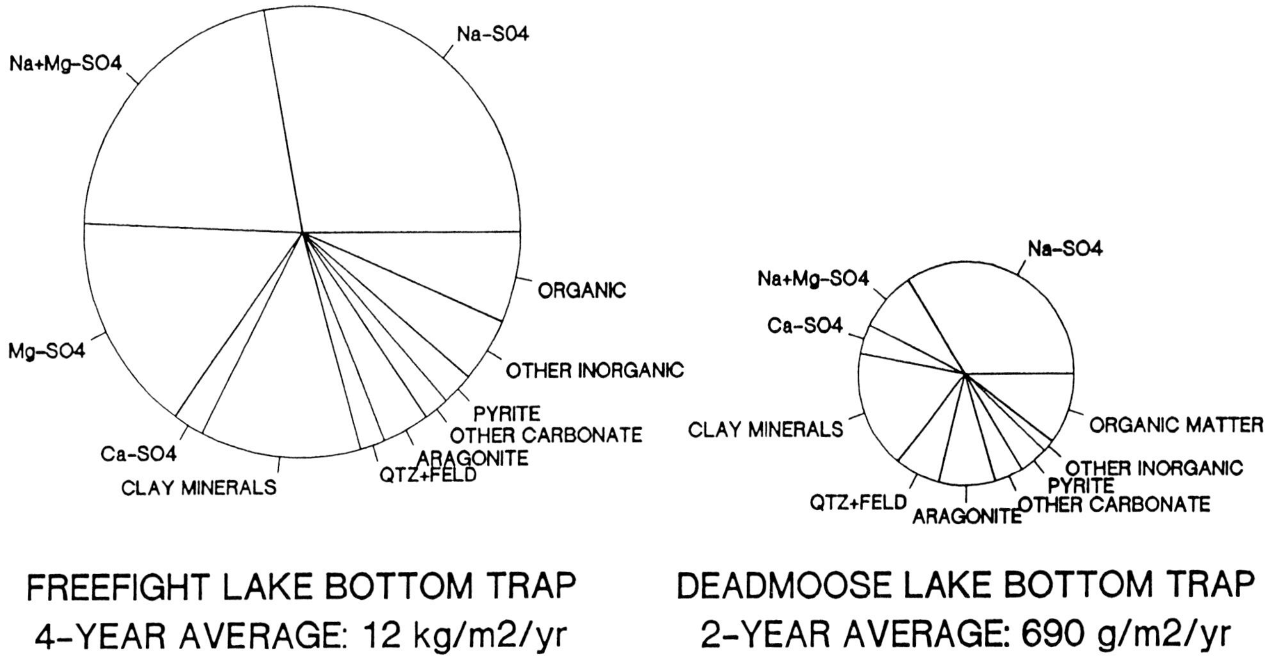

FIG. 7.—Pie charts summarizing the average composition and amount of material retrieved from sediment-water interface sediment traps in Freefight and Deadmoose lakes.

northern Great Plains. The results show that the surficial sediments in the deep basinal areas of three lakes in this region are dominated by soluble salts. The composition of these salts varies as dictated by the major ion content of the brines: mirabilite is dominant in the lake with the highest proportions of Na^{+} and SO_4^{2-} ions; epsomite and bloedite dominate the salts in the lake with a high Mg^{2+} content.

Although winter freeze-out precipitation of soluble salts has been reported from numerous lakes in this region (e.g., Last and Schweyen, 1983; Hammer, 1978; Rawson and Moore, 1944), until now it has been tacitly assumed that these freeze-out salts redissolved during the ice-free season. Preliminary coring and augering of the deep basin sediments in Freefight, Deadmoose, and Little Manitou lakes reveals that relatively thick (greater than 1 m) sections of salt have accumulated. Although it is clearly misleading to generalize about long-term sedimentation and precipitation rates on the basis of the relatively short temporal observational results reported here, the data do show several interesting features. Modern sedimentation rates in Freefight Lake are exceptionally high. If the observed accumulation rates are maintained, this 25 m deep basin would be filled with salt in less than 2500 years. While sedimentation rates of this magnitude have not been reported from saline lakes elsewhere, the post-glacial stratigraphic records of other lacustrine basins in western Canada suggest that such high rates of evaporite mineral formation did occur, at least periodically, during the Holocene. For example, Ingebright Lake in western Saskatchewan contains in excess of 45 m of salt which was deposited in approximately 6000 years.

There is considerable scope for future work on the deep-water evaporites of the saline lakes of the Great Plains region. While significant advances have been made in our ability to recognize shallow-water/salt pan evaporites (e.g., Lowenstein and Hardie, 1985; Hardie and others, 1985; Rosen, 1991), it is essential that the deep-water deposits of modern evaporitic settings be examined from a petrographic and sedimentary structure perspective. It is commonly accepted that closed basin saline lakes provide the best record of changing hydrological and geochemical conditions in a continental region. However, if the salts of these basins are to be effectively used in paleoenvironmental reconstruction, it is particularly important to identify and evaluate criteria that can be used to differentiate shallow-water and salt pan sediments from true deep-water evaporites.

ACKNOWLEDGMENTS

My long-term research on salt lakes of the northern Great Plains has been supported by grants from the Natural Sciences and Engineering Research Council of Canada, the American Chemical Society, Energy, Mines, and Resources Canada, and the University of Manitoba. I wish to thank D. Bell, T. Ewing, L. A. Sack, and L. A. Slezak for their valuable assistance in the field and laboratory. An earlier version of this manuscript was greatly improved by the critical comments of Dr. T. K. Lowenstein and an anonymous reviewer. This is Palliser Triangle Global Change Project Contribution Number 9.

REFERENCES

BODEN, S. E., 1985, Chemical and sedimentological study of Deadmoose Lake, Saskatchewan: Unpublished B.Sc.(Hon.) Thesis, University of Manitoba, Winnipeg, 64 p.

DRONKERT, H., 1985, Evaporite Models and Sedimentology of Messinian and Recent Evaporites: Utrecht, GUA Papers of Geology, Series 1, No. 24, 283 p.

GHEBRE-EGZIABHIER, K., AND ST. ARNAUD, S. J., 1983a, Carbonate mineralogy of lake sediments and surrounding soils, I. Blackstrap Lake: Canadian Journal of Soil Science, v. 63, p. 245–257.

GHEBRE-EGZIABHIER, K., AND ST. ARNAUD, S. J., 1983b, Carbonate mineralogy of lake sediments and surrounding soils, 2. The Qu'Appelle Lakes: Canadian Journal of Soil Science, v. 63, p. 259–269.

HÅKANSON, L., AND JANSSON, M., 1983, Principles of Lake Sedimentology: New York, Springer-Verlag, 316 p.

HAMMER, U. T., 1978, The saline lakes of Saskatchewan. III. Chemical characterization: Internationale Revue der gesamtem Hydrobiologie, v. 63, p. 311–335.

HAMMER, U. T., 1986, Saline Lake Ecosystems of the World: Boston, Dr W. Junk Publishers, 616 p.

HAMMER, U. T., AND HAYNES, R. C., 1978, The saline lakes of Saskatchewan. II. Locale, hydrography and other physical aspects: Internationale Revue der gesamtem Hydrobiologie, v. 63, p. 179–203.

HAMMER, U. T., SHAMESS, J., AND HAYNES, R. C., 1983, The distribution and abundance of algae in saline lakes of Saskatchewan, Canada, *in* Hammer, U. T., ed., Saline Lakes: The Hague, Junk, p. 1–26.

HARDIE, L. A., LOWENSTEIN, T. K., AND SPENCER, R. J., 1985, The problem of distinguishing between primary and secondary features in evaporites, *in* Schreiber, B. C. and Harner, H. L., eds., Sixth International Symposium on Salt: Cleveland, Northern Ohio Geological Society, v. 1, p. 11–40.

HAYNES, R. C., AND HAMMER, U. T., 1978, The saline lakes of Saskatchewan. IV. Primary production by phytoplankton in selected saline ecosystems: Internationale Revue der gesamtem Hydrobiologie, v. 63, p. 337–351.

LAST, W. M., 1988, Salt lakes of western Canada: a spatial and temporal perspective, *in* Nicholaichuk, W., and Steppuhn, H., eds., Proceedings of the Symposium on Water Management Affecting the Wet-to-Dry Transition: Planning at the Margins: Regina, Water Studies Institute, p. 99–111.

LAST, W. M., 1989, Continental brines and evaporites of the northern Great Plains of Canada: Sedimentary Geology, v. 64, p. 207–221.

LAST, W. M., 1990, Lacustrine dolomite—an overview of modern, Holocene, and Pleistocene occurrences: Earth Science Reviews, v. 27, p. 221–263.

LAST, W. M., 1992, Salt lake paleolimnology: the facts, the fears, the future, *in* Robarts, R. D., and Bothwell, M. L., eds., Aquatic Ecosystems in Semi-Arid Regions: Implications for Resource Management: Saskatoon, National Hydrology Research Institute Seminar Series No. 7, p. 51–62.

LAST, W. M., AND SCHWEYEN, T. H., 1983, Sedimentology and geochemistry of saline lakes of the northern Great Plains: Hydrobiologia, v. 105, p. 245–263.

LAST, W. M., AND SLEZAK, L., 1988, The salt lakes of western Canada: a paleolimnological overview: Hydrobiologia, v. 153, p. 301–316.

LOWENSTEIN, T. K., AND HARDIE, L. A., 1985, Criteria for the recognition of salt-pan evaporites: Sedimentology, v. 32, p. 627–644.

RAWSON, D. S., AND MOORE, J. E., 1944, The saline lakes of Saskatchewan, Canadian Journal of Research, Part D, v. 22, p. 141–201.

ROSEN, M. R., 1991, Sedimentologic and geochemical constraints on the evolution of Bristol Dry Lake Basin, California, U.S.A.: Palaeogeography, Palaeoclimatology, Palaeoecology, v. 84, p. 229–258.

RUTHERFORD, A. A., 1970, Water quality survey of Saskatchewan. Surface waters: Saskatoon, Saskatchewan Research Council Report C70–1, 133 p.

SCHMALZ, R. F., 1969, Deep-water evaporite deposition: a genetic model: Bulletin American Association of Petroleum Geologists, v. 53, p. 798–823.

SCHREIBER, B. C., 1986, Arid shorelines and evaporites, *in* Reading, H. G., ed., Sedimentary Environments and Facies (2nd ed.): New York, Elsevier, p. 189–228.

SCHULTZ, L. G., 1964, Quantitative interpretation of mineralogical composition from X-ray and chemical data for the Pierre Shale: Washington, D. C., United States Geological Survey, Professional Paper 391-C, 31 p.

SCHWEGER, C. E., AND HICKMAN, M., 1989, Holocene paleohydrology of central Alberta: testing the general-circulation-model climate simulations: Canadian Journal of Earth Sciences, v. 26, p. 1826–1833.

SLEZAK, L., 1989, Modern sedimentary environments and early sediment diagenesis of Freefight Lake in southwestern Saskatchewan: Unpublished M.S. Thesis, University of Manitoba, Winnipeg, 177 p.

SONNENFELD, P., 1984, Brines and Evaporites: New York, Academic Press, Inc., 613 p.

TILL, R., 1978, Arid shorelines and evaporites, *in* Reading, H. G., ed., Sedimentary Environments and Facies: New York, Elsevier, p. 178–206.

WARREN, J. K., 1989, Evaporite Sedimentology: Englewood Cliffs, Prentice Hall, 285 p.

PRECIPITATION OF SALTS FROM Mg^{2+}-(Na^{+})-SO_4^{2-}-Cl^{-} PLAYA—LAKE BRINES: THE ENDORHEIC SALINE PONDS OF LA MANCHA, CENTRAL SPAIN

SALVADOR ORDÓÑEZ

Departamento de Petrología y Geoquímica, Facultad de Ciencias Geologícas, Universidad Complutense, 28040 Madrid, Spain (present address: Departamento de Ciencias de la Tierra y del Medio Ambiente, Facultad de Cienias, Universidad de Alicante, 03690 Alicante, Spain)

AND

SERGIO SÁNCHEZ MORAL, MARIA DE LOS ANGELES GARCÍA DEL CURA, AND EDUARDO RODRIGUEZ BADIOLA

Consejo Superior Investigaciones Científicas, Spain

ABSTRACT: More than 25 ephemeral saline lakes are found in La Mancha, an endorheic region with a semi-arid climate in central Spain. The annual rainfall is 400 mm/yr, with an evaporation rate of 1,400 mm/yr. Many of the lakes occupy the small closed or nearly closed hydrological basins, approximately 10 km^2 in extent. The area annually flooded by brines is approximately 0.5 km^2. The water depth in the lakes during the rainy season (winter) is <0.5 m. In some lakes, precipitation of salts during summer develops a "bull's eye" pattern of concentrically zoned salts, with bischofite crusts in the innermost zone of the lake, then a zone of halite-epsomite crusts, followed by hexahydrite-halite crusts, and finally, hexahydrite-gypsum crusts in the outermost zone.

The thin crust of the outer zone is always overlain by organic muds; some crusts commonly display tepee structures. Salts precipitated in summer are dissolved during winter, and mass-balance solute models show some playas to be nearly closed systems. An experimental pond was built in one of the lakes (Quero). This pond has allowed us to monitor temporal variations in the concentration and temperature (both upper and lower layers) of the brine, and the precipitation of saline minerals. The sequence of minerals precipitated in the experimental pond agrees with the observed natural sequence; most salts grow at the air-brine interface. A direct correlation has been recognized between the vertical temperature patterns, and various parameters of the brine concentration during periods of brine "stagnation."

INTRODUCTION

La Mancha is a semi-arid region in the Castilla-La Mancha district of central Spain. Within the region, there are many endorheic lakes that alternately experience phases of flooding and desiccation according to the prevailing weather. The term "endorheic" is used for closed basins where evaporation losses exceed inflow (cf. Sonnenfeld and Perthuisot, 1989).

The general geomorphologic and climatic features of the region have been described by Hernández-Pacheco (1949) and Pérez-González (1982). The lacustrine brines are dominated mainly by magnesium and sulfate ions, with minor quantities of chloride and sodium (Ríos, 1968). The origin of the solutes was attributed by Octavio de Toledo and López-Camacho (1976) to the leaching of Tertiary and Triassic evaporites. The general distribution and mineralogical composition of the salts in the lakes were described by Bustillo and others (1978). Ordóñez and others (1983) described the annual hydrochemical evolution of the lakes, and later presented a detailed description of the Alcahozo playa (Ordóñez and others, 1987).

The principle aims of this paper are to describe the mineralogical composition, textural features and areal distribution of the evaporite deposits of the La Mancha playa-lakes. We propose a nearly closed model for the evolution of brines, in which recycling of the solutes is nearly complete and is almost an annual event. We show that the physicochemical properties of the brines and the hydrologic patterns within the playa lakes are the key factors in explaining the general aspects of evaporite sedimentation. Although there are many playa lakes in the region, we focus mainly on the Quero and Alcahozo playa-lakes.

We also include a description of an artificial playa lake that we constructed to monitor the relations between concentration, density, temperature, and the thickness of the brine layers during salt precipitation. Some basic conclusions were reached by comparing the experimental lake with the natural playas.

Sediment samples were collected in equilibrium with the brine and were transported to the laboratory in isothermal bags. Mineralogical analyses were made using X-ray diffraction (XRD). Diffractograms were obtained for both wet and dry samples and commonly displayed different mineralogy.

Magnesium chloride-brine is actively being recovered from several playa lakes in the La Mancha region. At present, this brine is used by the sugar manufacturing industry. Several playa lakes were formerly a source of high quality magnesium sulphate that was used principally as a fertilizer.

GEOGRAPHICAL AND GEOLOGICAL SETTING

The playa lakes studied lie within the Guadiana River basin in the region between 39°40′ to 39°20 N, and 3°20′ to 2°40′ W (Fig. 1). This region is a Tertiary-Quaternary basin located between the Toledo Mountains to the west and the Iberian Range to the east. The Toledo Mountains are part of the Hercynian Shield of the West Iberian Peninsula, whereas the Iberian Range belongs to the Alpine Cordillera. The Toledo Mountains are composed of metamorphic and calc-alkaline plutonic rocks. Limestones, dolomites, mudstones, sandstones and gypsum are common in the Mesozoic sediments of the Iberian Range. The sedimentary fill of the Guadiana River basin is mainly continental lacustrine sediments, limestones, mudstones and gypsum, with some Neogene volcanic rocks (Calvo and others, 1990).

The basin is underlain by a major aquifer. The Guadiana River, the principal river in the basin, rises in a spring situated in the central part of the basin (Fig. 1A). In the lower part of its basin, the river flows through a marshy zone called Tablas de Daimiel. The longitudinal profile of the river is gently sloping and the drainage is not entrenched, thus the regional relief is low. At present, the phreatic level

Sedimentology and Geochemistry of Modern and Ancient Saline Lakes, SEPM Special Publication No. 50
 ISBN 1-56576-014-X

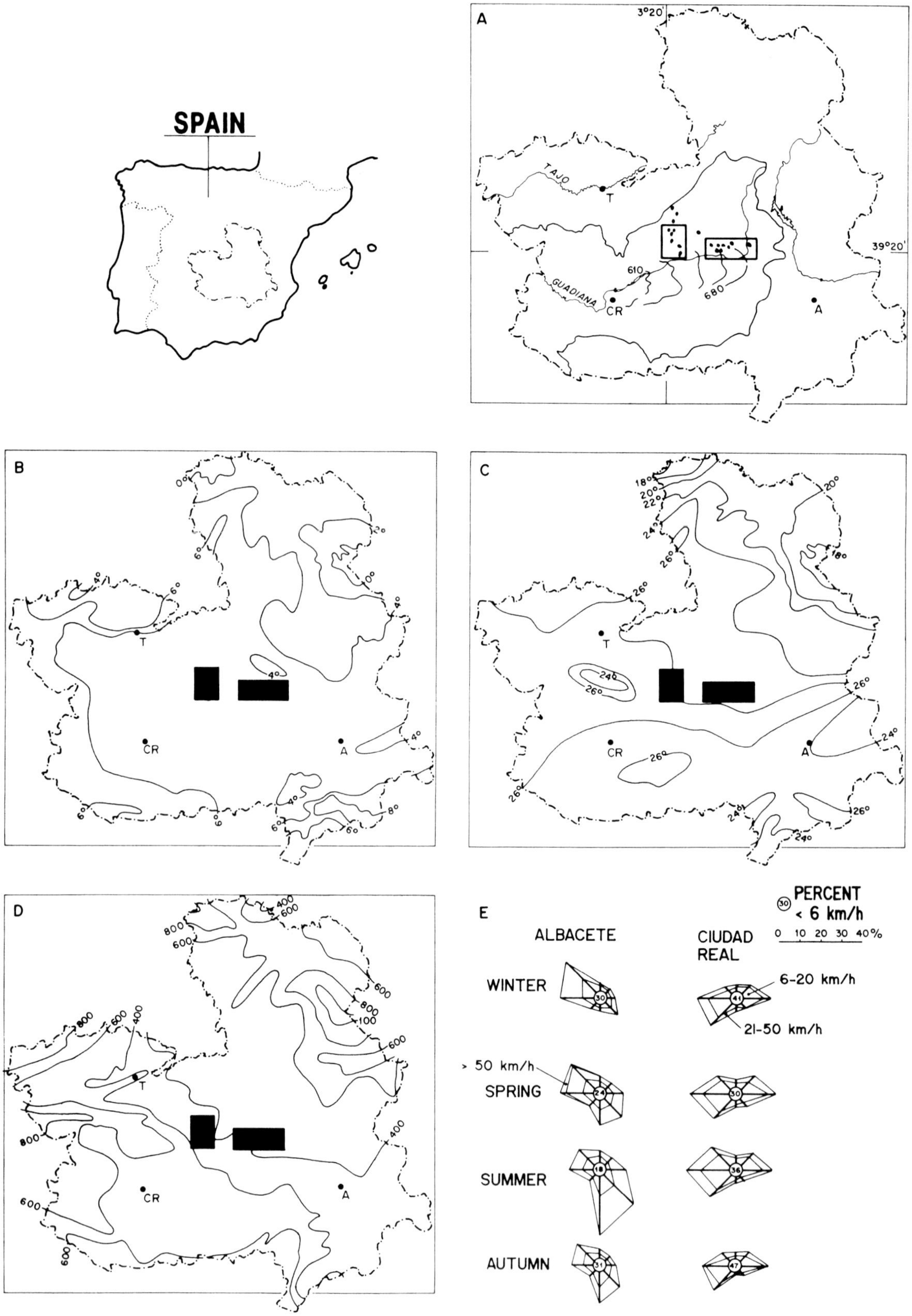

FIG. 1.—Location map of the study region. (A) Drainage basin of the playa lakes, (B) Isohyps map of the mean January temperatures (°C), (C) Isohyps map of the mean July temperatures (°C), (D) Isohyps map of annual rainfall (mm/yr), and (E) Rose diagrams of the wind direction and velocity (km/h). Locations of Albacete and Cuidad Real stations are shown on diagrams. T-Toledo. Boxes show location of the study areas, Cigüela River Valley (left) and Zancara River Valley (right).

has been depressed by the over-exploitation of groundwater and the Tablas de Daimiel marsh has been seriously damaged.

Two main regions with intermittent saline lakes are recognized. The first lies along the Cigüela River and is represented by the Quero saline lake (Fig. 2). The second region, near the Záncara River, is represented by the Alcahozo saline lake (Fig. 3). The Cigüela and Záncara are both tributaries of the Guadiana River.

The regional climatic conditions are summarized in Figure 1, B to E. The mean annual temperature is approximately 11 to 14.5°C, with a mean January temperature of 4 to 6°C, and a mean July temperature of 23 to 26°C. For

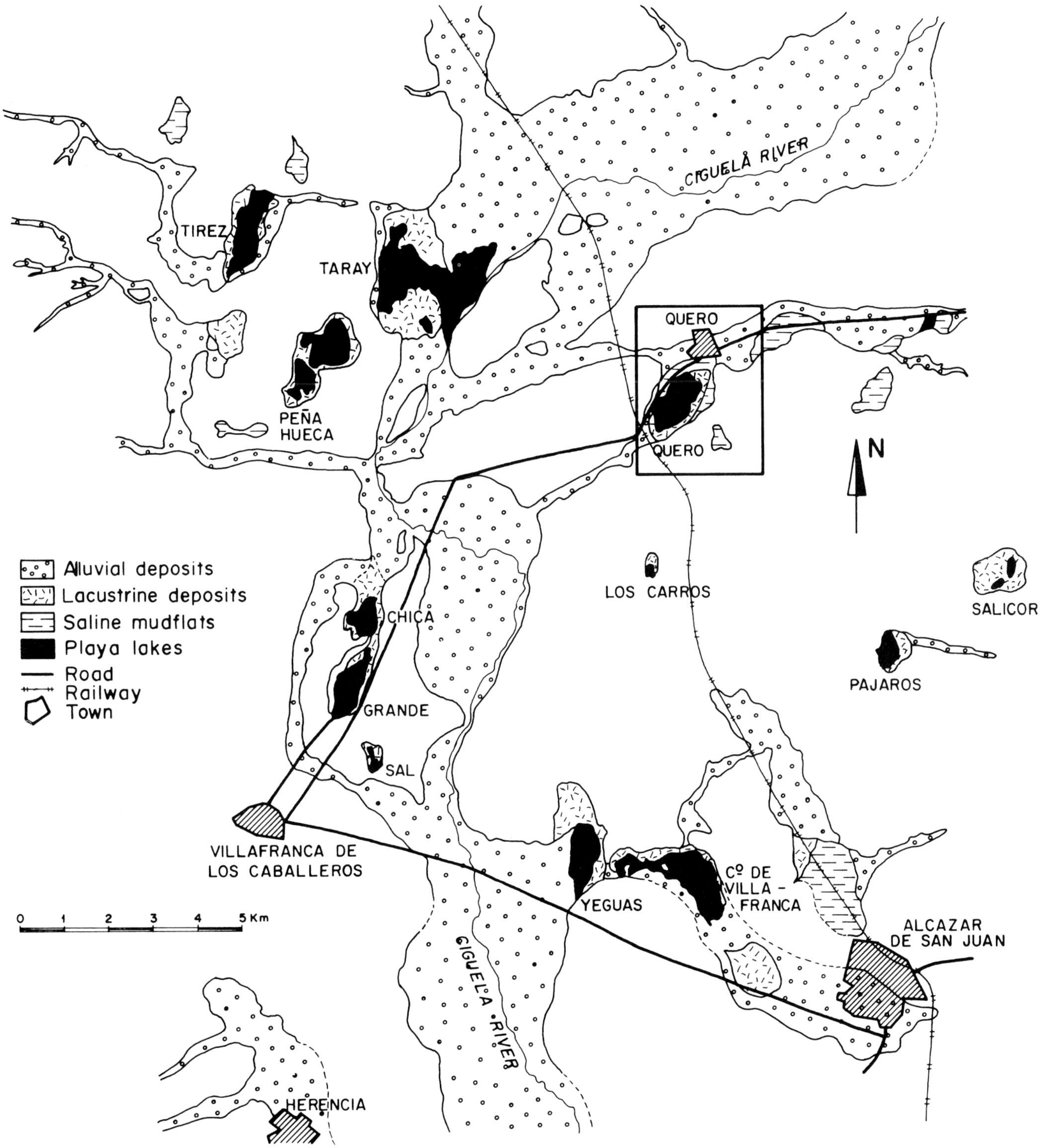

FIG. 2.—The playa lakes of the Cigüela River Valley district.

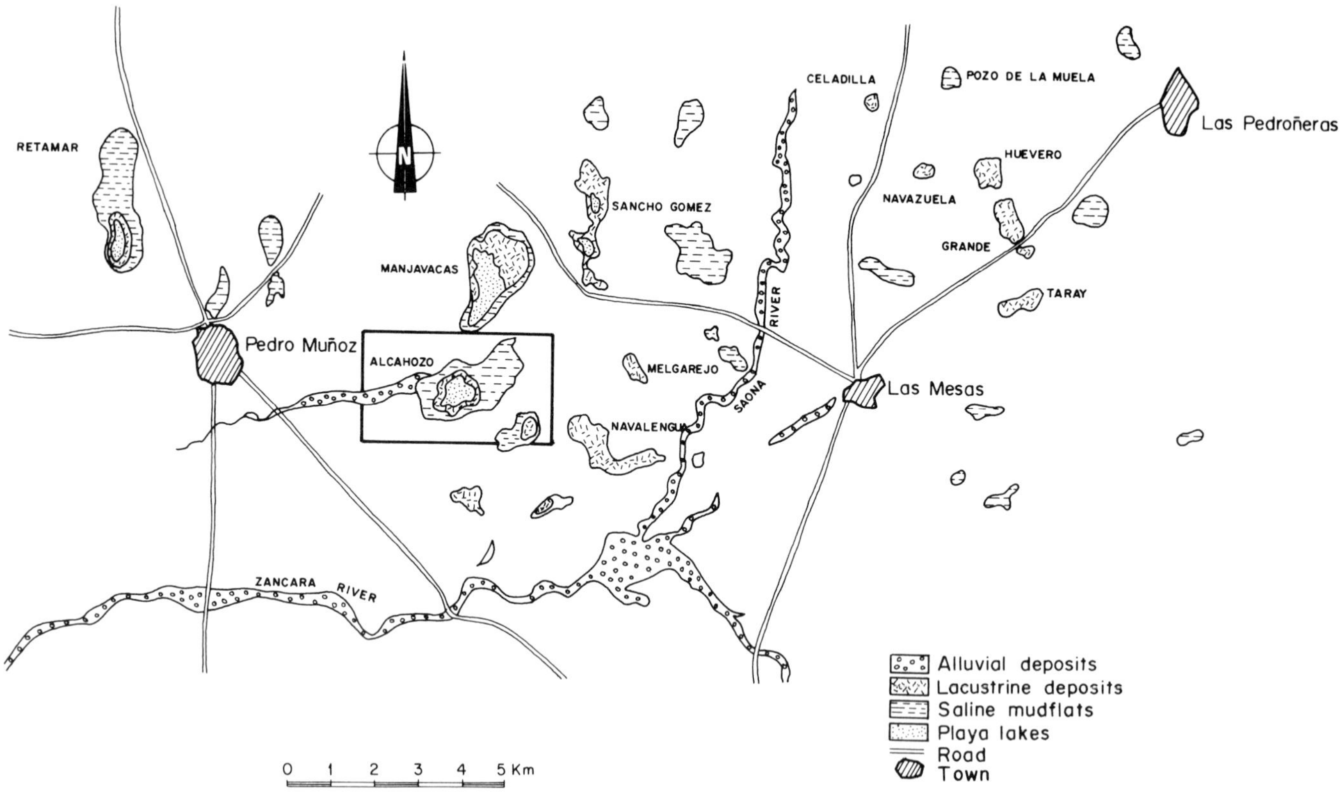

FIG. 3.—The playa lakes of the Záncara River Valley district.

four months of the year the region experiences sub-zero temperatures at night. The mean annual rainfall for the period 1930 to 1987 was approximately 415 mm, but the rainfall displays an irregular distribution, including a dry period of 17 years with 373 mm/yr and a 20 year period with a mean annual rainfall of 458 mm/yr. At present, we are in a dry period, with a mean annual rainfall of 362 mm/yr.

The mean summer evaporation rate in the region with saline lakes is 5.5 mm per day. This agrees with various model calculations, and is supported by Mayer's formula, corrected using Helgeson's water-activity approach (Ordóñez and others, 1991). The mean annual relative humidity is 65 percent and the evapotranspiration rate is approximately 850 mm per year.

The mean surface area of the playa lakes is 0.44 ± 0.35 km^2 ($n = 25$). Their mean elevation is 664 ± 17.12 m above sea level. Many of the playa lakes are no longer in their virgin state. The shape of several of the playa lakes, their inflow, and the chemical composition of brines have been modified by: artificial channels that connect them to rivers, economic recovery of salts and brines, urban water runoff, industrial waste disposal, inflow of fresh water, inadvertent introduction of agricultural herbicides and pesticides, and other local factors.

EVAPORITE SEDIMENTATION

Alcahozo Playa Lake

The Alcahozo playa lake (Fig. 4) occupies the lowest part of a closed drainage basin with an area of 9 km^2. Lacustrine evaporites and muds extend over an area of 0.94 km^2, and after major flooding water covers approximately 0.86 km^2 of the lake basin floor. The lake and its drainage basin may have originated by former karstification of the underlying Mesozoic or Tertiary limestones and eolian processes. Deflation is locally active in saline depressions lacking vegetation, which may explain why the lake is oriented nearly parallel to the dominant wind direction.

The inflow waters derive mainly from direct precipitation (0.423 Hm^3/year) with minor additions from groundwater (0.171 Hm^3). The annual evaporation rate probably exceeds 1,000 mm/yr. Maximum water depth during winter floods may reach 0.5 m. At this time, the evaporites that formed during the previous summer are partially dissolved, and salts may only be visible as saline efflorescences in the peripheral mudflats. These characteristics define Alcahozo as an ephemeral playa-lake. The brine during flooding has a composition of Na-Mg-SO_4-Cl and a salinity of ~180 g/l.

Commonly, during summer, a saline crust with tepee structures and a puffy surface (cf. Hardie and others, 1978) develops in the central part of the lake. The mineralogical composition of this crust during the summers of 1982 and 1983 (Fig. 5) was bloedite ($Na_2SO_4 \cdot MgSO_4 \cdot 4H_2O$), but in 1984, it was bischofite ($MgCl_2 \cdot 6H_2O$) with minor halite (NaCl). The brine temperature in equilibrium with bloedite was 49°C, while the air temperature at that time was 32°C. Surrounding the puffy saline crust was a microbial blister mat, more than 20 m wide. Efflorescent salts in the mat were mainly bischofite, thenardite (Na_2SO_4), gypsum ($CaSO_4 \cdot 2H_2O$) and halite.

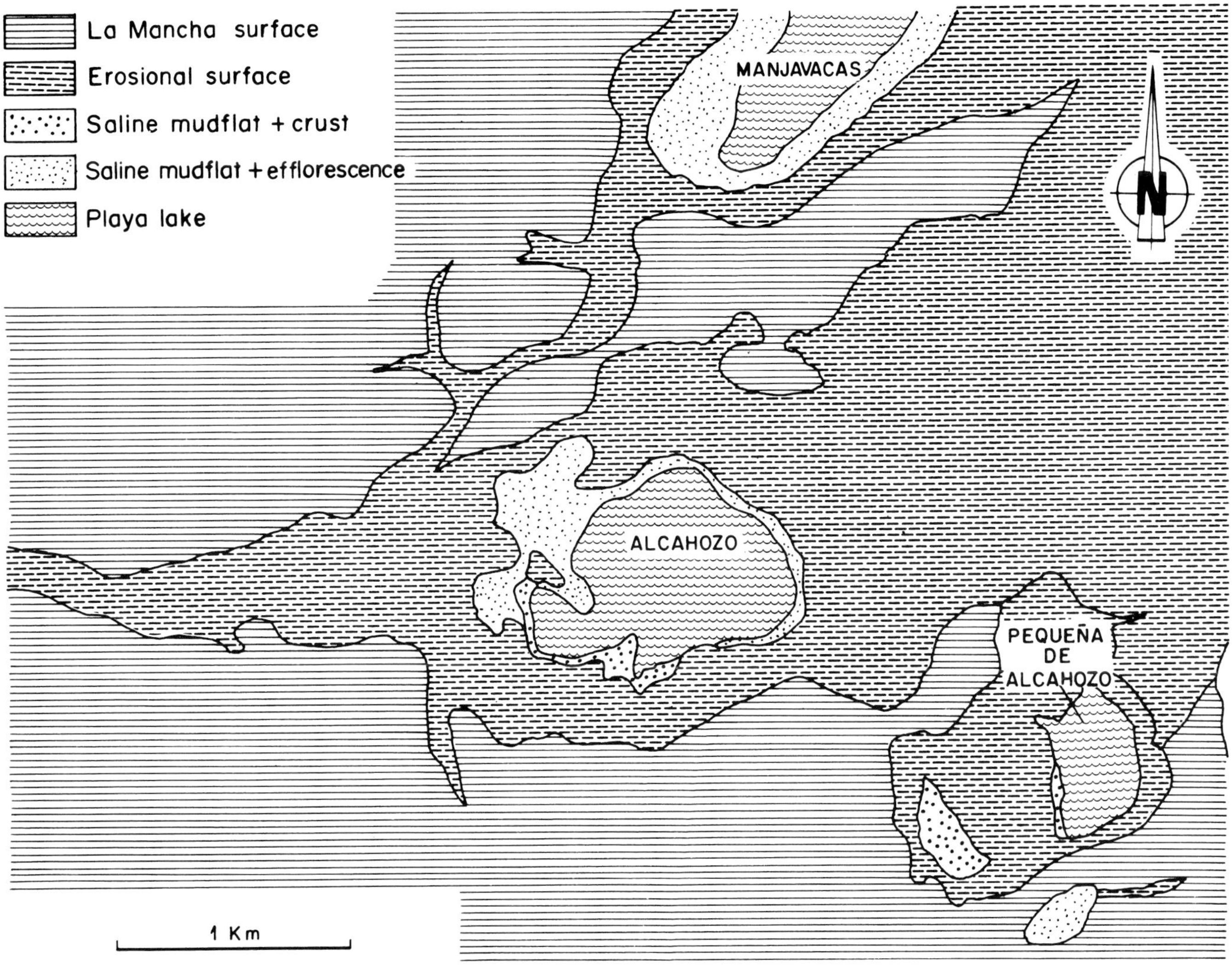

FIG. 4.—Geomorphologic map of the Alcahozo playa lake.

In the marginal areas of the salt crust and on the surface of the peripheral mudflat, we found a thin flat crust of gypsum-hexahydrite ($MgSO_4 \cdot 6H_2O$). The upper part of this crust displayed small wind-blown lineations.

Quero Playa Lake

The Quero playa lake occupies the lowest part of a closed basin with an area of 13 km^2. The drainage basin lies within a former valley of the Cigüela and Riansares Rivers (Fig. 2). Reddish Triassic gypsiferous mudstones form the bedrock and the regional aquiclude, and are probably the main source of solutes in the brine. Muddy sandstones, gypsiferous marls, and limestones of Upper Tertiary age crop out in the gentle slopes surrounding the playa lake (Fig. 6).

A bathymetric map of the playa lake (Fig. 7) displays a flat bottom oriented in a NE-SW direction, with the lowest point (−5 m) in the NE corner of the lake. The southeast and northwest sides of the lake have relatively steep margins. Maximum water depth during flooding may reach 0.2 m. Magnesium chloride brine has been recovered from the playa lake. Table 1 gives chemical analyses of the brines.

During the summer of 1989, the evaporite deposits of the Quero playa-lake displayed a bull's eye pattern. The outer zone of the summer evaporite deposits was formed by gypsum, bloedite, epsomite ($MgSO_4 \cdot 7H_2O$) and halite efflorescences over the sandy-muddy sediments of the saline mudflat. The inner zone was occupied by a puffy saline crust of epsomite + halite + bloedite.

Investigations during summer 1990 allowed us to obtain a more detailed description of the salts, their precipitation sequence, and the saline crust. Observations were made and samples were collected from an artificial pond in the playa basin, with a surface area of over 25,000 m^2, that contained concentrated brine from the previous year's inflow. In this pool, it was possible to monitor the salt precipitation processes, and the density and concentration of brine. Figure 8 summarizes the precipitated salts and other features. All zones occupied by the brine were covered by a centimeter-thick crust from the inner to the outer margins. The crust showed wide mineralogical variations, not only from its base

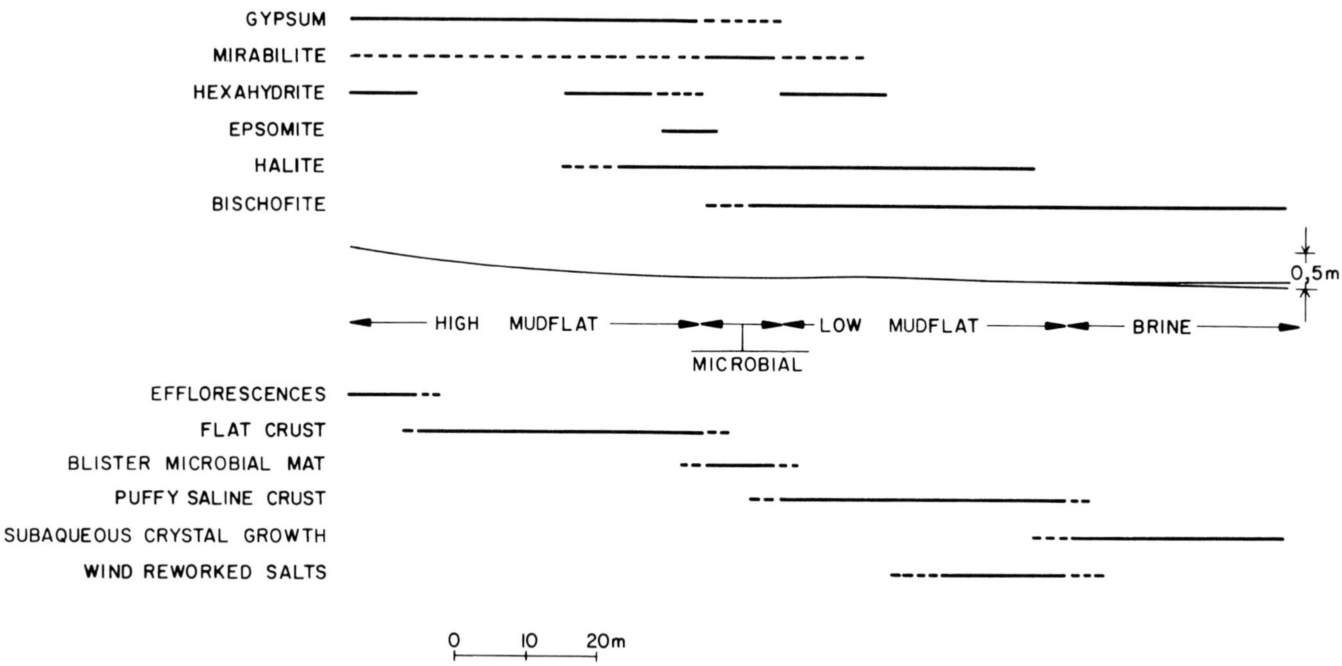

FIG. 5.—Distribution of saline minerals and sedimentary features of the evaporite deposits of the Alcahozo playa lake (summer, 1984).

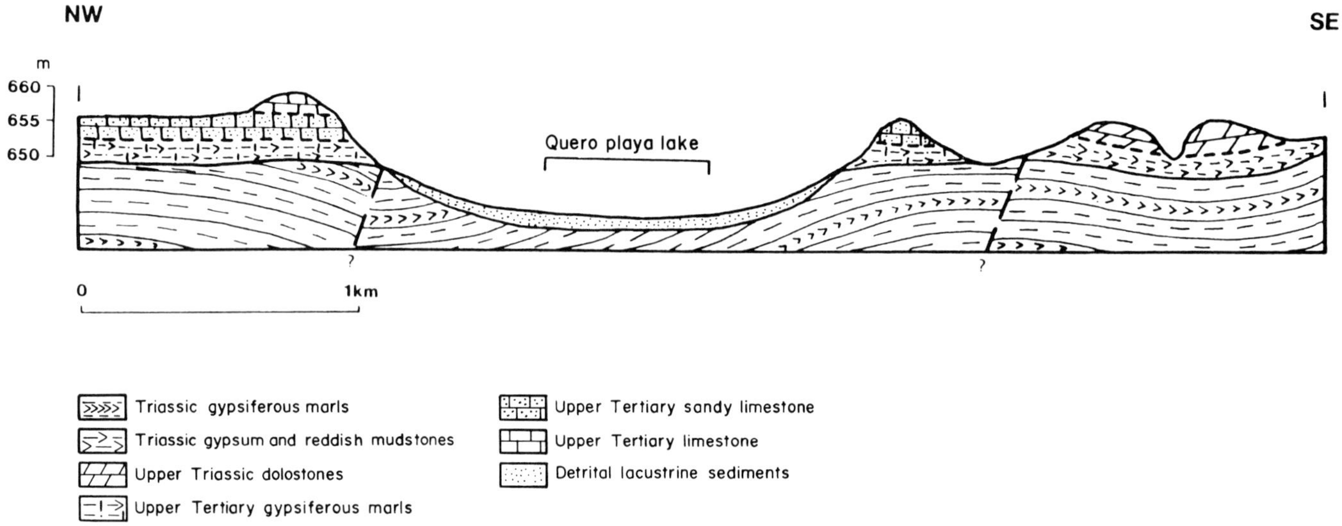

FIG. 6.—Geologic cross section of the Quero playa lake.

to the top, but also from the margins to the center of the lake. The main saline minerals identified in the crust were bloedite, epsomite and halite, with minor quantities of hexahydrite, gypsum, calcite, dolomite, pentahydrite ($MgSO_4 \cdot 5H_2O$) and bischofite (Table 2).

Gypsum is the principal mineral in the marginal deposits, and is also a common secondary mineral in the dry crusts and ridges of the tepees. *Calcite* and *dolomite* have a similar distribution to gypsum.

Bloedite is the most abundant mineral in the middle part of the crust. It occurs with a granular habit in hard layers, 0.5 to 1.5 cm thick. The color of the bloedite varies from vitreous translucent to brown, where rich in inclusions. The bloedite forms diagenetically during periods of brine "stagnation" (discussed below), rather than at the air-brine interface. The bloedite crystals in organic muds commonly display a tabular habit.

Two varieties of epsomite were recorded. *Type I Epsomite* is the most common mineral in the lower part of the crust. It forms short vitreous prisms from 0.5 to 0.75 mm long and probably grows from bottom nucleated crystals. *Type II Epsomite* is the most common mineral in the surface

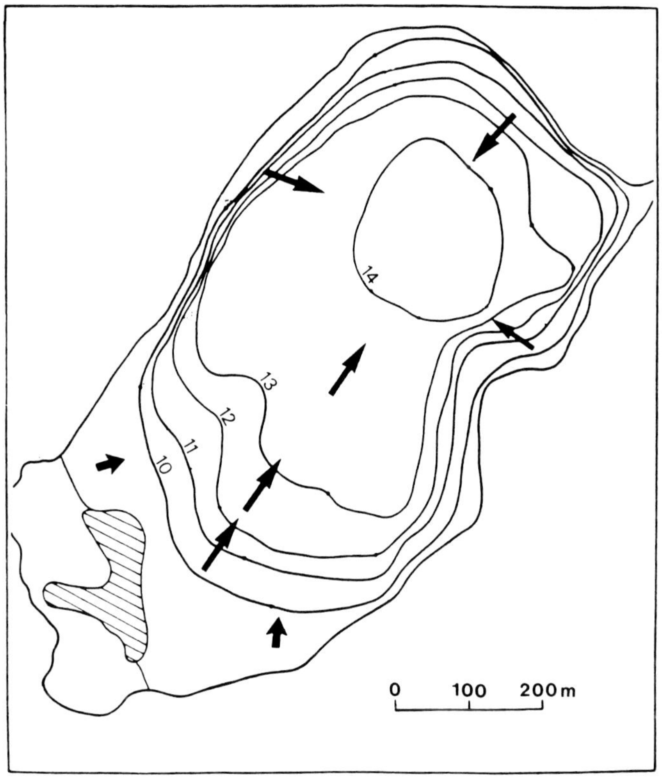

FIG. 7.—Bathymetric map of the Quero playa lake. Contour lines are given in cm. Arrows denote regional slope. The reference level for the bathymetric map was the upper limit of the mudflat deposits.

TABLE 2.—MAIN FEATURES OF EVAPORITE MINERALS IN QUERO PLAYA LAKE

1. AIR-BRINE INTERFACE MINERALS (EPHEMERAL SALTS)
- CALCITE
- GYPSUM
- THENARDITE, secondary after bloedite
- BLOEDITE, tabular habit (?)
- EPSOMITE, acicular
- HALITE, rafts of hopper crystals (3 mm crystals)
- HALITE RAFT + EPSOMITE, columnar prisms
- HALITE RAFT + HEXAHYDRITE, acicular

2. SUBAQUEOUS SALT CRUSTS, BOTTOM OF THE PLAYA LAKE (PERMANENT SALTS)
- CALCITE—lower part of the crust, outer zone
- GYPSUM—lower part of the crust, outer zone
- DOLOMITE—lower part of the crust, outer zone
- EPSOMITE: TYPE I—lower-middle crust, outer part of the inner zone
  Vitreous idiomorphic prisms, size 5–7.5 mm
  Growth from bottom nucleated crystals
- BLOEDITE, middle part of the crust, inner zone
  Hard layers of 5–15 mm
  Granular habit
  Vitreous to brown color
  Probably early diagenetic from epsomite
- EPSOMITE: TYPE II, surface of the crust, inner zone
  Growth from epsomite needles from the air-brine interface
  Prismatic habit (30 mm crystals)
- HALITE, surface of the crust, inner zone
  Chevron growth from hoppers
- BISCHOFITE, middle part of the crust
  Interstitial cement within bloedite
  Early diagenetic mineral

3. CRUST DESICCATION EFFLORESCENCES AND DESICCATED TEPEES
- HEXAHYDRITE
- PENTAHYDRITE
- STARKEYITE
- HALITE, cubic crystals in the ridges of tepees

TABLE 1.—CHEMICAL COMPOSITION OF QUERO PLAYA LAKE BRINES AND GROUNDWATER

	SO_4 mg/l	Cl mg/l	Ca mg/l	Mg mg/l	Na mg/l	K mg/l
Spring	44,952	31,188	82.6	15,456	11,106	433.6
Summer	239,995	62,440	130.2	52,973	38,834	1,285.1
Summer	163,995	94,908	82.5	48,321	46,982	2,046.8
Groundwater	60,058	145,003	140.3	54,375	21,492	1,992.2

of the crust and has a prismatic habit, with crystals up to 30 mm long. These probably develop from acicular crystals nucleated at the air-brine interface that sink to the bottom of the lake, where they continue to grow.

Halite is associated with Type II epsomite and always grows from rafts of hopper crystals. During the winter and spring of 1990, a two-layered structure of parallel vertical prismatic crystals of *hexahydrite* that measured 30 cm long appeared along the rim of the artificial pool, associated with halite. *Bischofite* forms a cement that grew interstitially between bloedite crystals. When saline crusts and tepees emerge from below brine, many efflorescent minerals are found, including *hexahydrite*, *pentahydrite* and *starkeyite* ($MgSO_4 \cdot 4H_2O$).

Experimental Scale Model.—

To investigate the characteristics of the brine during evaporite precipitation, an experimental model was created (Fig. 9). Two drums were buried in the playa sediments and filled with brine from the playa-lake. The initial depth of brine in the drums was similar to that of the lake. The data record for this experiment began on August 10, 1989. The following data were recorded daily: concentration of $MgCl_2$, density, depth of brine, surface brine temperature (ts), bottom brine temperature (tb), and the composition of the salts precipitated at the air-brine interface.

The $MgCl_2$ concentration displayed a continuous increase in concentration during the first 10 days, but remained almost constant for the following 12 days. This "stagnant period" of $MgCl_2$ concentration coincided with precipitation of salts at the air-brine interface. The precipitation points and mineralogical composition of the salts are shown in Figure 10A.

The temperature balance (Δt) between the bottom brine temperature (tb) and the surface brine temperature (ts), in the first 10 days of the experiment, remained almost constant (Fig. 10A). This can be explained by the cooling of the surface brine due to salt crystallization (cf. Sonnenfeld and Perthuisot, 1989). During the "stagnant period," the bottom temperature increased on occasions by as much as 4°C, probably associated with major growth of salts on the bottom. The "stagnant period" is always simultaneous with the genesis of bloedite on the lake floor.

Brine density and its thickness versus time show a similar pattern to that of $MgCl_2$ concentration (Fig. 10B). How-

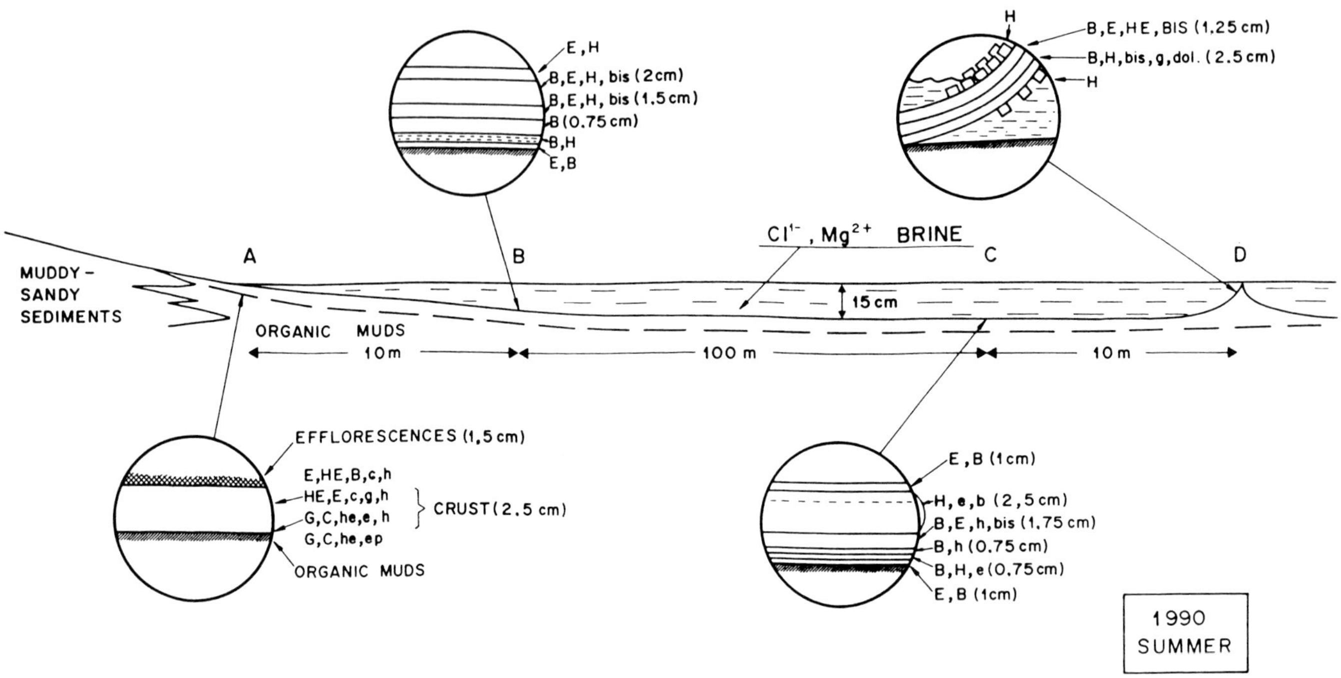

FIG. 8.—Cross section of saline deposits in the artificial pond located at Quero playa-lake. E/e-epsomite; B/b-bloedite; H/h-halite; G/g-gypsum; C/c-calcite; HE/he-hexahydrite; BIS/bis-bischofite (Upper case letters indicate dominant minerals).

ever, the brine density clearly decreases during the "stagnant period," while the thickness of the brine decreased steadily from the starting point of the experiment.

DISCUSSION AND CONCLUDING REMARKS

The annual hydrologic cycle of the playa lakes of La Mancha (Fig. 11) can be summarized as follows. During winter and probably early spring, the previous summer's salts are almost totally dissolved, and the salinity of the brine is approximately 150 mg/l. With sub-zero temperatures during winter, mirabilite ($Na_2SO_4 \cdot 10H_2O$) may precipitate both within the brine and across the mudflat. Both forms of mirabilite are ephemeral.

Direct rainfall is the dominant inflow, and groundwater recharge is commonly less than 25 percent of the total budget. Sporadic flash floods may also flood the playa lakes in the early fall. Some of the saline playa-lakes receive polluted water inflow from urban and industrial sources.

During early summer, evaporation and evaporative pumping increase the salinity of the brines. The mineral precipitation sequence, when plotted on a stability field diagram for the system $Mg^{2+}-Na^{+}-SO_42-Cl^{-}$ (Valyashko, 1972) is: gypsum−calcite−(dolomite)−bloedite−bloedite + halite−bloedite + halite + epsomite−halite + epsomite + hydrated magnesium sulphates + bischofite (eutonic point). Before the brine reaches the eutonic point in some of the playa-lakes, it is recovered for industrial uses as previously noted. Unpublished hydrochemical data for brines in the Quero playa lake show that the Cl^{-}/SO_4^{2-} ratio has decreased in accord with the amount of Mg-Cl brine recovered during the last 10 years (1980–90). This supports the hypothesis of a nearly closed hydrological model for the La Mancha playa-lakes. Deflation of salts may slightly alter the salt balance in the playa-lakes.

The evidence for bottom salt precipitation in the Quero playa-lake shows that bloedite is probably an early diagenetic mineral. The observed mineralogical sequence and textural features of the epsomite and bloedite, suggest that epsomite may alter to bloedite by the reaction:

$$\underset{\text{epsomite}}{2MgSO_4 \cdot 7H_2O} + 2Na^{+} = \underset{\text{bloedite}}{MgNa_2(SO_4)^2 \cdot 4H_2O} + 10H_2O + Mg^{2+}$$

According to this equation, for each 1 cm^3 of epsomite, we obtain 0.51 cm^3 of bloedite and 0.61 cm^3 of water released to the brine. The high salinity of the brine and low quantity of magnesium in solution favor the genesis of bloedite. The water released to the brine may be responsible for the "stagnant period" of concentration and density. The Mg^{2+} released may favor the genesis of bischofite as an interstitial cement mineral within bloedite to produce a hard crust layer several centimeters thick. The transformation of epsomite to bloedite may be the most logical explanation for the development of polygonal cracks observed in the bottom lake sediments, associated with a decrease in molar volume of the solid phases.

When the activity of magnesium in the brine increases, the direction of the reaction may reverse and epsomite may become the stable evaporite mineral. The thickness of the brine layer falls drastically at this time. In this sense, the highly concentrated brine layer, can be related to: (1) the water/crystallization balance, (2) the relative humidity of

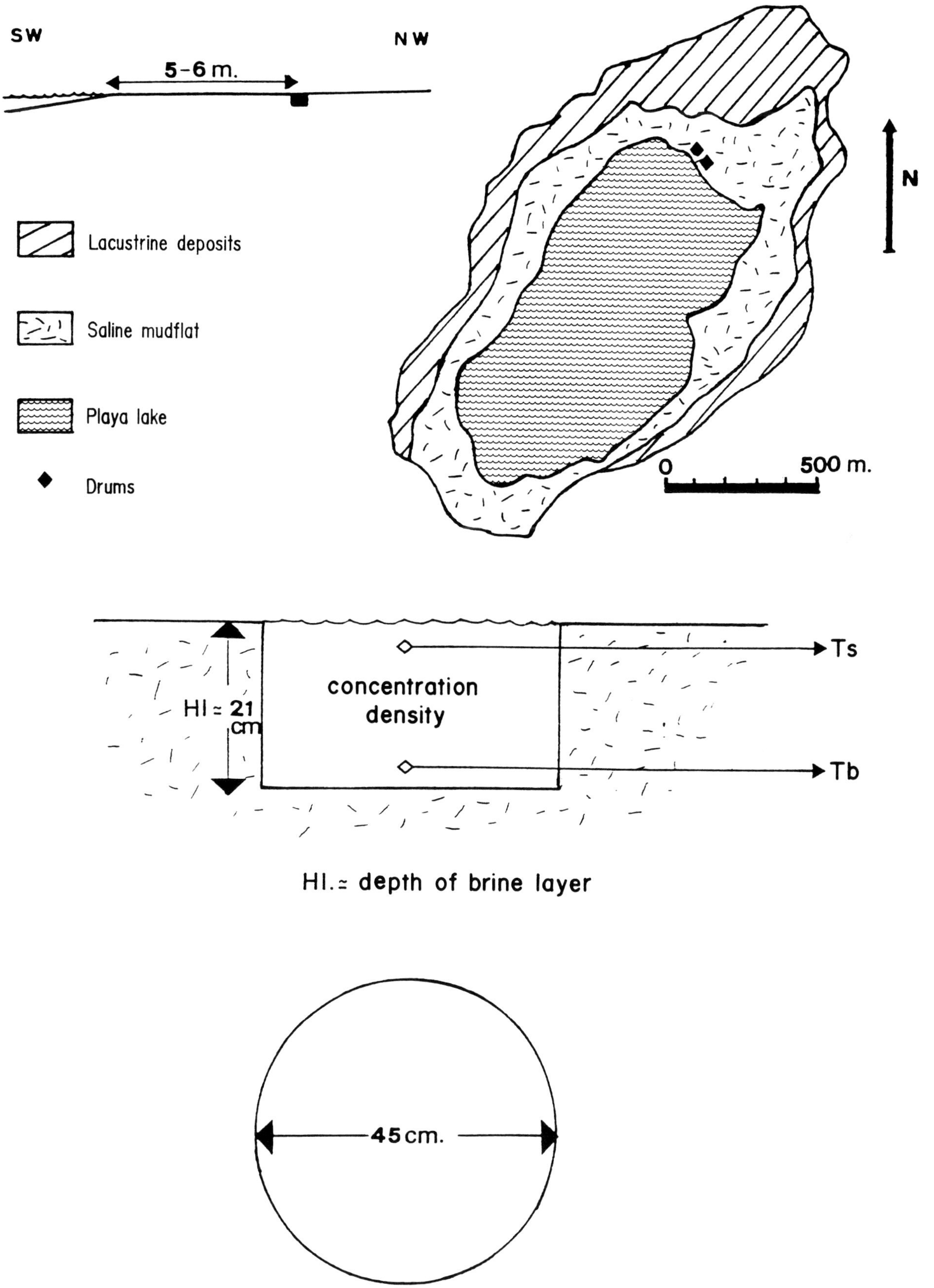

FIG. 9.—Location of drums at the Quero playa lake placed to investigate the "stagnant period" of brine concentration and density. Ts-surface temperature; Tb-bottom temperature.

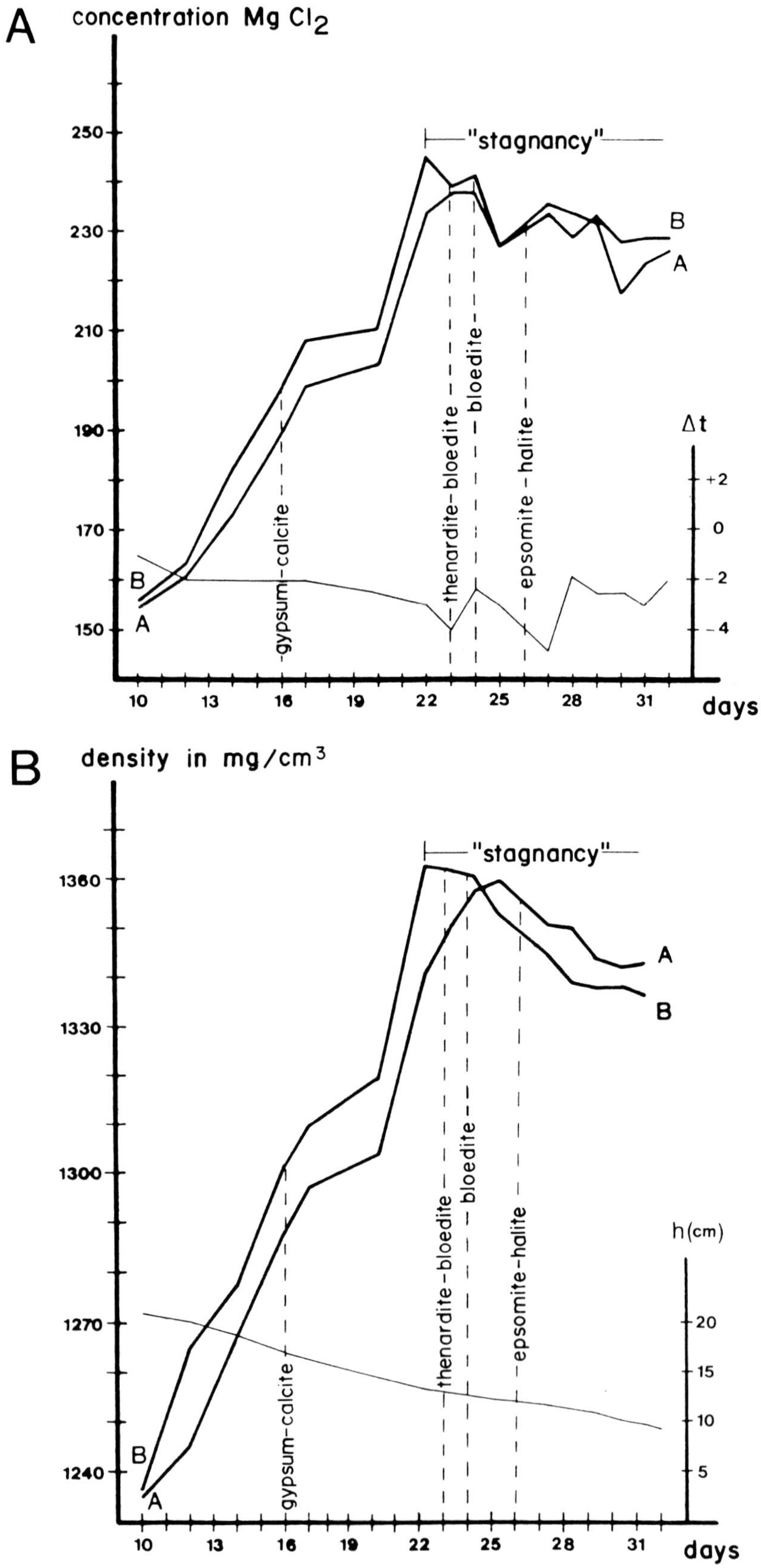

FIG. 10.—Plots of variations in brine concentration and density for the artificial drum ponds (A and B) at Quero playa lake during the period 10–31 August, 1989. (A) Concentration of $MgCl_2$ (mg/l). Δt-Difference (°C) between surface (Ts) and bottom (Tb) water temperature. (B) Brine density (mg/cm^3). h-depth of brine layer (cm).

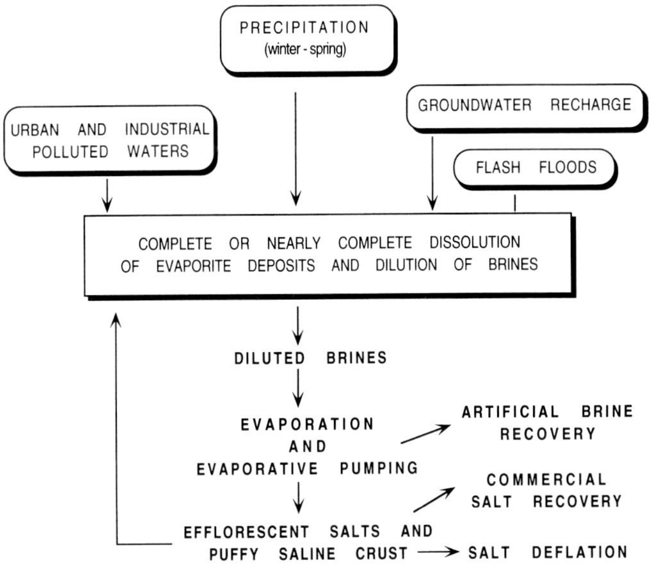

FIG. 11.—Flow diagram to illustrate annual brine evolution and evaporite sedimentation in the La Mancha playa-lakes.

the air over the playa lake, and (3) the evaporation of the brine layer. Under nearly complete desiccation, epsomite and halite crystallization may be responsible for the genesis of tepee structures. In the Quero playa-lake, the rate of dissolution of bloedite is lower than that of the other common saline minerals, and this is the only mineral that may survive the winter-spring dissolution processes.

In conclusion, the chemical composition of brines in the La Mancha playa-lakes may change annually associated with the dissolution rate of the previous summer's evaporites, the volume of artificial salt recovery, the chemical composition and volume of artificial brine recovery, the volume of salts deflated, and the inflow of polluting agents (waters, herbicides, pesticides, fertilizers).

ACKNOWLEDGMENTS

We thank two anonymous referees for suggestions and corrections that helped to improve the manuscript. CLORUR S.A. (J. L. Sabaté) is thanked for providing technical assistance during field work.

REFERENCES

BUSTILLO, M. A., GARCÍA DEL CURA, M. A., MARFIL, R., ORDÓÑEZ, S., AND PEÑA, J. A., 1978, Estudio sedimentológico de algunas lagunas de la Región Manchega, sector Lillo—Villacañas—Quero (provincia de Toledo): Estudios Geológicos, v. 34, p. 187–191.

CALVO, J. P., HOYOS, M., MORALES, J., AND ORDÓÑEZ, S., 1990, Neogene stratigraphy, sedimentology and raw materials of the Madrid basin: Paleontologia i Evolució, v. 2, p. 63–95.

HARDIE, L. A., SMOOT, J. P., AND EUGSTER, H. P., 1978, Saline lakes and their deposits: a sedimentological approach, *in* Matter, A., and Tucker, M. E., eds., Modern and Ancient Lake Sediments: Oxford, Special Publication of the International Association of Sedimentologists 2, Blackwell, p. 7–42.

HERNANDEZ PACHECO, F., 1949, La Mancha: Madrid, Real Academia de Ciencias Exactas, Físicas y Naturales, p. 29–61.

OCTAVIO DE TOLEDO, F., AND LOPEZ CAMACHO, B., 1976, Relaciones entre el flujo subterráneo y la calidad química de las aguas en formaciones continentales: caso del SE de Toledo: Valencia, Simposio Nacional de Hidrogeología, v. 2, p. 1078–1101.

ORDÓÑEZ, S., GARCÍA DEL CURA, M. A., MINGARRO, F., AND LOPEZ DE AZCONA, M. C., 1987, Mineralogénesis de sales sulfatadas-cloruradas magnésicas en la Laguna de Alcahozo (La Mancha-Ciudad Real): Boletín Sociedad Española de Mineralogía, v. 10, p. 219–229.

ORDÓÑEZ, S., MINGARRO, F., GARCÍA DEL CURA, M. A., AND LOPEZ DE AZCONA, M. C., 1983, Sedimentoquímica e hidroquímica de lagunas castellano-manchegas: Menorca, Cuenca del Cigüela y del Záncara, 10 Congreso Nacional de Sedimentología, p. 1.53–1.56.

ORDÓÑEZ, S., SANCHEZ MORAL, S., AND GARCÍA DEL CURA, M. A., 1991, Modelización de la hidroquímica y sedimentoquímica de una laguna tipo playa (Cl^--SO_4^{2-}-Mg^{2+}-Na^+): La Laguna Grande de Quero (Toledo): Estudios Geológicos, v. 47, p. 207–219.

PEREZ GONZALEZ, A., 1982, Neógeno y Cuaternario de la Llanura Manchega y sus relaciones con la Cuenca del Tajo: Ph.D. Thesis, Universidad Complutense, Madrid, 787 p.

RIOS, J. A., 1968, Saline deposits of Spain: Madrid, Geological Society of America Special Paper 88, p. 59–74.

SONNENFELD, P., AND PERTHUISOT, J. P., 1989, Brines and Evaporites: Washington D. C., Short Course Notes *in* Geology, v. 3, American Geophysical Union, 126 p.

VALYASHKO, M. G., 1972, Playa lakes: a necessary stage in the development of a salt-bearing basin, *in* Geology of Saline Deposits: Paris, Proceedings of the Hannover Symposium, May 1968, UNESCO, p. 41–51.

SEDIMENTARY FEATURES PRODUCED BY EFFLORESCENT SALT CRUSTS, SALINE VALLEY AND DEATH VALLEY, CALIFORNIA

JOSEPH P. SMOOT
United States Geological Survey, Box 25046, Denver Federal Center, Denver, Colorado 80225
AND
BARBARA CASTENS-SEIDELL
116 Runnymeade Court, Greenwood, South Carolina 29649

ABSTRACT: Efflorescent salt crusts composed primarily of halite dominate the saline mudflats of Saline Valley and Death Valley, California. These crusts form by the complete evaporation of saline groundwater at the sediment-air interface. Efflorescent crusts also form where halite dust is introduced by wind, then dissolved by rain and reprecipitated as the rainwater is evaporated. Wind-blown silt and clay adhere to thin hydroscopic water films coating crystals in the efflorescent crusts, and coarser sediment is trapped in surface depressions. The sediment is left as a lag deposit when the halite dissolves at the base of the crust in the undersaturated waters below the surface. Efflorescent crust deposits slowly aggrade producing irregular sand and silt lenses in poorly sorted porous mud. The sand and silt lenses have distinctive cuspate contacts, ragged edges, and irregular layering and grain-size distributions. Efflorescent crusts growing on sandy sediment distort the upper surface into polygonal bowl shapes, or deform ripples into hump-shaped lenses. Deposition during flooding over an efflorescent crust commonly produces local areas of solution collapse, which are filled with the sediment as it accumulates. The latter resemble load casts but do not have associated flame structures. Efflorescent crust fabrics similar to those in Saline Valley and Death Valley are documented in the lacustrine Blomidon Formation (Fundy basin, Nova Scotia, Canada) and Bigoudine Formation (Argana basin, Morocco). Efflorescent crusts composed of less soluble minerals, such as gypsum or borates, may leave humpy layers of broken crystals and plates. Powdery efflorescence of minerals, such as thenardite or thermonatrite, generally only disturbs the internal layering of sandy deposits. Puffy ground in dry mudflats, formed by powdery efflorescence growth in mudcracks, produces a distinctive granular fabric.

INTRODUCTION

Efflorescent salt crusts are an ubiquitous component of modern lacustrine saline mudflats and supratidal coastal sabkhas (e.g., Smoot and Lowenstein, 1991; Handford, 1991) and are an important feature of saline soils (Watson, 1983a, b). These crusts form by the complete evaporation of surface pore water producing rock-hard pavements to powdery microcrystalline aggregates. The mineralogy of the crusts reflects the total assemblage of solutes in pore water, and they commonly contain a variety of highly soluble minerals (Eugster and Hardie, 1978). The most common efflorescent crusts are composed mostly of halite (Hunt and others, 1966; Stoertz and Ericksen, 1974; Hardie and others, 1978; Smoot and Lowenstein, 1991), but crusts dominated by gypsum are also common (Stoertz and Ericksen, 1974; Watson, 1983a, b, 1985). Efflorescent crusts may consist predominantly of other minerals, such as trona (Hay, 1968; Papke, 1976), thermonatrite (Eugster, 1970, 1980), thenardite (Jones, 1965; Last, 1984), borate minerals (Muessig, 1959, 1966; Bowser and Dickson, 1966; Hunt and others, 1966; Barker and Lefond, 1985; Smith, 1985), and nitrate salts (Ericksen, 1981; Garrett, 1985). Unlike the coarsely crystalline crusts of saline pans (Lowenstein and Hardie, 1985), the highly soluble efflorescent salt crusts are rarely preserved beneath the surface. For this reason, efflorescent crusts have received little attention in studies of modern saline environments, even though they may cover most of the surface.

We have determined that efflorescent crusts produce a variety of distinctive sedimentary features that are well preserved at depth and that occur in the geologic record. In this paper, we will concentrate on those structures produced by halite efflorescent crusts in the continental saline playas of Saline Valley and Death Valley, California. We compare these structures to fabrics observed in the Triassic lacustrine rocks of the Fundy basin, Nova Scotia, and the Argana basin, Morocco. We also include a brief section describing additional efflorescent crust fabrics based on our observations in: Deep Springs Lake, California; Desert Dry Lake, Big Smoky Playa, and Rhodes Marsh, Nevada; the Bonneville Desert, Utah; Salt Flat Graben, Texas; Lake Eyre and Lake Frome, Australia; and the coastal sabkhas of Baja California, Mexico.

FORMATION OF EFFLORESCENT CRUSTS

Efflorescent salt crusts form under two fundamentally different hydrologic processes (Fig. 1). The first is the evaporation of a saline groundwater table near the surface, and the second is the surficial redistribution of eolian dust

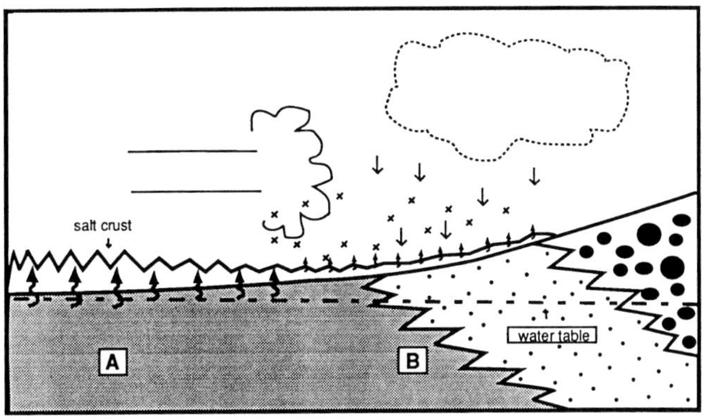

FIG. 1.—Processes that produce efflorescent salt crusts. In region A, evaporation of groundwater near the surface results in precipitation of a crust that thins as the groundwater table occurs at greater depth. In region B, solutes eroded from the flats by wind and deposited on the alluvial fan toe are washed into sediment by rain; subsequent evaporation produces a thin efflorescent crust that is independent of the groundwater table depth.

Sedimentology and Geochemistry of Modern and Ancient Saline Lakes, SEPM Special Publication No. 50

by rainfall. Both processes operate in similar settings, and overlap in certain conditions.

In saline mudflats, areas of brine-saturated sediments are at, or near, the surface in the hydrographic lows. The sources of these brines include groundwater flowing in from the basin margin, surface water that sinks into the sediment as it becomes concentrated by evaporation, and artesian spring water resulting from the intersection of groundwater flow and sinking surface brines (Smoot and Lowenstein, 1991). As the pore-filling brine at the surface is evaporated, additional brine moves to replace it primarily in response to the hydrostatic head of the inflow ground water (Macumber, 1983). Evaporative pumping (Hsu and Siegenthaler, 1969) may cause some vertical brine movement, but it is probably not very important (Macumber, 1983). The evaporation and mixing of these waters result in a lateral distribution of intrasediment evaporite crystals (Smoot and Lowenstein, 1991), which are commonly concentrically zoned in closed basins (Jones, 1965; Hardie, 1968).

The complete evaporation of brine at the sediment surface produces the efflorescent crusts. The mineralogy of efflorescent crusts reflects the chemistry of the underlying brine, but not the concentration of solutes. Thick crusts form where the surface sediments are completely saturated with brine, reaching more than a meter in thickness as brine is "wicked" upward along the crystalline lattice. Thinner crusts form where the surface sediments are less saturated and the phreatic brine table is deeper. The thickness of crusts is also controlled by the frequency of flooding with undersaturated water and by the rate of sediment accumulation. Efflorescent crusts dissolve rapidly in standing water, such as in a saline pan, and in contact with moving flood water or rainfall. Because they are commonly out of equilibrium with ground water, efflorescent crusts dissolve once they lose contact with surface evaporation. This loss of contact results from sediment accumulation on top of the crust, or by continued crust growth at the upper surface. Thick efflorescent crusts commonly show evidence of dissolution at their base (see below).

Efflorescent crusts also form where wind-blown evaporite minerals are dissolved by rainfall or flood waters that sink into the sediment surface and then evaporate reprecipitating the minerals as an efflorescent crust. This process is particularly important in saline soils (Watson, 1983a, b). Solutes may also be introduced by tiny water droplets in fog and mist blown over land surfaces from the ocean or a saline lake. Efflorescent crusts formed by these mechanisms may be as much as several meters thick and drape topography, independent of the depth or chemistry of ground water. They may overlie sediments of a variety of depositional environments and even bedrock. The thickness and mineralogy of these crusts are determined by the availability of eroded salts transported by wind and by the frequency of rainfall and flooding. They are commonly associated with groundwater-controlled efflorescent crusts and saline pans in both non-marine and marine settings.

EFFLORESCENT CRUSTS IN SALINE VALLEY AND DEATH VALLEY

Saline Valley and Death Valley are arid, hydrologically closed, fault-bounded intermontane basins in eastern California (Hunt and others, 1966; Hardie and others, 1978; Smoot and Lowenstein, 1991). Alluvial fans extend from the mountain fault-scarps, fining basinward into broad evaporitic saline mudflats. Saline Valley has a saline pan near the southwestern margin of the basin floor, whereas Death Valley has three saline pans within the long, narrow valley floor. Efflorescent crusts, composed primarily of halite, cover the saline mudflats and encrust sandflats and alluvial fan toes (Figs. 2, 3). The crusts in the latter environments are formed from reworked airborne material, rather than by ground-water concentration. In the saline pan, efflorescent crusts form polygonal ridges along fractures in the bedded halite that are dissolved during each flood stage (Lowenstein and Hardie, 1985). Efflorescent crusts on the saline mudflat directly adjacent to the saline pans are as much as a meter thick with 40 to 50 cm of surface relief comprised of irregular pinnacles and depressions. Toward the edges of the saline mudflats, the efflorescent crusts are progressively thinner, have lower surface relief, and generally have more rounded highs and broader depressions. These crusts may exhibit a polygonal ridge pattern with diameters ranging from 0.1 to 3.0 m, the smaller diameters more commonly associated with sandy sediment. The saline mudflat efflorescent crusts are commonly incised by sinuous channels with ripples and low-relief point bars. Thin efflorescent crusts (from a few centimeters to less than a millimeter thick) that characteristically encrust the channel deposits, are also common in sandflats, eolian interdune areas and sandsheets, and alluvial fan toes. The thin efflorescent crusts have an irregular humpy to warty surface with low relief (less than 10 cm). Older channels in the saline mudflat are covered by thick crusts and are only distinguished as gentle depressions that are easily seen in aerial photographs.

Saline mudflat efflorescent crusts made of halite are characterized by a porous framework of interlocking, submillimeter, euhedral crystals (Fig. 4). Crystal aggregates typically form nodular "popcorn" surfaces. Solution features are common, including fluted pinnacles, tube-like solution pits, and rubbly brecciation (Hunt and others, 1966; Stoertz and Ericksen, 1974). Coarse halite crystals fill voids near the base of the crusts as rain and floodwater dissolve the crusts at the surface, then pool at the crust-sediment boundary, precipitating halite as they evaporate. These dense cemented areas comprise layers 5 to 10 cm thick. Dissolution features are less well developed in thin crusts, and secondary crystal growth in pore space is less common. The crystals composing these crusts are commonly smaller and more sugary than those of the thicker crusts. Halite crystals in efflorescent crusts have thin hydroscopic films of water on their surface. Wind-blown dust adheres to these films causing the crusts to become muddy and brown. Following rain storms, the surfaces of saline mudflat efflorescent crusts are coated with white, sediment-free crust that becomes brown over the course of a year. Efflorescent crusts in frequently flooded areas, such as saline mudflat channels, fan toes, and sandflats, incorporate less mud than those formed in saline mudflats. Crusts formed in areas that are rapidly covered with sediment, such as interdune areas and eolian sandsheets, contain less mud. Surface irregularities of the efflorescent crusts are commonly partially filled with

FIG. 2.—Surfaces covered by efflorescent halite crusts. (A) View from saline pan (foreground) across a saline mudflat to an alluvial fan, Death Valley. Note polygonal efflorescent crust in the saline pan and relative relief of efflorescent crust in saline mudflat. (B) Flooded surface of saline pan in Saline Valley. Note the dark polygonal pattern where the efflorescent crust has dissolved. Remnants of efflorescent crust (white) are visible in the right foreground. (C) High relief efflorescent salt crust (foreground) in an area near the saline pan, Saline Valley. Note the remnant of a channel in right background. (D) Detail of efflorescent crust in Figure 2C, showing narrow ridges and deep pits. Knife is 35 cm long. (E) Low relief efflorescent salt crust (foreground) near the outer edge of a saline mudflat, Saline Valley. Note how ridges (dark) define crude polygonal pattern. (F) Detail of surface in Figure 2E showing "popcorn" surface on ridges (dark) and broad, flat, sediment-filled depressions. Scale is 15 cm long.

FIG. 3.—Surfaces covered by thin efflorescent salt crusts. (A) Polygonal crust on sandy saline mudflat adjacent to an eolian sandsheet, Saline Valley. Crests (light) are narrow and troughs are broad. Shovel is 1.4 m long. (B) Low relief efflorescent crust in the outer part of a saline mudflat that is cut by a salt-encrusted channel (dark), Saline Valley. The dark area bordering the channel is overbank mud on the efflorescent crust. (C) Ripples in a channel in the saline mudflat that are encrusted with a thin efflorescent crust, Saline Valley. Note the "popcorn" surface of the crust. Scale is 15 cm long. (D) Eolian sandsheet with coppice dunes encrusted with thin warty efflorescent salt crust, Saline Valley. Mattock is 1.2 m long.

sediment (Fig. 5), including mud left behind after surface dissolution by rain, mud to sand washed over the crust surface, and eolian silt to sand.

Efflorescent crusts on the saline mudflat are more sediment-rich at the base, and the crystalline structure at the base is more corroded. This relationship is interpreted as indicating dissolution of the base of the crust and concentration of the incorporated sediment. This process results in the net aggradation of poorly sorted mud (Fig. 6). The aggradation of this mud by efflorescent crusts is also indicated by the presence of patchy layers of dense pore-filling halite within the poorly sorted mud, and by the nature and distribution of patches of sediment accumulated in the surface irregularities of the efflorescent crusts. Thinner efflorescent crusts are underlain by lesser amounts of poorly sorted mud, and crusts formed on sandy sediment typically incorporate the upper few centimeters of sediment. The efflorescent crusts that are obviously composed of reworked airborne solutes are typically sediment-free, except for the incorporated underlying sediment and sediment trapped in surface irregularities. In all cases of the halite efflorescent crusts, halite is present only near the surface, even though the sedimentary features indicate that as much as 60 cm of the sediment below the efflorescent crust-sediment contact was accumulated by or modified by efflorescent crusts.

SEDIMENTARY FABRICS PRODUCED BY EFFLORESCENT CRUSTS

The sedimentary features of efflorescent salt crusts in Saline Valley and Death Valley can be divided into three categories (Fig. 7): (1) aggradational features, (2) surface deformation features, and (3) solution collapse features. The variability and distribution of these features are described below.

Aggradational Features

As described above, the halite efflorescent crusts of the saline mudflats trap wind-blown dust and produce a poorly

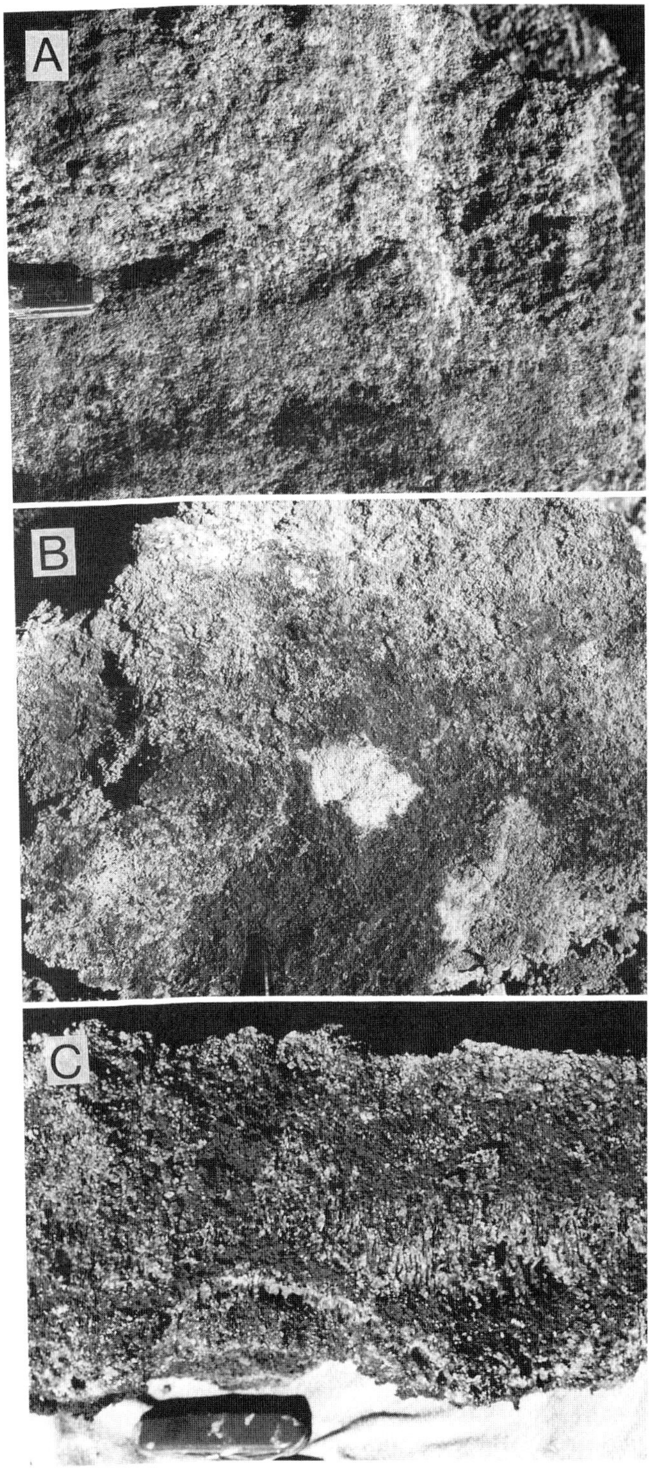

FIG. 4.—Efflorescent salt crusts in cross section. (A) Cross section of the upper part of the efflorescent crust in Figure 2D. Dark area in upper right is a pit lined with muddy "popcorn" surface. Note the increase of mud (dark) from top to base and cell-like division of mud by halite rings (light) near the base. Knife is 9 cm long. (B) Bedding plane-parallel cross section of efflorescent crust in Figure 4A showing lacy crystalline halite, mud (dark), and small patches of fine sand and silt that were trapped in the pits (very light). Field of view is 20 cm wide. (C) Cross section of lower part of the crust in Figure 2D. Note the abundance of mud and layers of halite-cemented crust (light). The cemented zones have ragged edges indicating that they are dissolving. Knife is 9 cm long.

sorted mud as the crusts dissolve at their bases. The mud is silty to sandy and, at least near the surface, has a porous, almost pelleted texture. This texture is produced by the dissolution of the last remnant halite, which commonly forms a honeycomb texture around clumps of mud. A distinctive suite of sedimentary features is formed by wind-blown or water-deposited sand and silt that accumulate within the surface irregularities of the crust and are left as lag deposits in the muddy matrix as the halite dissolves (Figs. 8, 9). These features vary laterally from the inner saline mudflat, near the saline pan, to the outer saline mudflat, where it grades into sandflat or eolian sandsheet deposits. These sand and silt patches are interpreted as lags because their size and shape in cross section directly reflect the length and width of the surface depressions and the amount of sediment entrapped in them, and because they can be seen in various levels in the crusts and partially in the crust and in the underlying mud. These sediment patches have several distinctive characteristics: (1) Wind-blown sediment patches are highly variable in grain size in adjacent patches and even within a single patch; (2) Patches commonly have jagged edges, contain laminae that differ in character and orientation from adjacent patches, and have numerous small internal faults reflecting solution during or after accumulation; (3) The margins of patches are commonly cuspate, with rounded protrusions of muddy matrix in the coarser patch (i.e., the opposite of load casts) reflecting the popcorn surface of the salt; and (4) Thin, dense mud layers occur at the base of patches or within patches that were deposited in the base of salt-crust depressions as rainfall or flood water dissolved the crust surface.

Efflorescent crusts that form during the desiccation stage in the saline pan are completely dissolved during the flood stage (Lowenstein and Hardie, 1985) contributing to solute concentration of the flood-stage lake. The muddy sediment that adhered to these crusts is dispersed into the standing water and contributes to the flood-stage muddy parting. The only record of efflorescent crusts in a saline pan deposit is polygonal clay fillings between upturned edges of the pan halite (Lowenstein and Hardie, 1985).

The efflorescent crust deposits of the saline mudflat immediately next to the saline pan are predominantly poorly sorted mud (Fig. 6). Rainfall, or occasional surface flooding, causes mud to accumulate in the surface depressions of the crust. The lag deposits appear as clots of denser mud or small, irregular lenses of vaguely laminated mud and silt in poorly sorted mud matrix. Small amounts of wind-blown silt accumulate in the steep narrow depressions that produce small (1 to 2 cm on the long axis) irregular lenses as lags in the poorly sorted mud. These textures are further modified by the growth of intrasediment evaporite minerals such as mirabilite, glauberite, and gypsum.

Toward the edges of the saline mudflat, the fine-grained muddy matrix becomes sandier and more poorly sorted. Wind-blown and water-deposited sand and silt compose larger patches that make up a larger proportion of the deposit than in the interior of the mudflat (Fig. 9). Toward the inner part of the saline mudflat, these patches are finer grained and are more irregular in shape and thickness, whereas toward the edge they tend to be more bowl-shaped

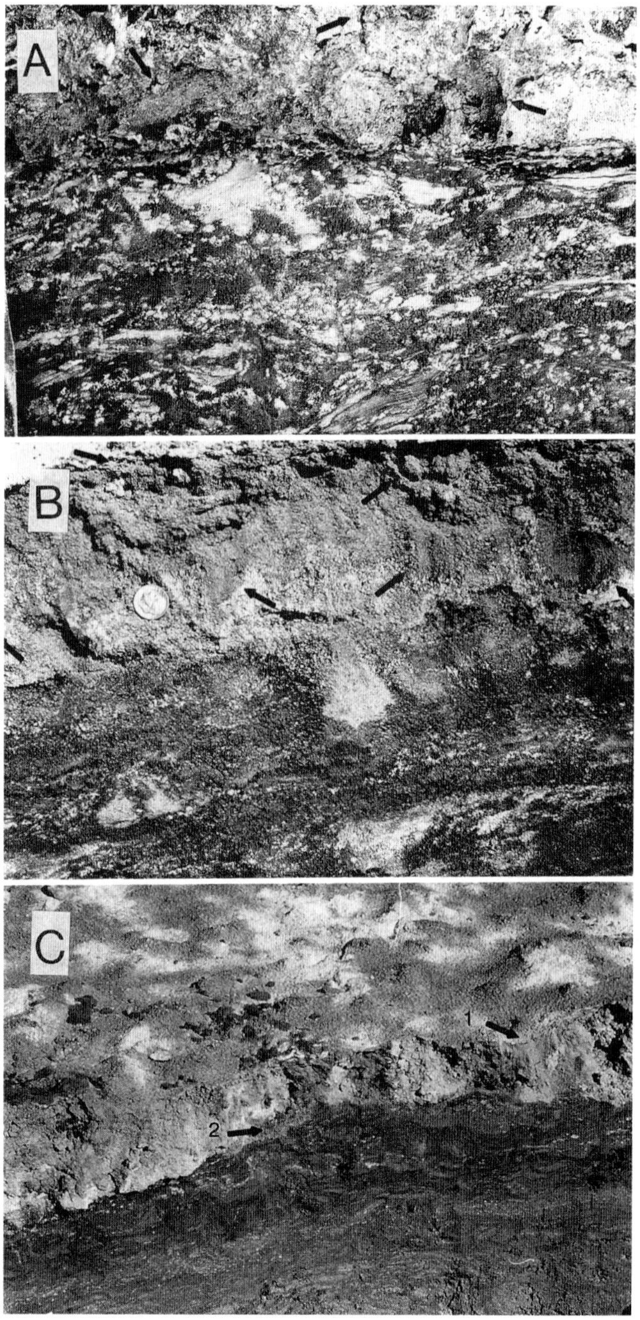

FIG. 9.—Variations of shape and size of sand and silt fillings of depressions in efflorescent salt crusts. (A) Sand and silt patches (light) are highly variable, reflecting the range of depression shapes in Figure 5A. Note the sand patches within efflorescent crust (arrows). Field of view is about 30 cm high. (B) Bimodal pattern of thick sand patches and thin, elongate silt and sand patches resulting from complete filling of large depressions and covering of warty upper surface. Thick sand pods are visible within the crust (arrows). Coin is 1.8 cm in diameter. (C) Bowl-shaped patches of wind-blown sand. Note that sand can be seen at the surface of the crust (1), at the contact with the underlying mud (2), and in the mud. The upturned edges are deformation due to continued crust growth at the edges of depressions. Trench is 45 cm deep.

FIG. 10.—Cross section of eolian sand flat in Figure 3D: Upper sand is massive except for muddy line defining former efflorescent salt crust (lines in sand are scrape marks). Eolian sand fills depressions in the old salt crust and is deformed by the present efflorescent crust. Lower sands have climbing ripple cross-lamination of sheet deltas. Muddy partings contain aggradational efflorescent crust fabrics and deformed ripples. Note bulbous basal contacts of sands, reflecting solution collapse during deposition. Pen is 14 cm long.

variable from tens of centimeters to several meters. The larger scale is particularly evident in sandy mud of the outer saline mudflat where aggrading efflorescent crust fabric is deformed into polygons as much as 3 m in diameter.

Efflorescent crust growth on ripples in channels and on sheet deltas (Smoot and Lowenstein, 1991) produce a distinctive deformation fabric (Fig. 12). Warty efflorescent crusts are thickest on the ripple crests causing them to become bulbous and to separate from the underlying sediment. The preserved fabric is a hump-like convex lens with oversteepened or convoluted internal lamination. Thin, warty efflorescent crusts may also deform planar laminae into irregular convex-upward bumps.

Solution Collapse Features

Efflorescent crusts in channels, sheet deltas, and saline pan margins are commonly inundated by floodwaters. The rapid dissolution of the crust at the surface results in the deformation of sediment within the crust. The small faults and fractures in water-deposited sediment patches within efflorescent crust irregularities are good examples of this process. These features are common in the areas of the saline mudflat adjacent to the channels that are occasionally covered with overbank flows. Larger scale solution-collapse features are more common within the channels in the saline mudflat and on sheet delta surfaces (Fig. 13). In these settings, the efflorescent crusts dissolve while sediment is accumulating on surface. Dissolution is rarely uniform and commonly occurs in discrete pockets that widen through the flooding event. Where efflorescent crusts are thick, the dissolution will produce a compact clay layer with numerous "growth faults" that are filled with the flood sediment. Thinner efflorescent crusts or sandy efflorescent crusts pro-

FIG. 11.—Deformation of sand into polygonal patterns by efflorescent crusts. (A) Sandy mud with patches of eolian sand is deformed into tepees (crest is left of ruler) by surface polygons (see Fig. 3A). Scale (center right) is 35 cm long. (B) Channel sand with surface deformed by efflorescent crust (light upper part) producing bowl-shaped polygons. Zone below upper sand layer is an older surface deformed by efflorescent crusts. Section shown is 50 cm thick.

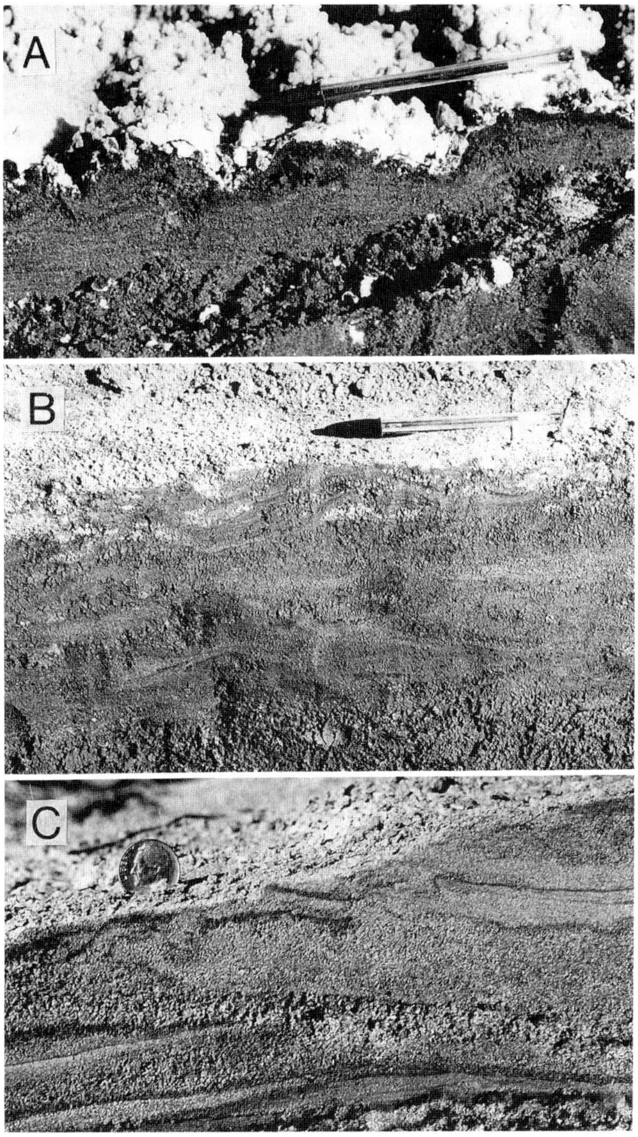

FIG. 12.—Deformation of sandy sediment by thin, warty efflorescent crusts. (A) Cross section of ripples in Figure 3C. Note how the upper surface is pulled up from the rest of the sand at the ripple crests. Pen is 14 cm long. (B) Ripples deformed by efflorescent crusts as seen deeper in a trench. Note the indistinct internal layering and steep hump-shaped upper surface (particularly below pen top). Pen is 14 cm long. (C) Alluvial fan-toe deposit, the upper surface of which is deformed by a thin, warty efflorescent salt crust. Note oversteepened clay laminae (dark) representing former crust and abrupt thickening of sand, which is mostly eolian fill of surface depressions. Coin is 1.8 cm in diameter.

duce fewer faults on dissolution, and the overlying layer may form local pods without the faults.

EFFLORESCENT SALT CRUST STRUCTURES IN THE TRIASSIC FUNDY AND ARGANA BASINS

Triassic lacustrine deposits of the Fundy basin (Canada) and Argana basin (Morocco) contain numerous fabrics similar to those formed by efflorescent crust in Saline Valley and Death Valley. The occurrence of these fabrics in hundreds of meters of section attests to the preservability of the textures and the importance of the processes in the geologic record.

The early Mesozoic Fundy basin is a half graben mostly filled with Upper Triassic to Lower Jurassic continental sediments. The exposed stratigraphic section is less than 1000 m thick, but equivalent strata in the subsurface are several thousand meters thick. The Fundy basin sedimentary fill consists of the basal fluvial Wolfville Formation, the evaporitic lacustrine Blomidon Formation, the North Mountain Basalt, then the fluvial-lacustrine Scots Bay and McCoy Brook Formations (see Chapter 9 in Olsen and others, 1989). The stratigraphy of the Argana basin is remarkably similar to that of the Fundy basin (Smoot, unpublished data). Brown (1980) estimated the thickness of strata as 2,500 to 5,000 m, although this thickness is probably exaggerated by fault repetition of the section. The

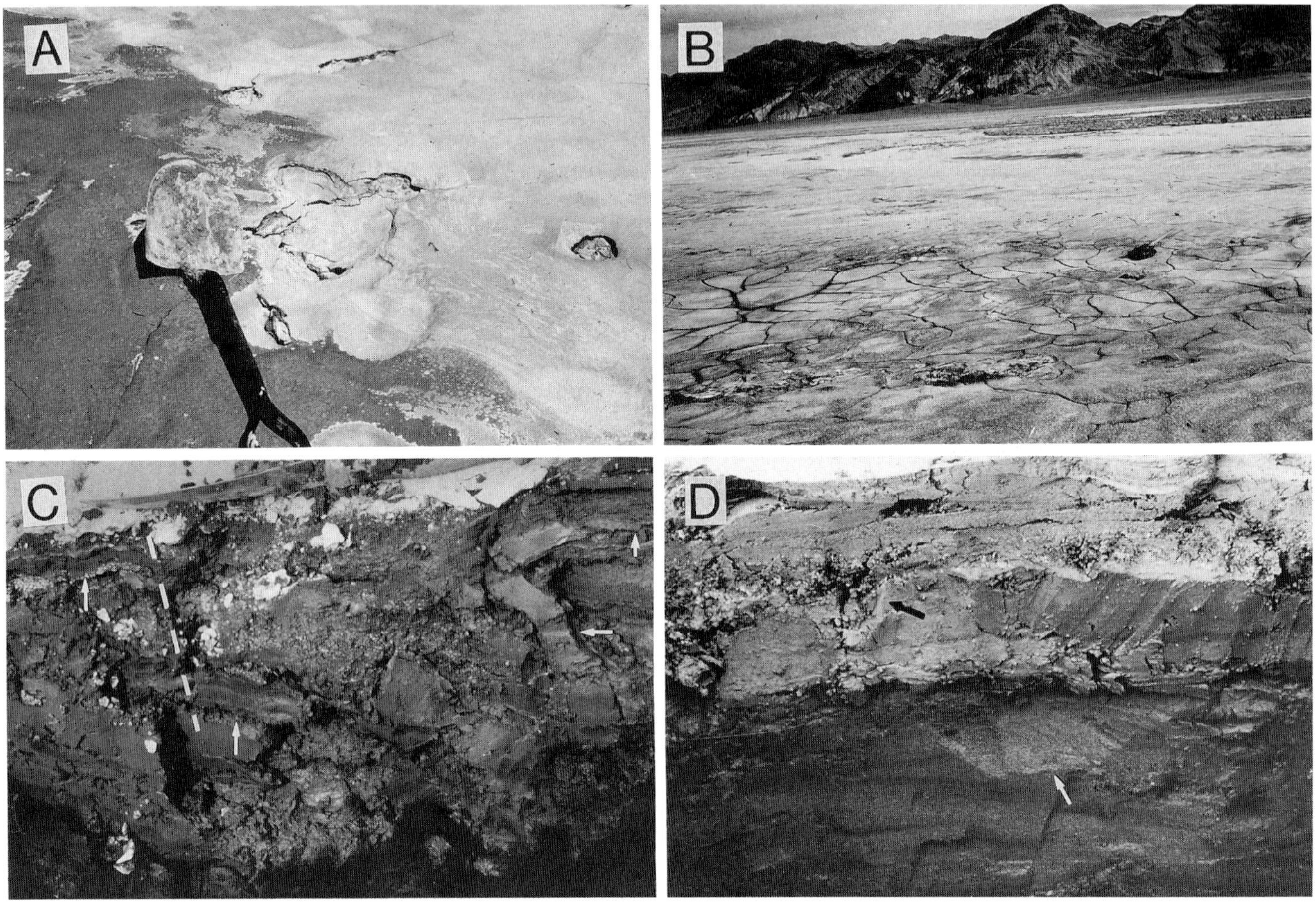

FIG. 13.—Solution-collapse features associated with efflorescent crusts, Death Valley. (A) Small-scale collapse structures on surface of sheet delta. Remnant patch of efflorescent crust in depression to the right. Light-colored surface of clay is bound by a thin microbial mat. Shovel is 1.0 m long. (B) Large-scale solution-collapse structure (area with large mudcracks) on surface of sheet delta. Undissolved efflorescent crust is visible in the background at the mouth of the delta. Shovel is 1.4 m long. (C) Growth-fault sand lobe formed by dissolution during deposition. Clay layer (arrows) that is faulted down (dotted line) was the surface of a crust at initiation of dissolution. Knife is 27 cm long. (D) Solution load structures (arrows) at the base of sheet-delta sands. Note the rotation of layering in lower sand pod and absence of flame structures. Flow was to the left. Field of view is 30 cm thick.

basal Ikakern and Timesgadiouine Formations of Brown (1980) are fluvial deposits overlain by the lacustrine Bigoudine Formation. The top of the Bigoudine Formation is a basalt, which is overlain by an unnamed fluvial lacustrine unit.

The uppermost Wolfville and Timesgadiouine Formations include abundant eolian strata (Hubert and Mertz, 1980, 1984; Smoot, unpublished data). The Blomidon Formation lacustrine strata contain abundant evaporite minerals and crystal casts (Hubert and Hyde, 1982; Mertz and Hubert, 1990), whereas no evaporites have been reported from the Bigoudine Formation, although halite diapirs in the Essaouira basin may be from correlative strata. The Blomidon and Bigoudine Formations are characterized by repeated 1 to 3 m thick lacustrine cycles (Smoot and Olsen, 1988) (Fig. 14). The base of a cycle typically consists of 20 to 30 cm of oscillatory ripple sand lenses with clay partings that are disrupted by polygonal cracks. This unit grades upward into 50 to 80 cm of sandy mud with irregular patches of sand and vugs resulting from the dissolution of evaporite minerals (gypsum?), common in the lower part. The upper 10 to 20 cm of a cycle may contain lenses of unidirectional-flow ripple cross lamination that are deformed into hump shapes. The base of a few cycles consists of 30 to 50 cm of red shale with pinch-and-swell lamination, which may contain disarticulated fish fossils. The cycles represent alternations of depositional environments from shallow lakes (mud-cracked thin beds) or relatively deep freshwater lakes (laminated shale) to subaerial conditions. Hubert and Hyde (1982) described sheetlike sandstone units, 1 to 5 m thick, in the Blomidon Formation that are dominated by climbing ripple cross-lamination. These sheet sandstones intertongue with the lacustrine beds locally. Hubert and Hyde (1982) interpreted these beds as sheetflood deposits, although the sandstone beds are better explained as sheet delta deposits (Smoot and Lowenstein, 1991). The sheet delta sandstones are laterally equivalent to the ripple cross-laminated sandstone beds in the uppermost part of the lake cycles. Similar sheet sandstones were not observed in the Bigoudine Formation.

FIG. 14.—Triassic lacustrine cycles in the Blomidon Formation, Fundy basin (A) and the Bigoudine Formation, Argana basin (B). Cycles in both exposures are 1 to 2 m thick composed of distorted ripple beds with mudcracked mudstone partings grading up into sandy mud with efflorescent crust fabrics. Note the channel in (A) filled with inclined fluvial and eolian deposits, and thick clay beds at the base and top of the outcrop (dark). Outcrop in (A) is about 15 m thick and in (B) is about 40 m thick.

Hubert and Hyde (1982) interpreted the irregular patches of sand in the upper parts of Blomidon lake cycles and the associated hump-shaped ripple lenses as adhesion ripples by analogy with Glennie (1970). Brown (1980) interpreted similar fabrics as bioturbation of mudflats on a deltaic plain. We believe that irregular lenses (Fig. 15) are lags of eolian sand trapped in surface irregularities of moderately thick efflorescent halite crusts (i.e., sand-patch fabric of Smoot and Olsen, 1988; Smoot, 1991). This interpretation is supported by the fact that: (1) the grain sizes of the sandy patches are highly variable both within and between patches; (2) many patches have jagged boundaries and internal small-scale faults; (3) most patches have cuspate boundaries; and (4) many patches have thin dense mud layers at the base or internally at grain-size contrasts. The association with intrasediment evaporites in the Blomidon Formation is also consistent with a saline mudflat environment. In the Bigoudine Formation, cavities 1 to 2 cm in diameter within the mudstone may represent molds of former evaporite minerals. Hump-shaped ripple lenses (Fig. 16) are similar to ripples deformed by efflorescent salt crust growth, and poorly sorted sandy mud is similar to the muddy matrix formed by dissolution of efflorescent crusts.

Near the area of the sandstone sheets in the Blomidon Formation, many sand-patch mudstone layers are deformed into a series of polygonal "tepees" (Fig. 17). These deformation features were noted by Mertz and Hubert (1990), who attributed them to the intrasediment growth of gypsum. There are no modern analogs or mechanisms for this type of deformation from intrasediment crystal growth. We believe these polygonal structures reflect efflorescent salt-crust deformation. The scale of the features and the association with the sand-patch fabric compare favorably to modern occurrences.

Where the thick sandstone sheets are interbedded with the lacustrine sand-patch fabric, the sandstone sheets have bulbous projections into the mudstone similar to load casts (Olsen and others, 1989, Stop 9.1) (Fig. 18). These sandstone projections, however, have no associated fluidized flow layering or grain alignment, more like the structures formed by solution collapse. In some cases, these sandstone bulbs are associated with small growth faults similar to those noted in our proposed modern analogs.

OTHER EFFLORESCENT CRUST FEATURES

We have made limited observations of efflorescent crust fabrics in other settings. In most of these, including the supratidal flats near San Felipe in Baja California, Mexico, the halite crusts are thin and mud-free, producing deformation fabrics and/or small sandy patches. In a marine coastal setting, seawater-derived brine saturates surficial sediments in the upper intertidal and supratidal zones, both by groundwater incursion and by intermittent surface flood-

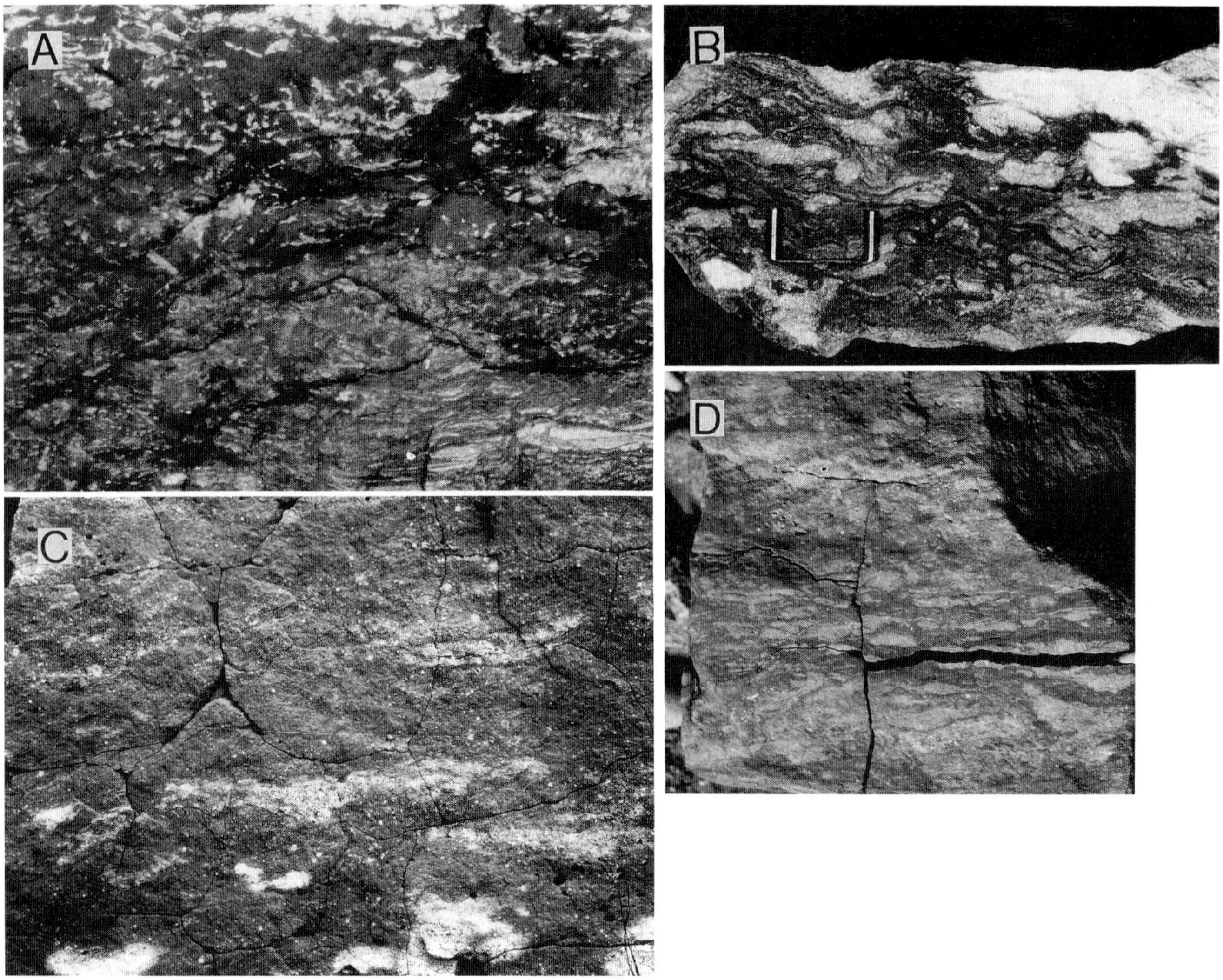

FIG. 15.—Aggradational fabrics of efflorescent crusts in the Triassic Blomidon and Bigoudine Formations. (A) Irregular sand patches (light) in a sandy mud matrix (dark), Blomidon Formation. Note the variability in size and shape of sand lenses and cuspate contacts. Outcrop area shown is 30 cm thick. (B) Slabbed sample of efflorescent crust fabric in Blomidon Formation showing irregular sand lenses with cuspate contacts (light) and thin clay linings along the base (very dark). Scale is 1 cm long. (C) Detail of sand lens from Blomidon Formation showing cuspate contact and cavities (black) from dissolved gypsum crystals. Outcrop area shown is 30 cm thick. (D) Irregular sand lenses (light) in sandy mud matrix (dark), Bigoudine Formation, similar to those in Figure 15A. Small cavities near top (dark blebs) are interpreted as former evaporite minerals. Sample thickness is about 25 cm.

ing. The efflorescent crusts are thin, perhaps because surface flooding is a more important source of brine than groundwater.

Saline Valley and Death Valley also have efflorescent crusts dominated by gypsum (Hunt and others, 1966; Smoot and Lowenstein, 1991). In Saline Valley, the gypsum efflorescent crust is developed near a deposit interpreted as a former gypsum saline pan. In Death Valley, the gypsum efflorescent crusts are developed in areas that may have been influenced by artesian springs or a pre-existing pan. Gypsum efflorescent crusts are composed of tiny euhedral to anhedral gypsum crystals and powdery bassanite. The crusts are typically less muddy than halite crusts, and many contain significant amounts of halite. The crusts form a low-relief humpy surface with a vague polygonal pattern. In cross section, the crusts are composed of irregular clumps of crystals, and broken and disoriented plates with some muddy sediment and coarser sediment that was trapped in the surface irregularities (Fig. 19A). These clumps define crude hummocky patterns similar to the modern surface. Crusts containing significant amounts of borate minerals in Death Valley and Rhodes Marsh, Nevada, produce fabrics similar to the gypsum crusts, including irregular clumps of crystals and little mud (Fig. 19B).

Efflorescent crusts dominated by thenardite and thermonatrite are typically thin and powdery. These crusts do not appear to accumulate much muddy material, and the surface irregularities lack the strength to trap significant coarser material. These efflorescent crusts mostly disturb the underlying sediment by incorporating the upper part.

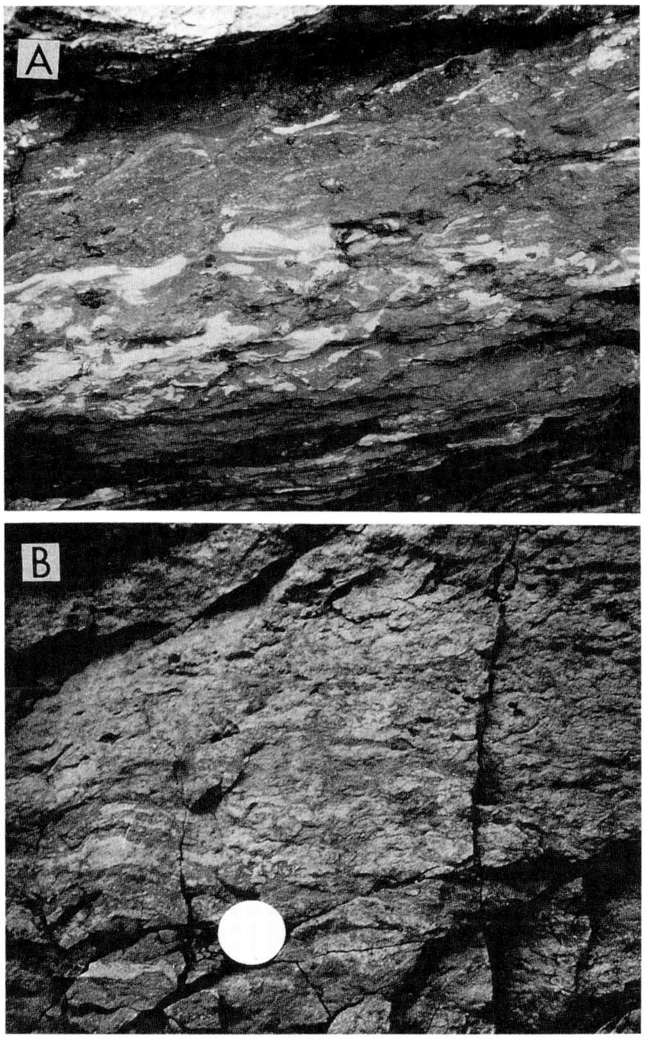

FIG. 16.—Ripple structures deformed by efflorescent crusts in the Triassic Blomidon and Bigoudine Formations. (A) Ripple lenses (both white and light gray) with vague layering and hump-shaped crests (particularly to the upper left) overlying mudcracked red shale, Blomidon Formation. Vugs (dark) are probably dissolved gypsum crystals. Outcrop shown is about 20 cm thick. Compare to Figure 12B. (B) Ripple lenses (light) with vague layering and hump-shaped crests (particularly to the left of coin) intermixed with some irregular sand lenses, Bigoudine Formation. Vugs (dark) are interpreted as dissolved evaporite minerals. Outcrop shown is about 20 cm thick.

FIG. 17.—Polygonal deformation structures associated with efflorescent crust fabrics in the Triassic Blomidon Formation. Tepee-like structures (crests are indicated by arrows) occur in sandy mud with large sand patches. Compare this to Figure 11A. Hammer is 35 cm long.

More permeable sediment is disturbed to a depth of several centimeters, whereas dense muddy sediment is barely affected.

Halite (and many other minerals) form powdery efflorescent crusts in areas where only small amounts of the minerals are introduced by wind or where groundwater tables are relatively dilute. The efflorescence grows in cracks and fenestrae in mud causing them to separate into sand- to silt-sized clumps separated by pedestals of salt. The resultant surface is referred to as puffy ground or self-rising ground (Snyder, 1962; Motts, 1965; Neal, 1969). These clumps are easily eroded by wind and are important to the deflation of mudflats and the formation of clay dunes (Bowler, 1973; Young and Evans, 1986). On dry mudflats (see Smoot and Lowenstein, 1991), puffy ground may form where floodwater is ponded, localizing dissolved salts (Fig. 20A). The clay clumps of the puffy ground may be preserved as a series of polygonal crack fillings producing a breccia-like fabric (Fig. 20B). The clumps are typically well rounded, perhaps due to dissolution of the salt along the edges during surface floods.

DISCUSSION

Several other papers describe sedimentary fabrics produced by efflorescent crusts. Nagtegaal (1973) described efflorescent crusts in the coastal Namib Desert that are covered by wind-blown sand that is partially fixed to the surface during diurnal cycles of partial solution (see also McKee, 1982, and Kocurek and Fielder, 1982). These crusts produce an irregular bedding, such as we describe for sediment accumulation on warty efflorescence, with some deformation by incorporative growth. The salt-ridge fabric of Fryberger and others (1983, 1984) is also formed by a combination of these processes. Figure 9 in Fryberger and others (1984), however, appears to show patch fabrics in a muddy matrix that are more indicative of a thicker more irregular efflorescent crust surface than the polygonal crust in their Figure 4D. Ahlbrandt and Fryberger (1981) illustrated two interesting efflorescent crust textures. Their Figure 13E shows a diffuse irregular bedding in a trench in an alkali lake deposit (Twin Lakes) in the Nebraska Sand Hills that is consistent with powdery efflorescent crusts, and their Figures 14C and 14D from a trench in Sabkha Ar Riyas in Saudi Arabia illustrate rubbly gypsum mounds with small sandy patches and a sandy drape that are consistent with the structures formed by efflorescent gypsum crusts. Watson (1983a, b, 1985, 1988) illustrated the textural characteristics of gypsum soil crusts. The presence of coarse gypsum crystals

semble dewatering structures (see Lowe, 1975). Dewatering structures show flow banding or grain alignment parallel to the upturned edges, which are not present in deformation by efflorescent crusts. Sediment pods formed in solution pockets resemble load casts (Dzulynski and Kotlarczyk, 1962; Kuenen, 1958). The solution-collapse pods differ from sediment loading on water-saturated mud in that the underlying layer is not deformed into flame structures. Solution-collapse breccias may be mistaken for earthquake-induced sandblows (see Obermeier and others, 1990). They differ in that the sandblow structures show upward sediment transport and common flow banding and grain alignment associated with the fluidized sediment, whereas the solution collapse is characterized by downward movement of sediment and no fluidized sediment flow features. The deformation fabrics of efflorescent crusts are also characterized by the association with the other efflorescence crust fabrics.

3. Glennie (1970) illustrated irregular sand lenses in a muddy matrix in the inland sabkhas of Libya, which he interpreted as adhesion ripples formed by wind-blown sand becoming trapped on a damp surface. These sand lenses (his figures 55, 62, 111, and 113) are very similar to the fabrics produced by aggradation and by deformation in efflorescent crusts. This similarity may indicate that there are serious problems in distinguishing between the fabrics, particularly because the features shown by Glennie (1970) also exhibit the mud-into-sand cuspate contacts. Alternatively, it is possible that Glennie misinterpreted the structures he observed. We suspect the latter for the following reasons: (a) all examples of modern adhesion ripples and experimentally formed adhesion ripples show convex-upward layering of sand on sand (e.g., Hunter, 1973, 1980; Kocurek and Fielder, 1982); (b) the basal contacts of the features shown by Glennie (1970) are composed of cuspate mud-into-sand, but there is no mechanism for deforming water-saturated mud in the adhesion ripple hypothesis; and (c) the localities where Glennie observed these features are covered by efflorescent crusts.

We believe that in addition to the fabrics described in the Fundy and Argana basins, structures produced by efflorescent salts are well represented in the geologic record. Distorted patchy sand lenses in the Permian Rotliegendes Formation, which have been interpreted as adhesion ripples (Glennie, 1972, 1983a, b, 1987; Nagtegaal, 1973), are probably fabrics produced by efflorescent crusts. Similar features in the Pennsylvanian and Lower Permian Minnelusa Formation (Fryberger and others, 1983) are also probably the result of efflorescent crusts. Chakraborty (1991) illustrated polygonal deformation structures in Proterozoic eolian sandstone in India that he attributes to salt-crust deformation. One of us (JPS) observed features that suggest the deposits of efflorescent salt crusts in the Late Proterozoic Bitter Spring Formation in Australia and in the Lower Jurassic Towaco Formation of the Newark basin of New Jersey. Smoot (1977) illustrated brecciated mudstones with diffuse layering and rounded mudstone clumps in the Wilkins Peak Member of the Eocene Green River Formation in Wyoming (his Figures 41 and 42) that resemble the textures produced by powdery efflorescent crusts forming puffy ground. This fabric is associated with sodium carbonate evaporite minerals, so a powdery thermonatrite efflorescence may be expected.

SUMMARY

Efflorescent salt crusts form by the complete evaporation of saline ground water at the sediment surface or by recycling of airborne minerals through rainfall and evaporation. Halite efflorescent crusts are composed of tiny crystals on which hydroscopic water films trap wind-blown dust. The dissolution of the halite in undersaturated groundwater leaves the poorly sorted mud as a lag deposit that slowly aggrades. Wind-blown and water-deposited sand and silt that is trapped in surface irregularities of the crust are also left as lags producing a sand-patch fabric. Efflorescent crusts formed on sandy sediment incorporate the upper layer and deform it into the shape of the crusts' surface. This deformation is commonly in the form of polygons. Where efflorescent crusts develop on ripple crests, they are deformed into hump-shaped lenses. Solution collapse of efflorescent crusts produces bulbous sand bodies along the base of channel deposits and sheet deltas that resemble load casts, but lack fluidization layering or aligned grains. The fact that all of these fabrics have been observed in the Triassic Blomidon Formation demonstrates the geologic preservability of these features and their utility for interpreting ancient environments. Although these fabrics may be associated with evaporites or their pseudomorphs, in some cases they may be the *only* indicator of saline conditions.

ACKNOWLEDGMENTS

Research of this project was partially funded by American Chemical Society Petroleum Research Fund grant #11086-GA, National Science Foundation grant #EAR-8020378, and the U. S. Geological Survey. We express our gratitude to Lawrence Hardie for introducing us to Saline Valley and making available his data, Bill Means for providing accommodation during our work in Saline Valley, and the National Park Service for allowing us to trench and core the Death Valley playas. The senior author would also like to thank Paul Olsen and Dave Brown for leading him through the Fundy basin, and the Government of Canada for allowing him to study these deposits. The senior author also thanks Ben Said and the Moroccan Directorate of Geology for allowing him to examine the Argana basin and for providing assistance in his work there. We thank Chris Schenk and Sherilyn Williams-Stroud for their reviews of this paper, which greatly increased its clarity.

REFERENCES

AHLBRANDT, T. S., AND FRYBERGER, S. G., 1981, Sedimentary features and significance of interdune deposits, *in* Ethridge, F. G., and Flores, R. M., eds., Recent and Ancient Nonmarine Depositional Environments: Models for Exploration: Tulsa, Society of Economic Paleontologists and Mineralogists Special Publication 31, p. 293–314.

BARKER, C. E., AND LEFOND, S. J., 1985, A re-evaluation of the origin and diagenesis of borate deposits, Death Valley region, California, *in* Barker, C. E., and Lefond, S. J., eds., Borates: Economic Geology and Production: New York, Society of Mining Engineers of the American Institute of Mining, Metallurgy, and Petroleum Engineers, p. 13–34.

BOWLER, J. M., 1973, Clay dunes: Their occurrence, formation and environmental significance: Earth Science Reviews, v. 9, p. 315–338.

BOWSER, C. J., AND DICKSON, F. W., 1966, Zonation of borates of Kramer, California, *in* Rau, J. L., ed., Second Symposium on Salt, I: Cleveland, Northern Ohio Geological Society, p. 122–132.

BROWN, R. H., 1980, Triassic rocks of Argana Valley, southern Morocco, and their regional structural implications: Bulletin of the American Association of Petroleum Geologists, v. 64, p. 988–1003.

CHAKRABORTY, T., 1991, Sedimentology of a Proterozoic erg: the Venkatpur Sandstone, Pranhita-Godavari Valley, south India: Sedimentology, v. 38, p. 301–322.

DROSER, M. L., AND BOTTJER, D. J., 1986, A semiquantitative field classification of ichnofabric: Journal of Sedimentary Petrology, v. 56, p. 558–559.

DZULYNSKI, S., AND KOTLARCZYK, J., 1962, On load-casted ripples: Annales de la Societé Géologique de Pologne, v. 32, p. 148–159.

EKDALE, A. A., BROMLEY, R. G., AND PEMBERTON, S. G., 1984, Ichnology, The Use of Trace Fossils in Sedimentology and Stratigraphy: Tulsa, Society of Economic Paleontologists and Mineralogists Short Course 15, 317 p.

ERICKSEN, G. E., 1981, Geology and origin of Chilean nitrate deposits: Washington, D. C., United States Geological Survey Professional Paper 1188.

EUGSTER, H. P., 1970, Chemistry and origin of brines of Lake Magadi, Kenya: Washington, D. C., Mineralogical Society of America, Special Paper 3, p. 215–235.

EUGSTER, H. P., 1980, Lake Magadi, Kenya, and its precursors, *in* Nissenbaum, A., ed., Hypersaline Brines and Evaporitic Environments: Amsterdam, Elsevier, p. 195–232.

EUGSTER, H. P., AND HARDIE, L. A., 1978, Saline lakes, *in* Lerman, A., ed., Lakes: Chemistry, Geology, Physics: New York, Springer, p. 237–293.

FRYBERGER, S. G., AL-SARI, A. M., AND CLISHAM, T. J., 1983, Eolian dune, interdune, sand sheet, and siliciclastic sabkha sediments of an offshore prograding sand sea, Dhahran area, Saudi Arabia: Bulletin of the American Association of Petroleum Geologists, v. 67, p. 280–312.

FRYBERGER, S. G., AL-SARI, A. M., CLISHAM, T. J., RISVI, S. A. R., AND AL-HINAI, K. G., 1984, Wind sedimentation in the Jafurah sand sea, Saudi Arabia: Sedimentology, v. 31, p. 413–431.

GARRETT, D. E., 1985, Chemistry and origin of the Chilean nitrate deposits, *in* Schreiber, B. C., and Harner, H. C., eds., Sixth Symposium on Salt, v. 1: Alexandria, The Salt Institute, p. 285–302.

GLENNIE, K. W., 1970, Desert Sedimentary Environments: Amsterdam, Elsevier, 222 p.

GLENNIE, K. W., 1972, Permian Rotliegendes of northwest Europe interpreted in light of modern desert sedimentation studies: Bulletin of the American Association of Petroleum Geologists, v. 56, p. 1048–1071.

GLENNIE, K. W., 1983a, Early Permian (Rotliegendes) paleowinds of the North Sea: Sedimentary Geology, v. 34, p. 245–265.

GLENNIE, K. W., 1983b, Lower Permian Rotliegend desert sedimentation in the North Sea area, *in* Brookfield, M. E., and Ahlbrandt, T. S., eds., Eolian Sediments and Processes: Amsterdam, Elsevier, p. 521–542.

GLENNIE, K. W., 1987, Desert sedimentary environments, present and past—a summary: Sedimentary Geology, v. 50, p. 135–165.

HANDFORD, C. R., 1991, Marginal marine halite: sabkhas and salinas, *in* Melvin, J. L., ed., Evaporites, Petroleum and Mineral Resources: Amsterdam, Elsevier, p. 1–66.

HARDIE, L. A., 1968, The origin of the Recent non-marine evaporite deposit of Saline Valley, Inyo County, California: Geochimica et Cosmochimica Acta, v. 32, p. 1279–1301.

HARDIE, L. A., SMOOT, J. P., AND EUGSTER, H. P., 1978, Saline lakes and their deposits: A sedimentological approach, *in* Matter, A., and Tucker, M. E., eds., Modern and Ancient Lake Sediments: Oxford, International Association of Sedimentologists Special Publication 2, p. 7–41.

HAY, R. L., 1968, Chert and its sodium precursors in sodium carbonate lakes of East Africa: Contributions to Mineralogy and Petrology, v. 17, p. 255–274.

HSÜ, K.J., AND SIEGENTHALER, C., 1969, Preliminary experiments on hydrodynamic movement induced by evaporation and their bearing on the dolomite problem: Sedimentology, v. 12, p. 11–25.

HUBERT, J. F., AND HYDE, M. G., 1982, Sheet-flow deposits of graded beds and mudstones on an alluvial sandflat-playa system. Upper Triassic Blomidon redbeds, St. Mary's Bay, Nova Scotia: Sedimentology, v. 29, p. 457–474.

HUBERT, J. F., AND MERTZ, K. A., 1980, Eolian dune field of Late Triassic age, Fundy basin, Nova Scotia: Geology, v. 8, p. 516–519.

HUBERT, J. F., AND MERTZ, K. A., 1984, Eolian sandstones in Upper Triassic-Lower Jurassic red beds of the Fundy basin, Nova Scotia: Journal of Sedimentary Petrology, v. 54, p. 798–810.

HUNT, C. B., ROBINSON, T. W., BOWLES, W. A., AND WASHBURN, A. L., 1966, Hydrologic basin, Death Valley, California: Washington, D. C., United States Geological Survey Professional Paper 494-B, 138 p.

HUNTER, R. E., 1973, Pseudo-crosslamination formed by climbing adhesion ripples: Journal of Sedimentary Petrology, v. 43, p. 1125–1127.

HUNTER, R. E., 1980, Quasi-planar adhesion stratification—an eolian feature in wet sand: Journal of Sedimentary Petrology, v. 50, p. 263–266.

JONES, B. F., 1965, The hydrology and mineralogy of Deep Springs Lake, Inyo County, California: Washington, D. C., United States Geological Survey Professional Paper 502-A, 56 p.

KOCUREK, G., AND FIELDER, G., 1982, Adhesion structures: Journal of Sedimentary Petrology, v. 52, p. 1229–1241.

KUENEN, P. H., 1958, Experiments in geology: Transactions of the Geological Society of Glasgow, v. 23, p. 1–28.

LAST, W. M., 1984, Sedimentology of playa lakes of the northern Great Plains: Canadian Journal of Earth Sciences, v. 21, p. 107–125.

LOWE, D. R., 1975, Water escape structures in coarse-grained sediments: Sedimentology, v. 22, p. 157–204.

LOWENSTEIN, T. K., AND HARDIE, L. A., 1985, Criteria for the recognition of salt-pan evaporites: Sedimentology, v. 32, p. 627–644.

MACUMBER, P. G., 1983, Interactions between ground water and surface systems in northern Victoria: Unpublished Ph.D. Dissertation, University of Melbourne, Melbourne, 506 p.

MCKEE, E. D., 1982, Sedimentary structures in dunes of the Namib Desert, Southwest Africa: Boulder, Geological Society of America Special Paper 186, 64 p.

MERTZ, K. A., AND HUBERT, J. F., 1990, Cycles of sand-flat sandstone and playa-lacustrine mudstone in the Triassic-Jurassic Blomidon redbeds, Fundy rift basin, Nova Scotia: implications for tectonic and climatic controls: Canadian Journal of Earth Sciences, v. 27, p. 442–451.

MOTTS, W. S., 1965, Hydrologic types of playas and closed valleys and some relations of hydrology to playa geology, *in* Neal, J. T., ed., Geology, Mineralogy and Hydrology of U.S. Playas: Cambridge, United States Airforce Cambridge Research Laboratories, Environmental Research Paper 96, p. 73–104.

MUESSIG, S., 1959, Primary borates in playa deposits: Economic Geology, v. 54, p. 495–501.

MUESSIG, S., 1966, Recent South American borate deposits, *in* Rau, J. L., ed., Second Symposium on Salt, I: Cleveland, Northern Ohio Geological Society, p. 151–159.

NAGTEGAAL, P. J. C., 1973, Adhesion ripple and barchan-dune sands of the recent Namib (SW Africa) and Permian Rotliegend (NW Europe) deserts: Madoqua, v. 2, p. 5–19.

NEAL, J. T., 1969, Playa variation, *in* McGinnis, W. G., and Goldman, B. J., eds., Arid Lands in Perspective: Tucson, University of Arizona Press, p. 13–44.

OBERMEIER, S. F., JACOBSON, R. B., SMOOT, J. P., WEEMS, R. E., GOHN, G. S., MONROE, J. E., AND POWARS, D. S., 1990, Earthquake-induced liquefaction features in the coastal setting of South Carolina and in the fluvial setting of the New Madrid seismic zone: Washington, D. C., United States Geological Survey Professional Paper 1504, 44 p.

OLSEN, P. E., SCHLISCHE, R. W., AND GORE, P. J. W., eds., 1989, Tectonic, Depositional, and Paleoecological History of Early Mesozoic Rift Basins, Eastern North America: Field Trip Guidebook T351, 28th International Geological Congress: Washington, D.C., American Geophysical Union, 174 p.

PAPKE, K. G., 1976, Evaporites and brines in Nevada playas: Nevada Bureau of Mines and Geology Bulletin 87, 35 p.

SMITH, G. I., 1985, Borate deposits in the United States: Dissimilar form, similar in geologic setting, *in* Barker, C. E., and Lefond, S. J., eds.,

Borates: Economic Geology and Production: New York, Society of Mining Engineers of the American Institute of Mining, Metallurgy and Petroleum Engineers, p. 37–51.

SMOOT, J. P., 1977, Sedimentology of a saline closed basin: the Wilkins Peak Member of the Green River Formation (Eocene), Wyoming: Unpublished Ph.D. Dissertation, Johns Hopkins University, Baltimore, 296 p.

SMOOT, J. P., 1991, Sedimentary facies and depositional environments of early Mesozoic Newark Supergroup basins, eastern North America: Palaeogeography, Palaeoclimatology, Palaeoecology, v. 84, p. 369–423.

SMOOT, J. P., AND LOWENSTEIN, T. K., 1991, Depositional environments of non-marine evaporites, *in* Melvin, J. L., ed., Evaporites, Petroleum and Mineral Resources: Amsterdam, Elsevier, p. 189–347.

SMOOT, J. P., AND OLSEN, P. E., 1988, Massive mudstones in basin analysis and paleoclimatic interpretation of the Newark Supergroup, *in* Manspeizer, W., ed., Triassic-Jurassic Rifting, North America and North Africa: Amsterdam, Elsevier, p. 249–274.

SNYDER, C. T., 1962, A hydrologic classification of valleys in the Great Basin, western United States: International Association of Scientific Hydrology, v. 7, p. 53–59.

STOERTZ, G. E., AND ERICKSEN, G. E., 1974, Geology of salars in northern Chile: Washington, D. C., United States Geological Survey Professional Paper 811, 65 p.

WATSON, A., 1983a, Gypsum crusts, *in* Goudie, A. S., and Pye, K., eds., Chemical Sediments and Geomorphology: London, Academic Press Inc., p. 133–161.

WATSON, A., 1983b, Evaporite sedimentation in non-marine environments, *in* Goudie, A. S., and Pye, K., eds., Chemical Sediments and Geomorphology: London, Academic Press Inc., p. 163–185.

WATSON, A., 1985, Structure, chemistry, and origins of gypsum crusts in southern Tunisia and the central Namib Desert: Sedimentology, v. 32. p. 855–875.

WATSON, A., 1988, Desert gypsum crusts as paleoenvironmental indicators: A micropetrographic study of crusts from southern Tunisia and the central Namib Desert: Journal of Arid Environments, v. 15, p. 19–42.

YOUNG, J. A., AND EVANS, R. A., 1986, Erosion and deposition of fine sediments from playas: Journal of Arid Environments, v. 10, p. 103–115.

SEASONAL EVAPORITE SEDIMENTATION IN DESERT PLAYA LAKES OF THE KARINGA CREEK DRAINAGE SYSTEM, CENTRAL AUSTRALIA

ARO V. ARAKEL AND TIAN HONGJUN
P.O. Box 481, West Ryde, New South Wales 2114, Sydney, Australia

ABSTRACT: The linear chain of playa lakes in the Karinga Creek drainage system, central Australia, hosts seasonal efflorescent crusts characterized by laterally intergrading halite and mirabilite facies. The sedimentary features of these evaporite facies and chemical compositions of the playa brines collectively reflect aperiodic thermodynamic equilibrium, with respect to sulfate and chloride minerals, due to repeated evaporation and cooling induced by desert climatic conditions. The coexistence of halite and mirabilite facies in the efflorescent crusts is largely a consequence of the groundwater discharge pattern and the hydrochemical setting of brine pools at a given point, rather than a true representation of concurrent episodes of chemical sedimentation.

These findings agree with results from isothermal evaporation experiments and reflect a pattern of playa sedimentation that differs significantly from that of simple evaporation. An assessment of the thermodynamic requirements for variations in the patterns of chemical sedimentation of ancient shallowing-upward lacustrine sequences may provide valuable paleoenvironmental information.

INTRODUCTION

Playa lakes, acting as local outlets for discharge from regional shallow aquifers, are common in the Central Australian Discharge Zone (Fig. 1; Jacobson, 1988; Arakel and others, 1990). These lakes occur in several discrete shallow drainage basins and, because of the arid conditions, they host a range of evaporite minerals. Facies variation is commonly attributed to changes in local hydrologic conditions and physiographic setting (Jacobson and others, 1988; Arakel and others, 1988). Sedimentary signatures, particularly evaporite facies changes, have also been used to assess the hydrogeochemical evolution of ancient lacustrine deposits, especially where shallowing-upward lacustrine sequences include evaporite caps dominated by sulfate and chloride mineral facies (Warren, 1989).

The literature on ancient lacustrine evaporites commonly mentions groundwater influences on the paragenesis of the sulfate and chloride facies (e.g., Orti Cabo and Salvany, 1990). For example, gypsum and glauberite have been identified as the most stable sulfate minerals in terminal lake facies; their coexistence is commonly attributed to early diagenetic replacement under changing hydrologic regimes (Orti Cabo and others, 1986; Arakel and Cohen, 1991). In contrast, there is little reference to the occurrence and distribution of temperature-sensitive and, thus, less stable evaporite minerals in both modern and ancient closed-basin lakes. This probably reflects masking effects by early diagenesis, thermodynamic instability, and repeated mineral phase transformations (Last and Slezak, 1986). Temperature is a fundamental control in commercial production of sulfate and chloride salts from brines by repeated evaporation and cooling (Barry, 1986). However, the significance of evaporation and cooling on the genesis and diagenesis of evaporite facies in natural playas is poorly documented.

In this paper, we use field observations from playa lakes in the Karinga Creek drainage system (Fig. 2) and experimental data to assess the fundamental controls on the occurrence and facies distributions of playa efflorescent crusts under prevailing climatic conditions. Most field information discussed here relates to systematic observations in the Calatta Lake system (Fig. 2), where seasonal development of a well-defined halite-thenardite crust offers a modern analog for assessing similar terminal evaporite facies in ancient lacustrine deposits.

LOCATION AND CLIMATE

Karinga Creek drainage system forms the easternmost component of the Central Australian Groundwater Discharge Zone, extending more than 120 km and covering a catchment of more than 1,280 km^2 (Fig. 1). Calatta Lake (Fig. 2) is one of many playa lakes in the drainage system. We selected it for this study because of its geomorphologic maturity and its seasonal development of a halite-mirabilite efflorescent crust.

The study area has an arid, desert climate. Mean annual precipitation varies from 226 mm (Curting Springs) to 333 mm (Ayers Rock). The measured minimum humidity varies from 43 percent at 9.00 AM to 24 percent at 3.00 PM, with an averaged daily sunshine incidence of about 9.6 hours for an average of 196 days per year. Prevailing winds are toward the southeast to east, mainly in the afternoon, and with a velocity generally less than 20 km/hr. The mean monthly maximum temperature for the recorded period is 33°C in January and 20°C in July, with mean monthly minimum temperatures of 20°C in January and 4°C in July. Nocturnal temperature measurements during winter (May to August) of 1989 and 1990 indicate a 78 percent probability for the occurrence of near-zero temperatures (2°C to −5°C), extending more than six hours. We recorded diurnal temperature variations of up to 28°C during winter; large fluctuations in diurnal and seasonal temperature ranges are reflected in significant variations in seasonal pan evaporation rates, which vary from 6.5 to 7.5 mm/day in summer, and 2.0 to 2.5 mm/day in winter. The mean annual evaporation potential in the study area exceeds 3,000 mm/year.

METHODS

We undertook fieldwork during seven extended trips and a few short visits between October 1987 and March 1991. The June-July 1989 fieldwork followed a major rainfall event that led to inundation of most playa lakes in the study area. Representative sediment samples at Calatta Lake and surrounding creeks were obtained by shallow coring using a 50 mm diameter PVC tube and a box corer. In total, 28 cores, up to 2.8 m long, and 47 grab and box core samples

Sedimentology and Geochemistry of Modern and Ancient Saline Lakes, SEPM Special Publication No. 50

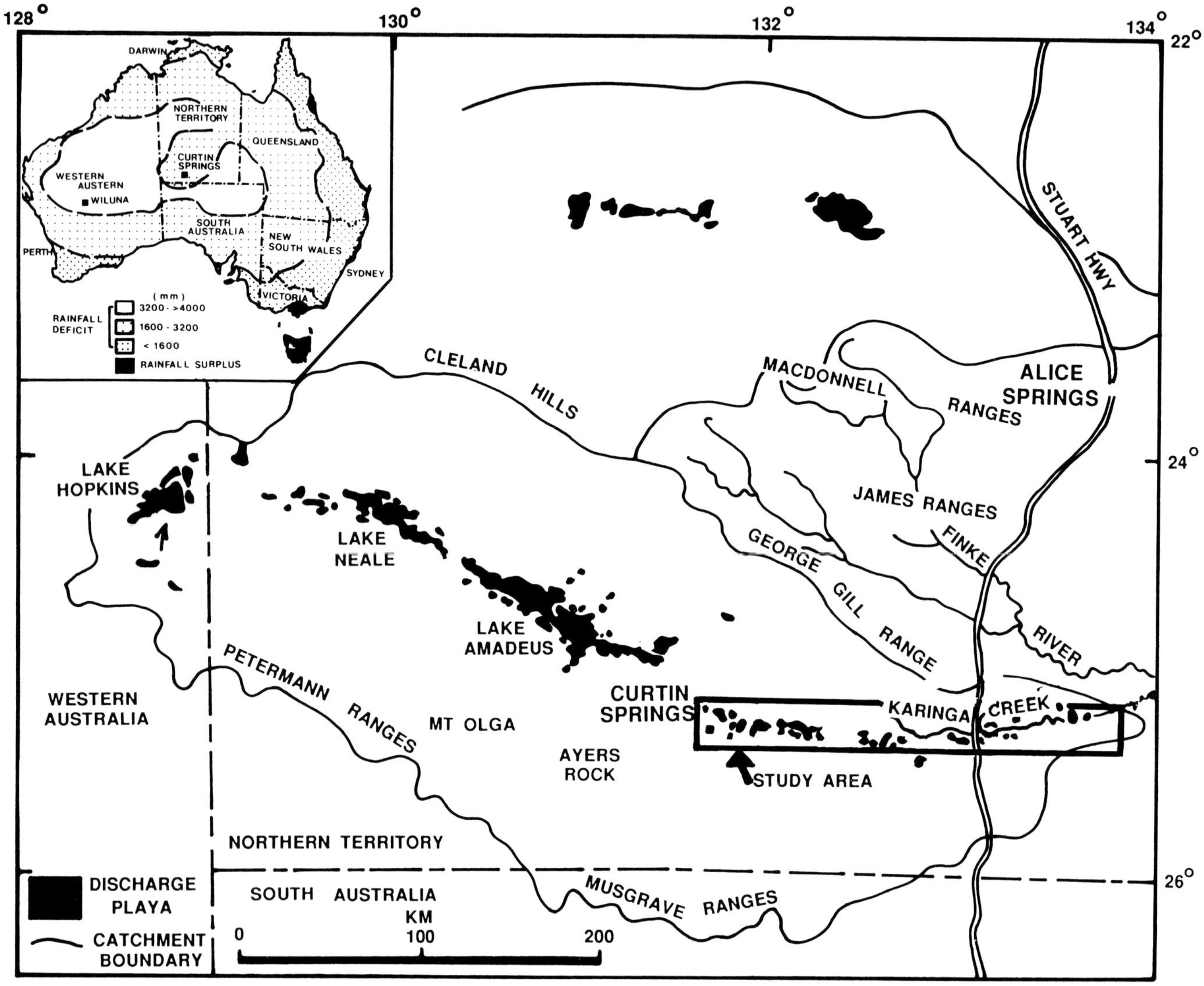

FIG. 1.—Regional map showing location of Karinga Creek drainage system in the Central Australian Groundwater Discharge Zone. The unshaded region on the map of Australia (inset) is characterized by desert climatic conditions with a rainfall deficit in excess of 3200 mm per annum.

were collected. The physico-chemical characteristics of the groundwaters (pH, Eh, TDS, conductivity, salinity, temperature) were determined in the field using portable instruments. Since 1988, 45 groundwater samples were collected for chemical analysis and experimental studies from different hydrologic zones of the lake, adjoining springs and seepage creeks. These samples, together with 550 other analyses of surface and ground-waters from the Karinga Creek drainage system, form the basis for comparative hydrochemical study of the playa lakes.

Laboratory investigations included sediment sample preparation, followed by microscopic studies, X-ray diffraction, scanning electron microscopy, and differential thermal determination of evaporite minerals, both from playa sediments and experimental precipitates. Groundwater samples were analyzed for major and trace elements by various titration, ion-selective electrode, atomic absorption spectroscopy, UV spectroscopy and ICP methods. Selected replicate samples were also analyzed as part of an interlaboratory verification process.

PHYSIOGRAPHIC SETTING AND REGIONAL HYDROGEOLOGY

The playa chain in the Karinga Creek system represents the down-drainage component of the Central Australian Groundwater Discharge Zone (Jacobson, 1988), which extends for more than 500 km, from Lake Hopkins in Western Australia to Lake Eyre in South Australia (Fig. 1). The regional groundwater system is two-layered, with thin Cenozoic sedimentary aquifers overlying fractured bedrock aquifers (Jacobson and others, 1988). Tertiary and Quaternary calcretes, alluvial clay-sand deposits and dunal sand bodies contain most of the shallow aquifers that discharge into the modern playa lakes.

The literature on arid interior drainage basins of Australia emphasizes the significance of past climatic regimes as

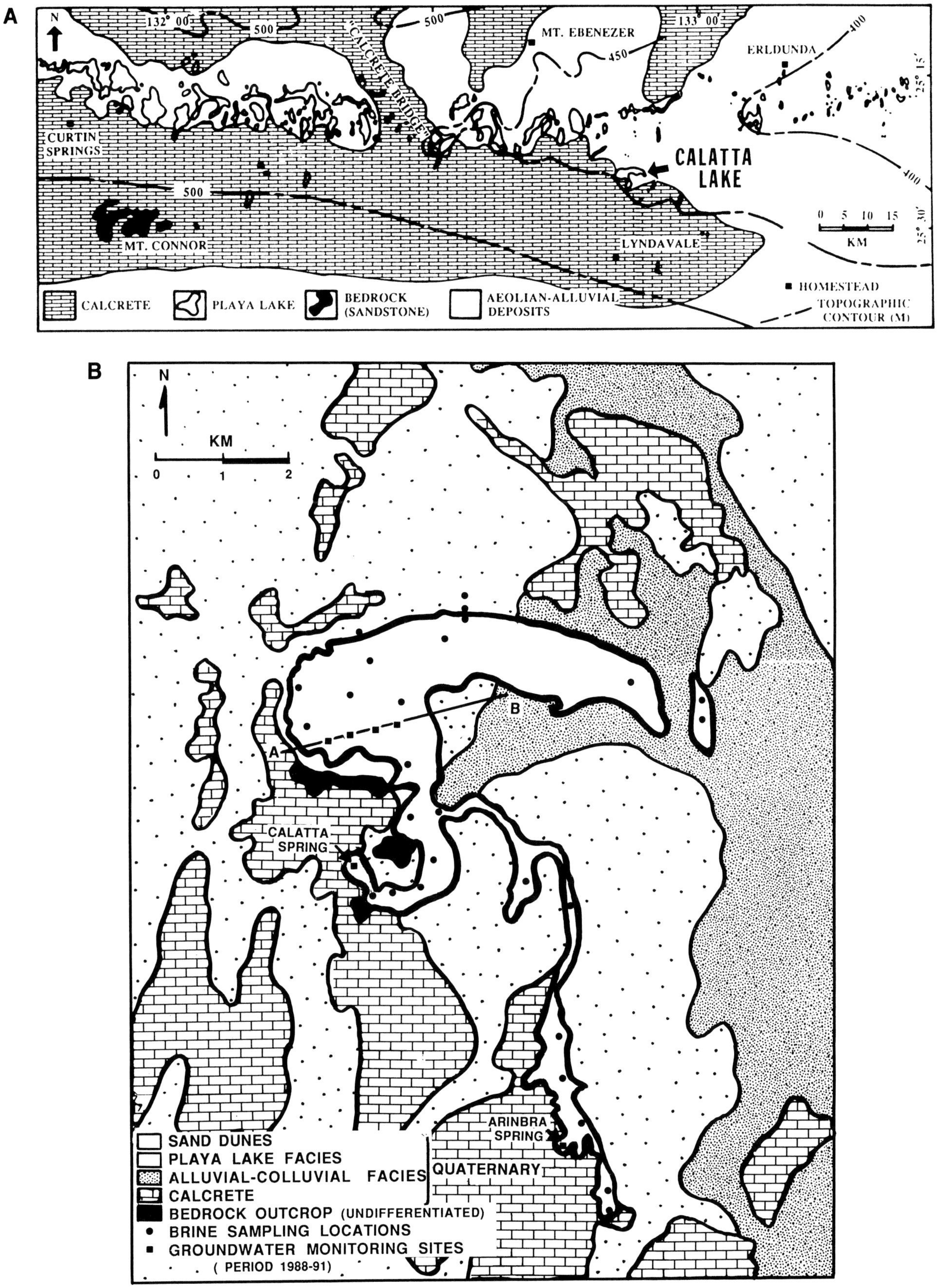

FIG. 2.—Maps showing (A) physiographic and geologic features of the Karinga Creek drainage system, and (B) surface geology and location of groundwater sampling and monitoring sites in the Calatta Lake system.

modifiers of landforms and groundwater hydrochemistry since the Tertiary time (e.g., Macumber, 1979; Bowler, 1986; Teller and others, 1982). Hydrologic adjustments have played a major role in the development of Quaternary duricrusts, especially at groundwater discharge and recharge areas (Arakel, 1991). Playa abandonment probably results from such hydrologic modifications during the Quaternary (Jacobson and others, 1988).

The Karinga Creek drainage system accommodates more than 100 playa lakes of different shapes and sizes (Fig. 2). Most lakes have long axes with an east-west orientation, probably indicating the influence of subsurface flow along an ancestral paleochannel. Regionally, a calcrete-capped basement high ('calcrete bridge' of Arakel, 1991) divides the study area into two hydrologic domains (Fig. 2). During the last pluvial phase (34,000 to 27,000 yr BP), the drainage basin in the eastern domain was marked by actively flowing creeks (Arakel and Wakelin-King, 1991). Headward erosion of calcretes by the ancestral Karinga Creek extended west toward the "calcrete" ridge (Fig. 2).

Drilling and hydrologic investigations indicate that the subsurface flow is impelled by a low regional hydrostatic head in a general easterly direction that coincides approximately with the direction of the modern Karinga Creek. The overall groundwater flow in the western domain is relatively slow and impeded by stagnation and subsurface ponding effects (Arakel and Wakelin-King, 1991). The minor surface flow is limited to a few springs and intermittent seepage creeks that emanate from the playa marginal calcretes and discharge into the lakes. Most surface flow recorded during this study occurred with aperiodic thunderstorms of local significance (Arakel, 1991).

The groundwater table in the calcrete and shale aquifers is generally a few meters below the surface. In the playa lakes, the groundwater level is commonly within the upper meter of the playa beds where most of the evaporites are present. Hydrologic monitoring in several of the playa lakes indicates moderate to minor groundwater level fluctuations (less than 20 cm) through each hydrologic cycle, probably reflecting a stable equilibrium between influxes and evaporative reflux (Chen, 1989; Arakel and Cohen, 1991).

Within the Calatta Lake system, a slightly undulating calcrete-capped terrain extending 18 km southward forms the main catchment for the lake (Fig. 2). The barren playa has a surface area of about 11 km^2 and remains dry throughout the year. Lakeward groundwater flow is mainly by subsurface seepage from the marginal calcrete aquifers and a mound spring in the southern arm of the lake. Surface flow along the 15 km long southern creek, which connects the Calatta and Arinbra Springs to the lake, is largely ephemeral with two brief periods of significant flow during the past five years. The springs emanate from shallow aquifers in the Tertiary and younger siliceous calcretes to the south of the lake, and from similar calcretes on the western and northwestern edges. The terrain east of the lake is mainly alluvial and colluvial deposits, overlain by quartzose and gypsiferous sand dunes. Calcretes north of the lake are overlain by mobile longitudinal dunes.

RESULTS

Playa Geochemistry

The chemical characteristics of the groundwaters in the different hydrologic zones of the Karinga Creek drainage system are described by Arakel and Cohen (1991). From the calcrete and bedrock terrains to the playa lakes, a trend of increasing salinity compatible with progressive evaporation concentration of the shallow groundwaters is apparent (Fig. 3). The greatest variation from the general trend is seen for Ca^{2+}, most of which is removed by formation of various carbonate minerals when groundwater approaches the surface and becomes depleted in CO_2. Despite significant variations in individual playa lakes, the following order of ionic concentration is common to most lakes:

$$Na^+ > Mg^{2+} > K^+ > Ca^{2+}$$

$$\text{and} \quad Cl^- > SO_4^{2-} > CO_3^{2-} + HCO_3^-$$

When compared with brines from other arid-zone Australian lake basins (Williams, 1984), the main difference is a significantly higher K/Ca ratio in the Karinga Creek playa lakes.

Using the Valyashko classification system for natural waters, most subsurface brines in the playa lakes belong to the sulfate type. Of these, the Na-sulfate subtype represents 65.3 percent, while the Mg-sulfate subtype represents 29.5 percent. Four-year monitoring of Calatta Lake shows that playa brines remain Na-subtype (Table 1). However, both Na and SO_4 show significant temporal variations in concentration trends when data for winter 1989 samples, collected after an earlier major rainfall event, are compared with the same period for subsequent years (Fig. 3).

Spatially, playa brines both in the eastern and western domains are of Na-sulfate subtype (Fig. 4). Brines in the Karinga Creek channel and a few adjoining lakes are of Mg-sulfate subtype. Seepage waters from the calcrete aquifers belong to carbonate or chloride types.

Using Hardie's (1984) classification, playa brines in Calatta Lake are dominated by Na-(Mg)-Cl-(SO_4), Na-(Mg)-Cl, and Na-Cl-(SO_4) types (Fig. 4). Brines in the western domain are mostly of the Na-(Mg)-Cl-(SO_4) type, with a few examples of Na-Cl-(SO_4) type. In the eastern domain, brines are more variable, but with Na-Cl-(SO_4) and Na-(Mg)-Cl types dominant, reflecting the chemistry of the calcrete and dune aquifers surrounding the lakes. Brines in Calatta Lake are mainly Na-Cl-(SO_4) type (Fig. 4) and are fed directly from the surrounding calcrete aquifers. This contrasts with lakes in the drainage system that receive seepage waters following extensive filtration, evaporation and precipitation.

Experimental Geochemistry

To assess the hydrochemical evolutionary path of the playa brines, the thermodynamic requirements for precipitation of sulfate salts were evaluated by experimental isothermal evaporation of a brine sample from Calatta Lake at 25°C. The experiment was undertaken in an evaporation chamber of similar design to that of De La Pena and others (1982), equipped with spotlights and a thermoregulator. Precipi-

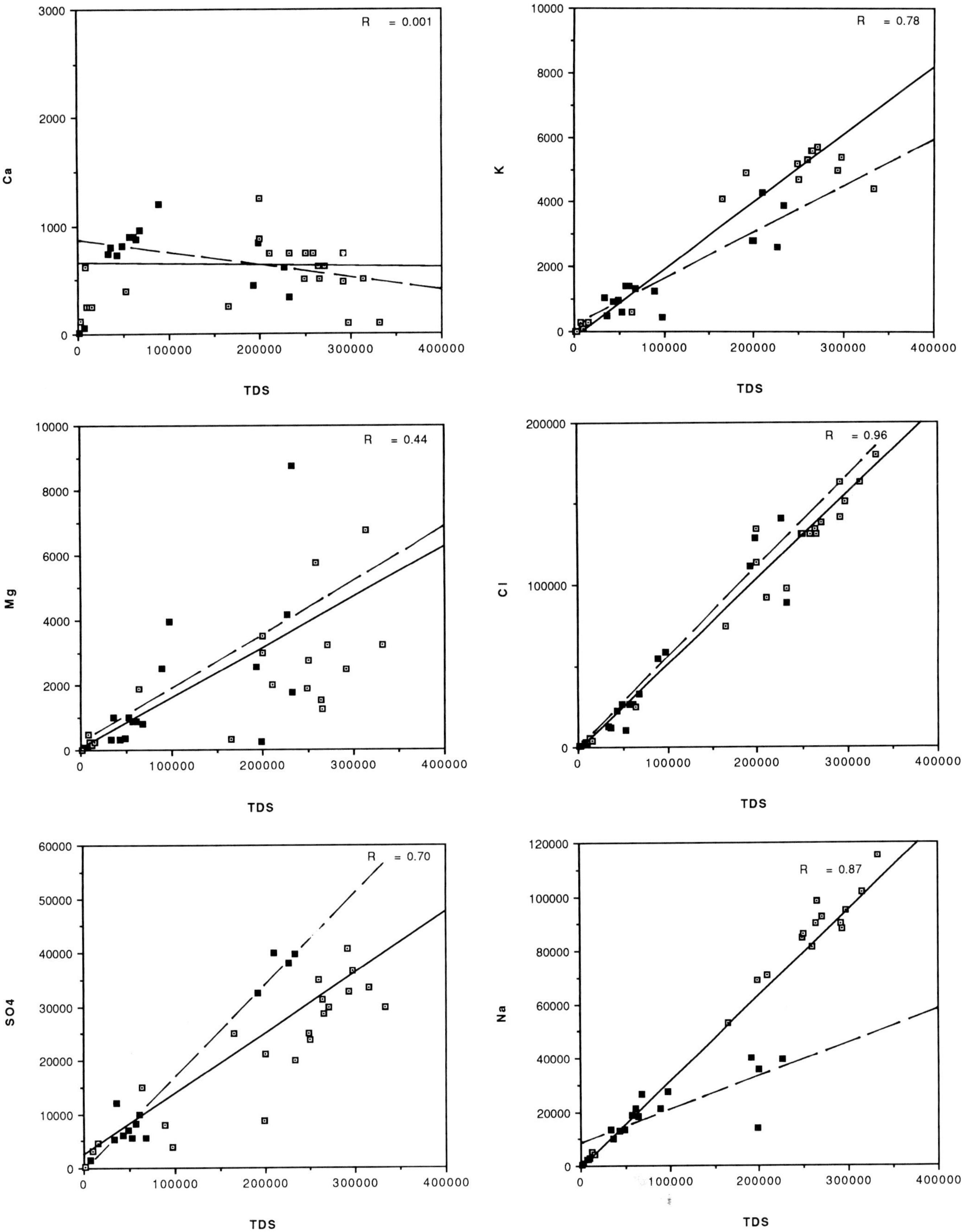

FIG. 3.—Relationship between salinity (expressed as Total Dissolved Solids: TDS ppm) and major-element concentrations for surface water samples collected from Calatta Lake during 1988–1991. Samples shown by full boxes are those following the June-July major rainfall event.

TABLE 1.—MAJOR ELEMENT CHEMISTRY OF CALATTA LAKE BRINES FOR THE PERIOD 1988–1992

Year	Parameter	Ca	Mg	Na	K	Cl	SO_4	$CO_3 + HCO_3$	Dominant brine
1988	Range:	102–481	2,481–9,070	88,000–115,000	4,400–5,400	150,850–180,070	29,883–36,725	16–169	Na-Cl-(SO_4)
(n = 18)	Mean:	228	4,931	99,334	4,933	164,805	33,136	106	Na-sulfate subtype
1989	Range:	90–2,800	600–8,777	4,910–108,790	190–8,343	5,587–220,667	2,270–53,781	55–1,176	As above
(n = 9)	Mean:	732	1,647	30,356	1,978	59,656	16,250	245	
1990	Range:	126–1,250	880–8,750	6,125–101,875	200–9,000	4,430–163,090	1,050–40,750	85–1,048	As above
(n = 12)	Mean:	610	2,413	59,277	3,767	85,750	18,580	348	
1991	Range:	94–887	1,215–3,108	88,812–121,600	887–7,807	11,250–178,850	2,262–38,334	101–987	As above
(n = 7)	Mean:	603	2,840	109,150	3,018	149,812	27,625	169	

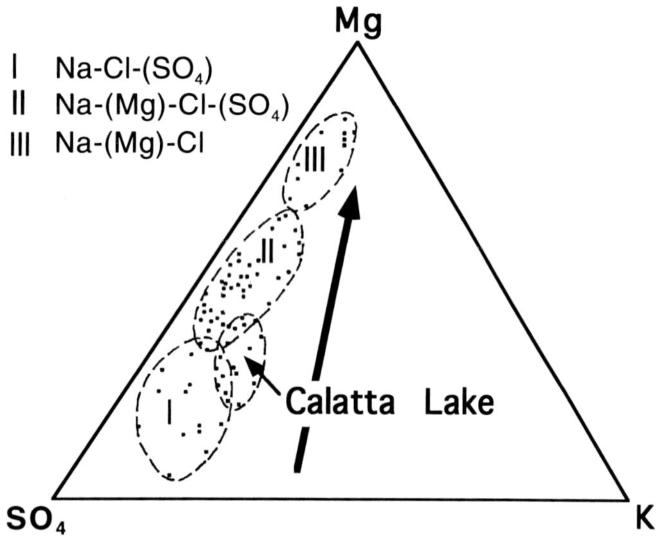

FIG. 4.—Ternary diagram showing spatial variation in chemistry of shallow groundwater solutions in the Karinga Creek drainage system, including Calatta Lake. Groups I and II represent brines in the western domain; Group III mainly represent the eastern domain.

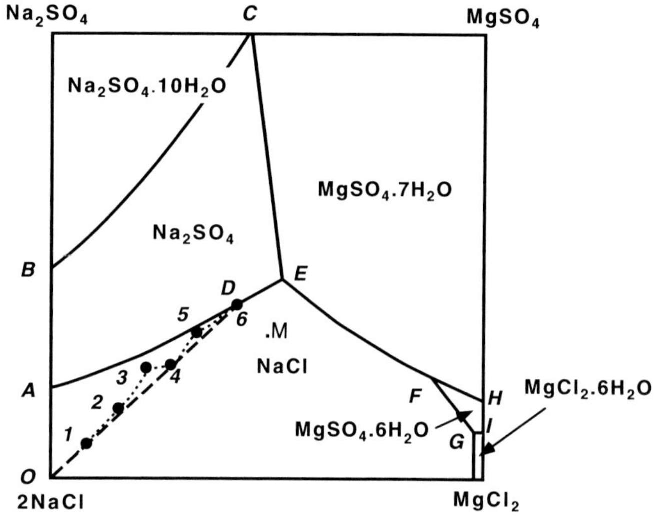

FIG. 5.—Evaporation path of a representative brine sample from the Calatta Lake in the Na-Mg-Cl-SO_4 system at 25°C, metastable equilibrium conditions.

tates were removed from residual solutions after completion of each main crystallization phase, thus a phase distribution in metastable equilibrium is used in preference to Janeke phase diagrams, which are commonly used to express the stable equilibrium system.

The composition of the liquid for single salt crystallization (NaCl-saturated) up to the thenardite stage in the metastable equilibrium system for Na-Mg-Cl-SO_4 at 25°C is shown in Figure 5. The measuring points, 1 through 6, follow the line OD (the theoretical evaporation path) until point D, where thenardite (Na_2SO_4) begins to crystallize. The position of groundwater solutions at point of discharge from the mouth of Calatta Springs is shown as point M, which at 25°C lies within the NaCl saturation field, whereas it falls within the mirabilite ($Na_2SO_4 \cdot 10H_2O$) saturation field at 0°C.

The experimental results show that by progressive evaporation, halite, and by cooling mirabilite, would be produced from Calatta Lake brine. With further evaporation, precipitation would move into the epsomite ($MgSO_4 \cdot 7H_2O$) crystallization field at 0°C, then into the schoenite ($K_2Mg(SO_4)_2 \cdot 6H_2O$) field.

Evaporite Sedimentation at Calatta Lake

Calatta Lake occupies an elongate low terrain confined between calcretes and alluvial-colluvial deposits (Fig. 2). Devonian shales of the Amadeus Basin (Edgoose, 1992; Arakel and Wakelin-King, 1991) occur on the margins and bottom of the lake, at shallow depths (Fig. 6). The shale outcrops on the western and northwestern lake margins are overlain unconformably either by massive bedded siliceous calcretes, or in places by bedded gypsite. The Quaternary bedded gypsite deposits are up to 4.5 m thick (Arakel and Cohen, 1991).

The playa facies of Calatta Lake are relatively thin and simple, and represented by a bright reddish to olive brown silty to sandy clay, 10 cm thick, that grades downward into a massive to weakly bedded, dark brown to red, clay unit (Fig. 6). The latter may be an alteration product of the Devonian bedrock (Edgoose, 1992). The overlying silty to sandy clay unit hosts several evaporite minerals, including halite, gypsum, glauberite, mirabilite, thenardite, epsomite, and polyhalite. Gypsum plus glauberite is the main evaporite assemblage in the playa lakes. The larger (>0.8 mm) gypsum crystals are mostly hemipyramidal and inclusion free; smaller discoid crystals fill interstitial pores between the larger crystals. The fine-grained glauberite ($Na_2Ca(SO_4)_2$)

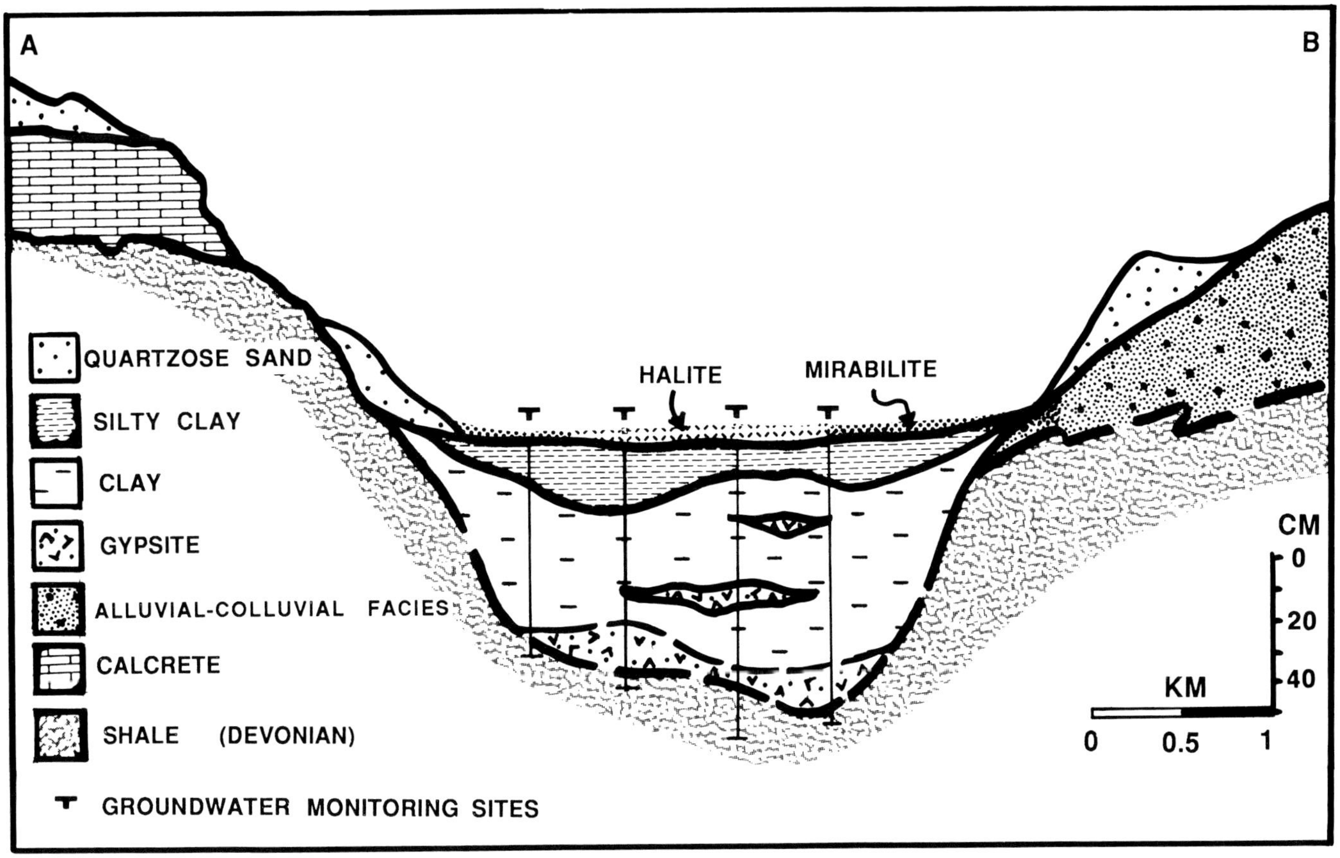

FIG. 6.—Cross-sectional view of Calatta Lake showing lithofacies distribution and location of cores and groundwater monitoring stations.

crystal assemblages found in the groundwater fluctuation zone are characteristically subhedral, 0.3 mm or less wide, and include textural and fabric features indicating replacement of a gypsum host (Arakel and Cohen, 1991). The standing groundwater level in the central part of the playa is from 5 to 25 cm below the surface, commonly marking the contact between the surficial silty to sandy clays and the underlying red clay unit. Single, large euhedral crystals of gypsum and glauberite, up to 1.5 cm wide, indicate the groundwater fluctuation zone.

Apart from halite, the presence of other evaporite minerals in efflorescent crusts depends largely on the extent of variations in groundwater discharge patterns and climate. Topography and brine chemistry play an overriding role in defining the type and distribution of evaporite mineral facies. During winter, mirabilite is dominant in most crusts, especially in playa-marginal areas that receive groundwater discharge from neighboring saline aquifers. Mirabilite crusts up to 15 cm thick are found in ponded areas of the playas (Fig. 7). Toward the playa centers, mirabilite grades laterally into a halite crust, several decimeters thick. The coexistence of mirabilite and halite is, however, short-lived because of the prevailing desert climate. The extreme diurnal and temporal variations in temperature and wind conditions during winter are reflected in efflorescent crusts that are subject to repeated dissolution, recrystallization and dehydration. These processes, together with continuous inflow of dilute groundwater from calcrete and dune aquifers, can result rapidly in a highly eroded crust composed of a white, powdery thenardite surrounding the playa that grades into a hard halite crust in the playa center. The dehydration of mirabilite to thenardite demonstrates the effectiveness of large temperature variations and eolian reworking during winter. The final product is commonly a white powder that eventually is dispersed by wind.

Field observations and thermodynamic criteria indicate the significance of temperature changes, both seasonally and daily, on the development and modification of efflorescent crusts. The greatest alterations occur during the two to four months of winter, especially when a significant rainfall event is followed by alternating daytime evaporation and nocturnal cooling of brines. The only significant variation in normal crust development was recorded following major rains in March and April 1989. Prismatic mirabilite crystallized across Calatta Lake from receding brines at temperatures below 2°C. By early August 1989, the playa surface was covered by a nearly continuous mirabilite crust with prismatic crystals up to 10 cm thick. In early September, daytime temperatures rose to more than 18°C, resulting in mirabilite dehydration to form thenardite pseudomorphs. We observed fibrous, radiating mirabilite associated with glauberite in vadose micropores and desiccation cracks of the silty mud below efflorescent crusts (Arakel and Cohen, 1991). Epsomite was found in small lenses and patches in

FIG. 7.—Close-up view of seasonal mirabilite, forming the bulk mineralogy of the efflorescent crusts on the playa-marginal flats, Calatta Lake system. The elongate mirabilite crystals (up to 12 cm long) occasionally form rosette-shaped assemblages that progressively undergo dehydration to powdery thenardite. Field of view is 50 cm wide.

the silty mud. By late September, the thenardite crust transformed to a loose, powdery layer while halite began to crystallize in the central playa.

Since 1989, the study area has experienced below average rainfall. Seasonal precipitation of thenardite in Calatta and most other lakes has mostly been localized near groundwater discharge areas. In winter 1991, even though surface brines were saturated with respect to mirabilite, no mirabilite crust formed in Calatta and neighboring lakes. Most mirabilite crystallization occurred in vadose playa sediments. X-ray diffraction of mirabilite-bearing sediments revealed minor diagenetic replacements by bloedite and epsomite. This attests to the relative geochemical stability of the playa waters, compared with the complex diagenetic alteration of evaporitic sediments in playa margins associated with groundwater level fluctuations (Arakel and Cohen, 1991).

DISCUSSION

Despite regional variations in the hydrochemistry of the eastern and western domains, each playa lake represents an independent hydrologic unit in which inflowing groundwaters are continuously concentrated by evaporation until reaching saturation with respect to halite. Brines in playa lakes, surrounded and/or underlain by Quaternary gypsite, commonly have high SO_4^{2-} concentrations. In contrast, Ca^{2+} and HCO_3^- are largely depleted due to early carbonate mineral precipitation. During winter, such brines easily acquire the thermodynamic requirements for mirabilite crystallization, notably sub-zero nocturnal temperatures. At Calatta Lake, mirabilite crystallization begins at sub-zero temperatures. With continuity of freezing temperatures, mirabilite precipitation extends from the marginal playa mudflats to the center, ultimately producing a continuous crust that incorporates earlier halite crusts in the central playa.

Minor variations in mineralogic associations and sequences of precipitation are found from lake to lake. The overall mineralogic sequence in the crusts, however, is in accord with results from the experimental studies, showing that precipitation of temperature-dependent sulfate minerals proceeds independently of halite crystallization in the ponded brines. Thus, despite a degree of commonality in brine sources and geochemical processes, the observed mineralogic zonation in efflorescent crusts mainly reflects the stages in thermodynamic equilibrium reached by the brines at different temperatures.

At Calatta Lake, which typifies most lakes in the region, the effects of groundwater seepage from the calcrete and dune aquifers are best seen in lake-marginal mudflats, where the gypsite facies exhibits significant mineralogic and textural modifications due to repeated wetting and drying (Arakel and Cohen, 1991). There, the salinity of shallow groundwater varies from 100–120 g/l TDS. Within the lake, both surface and subsurface brines reach salinities exceeding 200 g/l TDS, and their sequence of evaporite mineral precipitation depends largely on hydrochemical setting and climate.

Previous studies show that variations in groundwater chemistry in the study area are essentially a reflection of the cumulative effects of evaporation, sediment-water interactions, and mixing of brines, rather than the consequence of simple evaporative concentration (cf. Arakel and Cohen, 1991). Accordingly, in playa margins, most HCO_3^- + CO_3^{2-} and Ca^{2+} are removed from groundwater recharge solutions as carbonates (calcite, magnesite, dolomite) and sulfate minerals (gypsum, glauberite). However, SO_4^{2-} remains highly concentrated in both inflowing groundwater and playa brines, demonstrating influx from sources other than simple evaporative concentration. Concentrations of SO_4^{2-} up to 12 times, Mg^{2+} about four times, and Na^+, K^+ and Cl^- varying from seven to nine times that of seawater are common in the study area. The molar ratios, Mg/K and Mg/SO_4, for these brines fall between 0.8 and 4. Thermodynamically, the brines follow a seasonal mineral precipitation sequence seen in crusts of the Karinga Creek playa lakes (Fig. 8). Field monitoring shows that such salinity and temperature ranges are attained in the study area, both diurnally and seasonally.

In summary, the sedimentary and chemical signatures that characterize efflorescent crusts in the Karinga Creek drainage point to the role of temperature variations in triggering

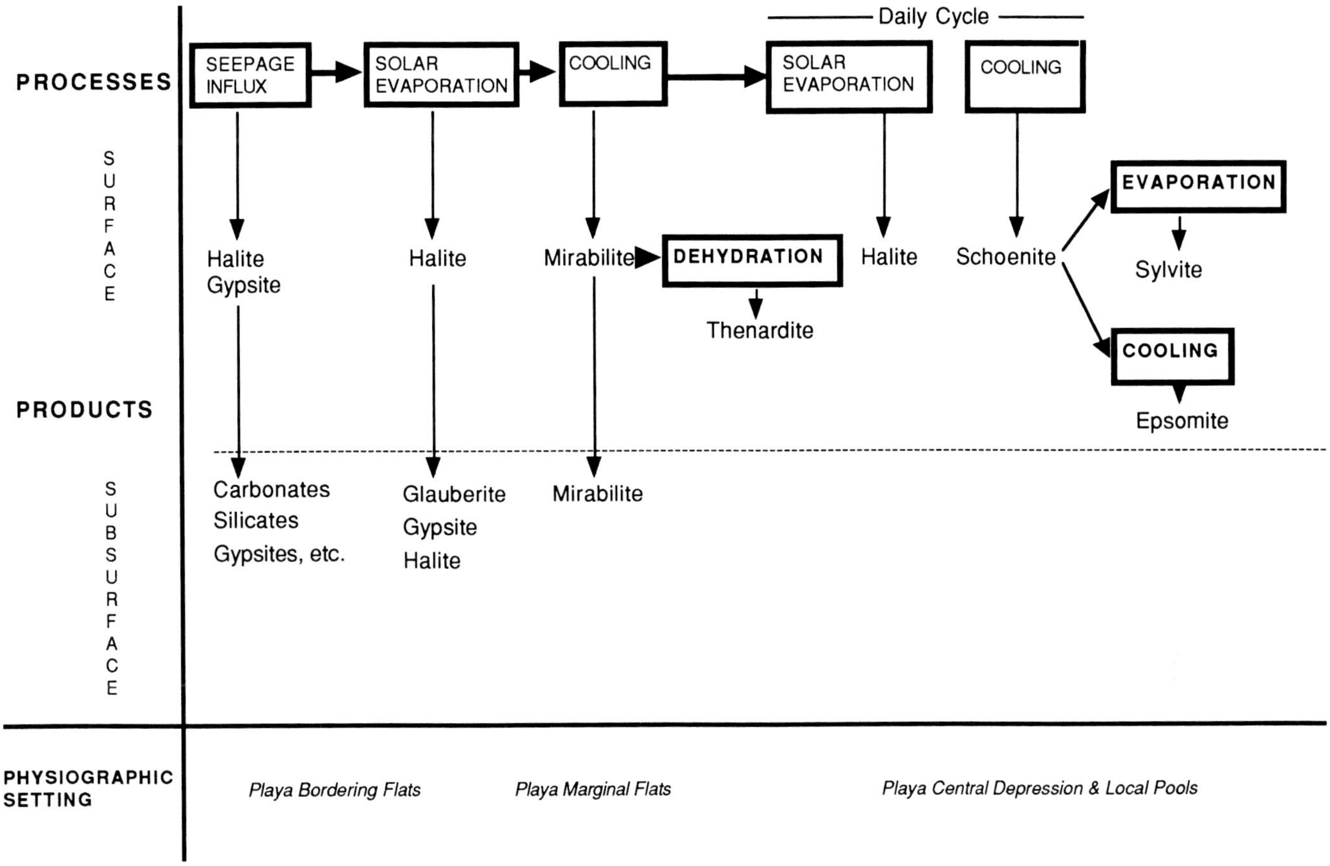

FIG. 8.—Simplified chart showing the evaporite precipitation sequence in Calatta Lake during a full hydrologic cycle. The recurrence of the mineralogic sequence largely depends on groundwater discharge pattern and prevailing climatic conditions. It is assumed that the brines have a salinity of 250–270 g/l TDS on entering the playa and temperatures of −2°C to 40°C.

precipitation of winter salts. During winter, repeated evaporation and cooling of surficial brines results in an evaporite mineralization pattern that, if preserved, shows lateral continuity with halite precipitated independently of ponded brines. The coexistence of halite and mirabilite in a single continuous crust that is progressively modified by the desert climate, clearly demonstrates playa evaporite sedimentation that differs significantly from that proposed previously for Australian arid-zone basins. Although climatic conditions and mineralogic associations changes from year to year, monitoring the lakes over four years indicates that most changes have a short-lived influence on the overall evolution of the brines, as shown by the recurrence of halite-mirabilite crusts in the Karinga Creek system.

When studying ancient lacustrine evaporites, problems may arise in assessing the role of climatic variability in chemical sedimentation—particularly where rapid lateral evaporite facies changes within an individual depositional unit cannot be correlated with physiographic boundaries (Dronkert, 1985). Halite and mirabilite are not unequivocal indicators of lacustrine evaporites, but their coexistence in the Karinga Creek playa lakes indicates periodic precipitation under repeated evaporation and cooling, a feature reflecting desert conditions. Small-scale evaporite facies associations, especially the thermodynamically incompatible mineral groups of halide, glauber and epsom salts, may provide useful information for interpreting depositional processes and paleoclimate (Tucker, 1978; Orti Cabo and Salvany, 1990).

CONCLUSIONS

Field observations of halite-thenardite efflorescent crusts in playa lakes of the Karinga Creek drainage system, together with an assessment of hydrogeochemical processes that are conducive to such facies relationships, demonstrate the effectiveness of brine recycling through seasonal mineral crystallization, phase transformations, and dissolution-precipitation under an arid climate.

The occurrences and distribution pattern of evaporite facies in the playa efflorescent crusts are primarily a reflection of variations in groundwater discharge, topography and brine chemistry. However, diurnal and seasonal temperature variations play an overriding role in defining the type and sequence of evaporite facies development. The observations agree with results from experimental studies that show that precipitation of temperature-dependent sulfate minerals may proceed independently of halite crystallization from ponded brines. Therefore, despite an apparent commonality in brine sources and geochemical processes, the mineralogic zonation observed in playa efflorescent crusts is most likely a consequence of significant variations in thermodynamic equilibrium under changing temperature conditions, rather than evaporative concentration and crystallization from ponded brines. Facies associations of thermodynamically incompatible evaporite mineral groups in ancient lacustrine sequences may provide useful information for reconstructing both depositional setting and paleoclimate.

ACKNOWLEDGMENTS

This study was supported by the Australian Research Council and Earthwatch, USA. We particularly thank Earthwatch volunteers for support with field monitoring and sampling. We thank the Builder and Kilgariff families for their highly valuable field support and "Territorian" hospitality. The support of Shirley Pickering and reviewers of the manuscript are gratefully acknowledged.

REFERENCES

ARAKEL, A. V., 1991, Evolution of Quaternary duricrusts in Karinga Creek drainage system, central Australian groundwater discharge zone: Australian Journal of Earth Sciences, v. 38, p. 333–347.

ARAKEL, A. V., AND COHEN, A., 1991, Deposition and early diagenesis of playa glauberite in the Karinga Creek drainage system, Northern Territory, Australia: Sedimentary Geology, v. 70, p. 41–59.

ARAKEL, A. V., JACOBSON, G., AND LYONS, W. B., 1990, Sediment-water interaction as a control on geochemical evolution to playa lake system in the Australian arid interior: Hydrobiologia, v. 197, p. 1–12.

ARAKEL, A. V., JACOBSON, G., SALEHI, M., AND HILL, C., 1988, Silicification of calcrete in inland drainage basins of the Australian arid zone: Australian Journal of Earth Sciences, v. 36, p. 73–89.

ARAKEL, A. V., AND WAKELIN-KING, G. A., 1991, Chemical sedimentation and facies relationships in the Karinga Creek drainage system, Amadeus Basin, Northern Territory, *in* Korsch, R. J. K., and Kennard, J. M., eds., Amadeus Basin: Canberra, Bulletin of the Bureau of Mineral Resources, Australia, 236, p. 581–589.

BARRY, G. S., 1986, Sodium sulphate: Canadian Mining Journal, v. 107, p. 77.

BOWLER, J. M., 1986, Aridity in Australia: age origins and expression in aeolian landforms and sediments: Earth Science Reviews, v. 12, p. 279–310.

CHEN, X. Y., 1989, Lake Amadeus, Central Australia: Modern processes and evolution: Unpublished Ph.D. Thesis, Australian National University, Canberra, 387 p.

DE LA PEÑA, J. A., GARCIA, J. M., MARFIL, R., AND PRIETO, M., 1982, Growth features of magnesium and sodium salts in a recent playa lake of La Mancha (Spain): Estudios Geológicos, v. 38, p. 245–257.

DRONKERT, H., 1985, Evaporite models and sedimentology of Messinian and recent evaporites: Amsterdam, GUA Papers in Geology, Series 1, v. 24, 283 p.

EDGOOSE, C., 1992, Kulgera SG 53–05: Explanatory Notes: Darwin, 1:250,000 Geological Map Series, Northern Territories Geological Survey.

HARDIE, L. A., 1984, Evaporites: marine or non-marine?: American Journal of Science, v. 284, p. 193–240.

JACOBSON, G., 1988, The hydrology of Lake Amadeus, a groundwater discharge playa in central Australia: Journal of Australian Geology and Geophysics, v. 10, p. 301–308.

JACOBSON, G., ARAKEL, A. V., AND CHEN, X. Y., 1988, The central Australian groundwater discharge zone: evolution of associated calcrete and gypcrete deposits: Australian Journal of Earth Sciences, v. 35, p. 549–565.

LAST, W. M., AND SLEZAK, L., 1986, Sodium sulphate deposits of Western Canada: geology, mineralogy, and origin, *in* Gilboy, C. F., and Vigrass, L. F., eds., Economic Minerals of Saskatchewan: Regina, Special Publication of the Saskatchewan Geological Society 8, p. 197–205.

MACUMBER, P. G., 1979, The influence of groundwater discharge on the Mallee landscape, *in* Storrier, R. R., and Stannard, M. E., eds., Aeolian Landscapes in the Semi-arid Zone of Southeastern Australia: Canberra, Riverine Branch, Australian Society of Soil Science, p. 76–85.

ORTI CABO, F., AND SALVANY, J. M., 1990, Formaciones Evaporíticas de la Cuenca del Ebro y Cadenas Periféricas, y de la Zona de Levante. Nuevas Aportaciones y Guía de Superficie: Barcelona, ENRESA-GPPG, Universidad de Barcelona, 307 p.

ORTI CABO, F., SALVANY, J. M., QUADRAS, M., AND PUEYO, J. J, 1986, Glauberite growth in playa-lake environments (Tertiary of Ebro Basin, Alcanadre-Arrubal zone) (abs.): Granada, Abstracts, International Symposium on Crystal Growth Processes in Sedimentary Environments, p. 142–143.

TELLER, J. T., BOWLER, J. M., AND MACUMBER, P. G., 1982, Modern sedimentation and hydrology in Lake Tyrrell, Victoria: Journal of the Geological Society of Australia, v. 29, p. 159–175.

TUCKER, M. E., 1978, Triassic lacustrine sediments from South Wales: shore-zone clastics, evaporites and carbonates, *in* Matter, A., and Tucker, M. E., eds., Modern and Ancient Lake Sediments: Oxford, International Association of Sedimentologists Special Publication, v. 2, Blackwell, p. 203–222.

WARREN, J. K., 1989, Evaporite Sedimentology: New Jersey, Prentice Hall, 285 p.

WILLIAMS, W. D., 1984, Australian lakes, *in* Taub, F. B., ed., Lakes and Reservoirs: Amsterdam, Elsevier, p. 499–518.

LAKE BOGORIA, KENYA RIFT VALLEY—A SEDIMENTOLOGICAL OVERVIEW

ROBIN W. RENAUT
Department of Geological Sciences, University of Saskatchewan, Saskatoon, Saskatchewan, Canada S7N OWO
AND
JEAN-JACQUES TIERCELIN
GDR 910 "Genèse et Evolution des Domaines Océaniques," Université de Bretagne Occidentale, F-29287 Brest, France

ABSTRACT: Lake Bogoria is a saline (100 g l^{-1} TDS), alkaline (pH- 10.3) meromictic lake that lies in an asymmetric half-graben in the semi-arid central Kenya Rift. The lake is fed by about 200 hot springs and ephemeral streams draining a volcanic catchment. Despite its shallow depth (11 m), the brine is stratified and stable meromixis prevails. Sedimentation along the border fault is dominated by major slope failures, many earthquake-induced, and youthful fan deltas. Small deltas on the ramp and axial platforms are slowly prograding lakeward. Within the lake, the littoral zone is dominated by siliciclastic sediments, many deposited by density flows. In the deeper, central part of the lake, sedimentation is mainly chemical-organic, with some detrital fines introduced as hypopycnal plumes. Cores spanning the last 30,000 years reveal organic muds, which formed during periods of intermediate to high lake levels and high microbial productivity, interbedded with evaporites (trona, nahcolite) that were precipitated during arid phases of low lake level. A large suite of authigenic minerals, including sodium silicates and zeolites, is present in the lake sediments. The evidence from Lake Bogoria demonstrates the complex interplay between climatic change and tectonics in a young rift lake, and confirms the important role played by hydrothermal springs in sedimentation in continental rifts.

INTRODUCTION

Saline lakes are common in the drier parts of the East African Rift, mainly in the volcanic regions of Ethiopia, Kenya and Tanzania. The overall rift structure is one of segments, characterized by alternating series of asymmetric half-grabens up to about 100 km long, linked by various transverse structures that commonly follow the fabric of pre-rift rocks (Rosendahl and others, 1986; Rosendahl, 1987; Daly and others, 1989). This architecture has produced numerous morphological barriers that, together with volcanic dams (central volcanoes and lava flows), intercept and modify axial and lateral drainage, contributing to the formation of many closed hydrological basins.

Except Lake Asal in Djibouti, which is fed partly by springs derived from seawater and has a Na-Ca-SO_4-Cl brine, waters in all the large saline rift lakes in East Africa are dominated by Na^+, CO_3^{2-} and HCO_3^- ions. Evaporites are common, including the 40-m thick trona deposits at Lake Magadi, Kenya (Eugster, 1980). All these lakes had more dilute precursors during the last 30,000 years (Street and Grove, 1979). Ancient saline, alkaline lakes (e.g., the Tambach, Ngorora and Kapthurin Formations west of Lake Baringo) have existed in the Kenya Rift since the Miocene epoch (Renaut and Owen, 1987; Renaut and Ego, unpublished).

Lake Bogoria is a perennial saline, alkaline lake in the central Kenya Rift Valley (Fig. 1). Its waters are stratified, with less dense saline waters lying upon a denser, anoxic brine. Similar meromixis has been inferred for many ancient saline, alkaline lakes including the precursors of Lake Magadi (Eugster, 1980, 1986; Damnati and others, 1992) and parts of the Eocene Green River Formation in the western United States (e.g., Desborough, 1978; Buchheim, this volume). Periodically, Lake Bogoria has very high microbial productivity with deposition of sapropelic muds below the chemocline. In contrast, sodium carbonate-bicarbonate evaporites precipitated during former drier climatic phases, together with a diverse authigenic mineral suite, including magadiite and many zeolites. Lake Bogoria may, therefore, provide a modern analog to help interpret aspects of many ancient saline lacustrine deposits, notably those where organic-rich shales are interbedded with trona and nahcolite.

During the late 1970's, Lake Bogoria was cored extensively (Tiercelin, 1981; Tiercelin and others, 1980, 1981, 1982). Concurrently, studies were made of the exposed sediments, hydrochemistry and hot springs (Young and Renaut, 1979; Renaut, 1982; Renaut and others, 1986). Further studies of the sedimentology, mineralogy, geochemistry, paleontology and palynology of the cores continued in the 1980's and are documented in a monograph edited by Tiercelin and Vincens (1987).

The purpose of this paper is: (1) to summarize the main depositional environments and facies in the Lake Bogoria basin and (2) to discuss some consequences of late Quaternary environmental changes on sedimentation in a meromictic saline, alkaline rift lake.

ENVIRONMENTAL SETTING

Geology

Lake Bogoria lies in a structurally complex part of the axial trough of the central Kenya Rift Valley between 0°11'N and 0°20'N (Figs. 1, 2). The lake occupies a narrow, strongly asymmetric, north-south trending half-graben, bounded to the east by 700-m high fault escarpments (Bogoria-Emsos escarpments; Figs. 3A, B), and on the ramp margin of the rift floor to the west by the northward-dipping Lake Bogoria Plateau, composed of closely grid-faulted Pleistocene lavas (Fig. 1B). The Marmanet and Laikipia Escarpments rise a further 700 m, about 10 km east of the Bogoria-Emsos Escarpment, to form the rift margin. The Bogoria half-graben formed during a Middle Pleistocene phase of volcano-tectonic activity.

Lying at about 20-m lower elevation, 22 km north of Lake Bogoria, Lake Baringo is a shallow (4 m deep) fresh lake with mainly siliciclastic sediments. The two lakes are separated by the Loboi Plain, a locally swampy alluvial surface underlain by up to 1 km of fluvial, deltaic and lacustrine sediments (Swain and others, 1981). The Loboi Plain may be the surface expression of a subsiding rhomb-graben

Sedimentology and Geochemistry of Modern and Ancient Saline Lakes, SEPM Special Publication No. 50

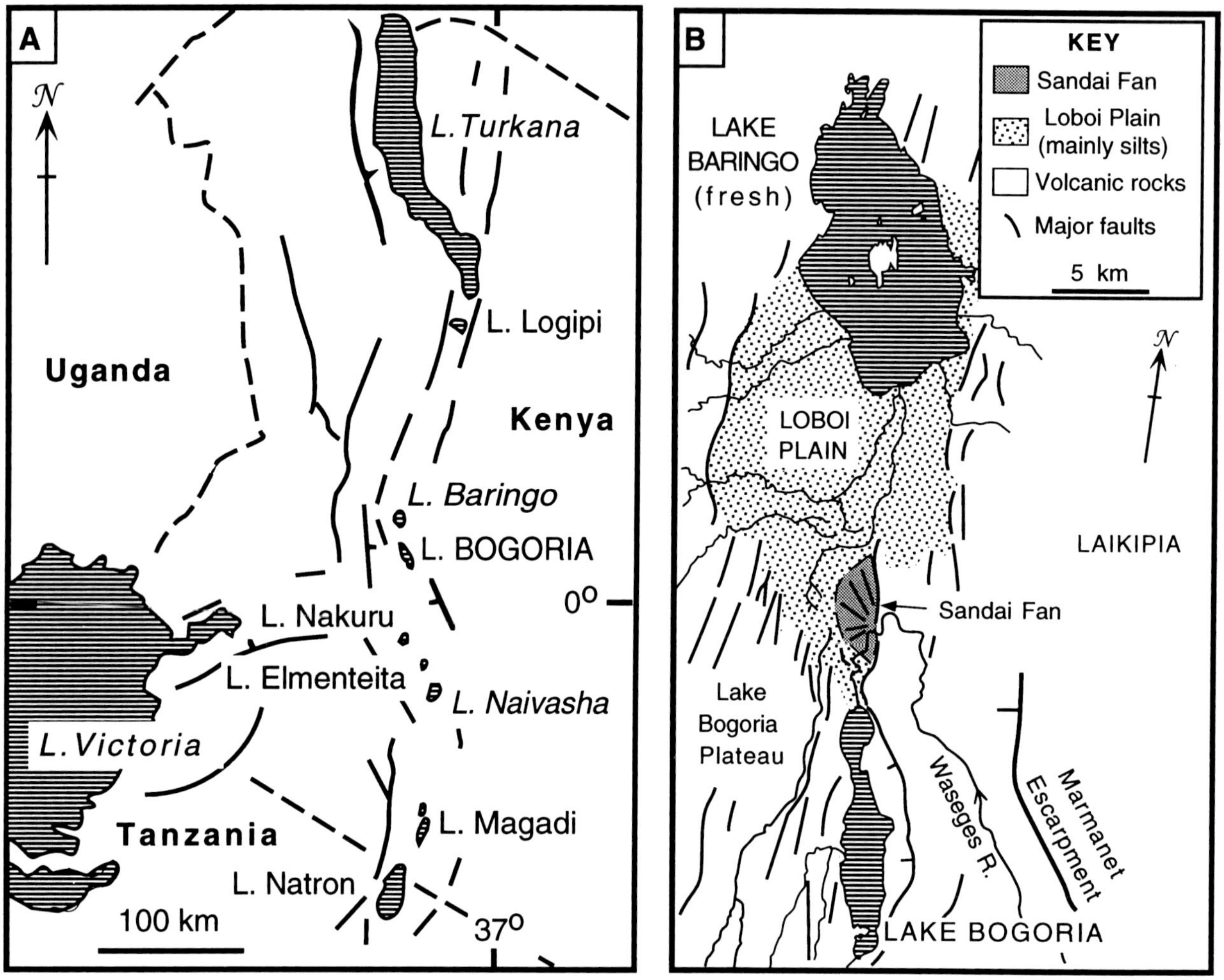

FIG. 1.—Location and setting of Lake Bogoria. (A) The Kenya Rift Valley, showing the major faults and main lakes. Fresh and subsaline (<5 g l^{-1} TDS) lakes are shown in italics; other lakes shown are saline and alkaline. (B) Lakes Bogoria, Baringo and the Loboi Plain, with location of Sandai Fan.

(Fig. 1B; Tiercelin and others, 1981). The drainage basins and morphology of both lakes are strongly controlled by three regional tectonic lineaments: North 0–20°, North 50°, and North 150° (Tiercelin and others, 1980; Richert and others, 1987).

Volcanic rocks and their weathering products constitute most of the catchment (McCall, 1967; Griffiths, 1977; Hackman, 1988). East of Lake Bogoria, bedrock is mostly basalts, trachytes and phonolites of Miocene and Pliocene age; the plateau to the west is mainly Pleistocene (1 Ma) trachyphonolites and basalts. The youngest lavas (0.3 Ma) were erupted mainly around the southern margins of the lake, including the Mwanasis peninsula (Fig. 2), an area of major geothermal activity.

The region remains seismically active (Young and others, 1991). Many historic earthquakes have been recorded locally, including a magnitude 6.9 event in 1928 (Ambraseys, 1991) that led to fissuring, slope failures, drainage diversion and changes in hydrothermal discharge. The lake reputedly became black from disturbed anoxic bottom muds (McCall, 1967). Seismites, including chaotic mud and sand layers mixed when in a thixotropic state, are locally interbedded within undisturbed Holocene fan-delta sediments east of the lake.

Drainage Basin

Lake Bogoria lies in a topographically, and probably hydrologically, closed drainage basin (700 km^2). McCall (1967) suggested that some lake-derived saline groundwater may flow northward, but evidence is lacking. The lake is recharged by rainfall, ephemeral streams, unchannelled runoff and hot springs. The Waseges River, termed the Sandai River in its lower course, drains 70% of the catchment and is the main affluent, discharging into the lake at the Sandai Delta (Fig. 2). During wetter years (e.g., 1978), inflow is perennial; in dry years (e.g., 1976), the river dries up before reaching the lake. The lake is also fed by about 100 minor ephemeral streams that only flow intermittently after heavy rains, of which the Emsos and Parkirichai are most important. Several ephemeral springs with cool (7–10° C), dilute waters discharge lakeward near the base of the alluvial fans east of the lake.

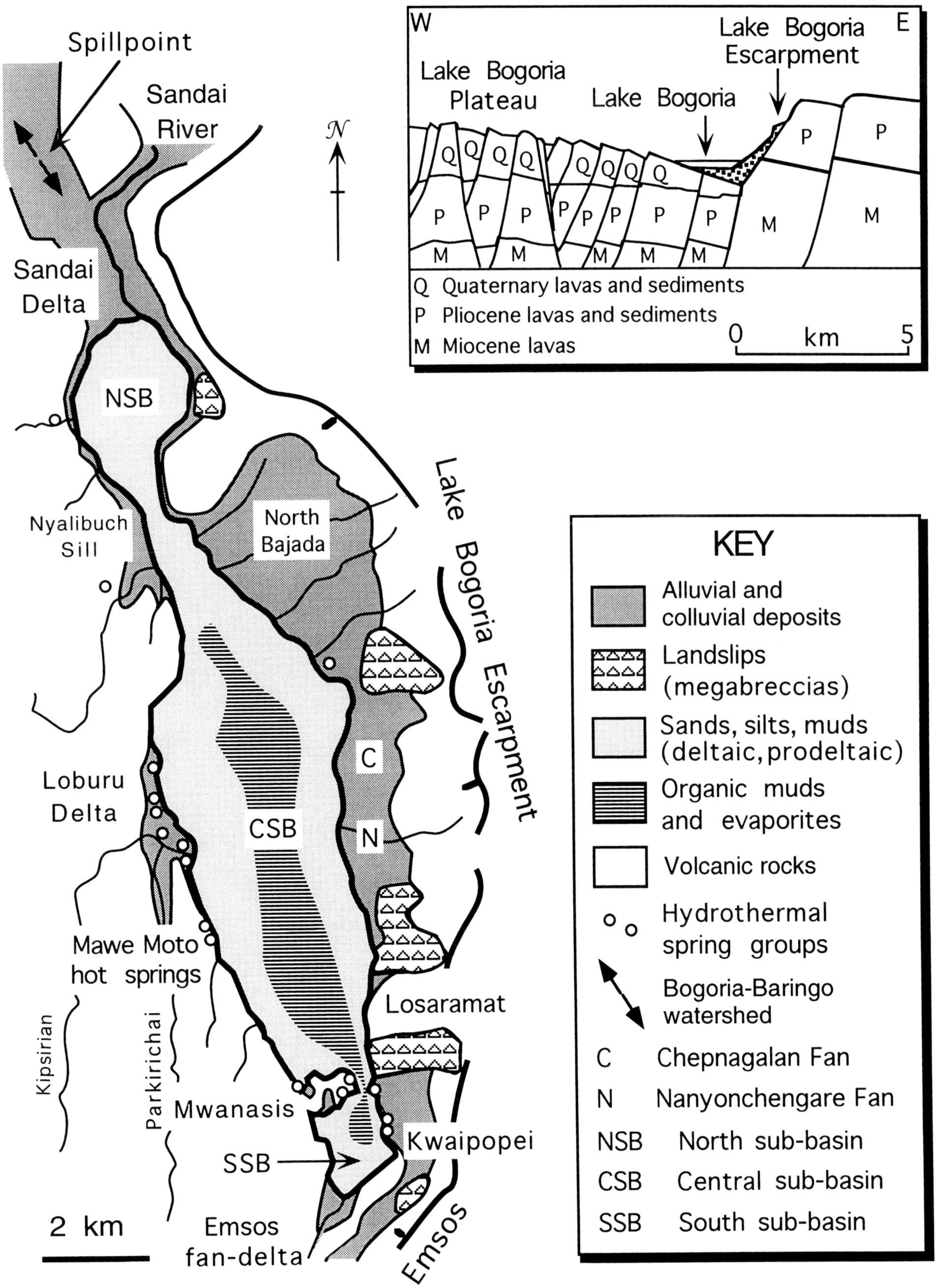

FIG. 2.—Sedimentary depositional environments of the Lake Bogoria half-graben. Inset shows a generalized west to east geological cross-section through the southern half of the central sub-basin (modified from Naylor, 1972 and Cioni and others, 1992).

FIG. 3.—Lake Bogoria. (A) The north sub-basin looking southeast toward the Lake Bogoria Escarpment; (B) The densely vegetated North Bajada and Chepnagalan alluvial fans below the Lake Bogoria Escarpment. Strong, late afternoon winds are moving saline water across the shoreline of Loburu Delta, inundating a pit (left) dug in the sediments an hour previously; (C) Perpetual spouting hot spring (Spring KL8a of Renaut, 1982) with geyserlike cyclicity at Loburu geothermal site. The mound draped in microbial mats is composed of fossil carbonates precipitated from former, less saline, spring waters; and (D) View across the lake from central Loburu Delta; hot springs discharge lakeward through *Cyperus laevigatus* marsh in the foreground.

At least 200 hot springs and a few small geysers discharge into Lake Bogoria (Figs. 3C and D; Renaut, 1982; Renaut and others, 1987; Cioni and others, 1992; Jones and Renaut, unpublished). Three hot-spring groups (64–99° C) lie along the shoreline, at Loburu, Mawe Moto, and in the southern sub-basin (Mwanasis-Kwaipopei-Losaramat Groups; Fig. 2). Sublacustrine hydrothermal springs have been confirmed by lake-bottom temperature surveys (Naylor, 1972) and sediments (Renaut and Owen, 1988).

Climate and Vegetation

Annual precipitation typically ranges from 600 to 900 mm yr^{-1} on the rift floor, increasing to more than 1,000 mm yr^{-1} on the adjacent escarpments. Most rain falls between March and August. Potential evaporation exceeds 2,500 mm^{-1}. The mean annual temperature is about 23–26° C, with little seasonal variation. Prevailing winds are from the east, but are funneled along the axis of the half-graben. Winds normally blow north to south by day, commonly reversing by night. The dominant vegetation in the catchment is *Acacia* savanna (Vincens, 1986). *Cyperus* marsh is found on the southern Loboi Plain and Loburu delta (Fig. 3D).

LIMNOLOGY AND HYDROCHEMISTRY

Morphometry and Lake Level

Lake Bogoria has a surface area of approximately 35 km^2 and consists of three sub-basins: north, central and south, respectively. The north and central sub-basins are separated by a shallow (<1.5 m), structurally controlled sill (Nyalibuch Sill in Fig. 2). The central and south sub-basins are separated by a narrow pass, about 5 m deep at the Mwanasis Peninsula. Maximum water depth (11.5 m in 1978) occurs in the southern central sub-basin.

Lake surface level fluctuates both seasonally and in response to changes in the precipitation/evaporation balance over a wide range of time scales. From 1970 until 1993, lake level fluctuated between 989 m and 993 m above sea level. Annual fluctuations are typically less than 1 m. After heavy rains in 1978, the lake rose nearly 3 m (to 993 m) above its 1976 level with a 30 percent increase in lake water volume (Tiercelin and others, 1980). By summer 1993, the lake had fallen close to its summer 1976 level (990 m).

Whenever the lake rises to its spillpoint, near Loboi village at 999 m elevation (Fig. 2), the waters will discharge northward toward Lake Baringo. The lake last approached this level (997 m) early this century (Johnston, 1902; Young and Renaut, 1979; Renaut and others, 1987). Long term fluctuations, of 10^2 to 10^3 years duration, are recorded in the late Quaternary sediments.

Hydrochemistry and Stratification

Representative analyses of the Bogoria basin waters are given in Table 1. The lake brine is of Na-CO_3-HCO_3-Cl composition. Recorded pH values have ranged from 9.3 to 10.6, averaging about 10.3 (Renaut and others, 1987). Surface water temperatures average 25–26° C. Bottom waters are slightly warmer (27° C), locally rising to about 60° C near sites of sublacustrine hot spring seepages.

In 1977, the salinity of the surface waters in the open lake exceeded 55 g l^{-1} TDS (Total Dissolved Solids), being lowest in the north sub-basin which receives the greatest dilute inflow. Although shallow, comparing surface and bottom waters at the same site (e.g., Site 85 in Table 1) shows clearly that the lake is chemically stratified or meromictic. The surface water (mixolimnion) lies upon denser anoxic brine (monimolimnion) with a salinity of up to 90 g l^{-1} TDS and a slightly lower pH. Several factors have likely contributed to meromixis, including sublacustrine saline spring seepage and dissolution of lake-floor evaporites. Although minor mixing may occur at the chemocline, the presence of shallow sills, particularly at Nyalibuch, limits exchange of bottom waters between the different sub-basins. The chemocline lies about 6 m below lake level in the central sub-basin, but its temporal vertical variability is unknown. Salinities exceeding 100 g l^{-1} TDS are found in shallow marginal lagoons. Cioni and others (1992) reported salinity of 135 g l^{-1} TDS during the mid-1980's.

The composition and evolution of the brine are broadly similar to that of the other saline lakes in volcanic sectors of the rift. The composition of the Lake Bogoria brine originates mainly from evaporative concentration, mineral precipitation and many other solute fractionation processes, but a detailed discussion is beyond the scope of this paper. After Na^+, K^+ is the subsidiary cation, while Ca^{2+} and Mg^{2+} are of low abundance. After the predominant carbonate-bicarbonate species, Cl^- is the subsidiary anion. Sulfate concentration is relatively low (<550 mg l^{-1}) and is exceeded by fluoride (up to 1,300 mg l^{-1}). Reported silica concentrations range from less than 100 to 1,210 mg l^{-1} (Baker, 1958; Renaut and others, 1987).

The lake is fed by dilute streams with a Na-Ca-HCO_3 composition, direct rainfall and about 200 hydrothermal springs. Cioni and others (1992) recognized two thermal aquifers at Bogoria. A shallow aquifer with Cl^- content of approximately 220 mg l^{-1} and temperature of about 100° C discharges at Loburu and Mawe Moto; a deeper reservoir with Cl^- content of 660 mg l^{-1} and a temperature of about 170° C underlies the southern hot spring group and con-

TABLE 1.—SELECTED CHEMICAL ANALYSES OF LAKE BOGORIA BASIN WATERS

Sample	Number	Date	pH	Temp. °C	Na^+ mg/l	K^+ mg/l	Ca^{2+} mg/l	Mg^{2+} mg/l	HCO_3^- mg/l	CO_3^{2-} mg/l	Cl^- mg/l	SO_4^{2-} mg/l	F^- mg/l	SiO_2 mg/l
Dilute inflow														
Sandai River	BW1	10–91	7.2	17	80	4.5	27	6.5	240	0	42	4.7	2.3	60
Parkirichai R. (Loburu)	KL1	8–77	7.5	19	101	6	26	8.0	306	0	46	3.0	2.6	45
Emsos River	EM1–77	8–77	7.4	22	84	3	21	7.0	306	0	46	3.0	4.2	61
Chepnagalan Fan	E 62 148	10–78	6.7	—	117	tr.	32	4.0	336	0	35	tr.	1.2	40
Hot springs														
N.W. Mwanasis	E64–023	10–77	7.10	42.0	161	15	15	5	500	0	18	tr.	—	175
Loburu	KL8–91	10–91	8.15	98.5	1480	28	1	tr.	2750	0	225	55	55	105
Loburu[1]	316	—	8.20	98.9	1270	26.2	1.60	0.97	2810	—	220	62.4	58.9	108
Mawe Moto	MM-4–91	10–91	8.75	98.0	1600	30	0.5	0	3480	0	305	58	65	130
Mawe Moto[1]	319	—	8.94	97.6	1590	34.7	0.54	0.33	3480	—	327	57.6	72.2	127
Kwaipopei[1]	219	—	9.42	96.3	4600	141	0.66	0.03	9760	—	1170	72.0	190	228
Lake waters														
Lake[2] (mix.)	K1	1–70	10.6	—	24400	387	0	2.2	5675	26165	6390	216	1064	260
Lake (mix.)	BW2	10–91	9.95	26.5	24600	435	12	<1	5500	24500	5680	330	110	225
Lake[1] (mix.)	321	—	10.10	27.3	39300	547	1.82	1.11	85400	—	7460	144	1350	378
N. sub. Site 68 (mix.)	E64 007	10–77	9.4	25.3	22700	432	24	0	1280†	23700	5850	222	nd	nd
N. sub. Site 68 (mon)	E64 010	10–77	9.3	27.2	41200	660	28	2	6100	43500	9400	340	nd	nd
C. sub. Site 85 (mix.)	E64 026	10–77	9.3	25.8	27800	428	24	0	0	31500	5850	434	nd	nd
C. sub. Site 85 (mon.)	E64 025	10–77	9.3	25.8	40900	644	34	2	6100	43500	9040	549	nd	nd
C. sub. Site 85 (mix.)	E62 181	10–78	10.2	33.5	10600	195	14	4	4270	8400	3900	68	2.5	nd
C. sub. Site 85 (mon.)	E62 182	10–78	10.1	28.9	26000	390	14	1	11900	19960	9570	300	2.5	nd
S. sub. Site 37 (mix.)	E64 016	10–77	9.4	27.8	26500	464	24	0	0	30000	5500	239	nd	nd
S. sub. Site 37 (mix.)	E62 173	10–78	10.2	nd.	10810	199	17	4	4030	8310	4430	175	3.0	nd
S. sub. Site 37 (mon.)	E62 174	10–78	10.1	nd.	22930	547	100	tr.	6410	18900	9220	354	<1	nd

nd.: not determined; tr.: trace; dashes indicate no data avilable; †: present as OH^-; mix: mixolimnion; mon: monimolimnion; N, C, S, sub.: north, central and south sub-basin.

[1]From Cioni and others (1992); [2]From Kilham (1971). Other analyses original or from Renaut and others (1987).

Note that several lake water analyses were taken at the same site to show difference in composition between surface and bottom waters.

tributes water to boiling springs at Mawe Moto. The salinity of the springs ranges from 4.5–25.8 g l^{-1} TDS and is highest in the southern group. Chemically, the springs are alkaline (pH- 8.0–9.8) and of Na-HCO_3 type with high concentrations of SiO_2 and F^-. They derive from steam-heated meteoric groundwater with a minor contribution (3–7 percent) of recycled lake water. Fumeroles are common, especially in the southern group. After steam, CO_2 is the main gas component with minor N_2 and CH_4. The reservoir PCO_2 is very high (up to 100 bar?; see Cioni and others, 1992). At least 12 dilute warm springs (33–49° C; pH- 6.5–7.5) with Na-Ca-HCO_3 waters also flow into the lake.

Assessing the relative importance of dilute inflow and hydrothermal sources in lake recharge is difficult, as many variables remain unknown or are ambiguous in interpretation. Changes in chemical composition of the lake waters between 1977 and 1978 offer some clues. Dilute inflow (stream) waters in 1977 and 1978 had a mean molar Na/Cl ratio of about 3.1 (range: 2.1–3.3). For the hot springs, the range was 4.3 to 13.4, with a mean Na/Cl ratio of about 7.0. In 1977, the mixolimnion waters had a Na/Cl ratio of about 6.6, twice that of dilute inflow. Thus if chloride behaves as a conservative ion in the lake until halite saturation is reached, evaporative concentration of dilute inflow waters alone could not produce the high Na/Cl ratio in the lake waters. The mixolimnion composition may thus reflect mixing of dilute inflow with hot spring waters or older lake water with a higher Na/Cl ratio. Following the heavy rains of 1978, the Na/Cl ratio of the mixolimnion decreased to a mean value of about 4.0 — a value closer to dilute inflow. These data indicate that the relative contributions of dilute inflow and hydrothermal springs in recharging the lake may vary reciprocally from year to year, and that the response to such changes is rapid. In very dry years, most recharge (perhaps 80–90 percent) derives from hydrothermal sources. In wet years, hot springs may contribute less than 30 percent.

Biology

Although modern lake waters have low species diversity, primary productivity is commonly very high (Melack, 1981). Cyanobacteria, notably *Oscillatoria (=Spirulina) platensis*, dominate the plankton (Beadle, 1932; Péniguel, 1987). Several diatoms and rotifers are present, but few other organisms inhabit the hypersaline waters. Lake Bogoria supports a major bird population, of which lesser flamingoes (up to two million birds) dominate. Hippopotamus and crocodile lived in the more dilute northern waters as recently as the 1920's. During late Quaternary humid phases, there was much greater faunal and floral diversity; fish, sponges, molluscs, ostracods, diatoms and other organisms were abundant (Carbonel and others, 1983; Tiercelin and Vincens, 1987).

MODERN SEDIMENTARY ENVIRONMENTS AND FACIES

The steep margins of the Lake Bogoria half-graben have limited development of subaerial depositional environments, thus most modern sedimentation is lacustrine (Fig. 2). Three tectono-geomorphic settings are recognized, each with a distinctive facies association: (1) boundary-fault margin, (2) axial and lateral littoral platforms, and (3) axial basin. Hot springs are a subsidiary, yet important, subenvironment.

Sedimentation along the Boundary Fault Margin

Slope Failures.—

The steep Bogoria Escarpment, an active normal fault with a throw of about 1 km, is a boundary fault of the half-graben, merging with the transverse Emsos Escarpment at its southern end (Fig. 2). Sedimentation along the piedmont zone below the escarpment is dominated by major landslips and coarse siliciclastics deposited by gravity and ephemeral streams (Table 2; Fig. 4). Large coalescent fan deltas, separated by large faulted blocks of lava bedrock and major slope failures, lie below the Bogoria-Emsos Escarpment (Figs. 2, 3B).

Major slope failures include large, back-tilted slump blocks of weathered lavas (some >1 km long) and megabreccias. Megabreccias are chaotic mixtures of blocks of lava, some exceeding 10 m across, in a boulder-gravel-sand-mud matrix, that form units tens of meters thick with erosional bases. They are commonly interbedded with coarse paraconglomerates and locally with minor imbricated, clast-supported, angular orthoconglomerates (sedimentary breccias). Megabreccias form during landslides and from very slow, highly viscous flows of coarse slope detritus. The paraconglomerates represent smaller-scale, viscous debris flows of lower density, generated during prolonged heavy rains (Tiercelin and Renaut, 1987). The angular orthoconglomerates are talus deposits (alluvial cones). Many slope failures are earthquake-induced (Tiercelin, 1990; Ambraseys, 1991). Reddish brown latosolic paleosols, up to 1 m thick, commonly separate depositional units; calcic paleosols are rare, reflecting the higher precipitation along the elevated rift margins.

Alluvial Fans and Fan Deltas.—

The coalescent North Bajada, Chepnagalan and Nanyonchengare alluvial fans (Fig. 2) issue from re-entrants in the Lake Bogoria Escarpment. Subaerial facies include paraconglomerates, orthoconglomerates, and stratified sands and silts (Table 2; Fig. 4; Renaut, 1982; Tiercelin and Renaut, 1987) that differ little from those described from fans elsewhere (e.g., Nilsen, 1982). Coarse paraconglomerates produced by debris flows dominate proximal locations, but are found throughout the fans. Streamflood and sheetflood deposits (orthoconglomerates, bedded sands and silts) dominate the mid- and toe slopes. All fans are segmented with deeply incised trunk channels near the apex (up to 15 m), reflecting both the relative subsidence of the rift floor and falling base (i.e., lake) level. Most proximal and medial fan sediments are poorly consolidated; intergranular micritic calcite (rarely, non-stoichiometric dolomite) and iron oxide cements (lepidocrocite, goethite and amorphous iron hydroxides) are found near the fan toes. Cemented zones are most common near dilute, cool spring seepages. Dense *Acacia* woodland covers the modern fan surfaces, with alkaliphilic grasses and reeds near the lake (Fig. 5A). Many

TABLE 2.—SUMMARY OF THE MAIN LITHOFACIES ASSOCIATED WITH THE BOUNDARY FAULT MARGIN OF LAKE BOGORIA

Lithofacies	Description	Other Features	Processes and Environments of Deposition
FA: Megabreccia	Chaotic mixtures of subangular rock fragments, some >10 m across, in poorly sorted, red to yellow-brown, muddy sand matrix; commonly in units tens of meters thick with irregular eroded bases.	Tree fragments; paleosols on top of units.	Landslides, talus and highly viscous flows of slope debris.
FB: Paraconglomerate	Polymictic reddish brown paraconglomerates with subangular lava clasts (1->100 cm) in a sandy (feldspathic) or muddy matrix; units <0.1 m to several m thick; erosive and non-erosive bases; mostly massive, some normally graded.	Rare tree fragments; rootlet horizons.	Debris flows and mudflows of variable viscosity; some deposited subaqueously.
FC: Orthoconglomerate	Polymictic angular and subangular gravels (>5 cm) with silty to muddy, reddish brown matrix in units 10 cm to several m thick; massive, imbricated and cross-stratified units present; many units are lenticular; erosional bases common.	Weakly lithified; local calcite cements; rare bone debris.	Subaerial channel fill and bar deposits on alluvial fans; some thin lenticular units are surficial lag gravels.
FD: Interstratified lenticular orthoconglomerate and sands	Thin (5–50 cm) lenticular units of orthoconglomerate interstratified with reddish brown to brown horizontal and cross-stratified sands in lenticular units 30–40 cm thick.	Weakly lithified; local calcite or Fe-oxide cements.	Sheetflood and shallow ephemeral wash deposits on mid- to lower fan slopes.
FE: Sands	FE_1: Planar and cross-stratified reddish brown to brown coarse to medium sands in units up to 1 m thick, commonly succeeding imbricate conglomerates in upward-fining packages. FE_2: Gravelly, black coarse sands; massive but with normal or inversely graded clasts, in units 10–20 cm thick. FE_3: Muddy, black to dark brown sands in massive or horizontally stratified beds, 5–200 cm thick. FE_4: Planar cross-stratified sands and silts in beds, 1–3 m thick.	Sands mainly K-feldspar and lithic grains; minor micritic calcite and rare dolomite cements; clays mainly smectite or illite-smectite.	FE_1: Stream channel fills and sand bars. FE_2 and FE_3: Deposition from turbidity currents (Bouma A, B and C?). FE_4: Homopycnal (Gilbert-type) fan-delta foresets deposited during high lake levels.
FF: Stratified sands and silts	Thin (1–70 cm) units of horizontally stratified brown sands and silts with thin lenticles of fine gravel.	Local calcite or Fe-oxide cements.	Distal fan and subaerial sandflat deposits.
FG: Silt and silty-clay	Reddish brown and grey black silts and clays in units 5–30 cm thick; silts mostly massive; clays may be laminated.	Clays mainly smectite; local vegetal detritus.	Distal turbidites (Bouma D?).
FH: Clay	Dark brown to black, massive or laminated clays with gas vesicles in units from 10 cm to >1 m thick; commonly weakly compacted.	Clays mainly smectite and illite; vegetal debris and insect fragments are common.	Muddy intervals of turbidites (Bouma E); sedimentation from hypopycnal plumes.

distal (toe zone) facies show extensive bioturbation by roots and termites. Calcite rhizocretions and incipient nodular calcretes are locally common in fan-toe sediments.

All fans discharge ephemeral runoff directly into the lake. To 6 m depth, short cores and exposed littoral sediments reveal partial to complete upward-fining cycles that resemble turbidite sequences (Fig. 4; Bouma, 1964), with a lower interval of massive gravelly sands, succeeded by deformed and horizontally stratified silts and fine sands, and capped by black or brown, massive to weakly laminated muds. Many sequences are incomplete (i.e., ABE, AE; see Walker, 1965). Both 'high density' and 'low density' turbidites (cf. Sturm, 1975; Sturm and Matter, 1978) have been recognized. Several cores and sections contain paraconglomerates with slightly erosive bases interbedded with lacustrine muds. These may have been deposited, and some turbidity currents generated, when debris flows entered the lake from steep fan deltas. Collapse of underwater slopes during earthquakes may also produce such deposits (Tiercelin, 1990).

Sedimentation along the Littoral Platforms

As at Lake Tanganyika (cf. Cohen, 1990; Tiercelin and others, 1992) depositional surfaces on the ramp side of half-grabens opposite border faults can be termed *lateral littoral platforms*, and those at the ends of the lake along the rift axis as *axial littoral platforms*. Sedimentation in both settings is mainly deltaic and progradational. Analogous platforms at Lake Bogoria are relatively minor features. Because of the steep transverse structural closure at the southern end of the half-graben, only one axial littoral platform is present.

Lateral Littoral Platforms.—

The western (ramp) margin of the Lake Bogoria half-graben consists of small deltas and narrow littoral platforms, separated by steep lava headlands, some extending tens of meters offshore. Headlands are mantled discontinuously by thin gravelly soils and colluvium. Along the shoreline, angular lava gravels form upward-coarsening littoral terraces with riser-tread couplets up to 999 m elevation. These degraded regressive terraces record former higher lake levels and higher littoral wave energy (Fig. 5B; Renaut and Owen, 1991).

All lateral deltas are very small. Loburu (Fig. 2), with two lobes and a marshy interdistributary embayment, is the largest (0.5 km^2). Most other deltas are very minor shore-

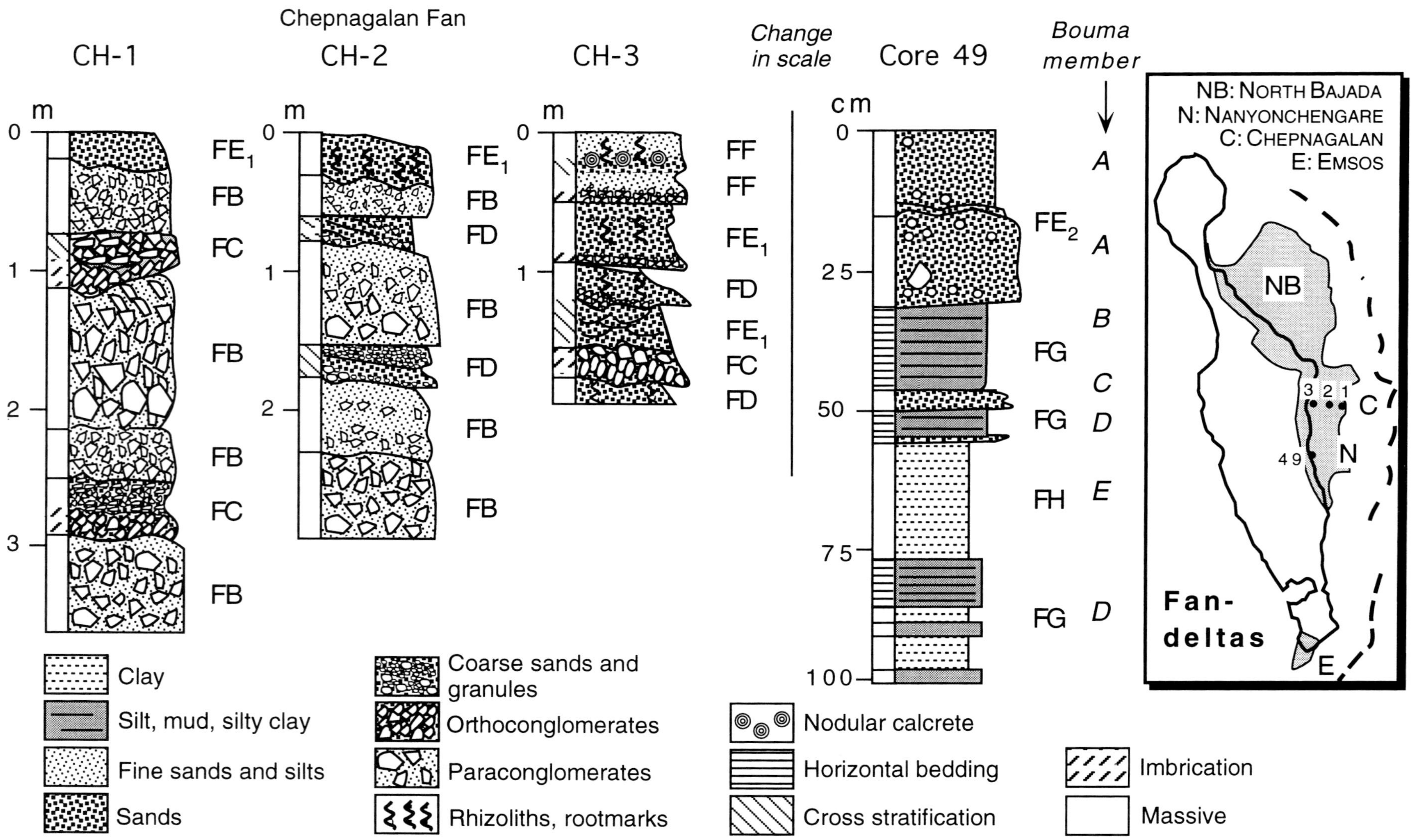

FIG. 4.—Selected lithological sections through fan-delta deposits east of Lake Bogoria. Code letters left of each section refer to the lithofacies in Table 2.

FIG. 5.——Sediments around margins of Lake Bogoria. (A) Distal, toe zone of Chepnagalan fan delta. In foreground, interbedded subhorizontal units of upward-fining sands and silts are onlapping steeply dipping, small-scale Gilbert-type foresets; (B) Rocky headland with degraded gravel terraces, north central sub-basin. Boulders are encrusted by stromatolitic limestones, 1 to 2 cm thick; (C) Holocene valley-fill sediments, Parkirichai River, 1 km west of Loburu Delta. Massive, reddish brown bioturbated silts and fine sands forming vertical cliffs were deposited during a late Quaternary high lake level; note poorly sorted gravels on floor of modern ephemeral stream bed; and (D) Sandai delta-plain encrusted by efflorescent trona, thermonatrite and halite; dust-devil in distance is redistributing efflorescent salts.

line promontories. The main axial rift drainage bypasses Lake Bogoria and flows north to Lake Baringo, thus all lateral deltas are fed by ephemeral streams with very small catchments, tightly confined within small N-S grabens. These streams have braided channels with very poorly developed medial and lateral bars, reflecting the erratic discharge. Channel sediments include poorly sorted imbricated gravels and planar and trough cross-stratified granular sands. Root-bioturbated and weakly laminated, feldspathic silts and fine sands represent overbank deposits. Near the lake, these affluent streams are incised within Holocene valley-fills of weakly lithified, reddish brown sands, silts, and fine gravels, weakly cemented by vadose and phreatic calcite and iron oxides (Fig. 5C). These fills were deposited subaqueously when late Quaternary high-stands of Lake Bogoria flooded the lower reaches of the valleys.

Modern delta plains are littered by poorly sorted gravels and granules, and are variably vegetated by scrub and grasses, with dense *Cyperus* marsh near the shoreline (Fig. 3D). Small (<1 m) wave-cut terraces lie parallel to paleoshorelines (Renaut and Owen, 1991). Modern channels are incised up to 3 m in the delta plains, and at Loburu bifurcate near the shoreline. Most sediments exposed in delta-plain sections were deposited during former higher lake levels and have since been modified by subaerial processes, including channelled and unchannelled wash, deflation, crys-

PART II
QUATERNARY SALINE LAKES

Sedimentology and Geochemistry of Modern and Ancient Saline Lakes, SEPM Special Publication No. 50

HOLOCENE SEDIMENTS OF THE CRATER LAKE AT MALHA, NORTHWESTERN SUDAN

FLORIAS MEES

Department of Mineralogy, Petrography and Micromorphology, Geological Institute, University of Ghent, B-9000 Ghent, Belgium

AND

NIGEL RICHARDSON

Department of Geography, University of Liverpool, Liverpool L69 3BX, United Kingdom

ABSTRACT: Two cores were recovered from deposits of the crater lake at Malha, in NW Sudan. One core was taken in the center of the lake (Core B), the other from a more marginal area (Core Y). In both cores, two main units are distinguished, each comprising the sediments of three distinct lacustrine phases. The first three lake stages (Periods I, II and III), all ending with lowstands or phases of complete dryness, are characterized by the deposition of mainly laminated sediments in a meromictic lake. During Periods IV, V and VI, the lake was holomictic and hyper- to mesosaline.

Mineral magnetic data for the cores allow us to trace several correlatable horizons that coincide with boundaries between the main lacustrine phases. Most peaks in magnetic susceptibility (χ) and saturation isothermal remanent magnetization (SIRM) are accompanied by shifts of the reverse field ratios to more negative values. These features indicate an increased supply of coarser grained detrital material to the coring sites, and correspond to periods when lake levels were either fluctuating or very low.

Only two radiocarbon dates are available for the entire Malha record: 8,290 ± 155 ^{14}C years BP for 921-cm depth in Core Y, and 2,940 ± 70 ^{14}C years BP for 643-cm depth in Core B. These two dates imply that the first lake desiccation, at the end of Period I, took place at ~7,000 years BP, and that the second lake desiccation, marking the boundary between Periods III and IV, occurred at ~2,200–2,300 years BP.

INTRODUCTION

The center of an explosion crater near the town of Malha in northwestern Sudan is occupied by a small saline lake. Two sediment cores were recovered from the lake. The first, Core B, was taken in the center of the lake. The second, longer core, Core Y, was taken about midway between the center and the present shoreline, at a distance of about 75 m from the first site. Core Y has been described elsewhere (Mees and others, 1991). In this paper, we describe Core B and present mineral magnetic data for both cores.

The village of Malha (15°08'N, 26°12'E) lies southwest of the Meidob Highlands in northern Sudan, 220 km north of El Fasher. The crater forms a small depression — 80 m deep and more than 1 km wide — surrounded by a 40-m high rim of pyroclastic materials. In 1985, the saline lake in the center of the basin had an average depth of 1 m and a mean diameter of 230 m. No data are available on lake water chemistry and groundwater composition.

Along the crater walls, the subsurface geology of the plains surrounding the Meidob Hills is exposed (Colchester, 1927; Nijs, 1986). Strongly kaolinized granites and metamorphic rocks of the Basement Complex occur at the base, overlain by sandstones of the Nubian Sandstone Formation, a thin basalt layer, and pyroclastic material that was ejected during the explosion that formed the crater. Undated diatomites and aragonitic travertines adhere to the crater walls, at levels of up to 50 m above the present lake level. K/Ar dates for the basalt layer, and for a lava flow that breaches the crater wall and dips into the depression, establish the period when the crater was formed between 250,000 and 100,000 years BP (P. De Paepe and P. Pasteels, pers. commun., 1991).

CORE STRATIGRAPHY

Core Y

The base of Core Y lies at 9.21 m below the mud-water interface (Fig. 1). The upper 1.2 m of the cored section consisted of brine-saturated sediments that could not be preserved and were sampled on site. Deeper drilling was prevented by a layer of pyroclastic material. The sediment in contact with this layer yielded an age of 8,290 ± 155 ^{14}C years BP (IRPA 623). Because this date was obtained for total organic carbon, it can only be considered to represent a maximum age. However, since any hard-water effects will be minimal during fresh-water stages, only a minor overestimation of the real age is expected.

The cored sequence comprises two distinct units. The Lower Unit (620–921 cm) consists mainly of laminated deposits, with alternating carbonate and clay laminae. The Upper Unit (0–620 cm) consists of a salt-rich dolomitic marl and contains only one laminated section.

We recognize three lithological subunits in the Lower Unit, each corresponding to a distinct lacustrine phase, termed Periods I, II and III. The sediments of Period I, including those in the strongly mineralized horizon at the top, are nearly continuously laminated. Only one unlaminated section occurs (878–875 cm), located within a zone with an important admixture of aragonite in the calcite-dominated deposits of this lake stage. Two more aragonite-bearing sections occur in the (dolomite-dominated) sediments of Period II. The deposits of the second lacustrine phase are unlaminated in the lower part and contain gaylussite ($Na_2Ca(CO_3)_2 \cdot 5H_2O$) at the base. Above this section, the sediments are laminated, followed by a zone with alternating laminated and unlaminated bands. These are succeeded by a section with several gaylussite layers, separated by bands with largely unlaminated sediments. The sediments of Period III are characterized by a regular lamination and by much lighter colors. The varved sequence is interrupted at several levels by wider bands that consist of an accumulation of small, discrete dolomite and clay aggregates.

The subdivision of the Upper Unit is based on the distribution of salt minerals. Three subunits are defined, corresponding to Periods IV to VI. Gaylussite is the only salt mineral in sediments of Period IV. The core section cor-

Sedimentology and Geochemistry of Modern and Ancient Saline Lakes, SEPM Special Publication No. 50

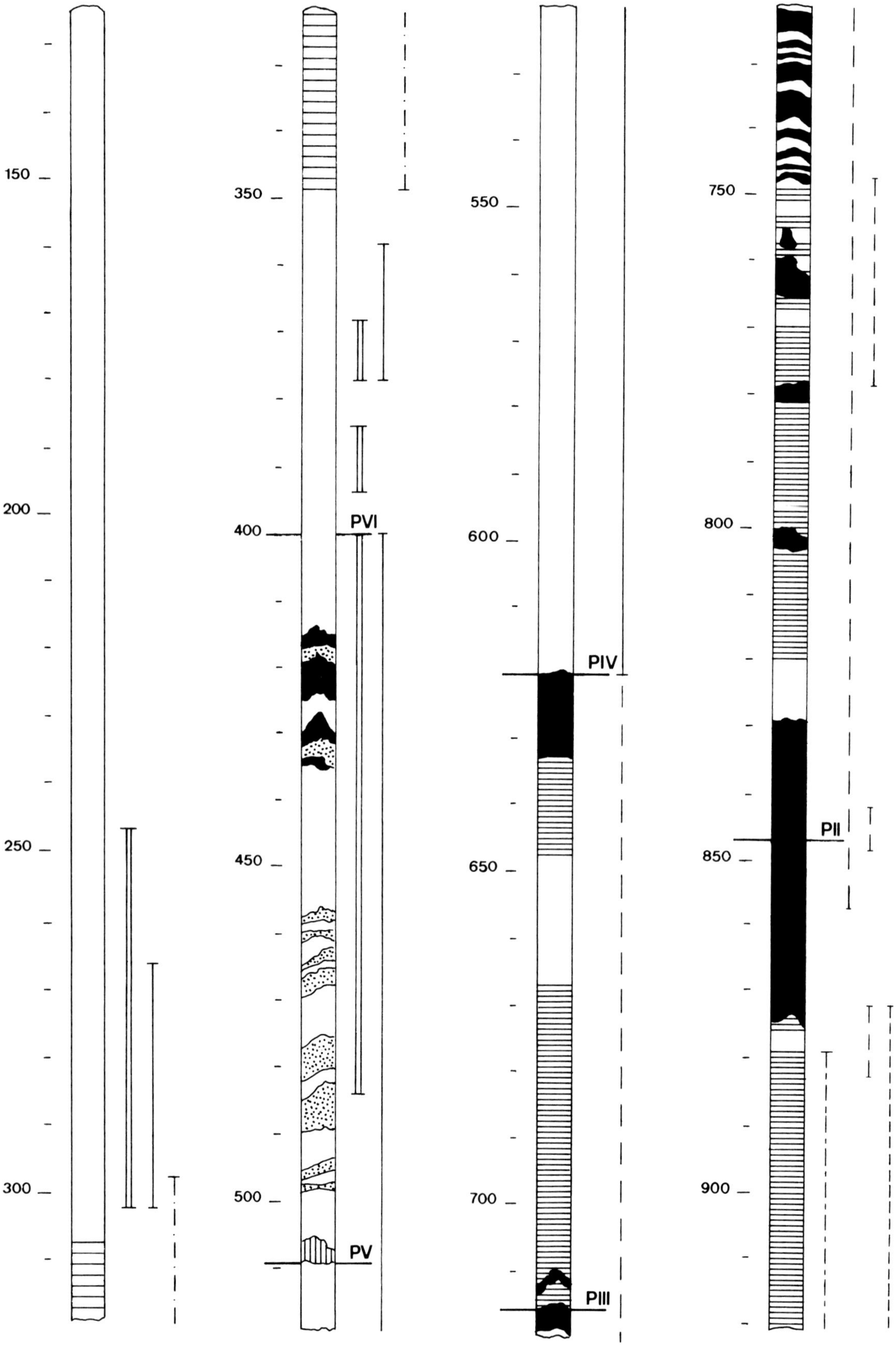

FIG. 1.—Core Y stratigraphy. See Figure 2 for explanation of symbols. Meso- and microscopic occurrences of nahcolite and gaylussite in sediments of Periods V and VI are presented in a simplified manner.

responding to the next lacustrine phase (Period V) has a trona ($NaHCO_3 \cdot Na_2CO_3 \cdot 2H_2O$) layer at the base and contains several nahcolite layers. Dark gaylussite-rich bands occur above and beneath two of the nahcolite layers. The sediments of Period VI comprise three separate zones with nahcolite ($NaHCO_3$) and gaylussite occurrences. In one intervening section, the sediments are laminated, due to differences in pigmentation and small variations in the dolomite content of the groundmass. This laminated section coincides with the presence of eitelite ($Na_2Mg(CO_3)_2$) in the core. The genesis and significance of the authigenic minerals are discussed by Mees and others (1991).

Core B

Core B extends to a depth of 7.8 m below the mud-water interface (Fig. 2). The brine-saturated upper 1.8 m of the core was not preserved as a continuous section. A piece of wood from 643-cm depth in the core was dated at 2,940 ± 70 ^{14}C years BP (UtC-1761).

The gaylussite-rich section at the base of the core correlates with the gaylussite horizon between 847- and 877-cm depth in Core Y; both display nearly identical macromorphologic characteristics, such as shape and size of the crystals. As a result, the section above this horizon in Core Y, with a much lower gaylussite content and with larger, better developed crystals, which was previously assigned to Period I (Mees and others, 1991), is now attributed to Period II. In Core B, the deposits of Period II lack gaylussite at the base, and, in further contrast with Core Y, the sediments are laminated from the onset of the lake stage. No section with alternating laminated and unlaminated bands occurs, and the unlaminated upper 40 cm lacks gaylussite layers. One aragonite-bearing section occurs at the base of the core, with important admixtures of calcite at the base of the section and around 735 cm. Another zone with aragonite (~665-cm depth) occupies an identical stratigraphic position to that at 765 cm in Core Y. The sediments of Period III are identical to those in Core Y. Besides the thick homogeneous interval between 536 and 547 cm, several wide laminae can also be correlated with corresponding features in Core Y — e.g., one layer around 572 cm, and three bands between 536 and 523 cm.

For the Upper Unit, the boundaries between Periods I, II and III can be easily traced, but finer correlations between cores are less evident. An important difference with Core Y is the lamination of Period IV deposits, which is similar to the 308–350 cm interval of Core Y. In the Period V section, no trona layer occurs at the base, and no gaylussite-rich bands are associated with any of the nahcolite layers. In sediments of Period VI, eitelite occurs in one narrow zone, but gaylussite seems to be absent.

MINERAL MAGNETIC MEASUREMENTS

A sequence of mineral magnetic measurements were completed on dried and powdered sediment samples from both cores. Table 1 summarizes the parameters and instrumentation used in the study. Magnetic susceptibility (χ), saturation isothermal remanent magnetization (SIRM), and reverse field ratios ($IRM_{-20mT}/SIRM$; $IRM_{-40mT}/SIRM$; $IRM_{-100mT}/SIRM$; $IRM_{-200mT}/SIRM$), plotted against depth, are presented in Figure 3. χ and SIRM values provide a fair indication of the changing concentration of ferrimagnetic minerals in the sediment. Reverse field ratios indicate changes in magnetic mineral grain size and/or magnetic mineral type. In general, more negative reverse field ratios indicate assemblages where "softer" ferrimagnetic minerals (e.g., magnetite and maghemite) and/or larger magnetic grain sizes dominate. The ratios become more positive with increasing proportions of "harder" antiferrimagnetic-type minerals, such as hematite, and/or finer ferrimagnetic grains. Volcanic rocks in the catchment are undoubtedly the main source of magnetic minerals. Some rocks contain up to 8 percent of Fe(Ti)-oxide minerals by weight (P. De Paepe, pers. commun., 1991). The deeply weathered Basement Complex granite and paleosols in the upper part of the Nubian Sandstone may also provide minor finer magnetic grains. Although the two commonly go together, magnetic grain size shifts are not necessarily linked to variations in sediment particle size.

We identify three potentially correlatable horizons in the magnetic sequence of Cores Y and B. A first feature, coinciding with the boundary between Periods II and III, is recognized by a shift in the reverse field ratios, from a peak with more negative values to an interval with more positive values. Also recognized is a major shift in the reverse field ratios to more negative values at the boundary between the Upper and Lower Units. Reverse field ratios for sediments of Period IV fluctuate throughout the interval, but there is an overall gradual shift to more positive values. Above a certain level, after a final stronger shift, the reverse field ratios remain relatively constant. This level, coinciding with the Period IV/V boundary, marks the position of a third correlatable horizon.

The sediments of the Lower Unit in Core Y display two broad peaks in χ and SIRM, centered on 880-cm and 740-cm depth, respectively. Sharp peaks in χ and SIRM occur at the base of the core and at about 880-cm depth, which are levels where gravel-sized fragments of pyroclastic material are present in the deposits. χ and SIRM values are generally low for sediments of Period III, which are also characterized by more positive reverse field ratios. The few small peaks in χ and SIRM that do occur, coincide with the unlaminated section between 667 and 648 cm, and with a band of identical structure and composition around 698 cm. χ and SIRM curves for the Upper Unit are generally constant and featureless. Higher values above 300 cm are probably artifacts of sampling or core preservation.

The mineral magnetic sequence for Core B is similar to that of Core Y, but χ and SIRM patterns do not correlate well with lithology. A sharp peak in χ and SIRM at about 735 cm coincides with an unlaminated section. Another peak, around 545-cm depth, correlates with the homogeneous interval in the sediments of Period III. Reverse field ratios tend to be more positive for normal Period III deposits, compared to sediments of Period II, indicating a "harder" magnetic mineral assemblage of finer ferrimagnetic grain size, while the section with higher X and SIRM values at ~545 cm shows a "softer" magnetic behavior indicating coarser magnetic grain sizes.

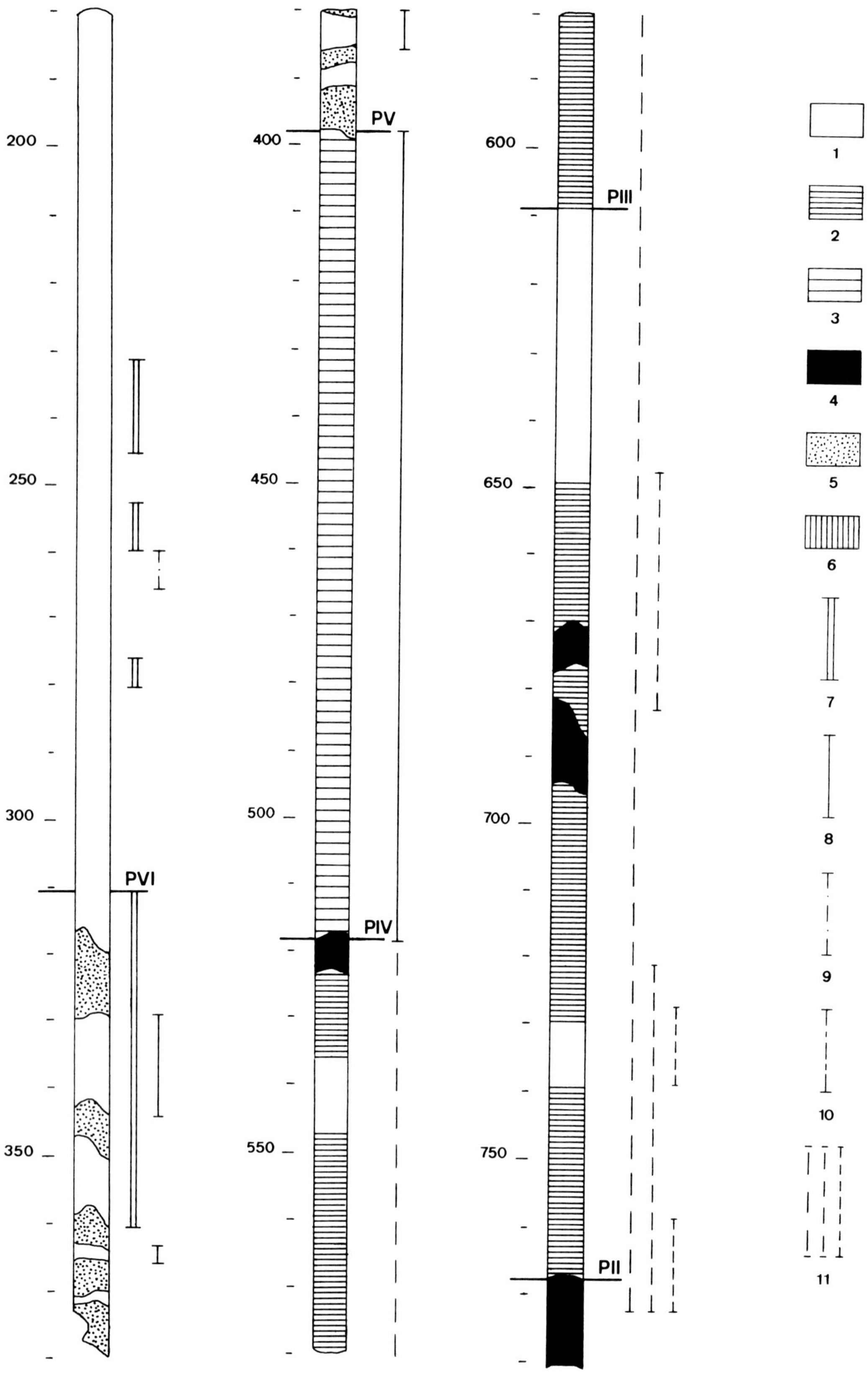

FIG. 2.—Core B stratigraphy. Legend : 1- unlaminated sediment; 2- laminated sediment, with clay/carbonate couplets; 3- laminated sediment, other than with clay/carbonate couplets; 4 to 6- macroscopic occurrences of gaylussite (4), nahcolite (5) and trona (6); 7 to 10- meso- (Core B) and/ or microscopic (Core Y) occurrences of nahcolite (7), gaylussite (8), eitelite (9) and magadiite (10); 11- distribution of dolomite (left), aragonite (center), and calcite (right) in sediments of the Lower Unit.

TABLE 1.—MINERAL MAGNETIC PARAMETERS AND INSTRUMENTATION

MAGNETIC SUSCEPTIBILITY (χ): The ratio of magnetization induced to intensity of the magnetizing field, measured within a low magnetic field with no induced remanence. Measured on a mass specific basis and roughly proportional to the concentration of ferrimagnetic minerals within a sample.
Units: m^3kg^{-1}
Instrumentation: Barington Instruments single sample susceptibility sensor.

SATURATION ISOTHERMAL REMANENT MAGNETIZATION (SIRM): The highest level of magnetic remanence that can be induced in a sample by application of a high field (1.0 T in this study). SIRM indicates the volume of magnetic minerals in a sample, but also responds to grain size variations.
Units: Am^2kg^{-1}
Instrumentation: Pulse magnetizer, fluxgate magnetometer.

DEMAGNETIZATION PARAMETERS (IRM_{-X}/SIRM): Obtained by applying one or more reversed magnetic fields to a previously saturated sample. The loss of magnetization is expressed as a ratio IRM_{-X}/SIRM, giving a result between +1 and −1.
Instrumentation: Pulse magnetizer, fluxgate magnetometer.

HISTORY OF THE CRATER LAKE AT MALHA

The history of the crater lake near Malha is summarized by the lake level curve in Figure 4, based in part on evidence from the study of Core Y (Mees and others, 1991). The highest lake levels were reached during Periods I, II and III. Meromixis probably occurred during highstands, which were bracketed between periods of lake desiccation or low lake levels. The lake became dry at the end of Periods I and III, resulting in protracted episodes of diagenetic gaylussite formation. During deposition of the Upper Unit, the lake was holomictic. Waters were mesosaline at first, followed by a hypersaline stage (Period V), that was followed in turn by waters with more moderate salinities. The new magnetic data and the study of Core B provide additional information on the lake's history.

The broad peak in *X* and SIRM in sediments of Period I relates to a period with high erosive input of magnetic minerals to the coring site. The material that was washed in, originating either from the surrounding catchment or from marginal areas of the lake, is of a coarser magnetic grain size than the normal sediments of Period I. The period of increased erosive input culminated with the deposition of a wide unlaminated band that might represent a complete overturn of the lake, which is supported by an absence of magadiite ($NaSi_7O_{13}(OH)_3 \cdot 3H_2O$) above this level in the deposits. This short-term event, which may correspond to a drop in lake level, is also evident from reverse field ratio patterns, which display a peak with more negative values at ~880-cm depth. After a short interval with more positive values, the reverse field ratios shift back toward more negative values, reflecting the deposition of coarser grained ferrimagnetic minerals at the end of Period I and beginning of Period II. χ and SIRM values in the corresponding core section are low, most likely due to diagenetic growth of gaylussite, which would dilute the concentration of ferrimagnetic minerals in the deposits. Post-depositional gaylussite formation also explains the local absence of alkaline-earth carbonate minerals in the upper part of Period I deposits.

The sediments of Period II provide a detailed record of a lake that rose to a highstand, followed by a drop in level. The more marginal coring site was located above the chemocline during a first part of Period II, while at the same time laminated sediments were being deposited and preserved in the center of the basin. After a highstand, recorded as sections with laminated deposits in both cores, lake level gradually fell. During this decline, short-term fluctuations in water supply occurred, with, in a first stage, vertical movements of the chemocline past the marginal coring site, resulting in the alternation of laminated and unlaminated sediments. In the shallower, more saline, holomictic lake that followed, fluctuations in water supply are recorded by their effects on lake salinity, resulting in periodic gaylussite crystallization in marginal areas. Three gaylussite occurrences between 750- and 830-cm depth in Core Y and two similar zones in Core B probably represent cuts through gaylussite nodules related to the lake desiccation that ended Period III.

A broad peak with higher χ and SIRM values and more negative reverse field ratios in the upper part of the deposits of Period II in Core Y is also attributed to an influx of coarser grained magnetic material. Because this core section corresponds with a stage of fluctuating lake levels, some of this influx is likely due to a redeposition of coarser marginal materials. A similar feature is not recorded in magnetic parameters for Core B, but, like Core Y, the sediments contain aragonite at the corresponding depth. This shows that this aragonite is detrital and is probably derived from the travertines. The origin of aragonite and calcite in the lower part of Core B is less clear. Intervals with more prominent, white carbonate laminae have a higher aragonite/dolomite ratio than intervening sections, suggesting that some aragonite might be a primary precipitate. The highest relative amounts of calcite are recorded at the base of Core B (primary?) and at ~735 cm — a level where an unlaminated band occurs that is also marked by a peak in χ and SIRM values (detrital origin?). The near-absence of these calcium carbonate minerals in the corresponding section of Core Y could be due to gaylussite formation.

Period III deposits are characterized by low concentrations of ferrimagnetic minerals of finer magnetic grain size. The localized peaks in χ and SIRM correspond with shifts in the reverse field ratios to more negative values, indicating a coarser magnetic assemblage. These peaks coincide with unlaminated sections in the deposits and relate to material that is probably derived from shallower areas of the lake. Since the most prominent bands occur in identical positions at both coring sites, deposition of these bands clearly corresponds to lake-wide (i.e., overturn) events.

For the Upper Unit, few conclusions can be added to those from the study of Core Y (Mees and other, 1991). Comparing cores shows that trona formation (507–510 cm in Core Y) was restricted to marginal areas of the lake. Dark, gaylussite-rich bands are only present in Core Y, supporting the notion that these layers might correspond to microbial mats. Eitelite formed lake-wide, and, as the corresponding sediments in Core Y are similar to the sediments of Period IV in Core B, the interval seems to correspond to a highstand period.

DISCUSSION AND CONCLUSIONS

During the Early- to Mid Holocene, most parts of northern Africa had a more humid climate than today. This hu-

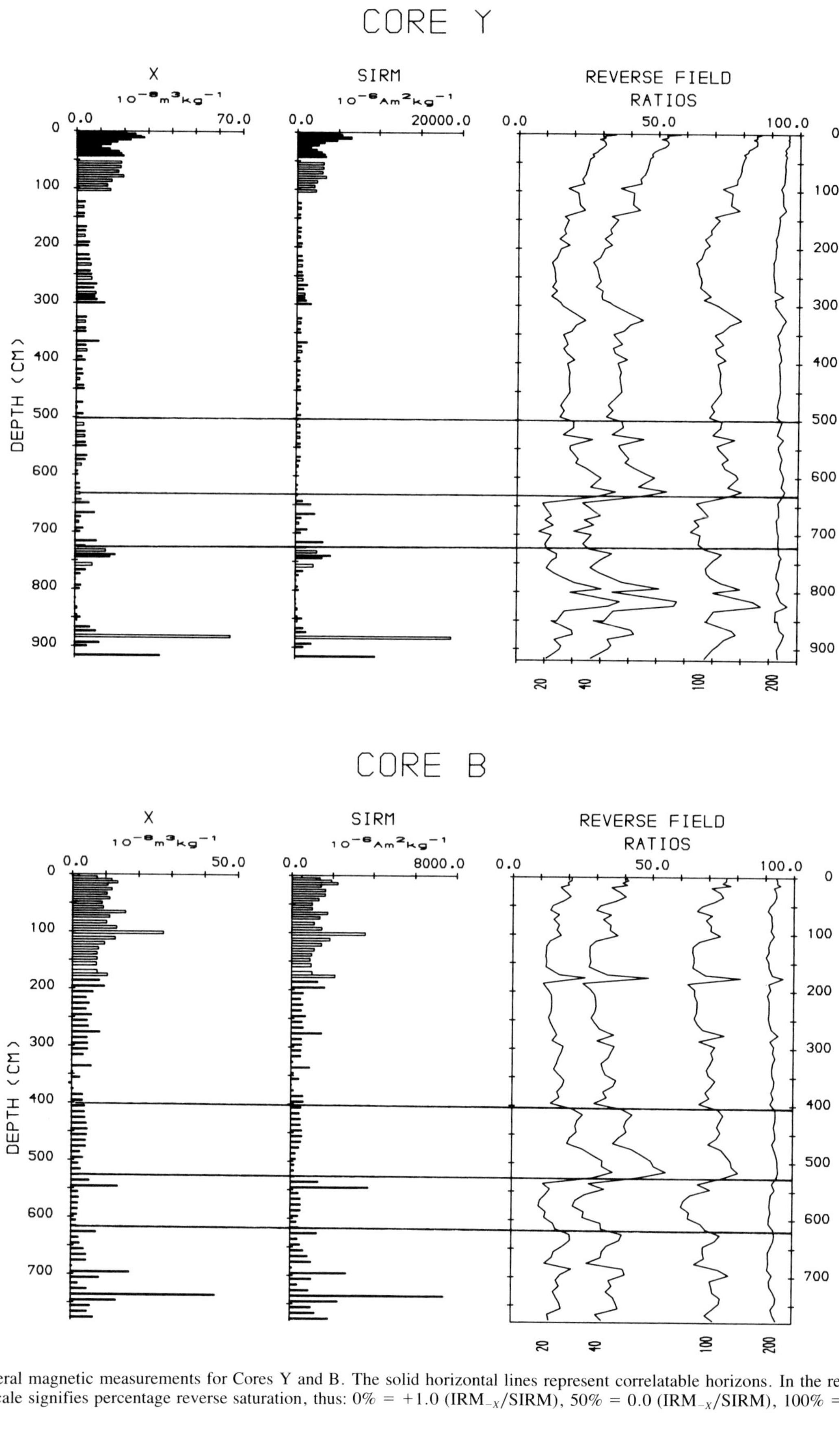

FIG. 3.—Mineral magnetic measurements for Cores Y and B. The solid horizontal lines represent correlatable horizons. In the reverse field ratio plot, the upper scale signifies percentage reverse saturation, thus: 0% = +1.0 (IRM_{-X}/SIRM), 50% = 0.0 (IRM_{-X}/SIRM), 100% = −1.0 (IRM_{-X}/SIRM).

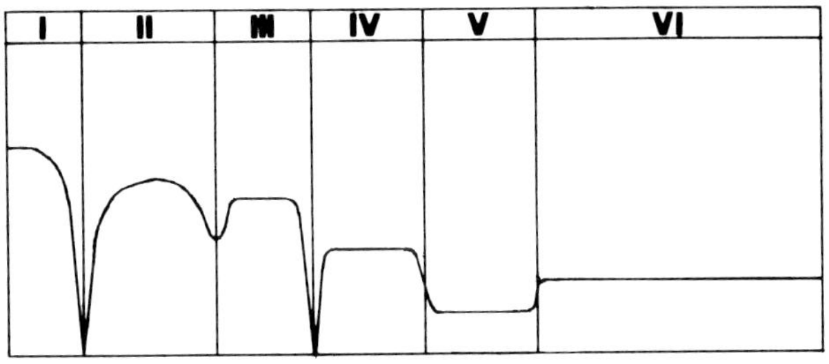

FIG. 4.—Relative lake level, plotted against depth in Core Y. Core sections corresponding with Periods I to VI are indicated at the top.

mid period seems to have been interrupted by periods of increased aridity. A few sites in northern Sudan and southern Egypt provide evidence for such oscillations. In northern Sudan, indications are only found at Wadi Shaw (Gabriel and Kröpelin, 1983, 1984), at Wadi Howar (Pachur and others, 1987), and possibly at Selima (Haynes, 1987). In southern Egypt arid intervals have been recognized at Nabta and Bir Kiseiba (Wendorf and Schild, 1980; Close, 1984). When the whole of northern Africa is considered, an arid interval at about 7,000 years BP has been clearly recognized for many basins (Street and Grove, 1976, 1979; Muzzolini, 1985a, b). In most parts of northern Sudan and southern Egypt, lacustrine sedimentation terminated at about 4,000 to 4,500 years BP. Climatic changes during the millennia following this phase are hardly documented for the region.

The base of the longest core taken at Malha, 80 cm below the Period I/II boundary, was dated at 8,290 ^{14}C years BP. The first lake desiccation, which ended the first lake stage, can therefore be assumed to coincide with the well-documented arid phase at about 7,000 years BP. The date of 2,940 ^{14}C years BP for 643 cm in Core B places the boundary between Periods II and III, marked by a lowstand, at around 2,700 years BP. Based on varve counts for the sediments of Period III, assuming an annual nature for the lamination of these deposits, the second lake desiccation is estimated to have occurred around 2,200 or 2,300 years BP. The date for the 643-cm level in Core B, 1.3 m above the Period I/II boundary, also implies that between the end of Period I and the onset of Period II, the lake remained dry for a prolonged period.

The crater near Malha forms a depression that cuts through the Nubian Sandstone Formation, which is the local aquifer, and reaches the underlying impermeable rocks of the Basement Complex. This favorable position in terms of groundwater supply allows the lake to survive during arid periods like the present. A lake desiccation at Malha around 7,000 years BP and its inferred long duration, is difficult to reconcile with the results of detailed studies at Selima (Haynes and others, 1989) and at Oyo (Ritchie and others, 1985), where this arid phase has not been recognized. Until the Malha cores have better radiometric age control, a discussion of these differences is inappropriate.

ACKNOWLEDGMENTS

The University of Liverpool offered financial support for N. R. to visit the University of Ghent. Gillian Yates provided invaluable assistance with sampling of the cores and with the completion of the magnetic measurements. Radiocarbon dating was performed by Mark Van Strydonck, Koninklijk Instituut voor het Kunstpatrimonium, Brussels. F. M. is research assistant with the Belgian National Fund for Scientific Research.

REFERENCES

CLOSE, A. E. , ed., 1984, Cattle-keepers of the Eastern Sahara: The Neolithic of Bir Kiseiba: Dallas, Southern Methodist University, 452 p.

COLCHESTER, C. V., 1925, Malha Crater, Darfur: Sudan Notes and Records, v. 10, p. 233–235.

GABRIEL, B., AND KRÖPELIN, S., 1983, Jungquartäre limnische Akkumulationsphasen im NW-Sudan: Zeitschrift für Geomorphologie Neue Folge, Supplementband 48, p. 131–143.

GABRIEL, B., AND KRÖPELIN, S., 1984, Holocene lake deposits in Northwest-Sudan: Palaeoecology of Africa, v. 16, p. 295–299.

HAYNES, C. V., 1987, Holocene migration rates of the Sudano-Sahelian wetting front, Arba'in desert, eastern Sahara, *in* Close, A. E., ed., Prehistory of Arid North Africa: Essays in honor of Fred Wendorf: Dallas, Southern Methodist University, p. 69–84.

HAYNES, C. V., EYLES, C. H., PAVLISH, L. A., RITCHIE, J. C., AND RYBAK, M., 1989, Holocene palaeoecology of the eastern Sahara, Selima Oasis: Quaternary Science Reviews, v. 8, p. 109–136.

MEES, F., VERSCHUREN, D., NIJS, R., AND DUMONT, H., 1991, Holocene evolution of the crater lake at Malha, Northwest Sudan: Journal of Paleolimnology, v. 5, p. 227–253.

MUZZOLINI, A., 1985a, Les climats au Sahara et sur ses bordures, du Pléistocène final l'aride actuel: Empúries, v. 47, p. 8–27.

MUZZOLINI, A., 1985b, Les pluviométries durant les 'Humides' holocènes au Sahara: Cahiers Ligures de Préhistoire et de Protohistoire, Nouvelle Série, v. 2, p. 163–179.

NIJS, R., 1986, The geology of Malha crater (Darfur, Sudan): Geo-ecotrop, v. 10, p. 1–10.

PACHUR, H. J., KÖPER, H. P., KRÖPELIN, S., AND GOSCHIN, M., 1987, Late Quaternary hydrography of the eastern Sahara: Berliner Geowissenschaftliche Abhandlungen 75, p. 331–385.

RITCHIE, J. C., EYLES, C. H., AND HAYNES, C. V., 1985, Sediment and pollen evidence for an early to mid-Holocene humid period in the eastern Sahara: Nature,v. 314, p. 352–355.

STREET, F. A., AND GROVE, A. T., 1976, Environmental and climatic implications of late Quaternary lake-level fluctuations in Africa: Nature, v. 261, p. 385–390.

STREET, F. A., AND GROVE, A. T., 1979, Global maps of lake-level fluctuations since 30,000 yr BP: Quaternary Research, v. 12, p. 83–118.

WENDORF, F., AND SCHILD, R., 1980, Prehistory of the Eastern Sahara: New York, Academic Press, 414 p.

HOLOCENE FLUCTUATIONS OF MONO LAKE, CALIFORNIA: THE SEDIMENTARY RECORD

MARK S. NEWTON

Earth Science Department, Los Angeles City College, Los Angeles, California 90029

ABSTRACT: Mono Lake is a large, hypersaline, alkaline water body, lying in the rainshadow of the Sierra Nevada. Unlike nearby Walker and Pyramid Lakes, Mono Lake has no major river input of water, but is fed by meltwater creeks. Since the early 1940's, part of this inflow has been diverted, causing the shoreline to fall from 1,957 m to 1,943 m, and the loss by evaporation of almost 50 percent of the lake's pre-1940 volume. The rate of decline closely reflects the regional climate. The concentration of major ions (Na^+, HCO_3^-, CO_3^{2-}, Cl^-, SO_4^{2-}) is remarkably conservative, showing an increase in salinity almost inversely proportional to the decrease in volume.

The first Holocene sedimentary column from Mono Lake has been obtained by coring and sedimentary facies have been mapped. Studies included the sedimentology, mineralogy and paleontology of the sediments by optical and SEM-EDAX methods and absolute dating by paleomagnetic, tephrochronology and ^{14}C methods, including AMS.

The results show: (1) a drastic recession from 350 m-deep Lake Lahontan to a 38 m-deep early Mono Lake sometime before 11,000 yr BP, and reaching a possible Holocene minimum about 10,000 yr BP; (2) a high stand beginning at least 9,000 years ago, when the lake was at least 16 m deeper than today; (3) a gradual recession starting about 6,500 yr BP in which shorelines eventually dropped to below the present level, reaching a minimum about 200 BP; meanwhile, the nearby Mono Craters were contributing large amounts of rhyolitic volcanic ash to the sediments at about 2,000 yr BP; (4) another high stand after 600 yr BP indicating a climatically induced influx of freshwater runoff peaking at about 300 yr BP, which probably caused long-term meromixis; and (5) a gradual recession, eventually resulting in overturn and the initiation of a monomictic lake, punctuated by brief interludes of meromixis—a regime that has prevailed until the present.

INTRODUCTION

Mono Lake is a large (178 km^2), saline (87‰) alkaline (pH = 10.0) lake located in the rainshadow of the eastern Sierra Nevada of California (Fig. 1). The Mono basin receives an average of 13 cm of precipitation annually and the evaporation rate exceeds 100 cm per year (Blevins and others, 1987). The residence time of the lake water is only 25 years; the lake loses 4 percent of its water to evaporation annually. Therefore, changes in lake volume take place swiftly due to natural or human-caused perturbations in seasonal runoff.

Mono Lake is fed primarily by two small meltwater streams draining the eastern Sierra Nevada—Rush Creek and Lee Vining Creek. Since the 1940's, much of this runoff has been diverted into the Los Angeles Aqueduct. Consequently, the shoreline elevation has dropped by more than 14 m to its lowest level in historic times, and the lake has lost almost half of its pre-1940 volume. As a result of evaporative concentration, the salinity and alkalinity of the lake have almost doubled. Only the Ca^{2+} content and pH, which are buffered by the high alkalinity, have remained relatively unchanged. The major-ion composition of the lake is shown in Figure 2. The 1940–1985 changes in dimensions and major ion concentrations are given in Table 1.

These changes have provoked an outcry from environmental groups. As shown by Stine (1987), the level of Mono Lake responds rapidly to oscillations in the Pacific weather system. Since the 1800's shoreline records have faithfully reflected the seasonal precipitation in San Francisco, similar fluctuations in level might be recorded in the lake's sedimentary record.

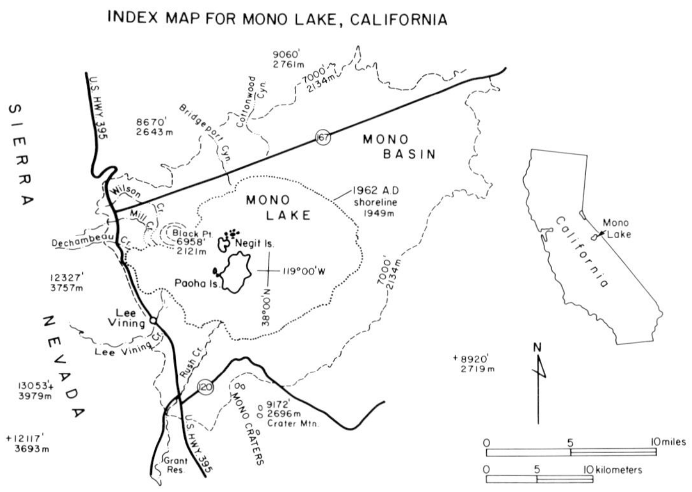

FIG. 1.—Location map of Mono Lake, California.

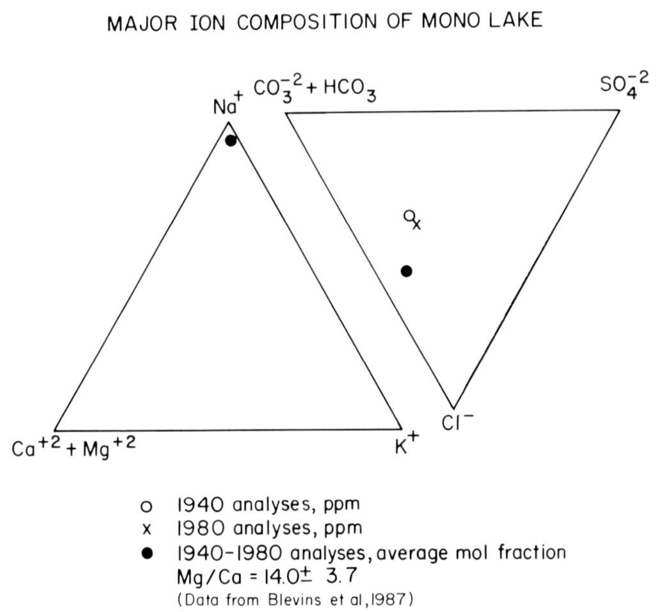

FIG. 2.—Major ion composition of Mono Lake (data from Blevins and others, 1987).

Sedimentology and Geochemistry of Modern and Ancient Saline Lakes, SEPM Special Publication No. 50

years ago at about 1,968-m elevation. Subsequently, the cores record the onset of drier conditions, evaporation of the epilimnion accompanied by increasing salinity, and the eventual overturn and restoration of a monomictic regime. Rather than a complicated, wave-like transgression/regression sequence in late Holocene time, the lacustrine sedimentary record obtained by coring shows a single post-600 yr BP high stand of Mono Lake that peaked around 300 years ago, and during which the lake increased to approximately 3.25 times its present volume.

ACKNOWLEDGMENTS

I thank Sandy Newton, Bob Newton, Marylou Remy, Janet Dodds, Brian Burnham, Alicia Thompson, Bill Ormerod, Mike Torresan, Larry Miller, Sally Henyey, Karen Young, Mike Prichard, Don Lowery, and Eric Stocker for their technical assistance. I am especially grateful to Holly Wagner, Steve Lund, Doug Hammond, Rick Forester, Platt Bradbury, Ed Theriot, Scott Stine, G. I. Smith, Bob Holmes, Bob Newton, Lou Scuderi, Bob Thompson, Roger Anderson, Andrei Sarna-Wojcicki and the two dedicated, anonymous reviewers, whose suggestions were scrupulously followed! Special thanks are due to Owen Davis for sharing his piston core and absolute dates, and to Professor John Melack and Gayle Dana of UCSB for providing the bottom sediment samples. Brian White and Susan Nakagawa of the LADWP generously donated copies of the Pelagos bathymetric maps of Mono Lake. The work was funded, in part, by grants from the Geological Society of America and the Graduate Research Fund at University of Southern California. The paleomagnetic work was supported by NSF grants.

REFERENCES

ADAM, D. P., AND WEST, G. J., 1983, Temperature and precipitation estimates through the last glacial cycle from Clear Lake, California, pollen data: Science, v. 219, p. 168–170.

ANDERSON, R. Y., DEAN, W. E., BRADBURY, P. J., AND LOVE, D. 1985, Meromictic lakes and varved lake sediments in North America: United States Geological Survey Bulletin 1607, 19 p.

BAILEY, D. K., 1980, Volcanism, earth degassing and replenished lithosphere mantle: Philosophical Transactions of the Royal Society of London, Series A, v. 297, p. 309–322.

BENSON, L. V., AND THOMPSON, R. S., 1987, Lake level variation in the Lahontan Basin for the past 50,000 years: Quaternary Research, v. 28, p. 69–85.

BENSON, L. V., CURRY, D. R., DORN, R. I., LAJOIE, K. R., OVIATT, C. G., ROBINSON, S. W., SMITH, G. I., AND STINE, S., 1990, Chronology of expansion and contraction of four Great Basin lake systems during the past 35,000 years: Palaeogeography, Palaeoclimatology, Palaeoecology, v. 78, p. 241–286.

BENSON, L. V., MEYERS, P. A., AND SPENCER, R. J., 1991, Change in size of Walker Lake during the past 5000 years: Palaeogeography, Palaeoclimatology, Palaeoecology, v. 81, p. 189–214.

BISCHOFF, J. L., HERBST, D. B., AND ROSENBAUER, R. J., 1991, Gaylussite formation at Mono Lake, California: Geochimica et Cosmochimica Acta, v. 55, p. 1743–1747.

BISCHOFF, J. L., FITZPATRICK, J. A., AND ROSENBAUER, R. J., 1993, The solubility and stabilization of ikaite ($CaCO_3 \cdot 6H_2O$) from 0° to 25°C: paleoclimatic implications for thinolite tufas: Journal of Geology, v. 101, p. 21–34.

BLANC, R. P., AND CLEVELAND, G. B., 1961, Pleistocene lakes of southern California — I: Sacramento, Mineral Information Service, California Division of Mines, 7 p.

BLEVINS, M. L., COUFAL, E. L., WATANABE, L. M., GARCIA, M. R., FOLEY, C. J., AND JIMENEZ, L. C., 1987, Mono Basin Geology and Hydrology: Los Angeles, Los Angeles Department of Water and Power, Aqueduct Division, Hydrology Section, 100 p.

BORN, S. M., 1972, Late Quaternary history, deltaic sedimentation, and mudlumps formation at Pyramid Lake, Nevada: Reno, Publications of the University of Nevada Desert Research Institute, 97 p.

BRADBURY, J. P., 1987, Late Holocene diatom limnology of Walker Lake, Nevada: Archive für Hydrobiologia/Supplementbande 79 (Monographische Beitrage), v. 1, p. 1–27.

BRADBURY, J. P., FORESTER, R. M., AND THOMPSON, R. S., 1988, Late Quaternary limnology of Walker Lake, Nevada: Journal of Paleolimnology, v. 1, p. 249–267.

BRADLEY, R. S., 1985, Quaternary Paleoclimatology: Boston, Allen and Unwin, 472 p.

BURSIK, M., AND SIEH, K., 1989, Range front faulting and vulcanism in the Mono Basin, Eastern California: Journal of Geophysical Research, v. 94, Part B, p. 15587–15610.

CHRISTENSEN, M. N., AND GILBERT, C., 1964, Basaltic cone suggests origin of some guyots: Science, v. 143, p. 240–242.

COUNCIL, C. C., AND BENNETT, P. C., 1993, Geochemistry of ikaite formation at Mono Lake, California: Implications for the origin of tufa mounds: Geology, v. 21, no. 11, p. 971–975.

DANA, G. L., AND JELLISON, R., 1989, Aquatic ecology: The Mono Lake Newsletter, v. 11, p. 11.

DAVIS, O. K., AND MORATTO, M. J., 1987, Evidence for a warm, dry early Holocene in the Western Sierra: Pollen and plant macrofossil analysis of Dinkey and Exchequer Meadows: Madrono, v. 35, p. 132–149.

FORESTER, R. M., 1987, Late Quaternary paleoclimate records from lacustrine ostracodes, *in* Ruddiman, W. F., and Wright, H. E., eds., North America and adjacent oceans during the last deglaciation (The Geology of North America, Volume K-3): Boulder, Colorado, Geological Society of America, p. 261–276.

GALE, H. S., 1914, Salines in the Owens, Searles and Panamint basins, southeastern California: United States Geological Survey Bulletin 580-L, p. 251–323.

GARRELS, R. M., AND MACKENZIE, F. T., 1967, Origin of the chemical compositions of some springs and lakes, *in* Equilibrium Concepts in Natural Water Systems: Washington, D. C., American Chemical Society, Advances in Chemistry No. 67, p. 222—242.

GILBERT, C. M., CHRISTENSEN, M., AL-RAWI, Y., AND LAJOIE, K. R., 1968, Volcanic and structural history of Mono Basin: Boulder, Geological Society of America Memoir 116, 58 p.

HAMMER, U. T., 1986, Saline Lake Ecosystems of the World: Dordrecht, Junk, 616 p.

HARDIE, L. A., SMOOT, J. P., AND EUGSTER, H. P., 1978, Saline lakes and their deposits: a sedimentological approach, *in* Matter, A., and Tucker, M. E., eds., Modern and Ancient Lake Sediments: London, International Association of Sedimentologists Special Publication 2, p. 7–41.

HERBST, D, 1986, Comparative studies of the population ecology and life history patterns an alkaline salt lake insect: *Ephydra (Hydropyrus) hyans* Say (Diptera: Ephydridae): Unpublished Ph.D. dissertation, Oregon State University, Corvallis, 222 p.

HERBST, D, AND BRADLEY, T., 1989, A Malpighian tubule lime gland in an insect inhabiting alkaline salt lakes: Journal of Experimental Biology, v. 145, p. 63–78.

JELLISON, R., AND MELACK, J. M., 1993, Meromixis in hypersaline Mono Lake, California, Part 1, Stratification and vertical mixing during onset, persistence and vertical breakdown of meromixis: Limnology and Oceanography, v. 38, p. 1008–1019.

JELLISON, R., MILLER, L. G., MELACK, J. M., AND DANA, G. L., 1993, Meromixis in hypersaline Mono Lake California, Part 2, Nitrogen fluxes: Limnology and Oceanography, v. 38, p. 1020–1039.

KISTLER, R. W., 1966, Geologic map of the Mono Craters Quadrangle, Mono and Tuolumne Counties: Washington, D. C., United States Geological Survey Map, GQ-462.

LAJOIE, K. R., 1968, Late Quaternary stratigraphy and geologic history of Mono Basin, California: Unpublished Ph.D. Dissertation, University of California, Berkeley, 271 p.

MASON, D. T., 1967, Limnology of Mono Lake, California: Los Angeles, University of California Publications in Zoology, v. 83, 110 p.

MELACK, J. M., 1983, Large, deep salt lakes: a comparative limnological analysis: Hydrobiologia, v. 105, p. 223–230.

MILLER, L. G., CULBERTSON, C. W., MAEST, A. S., OREMLAND, R. S., AND THORNE, J. H., 1989, Biogeochemistry of Mono Lake 1: Chemical response to stratification and a history of meromixis: Annual Meeting, Southern California Academy of Sciences, May 12–13, unpaginated.

MILLER, L. G., JELLISON, R., OREMLAND, R. S., AND CULBERTSON, C. W., 1993, Meromixis in hypersaline Mono Lake, Ca. III: Biogeochemical response to stratification and overturn: Limnology and Oceanography, v. 38, p. 1040–1051.

NEWTON, M. S., 1991, Holocene stratigraphy and magneto-stratigraphy of Owens and Mono Lakes, California: Unpublished Ph.D. Dissertation, University of Southern California, Los Angeles, 330 p.

NEWTON, M. S., AND GROSSMAN, E. L., 1988, Late Quaternary chronology of tufa deposits, Walker Lake, Nevada: Journal of Geology, v. 96, p. 417–434.

PAKISER, L. C., 1968, Seismic evidence for the thickness of Cenozoic deposits in the Mono Basin, California: Geological Society of America Bulletin, v. 79, p. 1833–1838.

PATTEN, D. T., ed., 1987, The Mono Basin Ecosystem: Effects of Changing Lake Level: Washington, D. C., National Academy Press, 272 p.

PUTNAM, W. C., 1949, Quaternary geology of the June Lake district, California: Geological Society of America Bulletin, v. 80, p. 1281–1302.

REED, W. E., 1977, Biogeochemistry of Mono Lake, California: Geochimica et Cosmochimica Acta, v. 41, p. 1231–1245.

RIND, D., PETEE, D., BROECKER, W., MCINTYRE, A., AND RUDDIMAN, W., 1986, The impact of cold North Atlantic sea surface temperatures on climate: Implications for the Younger Dryas cooling (11–10K): Climate Dynamics, v. 1, p. 3–35.

RUSSELL, I. C., 1889, Quaternary History of Mono Valley, California: Washington, D. C., U.S. Geological Survey 8th Annual Report, 394 p.

SARNA-WOJCICKI, A. M., CHAMPION, D. E., AND DAVIS, J. O., 1983, Holocene volcanism in the conterminous United States and the role of silicic volcanic ash layers in correlation of latest-Pleistocene and Holocene deposits, *in* Wright, H. E., ed., Late Quaternary Environments of the United States, Volume 2, The Holocene: Minneapolis, University of Minnesota Press, p. 52–78.

SCHOLL, D. W., VON HUENE, R., ST. AMAND, P., AND RIDLON, J. B., 1967, Age and origin of topography beneath Mono Lake, a remnant Pleistocene lake, California: Geological Society of America Bulletin, v. 78, p. 583–600.

SHEARMAN, D. J., AND SMITH, A. J., 1985, Ikaite, the parent mineral of jarrowite-type pseudomorphs: Proceedings of the Geologists Association, London, v. 96, p. 305–314.

SHEARMAN, D. J., MCGUGAN, A., STEIN, C., AND SMITH, A. J., 1989, Ikaite, $CaCO_3 \cdot 6H_2O$, precursor of the thinolites in the Quaternary tufas and tufa mounds of the Lahontan and Mono Basins, western United States: Geological Society of America Bulletin, v. 101, p. 913–917.

SIEH, K., AND BURSIK, M., 1987, Most recent eruption of the Mono Craters, eastern central California: Journal of Geophysical Research, v. 91, p. 539–571.

SIMPSON, H. J., 1970, Closed basin lakes as a tool in geochemistry: Unpublished Ph.D. dissertation, Columbia University, New York, 325 p.

SIMPSON, H. J., AND TAKAHASHI, T., 1979, Interstitial water studies, Leg 15, — chemical model of seawater and saline waters: New York, Lamont-Doherty Geological Observatory Contribution No. 1907, p. 877–886.

SMITH, G. I., 1976, Origin of lithium and other components in the Searles Lake evaporites, California, *in* Vine, J. D., ed., Lithium Resources and Requirements by the year 2000: Washington, D. C., United States Geological Survey Professional Paper 1005, p. 92–103.

SMITH, G. I., 1990, How chemically precipitated varves can add detail to a reconstructed lake history (abs.): Abstracts, Geological Society of America Penrose Conference on Stratigraphy of Deep Lakes, Lake Tahoe, California, unpaginated.

SMITH, G. I, AND PRATT, W. P., 1957, Core logs from Owens, China, Searles and Panamint Basins, California: United States Geological Survey Bulletin 1045A, p. 1–62.

SMITH, G. I, AND STREET-PERROTT, F. A., 1983, Pluvial lakes of the western United States, *in* Wright, H. E., ed., Late Quaternary Environments of the United States, Volume 1, The Late Pleistocene: Minneapolis, University of Minnesota Press, p. 190–214.

SMITH, R. S. U., 1978, Pluvial history of Panamint Valley, California: Los Angeles, Guide Book, Friends of the Pleistocene Field Trip, November 11, 1978, 36 p.

STEWART, J. H., 1977, Basin Range Structure in Western North America: a Review: Boulder, Geological Society of America Memoir 152, 31 p.

STINE, S. W., 1987, Mono Lake: The past 4,000 years: Unpublished Ph.D. Thesis, University of California, Berkeley, 614 p.

STINE, S. W., 1990, Late Holocene fluctuations of Mono Lake, eastern California: Palaeoclimatology, Palaeogeography, Palaeoecology, v. 78, p. 333–381.

VEROSUB, K. L., MEHRINGER, P. J., AND WATERSTRAAT, P., 1986, Holocene secular variation in western North America: The paleomagnetic record from Fish Lake, Harney Co., Oregon: Journal of Geophysical Research, v. 91, p. 3609–3623.

WOOD, S. H., 1977, Distribution, correlation and radiocarbon dating of late Holocene tephra, Mono and Inyo Craters, eastern California: Geological Society of America Bulletin, v. 88, p 89–95.

PALEOENVIRONMENTAL IMPLICATIONS OF LACUSTRINE CLAY MINERALS FROM THE DOUBLE LAKES FORMATION, SOUTHERN HIGH PLAINS, TEXAS

DANIEL M. WEBSTER AND BLAIR F. JONES
United States Geological Survey, 432 National Center, Reston, Virginia 22092

ABSTRACT: To investigate whether clay minerals reflect periodic variations in paleolake conditions, major-element chemical and X-ray diffraction analyses were done on ultrafine (<0.1 μm) fractions of Pleistocene-Holocene lacustrine sediments from the Double Lakes Formation, southern High Plains, Lynn County, Texas. Above a Cretaceous marine-shale substrate dominated by well-crystallized aluminous dioctahedral smectite, the clay-mineral assemblages of the younger lacustrine sediments show cyclic changes perhaps related to fluctuations in hydrologic environment and climate. The cycles apparently represent evaporatively induced salinity shifts from brackish to saline (perennial), or ephemeral-lake (playa) conditions, as inferred from the predominance of sepiolite, interstratified Mg-smectite, and palygorskite respectively. Two of at least four cycles show a complete sequence from low to high lake level and return to playa conditions. Between sediment units clearly dominated by a single clay-mineral species are intervals containing variable mixtures of chain-structure clays and smectite indicating change in the conditions of paleolake deposition. Chemically, the highest ratios of Si^{4+} to Al^{3+} are associated with sepiolite and the highest Mg^{2+} to Al^{3+} values are correlated with the abundance of interstratified smectite. Pulses of detritus associated with overland runoff and wind are indicated by high concentrations of illite and minor kaolinite, as evidenced by high K^{+} and Al^{3+} contents. Spring deposits (tufas) along a principal drainageway near the lake suggest inflow from groundwater has had a major effect on the nature of the clays.

INTRODUCTION

A number of studies have dealt with the question of the extent to which depositional conditions in lakes are reflected in the nature and diagenesis of clays in lacustrine sediments. The pioneering work in the field was that of Millot (1954, 1964) who recognized in ancient saline, lacustrine sediments an increase in silica and other cations, especially magnesium, relative to alumina in clay fractions from basin margin to center, often accompanied by a mineralogic change from smectite to a chain-structure clay. In their classic model of the development of alkaline, saline (so-called "soda") lakes by evaporative concentration of drainage from crystalline rock terranes, Garrels and Mackenzie (1967) proposed the direct precipitation of sepiolite from lake waters as the principal control on solute concentrations of silica and magnesium. Similarly, Gac and others (1977), following the indications of Carmouze (1976), used an evaporative concentration model to postulate the authigenic formation of smectite in Lake Chad in central Africa. Subsequently, Jones and Weir (1983) and Jones and Spencer (1985) have confirmed the sequence of more magnesian and siliceous smectite associated with more saline lacustrine conditions. Such an association has been further documented for the chain-structure clays, as well as Mg-smectite, by Trauth (1977) and Paquet and others (1987). Jones (1986) and Jones and Galan (1988) have suggested that the chain-structure clays are formed by direct precipitation from solutions more dilute (palygorskite) or aluminous-colloid-free (sepiolite) than waters forming stevensitic or Mg-rich smectite, which, in turn, is more likely to result from transformation of precursor 2:1 layer silicate. Thus the clay mineralogy of closed-basin lacustrine sediments might be used to indicate variations in hydrochemical conditions of deposition.

A number of well-defined, variably saline, closed basins on the southern High Plains of Texas, larger than 15 km^2 (5.8 mi^2) and incised 15.2–30.5 m (50–100 ft) below the regional land surface, have been described morphologically by Reeves (1966, 1976), Gustavson and Winkler (1988), Wood and Osterkamp (1984); hydrologically by Wood and Sanford (1990), and hydrochemically by Wood and Jones (1990). The clay mineralogy of Pleistocene to Holocene lacustrine sediments from several of these basins has been investigated by Parry and Reeves (1966, 1968), McLean and others (1972), Harris and others (1972), and McGrath (1984). The general stratigraphy of the upland areas around the basins consists of Permian to Cretaceous strata overlain by the Ogallala Formation of late Tertiary age strata which are locally topped by the Blanco Formation of Pliocene age and Blackwater Draw Formation of Quaternary age (Gustavson and Holiday, 1985). The clay minerals in these latter formations have been involved in calcic-soil development, and to a lesser extent, possibly modified by local lacustrine environments. A number of studies have associated chain-structure clays and smectites with specific soil, calcrete, playa, and lacustrine deposits common to the Llano Estacado and other similar localities in the semiarid southwestern United States. Parry and Reeves (1968), McLean and others (1972), Harris and others (1972), Reeves (1976), and McGrath (1984) have associated abundant palygorskite with the development of petrocalcic soils, and abundant sepiolite with lacustrine deposits. At the same time, smectites have provided no consistent characteristics for differentiating lake sediments from soils. Osterkamp and Wood (1987) reported an iron-rich smectite in sediment from a very small playa north of Lubbock, Texas, where smectite immediately below the playa was leached of iron and manganese oxides. Parry and Reeves (1968) indicated that most of the clay in lacustrine sediments in six of the larger saline-lake basins south of Lubbock, Texas, were detrital contributions from surrounding paleosols that antedated the Late Pleistocene Tahoka Formation (lacustrine deposits) of Evans and Meade (1945). Authigenic sepiolite was considered to be the most reliable indicator of lacustrine depositional conditions in the larger saline basins. McLean and others (1972) concluded that both palygorskite and sepiolite were lacustrine in origin and that the different chain-structure clays were related to variations in climatic conditions over the entire High Plains of Texas and eastern New Mexico. They noted that palygorskite was usually accompanied by calcite,

Sedimentology and Geochemistry of Modern and Ancient Saline Lakes, SEPM Special Publication No. 50

whereas sepiolite was associated with dolomite. The latter assemblage was attributed to increasing Mg^{2+} concentration with time.

Several studies, including the classic work of Vanden Heuvel (1966) on calcic soil horizons of southern New Mexico, have reported pedogenic sepiolite in the arid southwestern United States. Therefore, the presence of sepiolite alone cannot be used to infer lacustrine conditons. Hay and Wiggins (1980) recorded abundant sepiolite and only minor palygorskite coexisting in three well-developed pedogenic environments of the southwestern Great Basin. McGrath (1984) found sepiolite in the calcretes at West Mesa, Texas. Parry and Reeves (1968), and Bachman and Machette (1977) suggested a topotactic transformation process that formed sepiolite from precursor palygorskite; thus, the abundance of sepiolite was perhaps a direct function of the age of the calcrete. Most research has supported a solution-precipitation mechanism for chain-structure-clay formation (Vanden Heuvel, 1966; Gardner, 1972; McLean and others, 1972; Singer, 1977; Khoury and others, 1982; Jones, 1983, 1986; Jones and Galan, 1988), with their occurrence in surficial deposits most commonly resulting from evaporative concentration of vadose solutions. According to Jones (1983), with the parent material of most calcic horizons, palygorskite is favored initially and sepiolite can only precipitate after complete extraction of solute or colloidal alumina from solution. Sepiolite dominance is attributed to alkaline solutions with significant sources of solute Mg^{2+} and Si^{4+} (e.g., from dissolution of ancient dolomite and siliceous volcanics).

Objectives of the Double Lakes Formation study were twofold: (1) to examine the ultrafine mineralogy for type and relative abundance of clay species, and (2) to determine if differences in clay minerals can be related to variations in paleolacustrine conditions. Investigations were focused on the fine-clay (<0.1 μm) fraction because it was considered the most reactive with natural waters, and as McLean and others (1972) noted, chain-structure clays were the most abundant in <0.2-μm segregates.

GEOLOGIC AND HYDROLOGIC SETTING

The Double Lakes Basin is physiographically situated in the southern High Plains (Llano Estacado) of Texas. Reeves (1976), and Osterkamp and Wood (1987) describe the southern High Plains as a remnant of the alluvial plain formed by fluvial erosion of the southern Rocky Mountains to the west. Detrital transport and deposition of Ogallala sediments formed an alluvial plain during Miocene-Pliocene time. Wedge-shaped sediment aprons thin toward the eastern edge of the High Plains escarpment in Texas. The Ogallala Formation consists of fluvial gravels and sands interbedded with and overlain by aeolian sediments deposited as sand sheets and loess containing calcic paleosols (Gustavson and Winkler, 1988). The Ogallala Formation is topped by a thick, resistant calcrete (Reeves, 1970). A thin cover of fluvial sands, silts and clays, reworked by aeolian processes during Pleistocene glaciation, forms the Blackwater Draw Formation covering much of the Ogallala Formation in Texas. Blackwater Draw deposits are strongly marked, particularly at the top, by soil profiles that commonly contain calcic zones. Lacustrine sediments described by Evans and Mead (1945) and termed the Double Lakes and Tahoka Formations by Reeves (1976) were deposited during early and late Wisconsin time, respectively. Figure 1A shows the stratigraphic position of the Quaternary Double Lakes and Tahoka Formations in relation to underlying Tertiary-Cretaceous beds discussed by Reeves and Temple (1986).

The Ogallala and Blackwater Draw Formations compose the principal aquifer system of the Llano Estacado. Recharge is roughly 0.3 cm (1 inch) per year (Wood and Osterkamp, 1984). Recharge occurs primarily through numerous very small shallow (4 m, ~12 ft deep) playa-lake basins on the plains surface (Wood and Osterkamp, 1984).

Differences in basin size, sediment composition, and water chemistry distinguish recharge from discharge playas on the plains surface. Recharge playa waters are lost to the underlying aquifer, before significantly increasing solute concentration by evaporation (Harris, 1972; Wood and Osterkamp, 1984). Discharge playas are relatively large closed-basins intersecting the zone of saturation, thereby subjecting aquifer waters to evaporative concentration. Solute concentrations increase by one to three orders of magnitude over concentrations in aquifer waters (Wood and Jones, 1990). Evaporitic environments are indicated by the presence of abundant salts and authigenic clay minerals in the sediments (Parry and Reeves, 1966, 1968; McLean and others, 1972; Reeves, 1976; Reeves and Temple, 1986; Wood and Osterkamp, 1984, 1987; Wood and Sanford, 1990). Though several of the larger basins have been attributed to dissolution of evaporites at depth and subsequent collapse (Finley and Gustavson, 1981), Wood and others (1992) suggest a "modified deflation model" to explain the evolution of many of the more prominent saline-lake basins on the High Plains surface. The model reflects an increased hydraulic gradient and subsequent water table rise near Permian to Cretaceous bedrock highs associated with aquifer thinning. Under such conditions, groundwater discharges as springs forming tufas or marls at the head of local drainageways, but these points of discharge are small compared to the area of overall rise in the water table, leaving a significant area devoid of carbonate cement. Collectively, these conditions locally preclude calcrete formation, which, in turn, promotes deflation of loose sandy surface sediment, and basin enlargement.

LOCATION AND SAMPLE-COLLECTION METHODS

The Double Lakes area is located in Lynn county, 40 km (25 mi) south of Lubbock, Texas. According to Wood and others (1992), the basin is typical of the large saline-lake basins of the southern High Plains. The lakes are located within a local paleovalley incised into a bedrock high with small outcrops of Duck Creek and Kiamichi Formations on the northwestern margin of the lake. The basin is topographically closed, and its low point is ~25 m below the regional land surface. The basin is surrounded by a thick, continuous rim of calcrete caprock and has lee-side dunes. The floors are continually damp, indicating the discharge of groundwater to the lake. Groundwater beneath the lakes

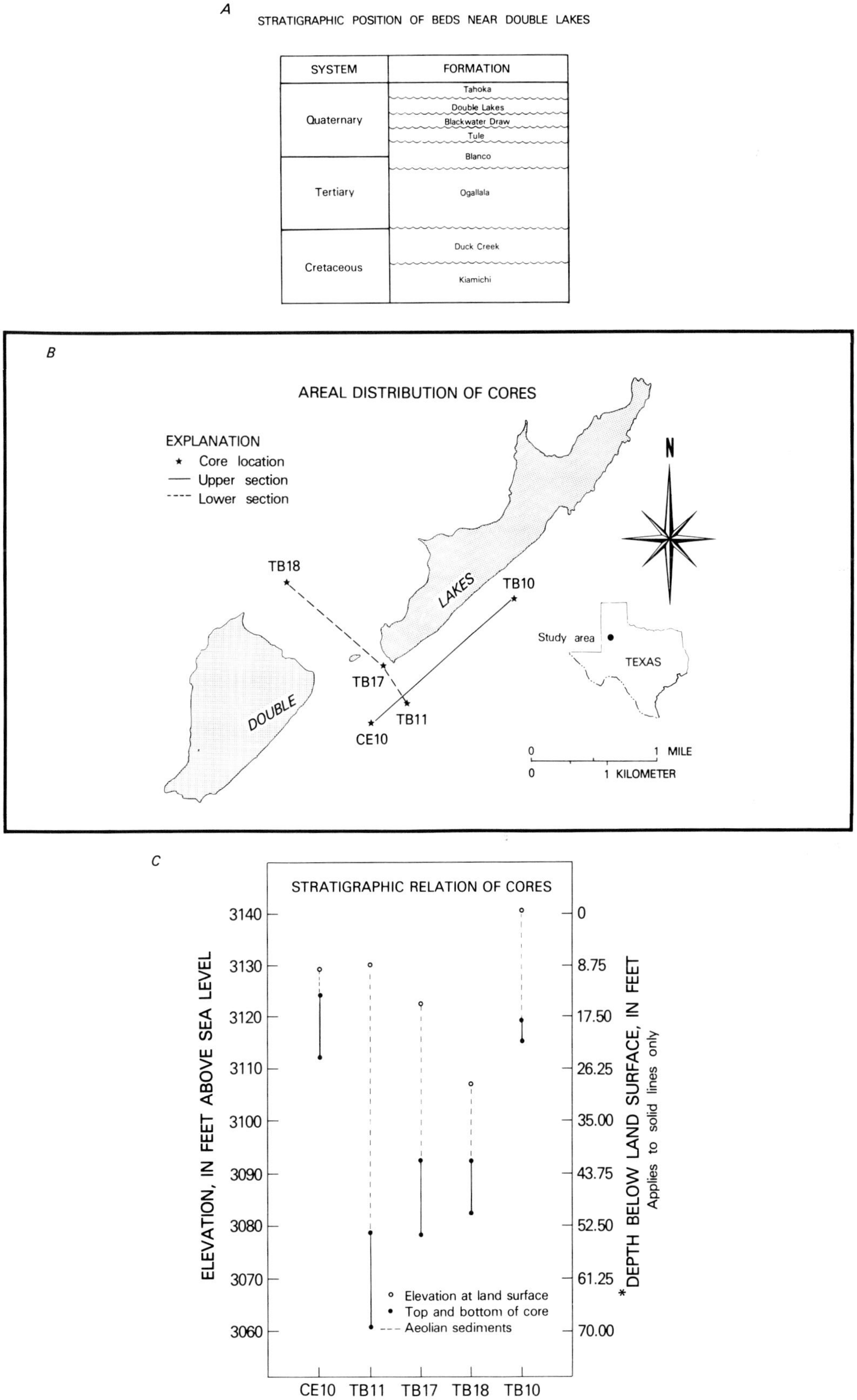

FIG. 1.—Location of Double Lakes area in the southern High Plains of Texas: (A) Stratigraphic position of Quaternary, Tertiary, and some Cretaceous beds below the Double Lakes and Tahoka Formations discussed by Reeves and Temple (1986); (B) Areal distribution of core locations; and (C) Stratigraphic relation of cores.

is of the sodium sulfate type; dissolved solids exceed 200,000 ppm.

Five cores were collected at specified locations around the modern saline-playa basin. The boreholes were drilled by the mud-rotary method through fluvial-aeolian sediments to depths where lacustrine deposits were identified by significant increases in clay content, and then cored until Cretaceous rocks were encountered. Samples of lacustrine sediments were collected at 0.61-m (2-ft) intervals from steel core barrels 7.62 cm (3 inches) in diameter. The sample-collection points ranged in depth from 1.52–21.34 m (5–70 ft) below land surface, spanning altitudes of 952.2 m (3,124 ft) to 932.7 m (3,060 ft) above sea level. Figure 1B shows the areal distribution of the five cores, whereas Figure 1C shows their stratigraphic relations. Sample depths, elevations, and sediment descriptions are given in Table 1.

The top, middle, and bottom parts of each 0.61-m (2-ft) long core segment were subsampled for X-ray diffraction (XRD) and chemical analysis. The five boreholes collectively provided 24.1 m (79 ft) of lacustrine sediment from outside the margin of the modern playa lake, from which a total of 85 samples were selected. Roughly 300 g of sediment were taken from each 0.23-m (0.75-ft) section. Final sampling intervals were rounded to the nearest 0.30 m (ft) and converted to elevations based on core location sites surveyed prior to drilling. All sediment described in Table 1 appears relatively homogeneous with respect to grain size, texture, and color.

SAMPLE-PROCESSING TECHNIQUES

Bulk samples were processed in the laboratory to remove coarse or soluble mineral components that interfere with XRD and chemical analysis of the clays. Eighty-two samples were treated with acetic acid to remove carbonate minerals through modification of the carbonate-digestion technique of Jackson (1969). Instead of adding buffer solution, the pH of the dispersed sample suspension was kept at 4.5 by continuously-monitored addition of acetic acid, and if necessary, NaOH. Three samples that contained sufficient gypsum to further inhibit fractionation of the ultrafines were treated with ethylenedinitrilotetraacetic acid (EDTA) to remove sulfate and carbonate minerals (Bodine and Fernalld, 1973). Samples with minor gypsum were placed in deionized water, ultrasonified, and centrifuged as many times as necessary to reduce the ionic strength of the sample suspension, resulting in dispersion of clay-sized particles. All samples were size-fractionated by centrifugation to obtain at least 250 milligrams of <0.1 μm particles for X-ray diffraction and chemical analysis.

X-RAY DIFFRACTION PROCEDURE

The ultrafine clay mineralogy was determined with an X-ray diffractometer using long-broad focused CuKa radiation spectrally isolated with a graphite monochrometor and a nickel filter. The scan rate was 1 degree 2-theta (1°2θ) per minute using a soller slit of 1° and a receiving slit of 1/4°. The ultrafine colloidal gel was dried on rectangular glass plates at laboratory temperature (25°C) to achieve maximum phyllosilicate orientation. At least two primary X-ray diffraction scans of each sample were obtained with the rate meter set at 2,000 and 5,000 counts per second (cps) and a 1-second time constant. Bulk random-powder scans were done at scale factors of 500 and 5,000 cps. The procedure used for mounting bulk random-powders was described by McCreery (1949), and Zen and Hammarstrom (1975). Layer silicate identification was based on X-ray diffractograms of samples that were untreated, ethylene glycol-solvated, and heated at 550°C. Randomly mounted ultrafine material was used to obtain (060) reflections for assessment of the octahedral occupancy of the clay structures.

BULK MINERALOGY

Bulk-mineral abundances were quantified by an external standard using designated reference peaks for both unknowns and standards (Schultz, 1964; Tatlock, 1966; Klug and Alexander, 1974; Webster, 1989). The reference-peak intensity was ratioed to that of a high-purity standard mineral, then normalized to 100 percent. Semiquantitative precision is about 10 percent. Table 2 shows the range in quantification for eight fundamental minerals composing a matrix common to all five cores. The matrix constituents are quartz, calcite, dolomite, sodic plagioclase, orthoclase, micas, gypsum, and total-clay (as defined by the *hk0* reflection at 19.9°2θ). The most useful quantification data are in the correlation of carbonates and clays. Generally, sepiolite and palygorskite are associated with dolomite and calcite, and are usually correlated with the highest carbonate values, especially palygorskite with calcite. Layers of illite, mixed with minor amounts of kaolinite, often contain calcite and only trace levels of dolomite. Interstratified Mg-smectite is always associated with dolomite. Interstratified Mg-smec-

TABLE 1.—GENERAL DESCRIPTION AND INFORMATION FOR DOUBLE LAKES CORE SEDIMENTS

Core	Elevation at land surface	Depth Interval below land surface	Elevation Interval	Sediment Description	Munsell Soil Color
TB11	3,130.30 ft	52–70 ft	3,078–3,060 ft		
	(954.12 m)	(15.85–21.34 m)	(938.17–932.69 m)	Very fine-grained clay-rich lake marl	5Y5/1–10Y8/2
TB17	3,122.50 ft	30–44 ft	3,092–3,078 ft		
	(951.74 m)	(9.14–13.41 m)	(942.44–938.17 m)	Very fine-grained clay-rich lake marl	5Y5/1–10Y8/2
TB18	3,107 ft	15–25 ft	3,092–3,082 ft		
	(947.01 m)	(4.57–7.62 m)	(942.44–939.39 m)	Very fine-grained clay-rich lake marl	10YR6/6–10YR8/1
CE10	3,129 ft	5–17 ft	3,142–3,112 ft		
	(953.72 m)	(1.52–5.18 m)	(921.72–948.54 m)	Very fine-grained clay-rich sandy marl	5YR6/2–7.5YR7/6
TB10	3,140.30 ft	12–25 ft	3,119–3,115 ft		
	(957.16 m)	(3.66–7.62 m)	(950.67–949.45 m)	Very fine-grained clay-rich sandy marl	7.5YR7/2–7.5YR8

TABLE 2.—RANGE IN SEMIQUANTITATIVE BULK MINERALOGY, IN WEIGHT PERCENT

Core	Quartz	Calcite	Dolomite	Plagioclase	K-Feldspar	Total Clay	Mica	Gypsum
TB11	5–45	tr–25	<5–60	tr–<5	tr–20	25–60	0–30	0–<5*
TB17	10–55	tr–10	<5–30	<5–20	<5–20	20–55	0–20	0–20
TB18	5–50	0–10	<5–25	<5–10	<5–10	30–55	0–20	0–tr*
CE10	10–50	5–50	0–10	<5–10	5–10	20–55	0–20	0–tr*
TB10	20–40	<5–20	tr–5	<5–10	<5–10	40–60	0–30	0–tr

Asterisk (*) = unidentified sulfate salt (glauberite and/or thenardite), tr = trace.

tite is also associated with high gypsum content, and probably with the alkali sulfates, glauberite and thenardite.

CLAY MINERALOGY

Initially it is necessary to define two terms associated with smectite. Aluminous smectite refers to an expandable clay mineral, with well-defined basal reflections, essentially a beidellite. Conversely, interstratified Mg-smectite refers to a 2:1 phase associated with high magnesium content, as well as very broad basal reflections indicative of significant interstratification. It has not been resolved whether the high magnesium content associated with the interstratified Mg-smectite is solely the result of natural lacustrine conditions or of an exchange process forced by laboratory treatment of bulk sediment with acetic acid, releasing magnesium from dolomite dissolution. At various times, hydrochemical conditions at Double Lakes have been suitable for the development of authigenic stevensite interstratification with dioctahedral smectite, as described by Jones and Weir (1983), and Banfield and others (1991).

X-ray diffraction analysis of ultrafine particulates (<0.1 μm) has identified four clay-mineral types in Double Lakes Formation sediment. These are aluminous smectite, sepiolite, interstratified Mg-smectite, and palygorskite. Each mixture is dominated by one clay-mineral phase but includes nearly equal to minor amounts of other clay species. The assemblage can contain minor to trace levels of illite, kaolinite, alkali feldspar, and quartz. Virtually all (060) measurements show only the reflections of dioctahedral 2:1 clays, though a minor trioctahedral component probably could not be detected in such low intensity peaks.

Aluminous Smectite

The dominant clay-mineral phase in Figure 2, is a well-crystallized dioctahedral smectite (~5°2θ, ~17Å) showing six orders of basal (001) reflections. This smectite is found in the top 0.61 m (2 ft) of the Lower Cretaceous Kiamichi Shale that occurs at the bottom of core TB11. Subordinate peaks of illite (8.8°2θ, 10.1Å) and kaolinite (12.4°2θ, 7.1Å) only occur in samples from above 933.3 m (3,060 ft). Smectite mixed-layering decreases with depth, as evidenced by decreased low-angle shoulder near 3°2θ as well as decreased principal peak full-width at half-maximum. In samples from 932.9 m (3,061 ft) and 933.3 m (3,062 ft), note the shift in the (002) peak for glycol-treated smectite from 10.4°2θ (8.51Å) to a position closer to the principal 10Å peak for illite. These reflections probably represent the resultant basal peak position for both end-members in the mixed-layered phase, following the illite-smectite interstratification model suggested by Mering (1949). All three samples show well-resolved (060) peaks at 1.51Å (61.4°2θ). The sample from 932.7 m (3,060 ft) probably is the best example of well-ordered smectite.

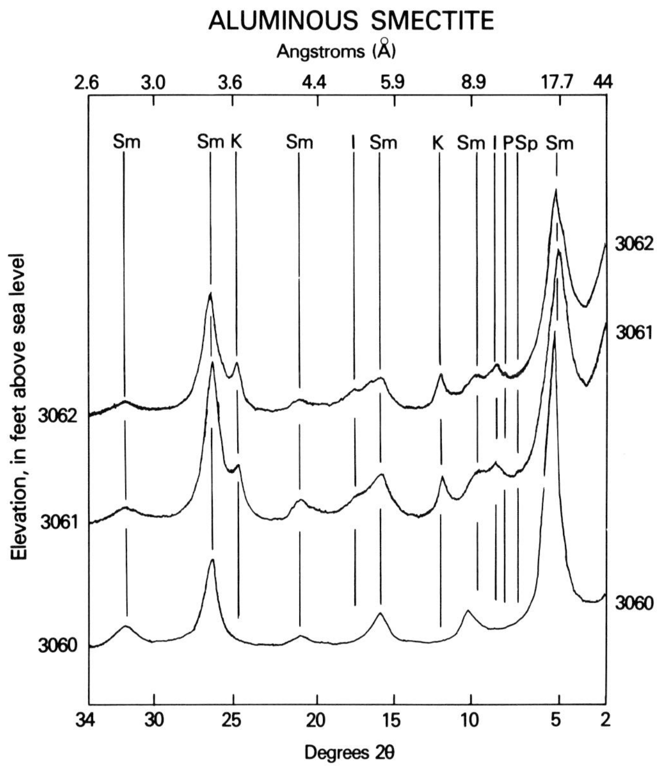

FIG. 2.—Well-crystallized aluminous smectite from the upper 0.61 m (2 ft) of the Cretaceous Kiamichi shale from the bottom of Core TB11. The following conditions apply to Figures 2 through 11: XRD patterns are for ethylene glycol-solvated <0.1 μm oriented mounts analyzed with CuKa radiation at 5,000 CPS and a 1-second time constant. Specific elevation above sea level and core locations are given on the vertical axes. Symbols identify minerals as follows: SM = smectite, SP = sepiolite, P = palygorskite, I = illite, K = kaolinite, and Q = quartz.

Sepiolite

XRD traces reflecting sepiolite as the most prominent clay-mineral species are presented in Figure 3. Sepiolite is identified by the principal (011) peak at 7.3°2θ (12.1Å) which does not shift on glycolation, and identification is supported further by resolution of the (060) reflection at 4.49Å (~20°2θ), as well as a (131) peak at 4.29Å (~21°2θ). These two overlapping peaks produce higher intensity from in-

SHAPIRO, L., 1975, Rapid analysis of silicate, carbonate, and phosphate rocks—revised edition: United States Geological Survey Bulletin, v. 1401, 76 p.

SINGER, A., 1977, Dissolution of two Australian palygorskites in dilute acid: Clays and Clay Minerals, v. 25, p. 126–130.

TATLOCK, D. B., 1966, Rapid model analysis of some felsic rocks from calibrated X-ray diffraction patterns: United States Geological Survey Bulletin, v. 1209, 41 p.

TRAUTH, N., 1977, Argiles Evaporitiques dans la Sédimentation Carbonatée Continental et Epicontinentale Tertiaire: Paris, Sciences Géologiques Mémoire 49, 195 p.

VANDEN HEUVEL, R. C., 1966, The occurrence of sepiolite and attapulgite in the calcareous zone of a soil near Las Cruces, New Mexico: New York, Clays and Clay Minerals, v. 13, p. 193–207.

WEBSTER, D. M., 1989, A semiquantitative X-ray diffraction method to determine mineral compositions in stream sediments with simlar mineralogy: Environmental Technology Letters, v. 10, p. 833–844.

WOOD, W. W., AND JONES, B. F., 1990, Origin of solutes in saline lakes and springs on the Southern High Plains of Texas and New Mexico, *in* Gustavson, T. C., ed., Geologic Framework and Regional Hydrology, Upper Blackwater Draw and Ogallala Formations, Great Plains: Austin, Texas Bureau of Economic Geology, p. 193–208.

WOOD, W. W., AND OSTERKAMP, W. R., 1984, Recharge to the Ogallala Aquifer from playa lake basins on the Llano Estacado (an outrageous proposal?), *in* Whetstone, G. A., ed., Proceedings of the Ogallala Aquifer Symposium II: Lubbock, Texas Tech University, p. 337–349.

WOOD, W. W., AND OSTERKAMP, W. R., 1987, Playa lake basins on the Southern High Plains of Texas and New Mexico, Part II, A hydrologic model and mass-balance argument for their development: Geological Society of America Bulletin, v. 99, p. 224–230.

WOOD, W. W. AND SANFORD, W. E., 1990, Ground-water control of evaporite deposition: Economic Geology, v. 85, p. 1226–1235.

WOOD, W. W., SANFORD, W. E., AND REEVES, C. C. JR., 1992, Large lake basins of the Southern High Plains: ground-water control of their origin?: Geology, v. 20, p. 535–538.

ZEN, E-AN, AND HAMMERSTROM, J. G., 1975, Quantitative determination of dawsonite in Green River Shale by powder-sample X-ray diffraction: effect of grinding: United States Geological Survey Journal of Research, v. 31, p. 21–30.

STRATIGRAPHY AND HOLOCENE HISTORY OF LAKE HAYWARD, SWAN COASTAL PLAIN WETLANDS, WESTERN AUSTRALIA

LEE COSHELL

National Key Centre for Resource Exploration, Curtin University of Technology, Perth, Western Australia, 6001

AND

MICHAEL R. ROSEN

Institute of Geological and Nuclear Sciences, Wairakei Research Centre, Private Bag 2000, Taupo, New Zealand

ABSTRACT: Lake Hayward is a small coastal hypersaline lake in the Clifton-Preston lakeland system of southwestern Australia. The lake is a Na-Cl-SO_4 brine, compositionally similar to seawater, with a salinity exceeding 200 g/l in summer. Although only 2.75 m deep, it is thermally and chemically stratified between April and January.

Groundwater tepees, desiccation features, crusts, gas escape structures, and degrading benthic microbial mats (BMC) occur around the lake margins. Pleistocene carbonate pinnacles acted as nucleation sites for carbonate-precipitating microbialite communities. In the lake center, the surface is covered with non-carbonate precipitating BMC communities, underlain by a thin gypsum layer.

The stratigraphy established from 20 cores shows a disconformable transition from Pleistocene eolian coastal-dune sedimentation at the base of the cores to a Holocene limnic swamp. The swamp phase is dated at about 8,000 years BP. Holocene deposition began when rising sea level caused the freshwater swamp to develop over the Pleistocene dune sand and calcretized calcarenite. Wetland conditions became predominant with progressively higher sea level and concomitant rise in water table. Dune barriers were breached and for a short interval Hayward became part of a complex marine inlet with a marine fauna.

Gradual lowering of sea level isolated Hayward from the rest of the system. Since 7,000 yr BP, a saline lacustrine environment has prevailed. The lacustrine sediments are laminated and dominated by chemically and biologically precipitated carbonates. Significant amounts of diagenetic dolomite are present. The lake-center cores contain ~25 percent TOC.

At Lake Clifton, the latest part of a pre-lacustrine marine inlet phase has been dated at between 4,670 and 3,890 yr BP, ending later than at Lake Hayward. Lake Hayward apparently became isolated from the ocean earlier than Lake Clifton and consequently has a longer lacustrine sedimentary record.

INTRODUCTION

The geochemistry of Holocene coastal evaporitic lakes or salinas has been studied by many authors, generally in areas of extreme aridity (Aharon and others, 1977; Arakel, 1980; Logan, 1981,1987; Kushnir, 1981; Warren, 1982a, b). These studies were concerned primarily with evaporite textures, sedimentology, and brine evolution. Except for the Coorong Region of South Australia, coastal saline lakes in semi-arid to temperate regions have received little attention. This probably reflects the difficulty in finding natural systems not significantly altered by human activity. This is unfortunate because closed saline coastal systems in areas of moderate rainfall may potentially provide detailed stratigraphic records of Holocene sea levels and climatic variations.

The lakes and surrounding wetlands of the Clifton-Preston Lakeland system in Western Australia are part of Yalgorup National Park (Fig. 1). Although not pristine, these wetlands offer an opportunity to study the stratigraphy, geochemistry and hydrology of a saline wetland system in a semi-arid to temperate zone that has had only limited human modification.

The aims of this paper are to describe the stratigraphy, sedimentology and Holocene history of Lake Hayward, one of the lakes in the Clifton-Preston Lakeland System (Fig. 2). Lake Hayward lies in the geographic center of the lake system and is the most saline of the group. The information gained from the study also adds insights into the Holocene sea-level changes and history of the Swan Coastal Plain region of Western Australia.

REGIONAL GEOLOGICAL AND GEOMORPHOLOGICAL SETTING

The Clifton-Preston Lakeland System lies on the coastal margin of the Swan Coastal Plain. This is a narrow north-south trending terrain flanked by the Precambrian Yilgarn Block to the east and the Indian Ocean to the west. Late Tertiary and Quaternary sediments of the plain form a thin veneer over Cretaceous strata (Playford and others, 1976). From east to west, there are four regional geomorphological divisions: (1) Ridge Hill Shelf and Pinjarra Plain, (2) Bassendean Dunes, (3) Spearwood Dunes (including Yoongarillup Plain), and (4) Quindalup Dunes (McArthur and Bettenay 1960; McArthur and Bartle, 1980; Gozzard, 1987; Semeniuk, 1990).

The lakes of the Clifton-Preston System (Fig. 2) lie in linear depressions parallel to the coast within the Yoongarillup Plain and Spearwood Dunes, a terrain composed of Pleistocene limestone mantled by quartz sands of variable thickness. The prominent ridge of the Spearwood Dunes, up to 80 m high, defines the eastern edge of the lakeland system. This sand-covered barrier abuts Pleistocene sands of the Bassendean Dunes. Another ridge of the Spearwood Dunes forms a barrier between Lake Preston and a chain of smaller lakes (the "lake chain") that includes Lake Hayward. The lake chain is separated from Lake Clifton by a low sand barrier. Along the western margin, the Holocene Quindalup dunes form a barrier to the ocean, 20 to 40 m high. This barrier is presently retrograding to the east and being eroded on its seaward face (Semeniuk and Meagher, 1981; Semeniuk, 1985; Semeniuk and others, 1989).

The Pleistocene limestone and quartz sands of the Yoongarillup Plain and Spearwood Dunes are ascribed to the Tamala Limestone of Playford and others (1976). The Bassendean Sand (Playford and others, 1976) is equivalent to the Bassendean Dunes, and the Safety Bay Sand (Passmore, 1970; Playford and Low, 1972) to the Quindalup Dunes. The Tamala Limestone and Bassendean Sand in-

Sedimentology and Geochemistry of Modern and Ancient Saline Lakes, SEPM Special Publication No. 50

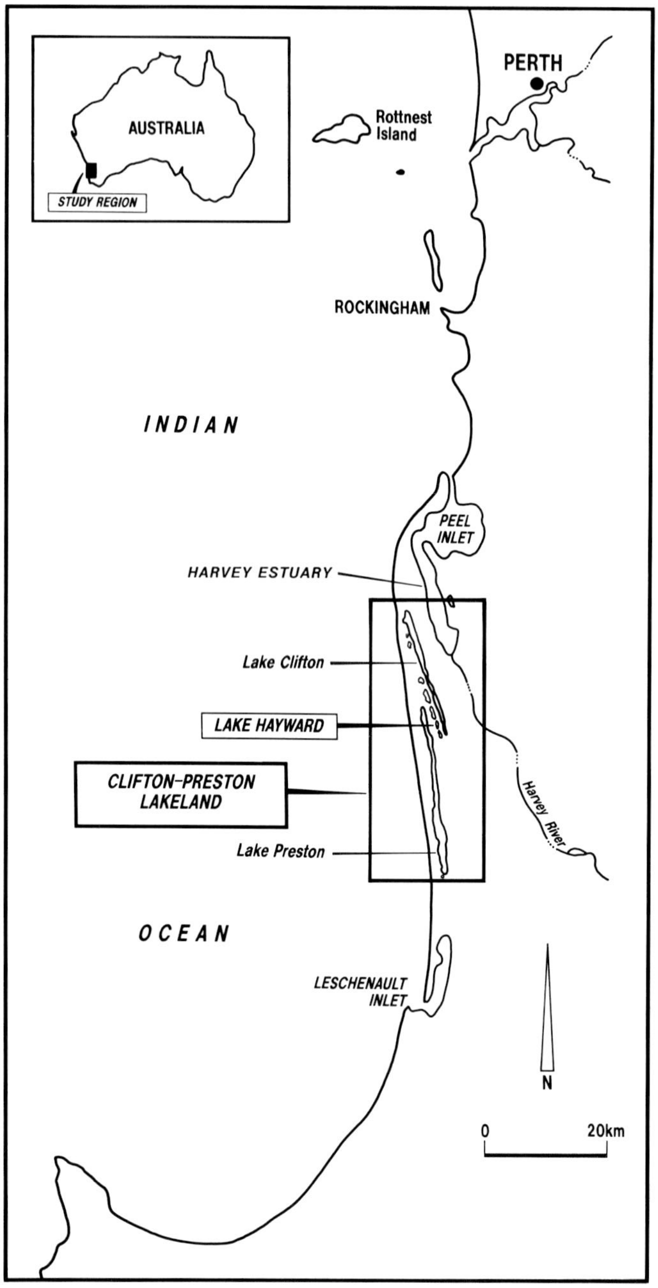

FIG. 1.—Location map of the Clifton-Preston Lakeland system. Map also shows the other areas where sea-level curves have been constructed.

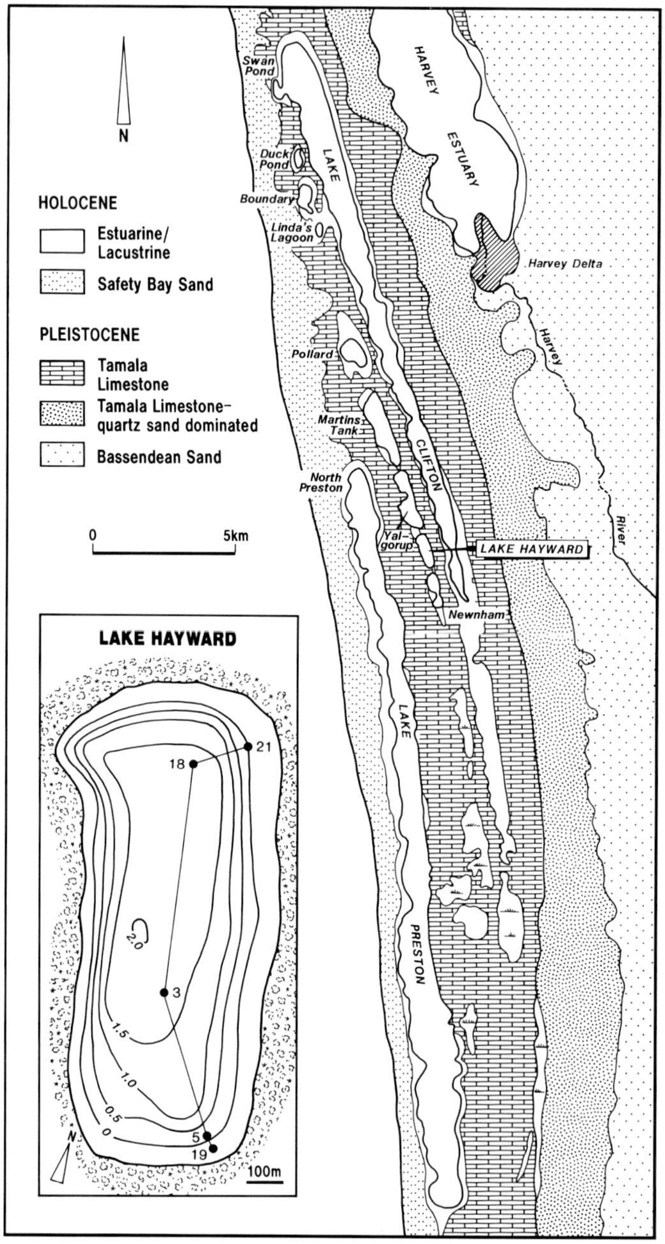

FIG. 2.—Regional geology of the Clifton-Preston Lakeland system. Modified from Wilde and Low (1980), Wilde and Walker (1982), and Gozzard (1987). Inset is Lake Hayward bathymetry (modified from Burke, 1989) and core locations (see Fig. 9 and 10).

terfinger along unconformity interfaces and the Bassendean Dunes contain isolated lenses of limestone (Semeniuk and Glassford, 1987, 1989). The yellow sands of the Bassendean Dunes and those associated with the Tamala Limestone are similar; both consist of yellow-stained quartz and minor feldspar (Commander, 1988). These sands form, in part, the basement of the Holocene sequence at Lake Hayward.

The yellow quartz sand of the Spearwood Dunes may be a residual product derived from leaching of the underlying quartzose limestone (McArthur and Bettenay, 1960; Playford and others, 1976; Commander, 1988). However, the very thick quartz sands in the Yalgorup area are not always underlain by limestone (Commander, 1988; Semeniuk, 1990). Semeniuk and Glassford (1987, 1989) suggested an eolian origin for these sands along the Spearwood and Bassendean Dune contact. Their model (following Fairbridge, 1964; Kukla, 1977; Sarnthein, 1978; Sprigg, 1979; and Glassford, 1980) invokes episodic sand incursions during periods of lower sea levels and associated aridity, with intervening periods of coastal dune building (see Fig. 10 in Semeniuk and Glassford, 1987). Thus, quartz sands near Lake Hayward may derive from reworked desert sands and residual quartz from weathering of calcarenite.

The Tamala Limestone consists of variably indurated eolian calcarenite and shelly beach deposits. High barrier ridges of eolian origin represent terminal stages of Pleistocene marine transgression, and overall, the unit is a complex beach-dune system (Playford and others, 1976; Semeniuk and Johnson, 1985). Semeniuk (1990) suggests that the fossiliferous limestone of the Yoongarillup Plain is a prograded beach-ridge plain that formed behind a barrier island limestone chain. The pre-Holocene framework of the Clifton-Preston area is thus one of multiple north-south trending dune ridges and depressions. These pre-existing features define, for the most part, present-day lake morphology.

CLIMATE

The region has a Mediterranean-type climate with hot, dry summers and wet, cool winters. Rainfall data have been collected at Yalgorup National Park by the Department of Conservation and Land Management since 1982. The rainfall is seasonal, most falling between May and September, with a maximum in June or July. Total annual rainfall values show a wide range (642 to 996 mm) with an average of 769 mm. Evaporation rates are nearly twice the average annual precipitation (Hodgkin and others, 1980).

HYDROLOGY

The lakes of the Clifton-Preston system receive almost no runoff, but are maintained by direct rainfall and groundwater inflow (Commander, 1988). Commander (1988) reported that the lakes are underlain by hypersaline groundwater (42–64 g/1) to a depth of 20 m. Consequently, lake-levels remain below sea level for most of the year. The marked seasonality produces a cyclic pattern of annual lake-level rises and falls. The September or October lake-level peak is three months later than the rainfall maximum, reflecting the groundwater influence on the system.

LIMNOLOGY

Most limnologic studies of the Clifton-Preston lakes have been concerned with comparative water chemistry (e.g., Williams and Buckney, 1976). The focus has been Lake Clifton, where there are actively growing microbialites (Moore and others, 1984; Moore, 1987; Moore and Turner, 1988). Burke and Knott (1989), and Burke (1990), described the limnology of four smaller lakes in the system, including Lake Hayward. All the lakes are saline, somewhat alkaline, perennial and shallow (<3 m). Lake Clifton is seasonally hyposaline, whereas all others are hypersaline.

Morphometry

At times of high water (i.e., when at sea level), Lake Hayward (Fig. 2) is 1.3 km long by 0.5 km wide, with a surface area of 0.6 km^2. Lake level falls approximately 1 m annually, reducing the area of water to approximately 0.45 km^2.

Water Chemistry

Lake Hayward is the most hypersaline lake in the Clifton-Preston system. The brine is similar in composition to seawater. The order of ionic dominance is $Na^+ > Mg^{2+} > Ca^{2+} > K^+$ (Ca^{2+} and K^+ may be reversed), and $Cl^- > SO_4^{2-} > HCO_3^-$. This pattern is the same as in the other lakes (Moore and Turner, 1988; Burke and Knott, 1989; Burke, 1990), and in many saline lakes elsewhere in Western Australia (Williams and Buckney, 1976; Geddes and others, 1981). Since 1988, Lake Hayward water has been at gypsum saturation but, as explained later, this may not be a permanent condition.

Lake Hayward is one of three lakes in the system known to have, at times, a chemically, thermally and density-stratified water column. The others are Lakes Yalgorup and North Newnham. Lake Hayward is stratified from autumn to early summer (April to January). The salinity of the lower layer in times of stratification may greatly exceed that of the upper layer. There is an accompanying inverse temperature regime and thermal gradients are sharply defined along the chemocline (Fig. 3). While scuba diving, we recorded a bottom temperature of 46°C. The chemocline is visible as a cloudy zone, a few centimeters deep, with clear water both above and below. Burke and Knott (1989) showed that

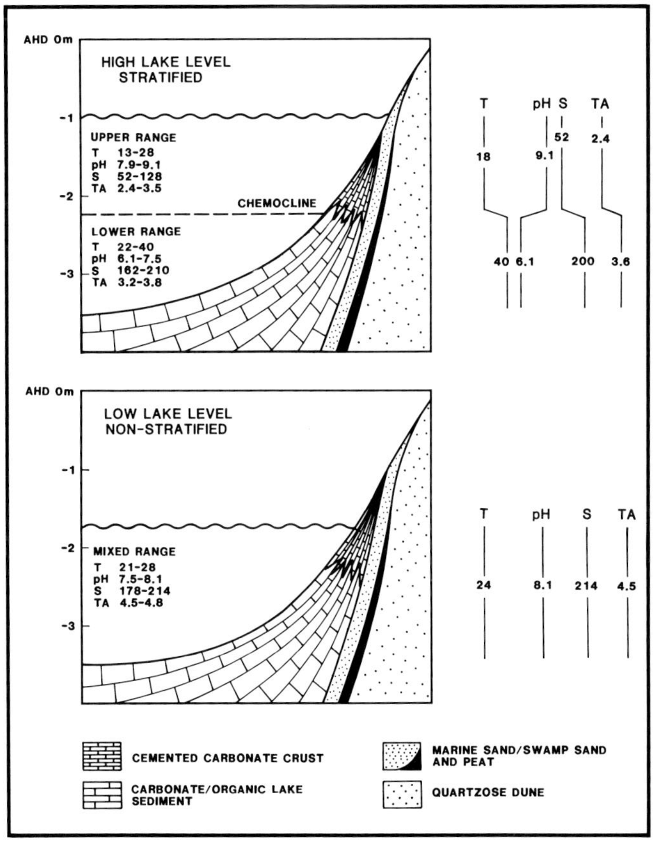

FIG. 3.—Relationship of simplified facies and stratified and non-stratified water chemistry of Lake Hayward, summarized for the period July 1985 to January 1986. T = temperature (°C), S = salinity (g/l), and TA = total alkalinity (meq/l). Pleistocene dune sand is overlain by carbonate lake sediment. A cemented crust zone surrounds the lake and has formed where sediments are seasonally exposed and periodically in contact with groundwater. The upper layer water chemistry reflects that of the inflowing ground-water at times of highest lake level. Chemical data summarized from Burke and Knott (1989); sedimentology, this paper.

in times of stratification there are significant physical and chemical differences between the bottom and overlying water masses, and that these differ from those in times of mixing.

BIOTA

A diverse biota exists within the lakes of the Clifton-Preston system (Moore and others, 1984; Moore, 1987; Burke and Knott 1989). Lake Clifton has an extensive reef of living microbialites. The spatial relationship of this reef to groundwater inflow is discussed by Moore and Turner (1988). Although fossil examples are present, no living microbialites occur in Lake Hayward or the other lakes. Some lake floors, notably Lake Hayward, have an extensive cover of cohesive, mucilaginous benthic microbial communities (BMC) consisting of cyanobacteria, bacteria, algae and meiofauna (Burke and Knott, 1989). Burke (1990) recognized four BMC types: "*Microcoleus*," "*Apanothece*," "*Chromatium*," and "Sulphate reducing" BMC.

The "*Microcoleus*" BMC, which only occurs at the seasonally exposed lake margin, comprises cohesive communities of *Microcoleus chthonoplastes* with *Chroococcus* sp. and diatoms. The other types occur in the permanent water zone and are unstable communities, waxing and waning according to the prevailing water chemistry. The "*Aphanothece*" BMC is a mucilaginous mat dominated by *Aphanothece* sp. This community covered the lake bottom in 1987 (Burke, 1990). However, in 1988 the onset of hypersaline conditions led to colloidal gypsum precipitation in the water column, which severely disrupted the "*Aphanothece*" BMC. Bottom water became sulfurous throughout 1988 and 1989. Purple sulfur bacteria, *Chromatium* spp., initially formed an extensive planktonic bloom and then coalesced as a BMC. Later during 1989, an orange "*Aphanothece*" BMC was re-established. By diving in December 1989 (near core site 3, Fig. 2), we found that about 60 percent of the sediment surface was a pale gypsum layer with degenerating "*Chromatium*" and some "*Aphanothece*" BMC. About 30 percent was covered by "*Aphanothece*" BMC with the remainder black and sulfurous ("Sulphate reducing" BMC).

Brine shrimp (*Artemia* sp.) commonly occur in large numbers in the littoral zone. In January 1990, we observed abundant *Artemia* concentrated above the chemocline. This brine shrimp is an exotic species that, according to Alan Savage (pers. commun.), has displaced ostracods in the lake because of its prolific breeding and more aggressive feeding habit. A lack of ostracod tests in the surficial sediment compared with abundant ostracod fossils in the underlying sequence supports this conclusion. However, the recent increase in salinity could also be invoked as a cause.

SURFICIAL SEDIMENTS AND STRUCTURES

The distribution of sediment types and sedimentary structures at Lake Hayward has a concentric pattern (Fig. 4). The surficial sediment of the innermost zone, of mainly permanent lake water, consists of non- or weakly lithified organic-rich carbonate. A thin gypsum layer is common in the upper few centimeters. The innermost zone is surrounded by a marginal environment that is cyclically exposed and inundated by rises and falls of lake level. The outer margin is characterized by a hard, strongly cemented carbonate crust. The inner and outer lake margin is transitional. Surrounding the outer crust margin is incipient soil and reworked sand containing Holocene marine fossils. The overall sedimentological pattern is ultimately attributed to the strong seasonal cycle of rainfall (and hence groundwater inflow) and evaporation in southwestern Australia. A predominant southwesterly wind vector is evident from the distribution of non-lithified surficial sediments overlying the crust.

FIG. 4.—View of the southern edge of Lake Hayward showing the apron of crusts and the tall eucalyptus (Tuart, Marri, and Jarrah) forest surrounding the lake.

Permanent Water Zone

Surficial sediment within the zone of permanent water consists of flocculant muddy and peloidal carbonate. The upper 2 to 5 cm is black and organic-rich, and commonly overlain by a thin, probably impermanent, gypsum crust. As described, the sediment has a cover of degraded and living BMC. The BMC mat, made buoyant by trapped gas, may lift from the sediment surface. It may "float" for a time at the chemocline (Fig. 5). Shoreward movement of mat fragments also occurs across the surface of the chemocline, possibly due to development of a heat-flow cell. On reaching the surface, mat fragments are carried rapidly shoreward by wind waves. Nevertheless, organic matter considered to be BMC-derived is preserved throughout the deeper lake sediments.

Gypsum

Lake Hayward water reached gypsum saturation in 1988 (Burke, 1990), and gypsum was still part of the surficial sediments until at least January 1992. The gypsum sheet occurs as a package of millimeter-thick laminae, indicating multiple precipitation episodes. This is consistent with the observations of Burke (1990), who reported episodic blooms of the sulfur bacteria *Chromatium* spp. These bacteria reduce colloidal gypsum suspended in the water column before settling as a BMC. The gypsum sheet is not laterally continuous but separated into plates, with a thickened "box-

FIG. 5.—Underwater photograph of benthic microbial mats (BMC) floating along the chemocline in Lake Hayward. Areas of mat may lift away from the sediment surface due to gas build-up in the mucilaginous fabric. This causes the mats to become less dense than the bottom water. The mats do not reach the surface because the fresher upper layer has a higher density, thus the mats float only to the sharply defined chemocline. Diver in the background for scale.

work" structure at plate margins. Recrystallization of gypsum takes place along plate boundaries, indicating interaction between underlying sediment pore water and lake water. The body of the plate acts as a permeability barrier, concentrating this interaction along plate contacts. The preservation potential of the gypsum is unknown but the mineral is absent in the deeper sediments. Gypsum saturation may occur periodically, but the mineral may dissolve in times of lower salinity, or associated with sulfate reduction, as reported in Solar Lake, Israel (Krumbein and others, 1977).

Crust Zone

The periodically exposed littoral margin has a hard, strongly cemented carbonate crust that extends from near the tree-line to just beyond the low water line (0-m contour in Fig. 2). Several superficial features that are mostly uncemented and have low preservation potential locally overlie this surface. These include sapropelic ooze, berms, desiccated sheets of BMC/carbonate lag, and gas-escape structures. The crust has well-developed tepee structures, degraded fossil microbialites, desiccation cracks and breccias, and iron-stained groundwater seeps. The crust is karstified in places, particularly where the tree-line has encroached upon the lake basin. Scattered shells, mainly the bivalve *Katelysia scalarina*, overlie the crust in the upper littoral zone. In many places, these shells are weathering from the carbonate crust. They are also scattered throughout the adjacent dune sand. As discussed below, *K. scalarina* originate from a Holocene marine phase and were incorporated into lacustrine carbonates during higher lake levels.

SUPERFICIAL FEATURES

The breakup and shoreward transportation of BMC organic matter cause most of the superficial deposits overlying the crust at Lake Hayward. Gypsum can be rafted ashore, attached to cohesive and mucilaginous BMC fragments. As lake level drops, waves periodically deposit a lag of coalesced mat fragments. This sheet later desiccates leaving millimeter-thick "biscuits" of organic matter and precipitated carbonate and salts. Desiccation polygons formed in this manner have a convex-upward form. Break-up of the "biscuits" produces gravel and sand-size particles. Although mainly on the eastern margins, deposits of this material occur as shore-parallel berms up to 10 cm thick and several meters wide. In the northeastern part, a substantial deposit of BMC-derived sapropelic ooze up to 30 cm deep has accumulated. In places, a superficial layer of carbonate forms a semi-lithified cap over decaying organic matter. This cap frequently develops a pustular character as expelled gases force up the less-permeable surface layer.

Pinnacles

Along the eastern edge of Lake Hayward, isolated pinnacles of calcretized and karstified fossiliferous limestone protrude through the crust pavement. The quartz-bearing limestone contains marine fossils similar to those in the Pleistocene Tamala Limestone. The structure of individual pinnacles varies, although near-vertical indurated skins indicating solution piping are common. Similar pinnacles occur in Pleistocene dunes throughout the Yalgorup area. In nearby quarries, deeply exposed pinnacles appear to be solution residuals developed between multiple solution pipes. The apex of each structure is calcretized but, at depth, eolian cross-stratification is preserved. The pinnacles of the Yalgorup area resemble those of the Nambung Pinnacle Desert, north of Perth, whose origin is discussed by McNamara (1983).

Tepee Structures

Tepee structures commonly disrupt the crust surface (Fig. 6). The general crack pattern is polygonal, with larger overthrust megapolygons occurring close to the lake. The dominant fracturing aligns perpendicular to the shoreline, as in other marginal carbonate crust zones of the Clifton-Preston System. Trees have encroached the crust by way of tepee fractures, producing further disruption. Away from the tree-line, samphire (glasswort) distribution patterns indicate tepees below a surficial cover. Using the criteria of Kendall and Warren (1987), there is good evidence for a relationship between the seasonal hydrologic cycle and tepee formation at Lake Hayward. The crust is underlain by a permeable sand that is iron-stained, containing shells with evidence of partial dissolution. Groundwater seeps occur along tepee structures. Isopachous aragonite rim cements with bright orange luminescence are abundant, with vadose and groundwater pisoliths common throughout the crust.

Fossil Microbialites

Fossil microbialites occur locally around the lake margin, associated with tepee fractures and attached to limestone pinnacles (Fig. 7A). We have also discovered fossil microbialites associated with tepee structures at other lakes in

FIG. 6.—Tepee structure just above low-water mark. Grass and shrubs preferentially grow in and around the broken crust areas indicating groundwater resurgence. Pen at arrow for scale.

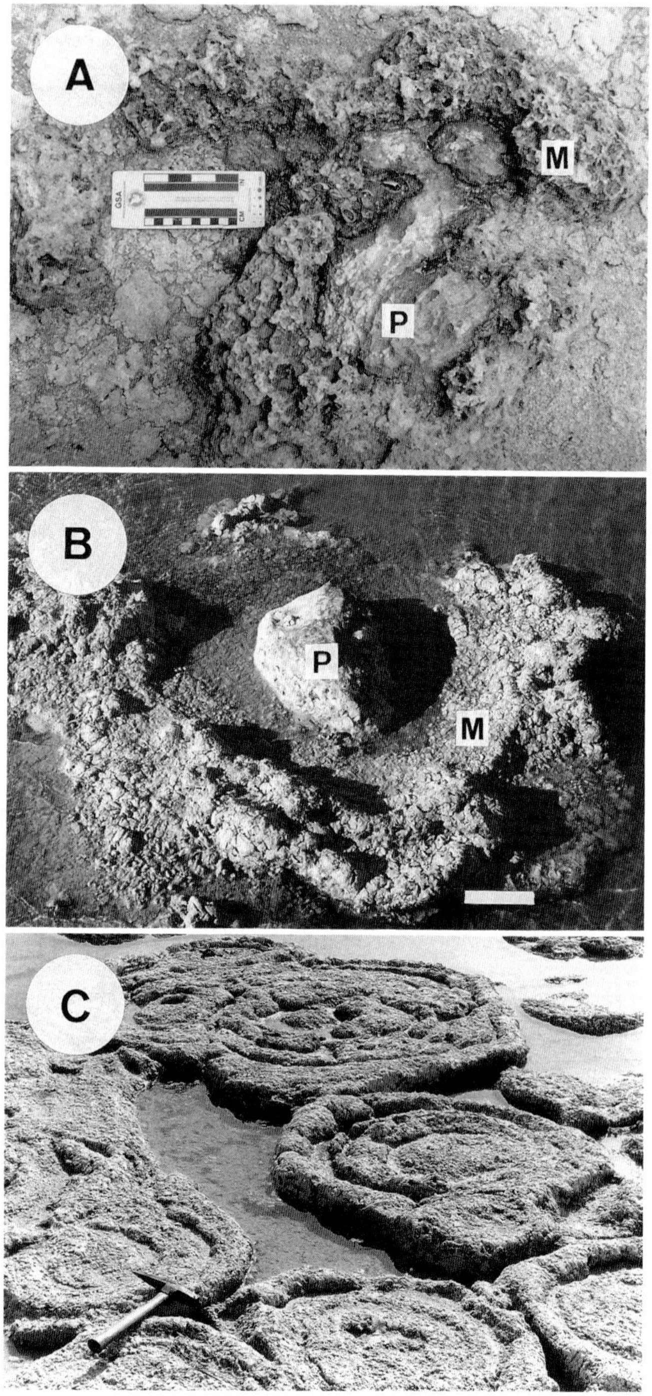

FIG. 7.—(A) Photograph of Pleistocene pinnacle core (P) encrusted by a Holocene fossil microbialite (M) on the eastern shore of Lake Hayward; (B) Modern eroded microbialite from Lake Clifton showing microbialite coating (M) eroded from a Pleistocene pinnacle core (P). (C) Living microbialites from Lake Clifton. Hammer for scale.

the system. Lake Preston has extensive marginal crust pavement with microbialite domes aligned along megapolygon contacts. Both pinnacle and tepee microbialites are well indurated but rather poorly preserved at Lake Hayward, which has implications for their age and the age of the lake. Living microbialites at Lake Clifton apparently do not require a hard substrate, although attachment to shell material is recorded (Moore and Burne, unpublished data). However, we have found many examples of attachment to pinnacles at Lake Clifton, where microbialites are being actively eroded (Fig. 7B). A nearshore to deeper water traverse reveals the following transition: (1) pinnacles with no attached microbialites; (2) pinnacles with relict microbialites; (3) increased microbialite preservation and degree of attachment; (4) coalesced microbialite structure with rare pinnacle exposure (Fig. 7C); and finally, (5) the apparently unattached, solitary, deep-water forms recognized by Moore (1987). Sediment surrounding nearshore pinnacles and extending beyond the microbialite "reef" is composed predominantly of microbialite debris. Larger nearshore microbialite fragments and dead microbialites attached to pinnacles show significant induration; they are compositionally and texturally similar to fossil microbialites from other lakes in the system.

The existence of living microbialites at Lake Clifton, a hyposaline lake, and only fossil examples in Lake Hayward and other hypersaline lakes in this system, implies that the water chemistry in these lakes has undergone significant change over time. It is also evident from Lake Clifton that microbialites have low preservation potential. The living structures are very poorly lithified and rapidly eroded. Lowering of lake level has subjected more marginal structures to increasing periods of exposure and, therefore, greater erosion. Cementation of the microbialites does not guarantee preservation in an environment that offers opportunities for meteoric dissolution. Multiple calcified crust pavements and strongly indurated microbialitic erosional remnants suggest that Lake Hayward may have a longer hypersaline history than the other lakes.

STRATIGRAPHY

We obtained 21 cores from Lake Hayward by pushing 50-mm diameter PVC pipe into the sediment. All littoral cores bottomed in Pleistocene dune sands (Fig. 8). In cores from the lake's center (3H and 18H; Fig. 2), a hard layer, interpreted to be calcretized calcarenite, limited penetration. Nevertheless, a complete or near complete Holocene sequence was obtained.

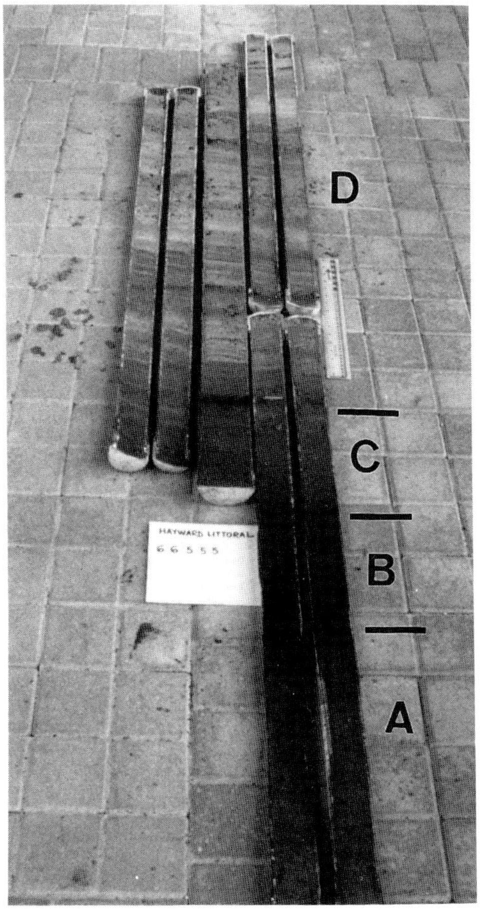

FIG. 8.—Littoral core from Lake Hayward showing four main units: Unit A- siliciclastic Pleistocene dune sand; Unit B- dark organic rich limnic swamp phase; Unit C- marine inlet phase; Unit D- well-laminated lake sediment. Ruler is 30 cm.

The total thickness of Holocene strata is approximately 2.5 m, assuming an average 30 percent compaction. We recognize four stratigraphic units (Fig. 9): (1) a basal Pleistocene unit dominated by quartz sand; (2) a 40-cm unit of peat and peaty carbonate mudstone with freshwater fauna; (3) a 35-cm interval of sandy, carbonaceous wackestone with abundant foraminifers and other marine or estuarine fauna, and (4) a 1.75-m thick lacustrine unit consisting of laminated wackestone and organic-rich wackestone. The depositional environments of these four units are: (1) reworked eolian dune, (2) freshwater swamp or lake, (3) marine inlet, and (4) saline lake (Table 1).

Unit A

The basal portion of the sequence comprises variably indurated calcarenite (facies Ab) or massive quartz sand (facies Aa). Both facies are ascribed to the Pleistocene Tamala Limestone.

Facies Ab.—

A karstified calcrete underlying at least the southeastern portion of Lake Hayward is inferred from sediment probing and piezometer bore intersections. This uneven surface protrudes above lake sediments as the limestone pinnacles along the eastern margin. The grey (5Y5/1), massive and laminar calcrete caps a light grey (5Y7/1) variably indurated, quartzose skeletal grainstone. Skeletal grains include marine molluscs, foraminifers, and echinoderm and algal-derived particles. The thickness of the calcarenite beneath the lake is unknown, but it is at least 10 m thick below the western shore of Lake Clifton.

Facies Aa.—

A quartz sand sheet at least 4 m thick underlies both the northern and southern marginal areas of Lake Hayward. This sand overlies calcretized calcarenite locally along the eastern shore. In cores, the lower part of the facies consists of yellow (5Y7/6) massive quartz sand with traces of woody plant matter. Yellow-stained quartz and rare grains of jasper and feldspar are predominantly fine and angular. Organic matter content increases gradationally toward the top of the unit, and the uppermost sand is dark olive grey (5Y5/2) with root structures. Except for a few sand-sized shell particles, no whole shell fossils were found.

Unit B

Unit B is a 40-cm thick interval comprising a basal peat (facies Bc), up to 20 cm thick, that grades upward into peaty carbonate mudstone (facies Bb). Contact with the underlying sand is generally sharp. The unit thins rapidly in the upper littoral zone, becoming increasingly sandy. In the uppermost littoral zone, the peat layer is absent. This littoral equivalent of facies Bc and Bb is designated "facies Ba."

Facies Bc.—

The black (5Y 2.5/1) peat facies of Unit B consists predominantly of higher plant remains (reeds and wood), minor fine quartz sand, and a trace of disseminated calcitic

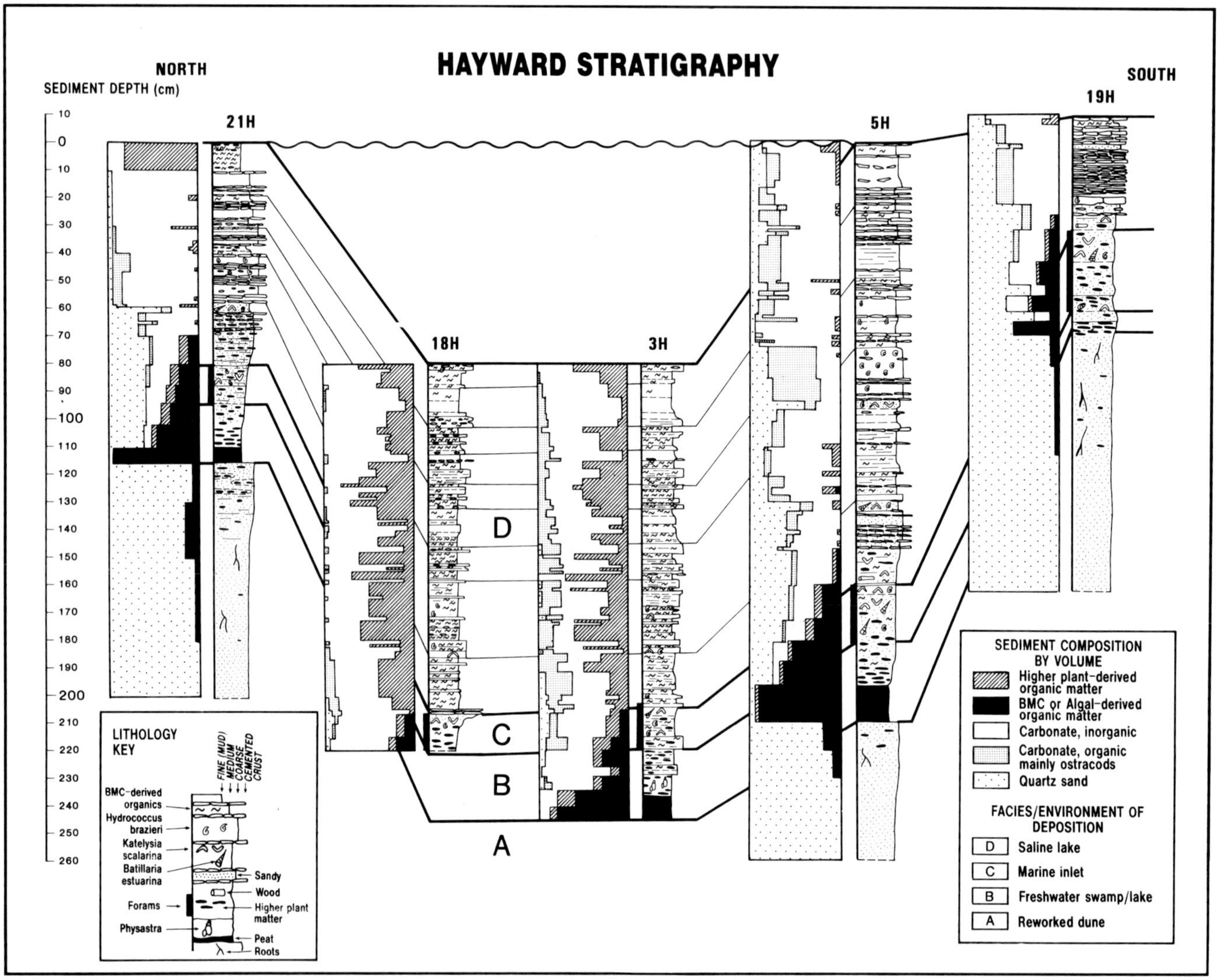

FIG. 9.—Lake Hayward stratigraphic correlation diagram. Sediment compositional variation is based on volume estimates. Also shown are major facies and inferred environments of deposition. Core sampling locations are given in Figure 2. Relative water depth shown is half actual depth. Core depths are not corrected for compaction.

mud. Microbial organic matter is intimately associated with the higher plant matter although the latter is far more abundant (Fig. 9). This facies contains the rare sinistrally coiled freshwater snail, *Physastra georgiana*. This extant planorbid gastropod inhabits reedbeds in still freshwater ponds. The shells at Hayward are delicate but unbroken. During deposition, the Hayward depression would have been fed mainly by groundwater inflow and direct rainfall. It is highly unlikely that *P. georgiana* was transported from outside the Hayward depression. The depositional environment for facies Bc is interpreted as a still freshwater swamp.

Facies Bb.—

The upper part of Unit B is a gradational continuation of facies Bb, but organic matter is subordinate to carbonate. The facies is an olive grey (5Y4/2), fine sandy, organic-rich calcitic mudstone. Organics are dominated by higher plant remains with lesser algal-derived material. Alternations in organic abundance impart a weak lamination to the sediments. The higher plant matter is finer and generally more degraded than in facies Bc. XRD analysis shows calcite to be the only carbonate phase although rare unidentified ostracod tests are also present (Fig. 10). Fine quartz sand forms less than 5 percent of the inorganic component, but may dominate in littoral areas. Remains of calcified *Charaphycae*, including oogonia, are also found mainly in the littoral parts. The depositional environment is interpreted to be a freshwater swamp or lake.

Unit C

Unit C, which is approximately 35 cm thick, has a diverse fossil assemblage of marine affinity, including abundant foraminifers. It comprises two facies: basinal, sandy calcareous fossiliferous mudstone or wackestone (facies Cb), and littoral, peloidal fossiliferous quartz sand (facies Ca).

TABLE 1.—LAKE HAYWARD DEPOSITIONAL FACIES

Unit	Facies/Criteria for recognition	Maximum Thickness	Environment of deposition	Age
D	(Da) Littoral Carbonate Crust Package White (5Y8/1) to light grey (5Y7/2) cemented carbonate laminae interlayered with unlithified or weakly lithified sandy grainstone. Overall wedge-shaped package. Karstified in part, tepee structures, degraded fossil microbialites, desiccation cracks and breccia. Predominantly aragonite with lesser calcite, Mg-calcite, and rare dolomite. *Cyprideis australiensis* common throughout, *Hydrococcus brazieri* in places abundant. (Db) Laminated, wackestone/packstone Repetitive laminae comprising: 1) dark olive grey (5Y3/2) to dark greyish brown (2.5Y5/2) organic-rich wackestone (30–70 vol% organics), 2) white (5Y8/2) to light olive grey (5Y6/2) organic-bearing wackestone or packstone (5–30 vol% organics), and 3) Olive grey (5Y5/2) to pale olive (5Y6/3) organic-bearing packstone or grainstone (20–40 vol% organics). Chemically precipitated carbonates with lesser contribution from shell material, mainly *Cyprideis australiensis*. Carbonates consist of aragonite, calcite, Mg-calcite, and dolomite. Recrystallized sheet gypsum lamina caps sequence.	1.75 m	Saline to hypersaline lake	Early Holocene to Present
C	(Ca) Littoral calcareous quartz sand Dark grey (5Y4/1) organic-bearing, foraminiferal, calcareous, quartz sand. Fine to medium quartz grain-size. Fossils are of marine affinity. (Cb) Foraminiferal mudstone/wackestone Dark grey (5Y4/1) to olive grey (5Y5/2) sandy, organic-bearing, fossiliferous mudstone or wackestone. Poorly laminated, organic matter both higher plant and algally derived. Diverse foraminifers, mollusc, and ostracod assemblage of marine affinity. Dominant carbonate is calcite.	0.4 m	Marine inlet	Early Holocene
B	(Ba) Littoral peaty quartz sand Olive grey (5Y4/2) peaty quartz sand and sandy peat. Abundant *Charaphycae* oogonia. (Bb) Peaty mudstone Olive grey (5Y4/2) fine sandy, organic-rich (higher plant matter), mudstone. Gradational basal contact with Bc. Poorly laminated. Main carbonate is calcite. (Bc) Peat Black (5Y2.5/1) peat. Higher plant (reedy monocot) dominated with rare quartz grains, minor calcite, and a sharp basal contact. Freshwater planorbid gastropod *Physastra georgiana*.	0.35 m	Freshwater swamp/lake	Early Holocene
A	(Aa) Quartz sand Yellow (5Y7/6) grading into dark olive grey (5Y5/2) quartz sand. Unfossiliferous, fine-grained and massive becoming rooting structured and increasingly organic rich (higher plant matter) towards the top. (Ab) Calcretised calcarenite: 'Pinnacles' Light grey (5Y7/1) variably indurated quartz skeletal (marine) grainstone. Massive and laminar, mottled, bored calcrete. Calcretised pinnacles, infilled solution pipes.	>4 m	Eolian dune	Pleistocene (Tamala Limestone; Playford and others, 1976)

There is a general trend of decreasing organic content with concomitant increase in quartz and carbonate content from units D to C. The base is identified by first appearance of marine foraminifers, the top by a sharp decrease in foraminifera (as well as ostracod) abundance and diversity. This coincides with a hard dolomite-bearing crust in the northern part of the basin. Calcite is the only carbonate phase indicated by XRD analysis.

The abundant ostracods include *Cyprideis australiensis, Phylctenophora zealandica*, and various unidentified types. Foraminifera include, amongst others, *Elphidium, Spiroloculina, Quinqueloculina* and *Triloculina*, of which *Quinqueloculina* spp. are the most abundant. The bivalve, *Katelysia scalarina*, of marine affinity (cf. Wells, 1984), is common throughout the unit. Gastropods include *Batillaria estuarina, Hydrococcus brazieri, Coxiella striata, Salinator fragilis* and *Liloa brevis*.

Facies Cb.—

Facies Cb is a dark (5Y4/1) to olive grey (5Y5/2) sandy, organic-bearing fossiliferous calcitic mudstone or wackestone to peloidal grainstone. It is unlithified or weakly lithified and may have poorly developed laminae. Organic matter constitutes up to 40 percent of the sediment volume, both as higher plant matter and algae-derived material. The inorganics comprise mainly carbonate mud and peloids in subequal abundance, quartz sand, ostracods, and other minor fossil material. Ostracods predominate in the basal part of the facies.

Facies Ca.—

Toward the littoral zone, quartz succeeds calcite as the dominant mineral, and in the outermost littoral area, unit C is a dark grey (5Y4/1) fossiliferous organic-bearing quartz sand. Quartz shows upward-increasing grain size from fine to medium. Organic matter content ranges from 30 percent in the central parts to less than 4 percent in the outer littoral area.

Most is derived from higher plants with a minor algal component. Although ostracods are less abundant than in the basinal facies, larger fossils like *Katelysia scalarina* are more common. Rare *Sanguinolaria biradiata*, a delicate

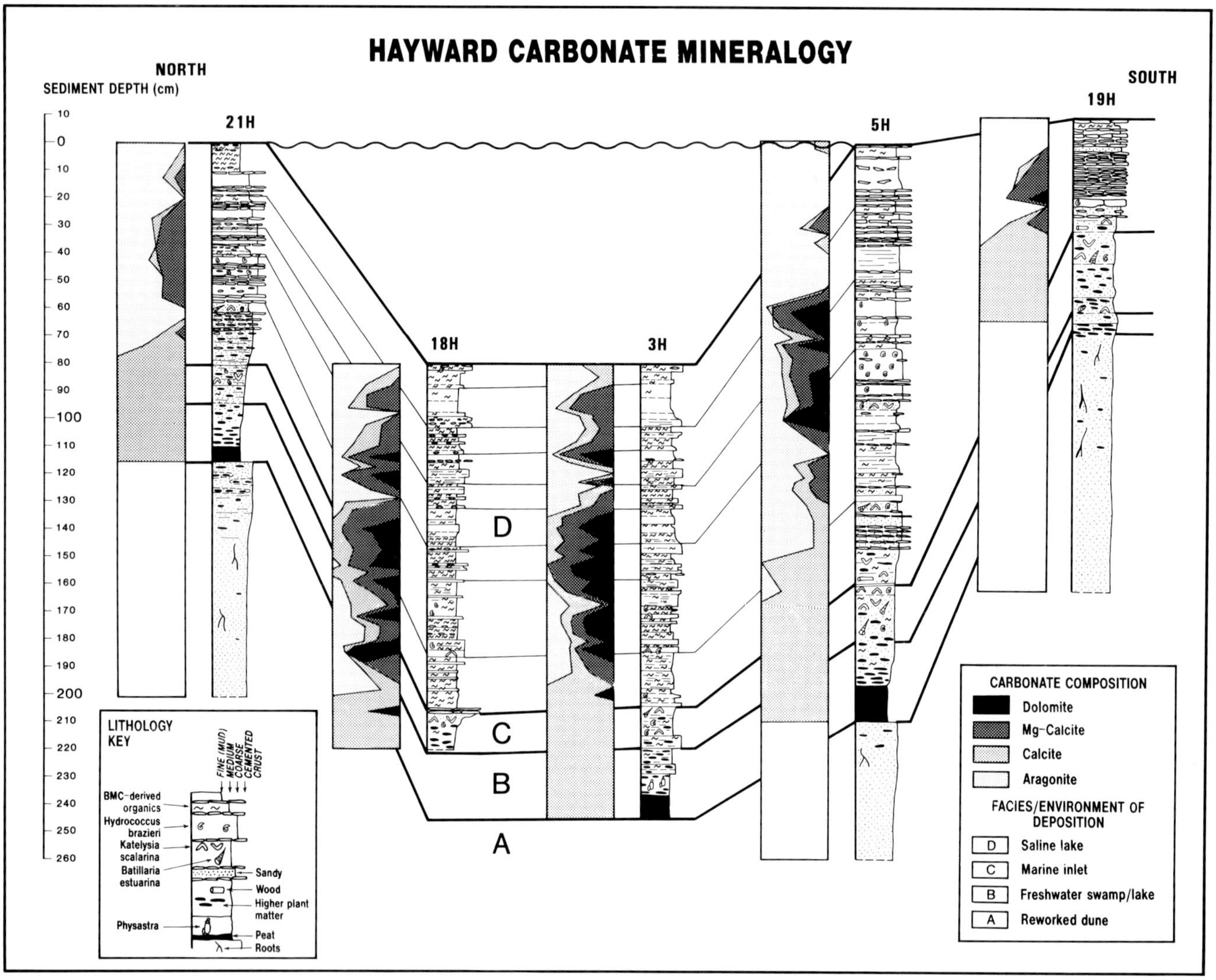

FIG. 10.—Correlation diagram showing vertical and lateral distribution on carbonate compositions in Lake Hayward. Features of note are the absence of carbonate minerals in Unit A, ubiquitous calcite in units B and C, and up to four concomitantly occurring carbonate phases in unit D. Relative water depth shown is half actual depth. Core depths are not corrected for compaction.

shelled bivalve, occur unbroken. This marine bivalve lives in lower intertidal and subtidal sand (Wells, 1984).

Unit D

In the basinal zone of Lake Hayward, Unit D is a 1.75-m thick sequence of laminated, variably organic-bearing wackestone (facies Db). Toward the littoral area, it grades laterally into a carbonate crust package (facies Da). The unit consists predominantly of chemically precipitated carbonate with lesser shell material, mainly ostracods (Fig. 10). These carbonates comprise aragonite, calcite, Mg-calcite and dolomite (Figs. 11A–D). The dolomite content tends to increase antithetically with other carbonate phases and is considered diagenetic (Rosen and Coshell, 1992). Quartz is rare across most of the basin, except in the littoral zone. Quartz decreases generally toward the top of the unit. Clays were not recorded from XRD analysis. The nature of modern Lake Hayward and sedimentological evidence outlined below indicates a saline lacustrine depositional environment for Unit D.

Facies Db.—

The basinal carbonate sequence is characterized by distinctive fine-scale alternations in organic abundance. Indeed, the overall organic content of this facies is high throughout the basin. Burke (1990) reported an average total organic carbon content (TOC) of 25 percent in this facies from one core. It is also possible to correlate this facies over the whole basin (Fig. 9), based upon visual estimates of organic abundance. Individual laminae (commonly on a millimeter scale) and thinly bedded laminae packages have a distinctive character (color, composition, pattern, etc.) that can be recognized easily in individual cores. However, decreased organic-matter content and concomitant development of cemented carbonate crusts in the transition to the littoral crust zone (facies Da) make fine-scale correlation more difficult. The organic matter throughout the facies is mainly BMC-derived with traces of higher plant matter, including charcoal. Fossils include abundant tests of the ostracod *Cyprideis australiensis*, minor *Hydrococcus brazieri*, and rare *Coxiella striata*. Ostracods generally comprise 20 percent of the carbonate fraction lower in the facies, with a decreasing trend toward the top. Apart from the upper few centimeters of sediment, ostracods are ubiquitous (Fig. 11A). Quiet depositional conditions are indicated by abundant articulated tests. In contrast to the marine phase, ostracod diversity is extremely low. Rare *Katelysia scalarina* and *Batillaria estuarina* are considered to be reworked from the marine inlet phase.

At least three subfacies occur repetitively throughout the sequence:

1. Dark olive grey (5Y3/2) to dark greyish brown (2.5Y5/2) organic-rich wackestone. Organic matter constitutes 30 to 70 percent of the sediment volume. Carbonate is uniformly distributed, with peloids and mud in about equal abundance.
2. White (5Y8/2) to light olive grey (5Y6.2) wackestone or packstone with 5 to 30 percent organic matter. Peloids are dominant, with minor cemented aggregates (Fig. 11B).
3. Olive grey (5Y5/2) to pale olive (5Y6/3) packstone or grainstone with 20 to 40 percent organic matter. The organics occur as distinctive yellowish brown (10YR5/4) to light olive brown (2.5Y5/4) clumped laminae. Interlayered carbonate is mainly peloidal with cemented crust-like aggregates formed along bedding (Fig. 11C).

Although no cyclic pattern is evident, type 1 commonly grades upward into type 2, perhaps indicating loss of organic matter in the latter. Nearer the littoral area the top of the type 2 sediment is commonly capped by a crust. Type 3 mainly occurs in the upper part of the sequence and may be equivalent to the type of sediment forming today. De-

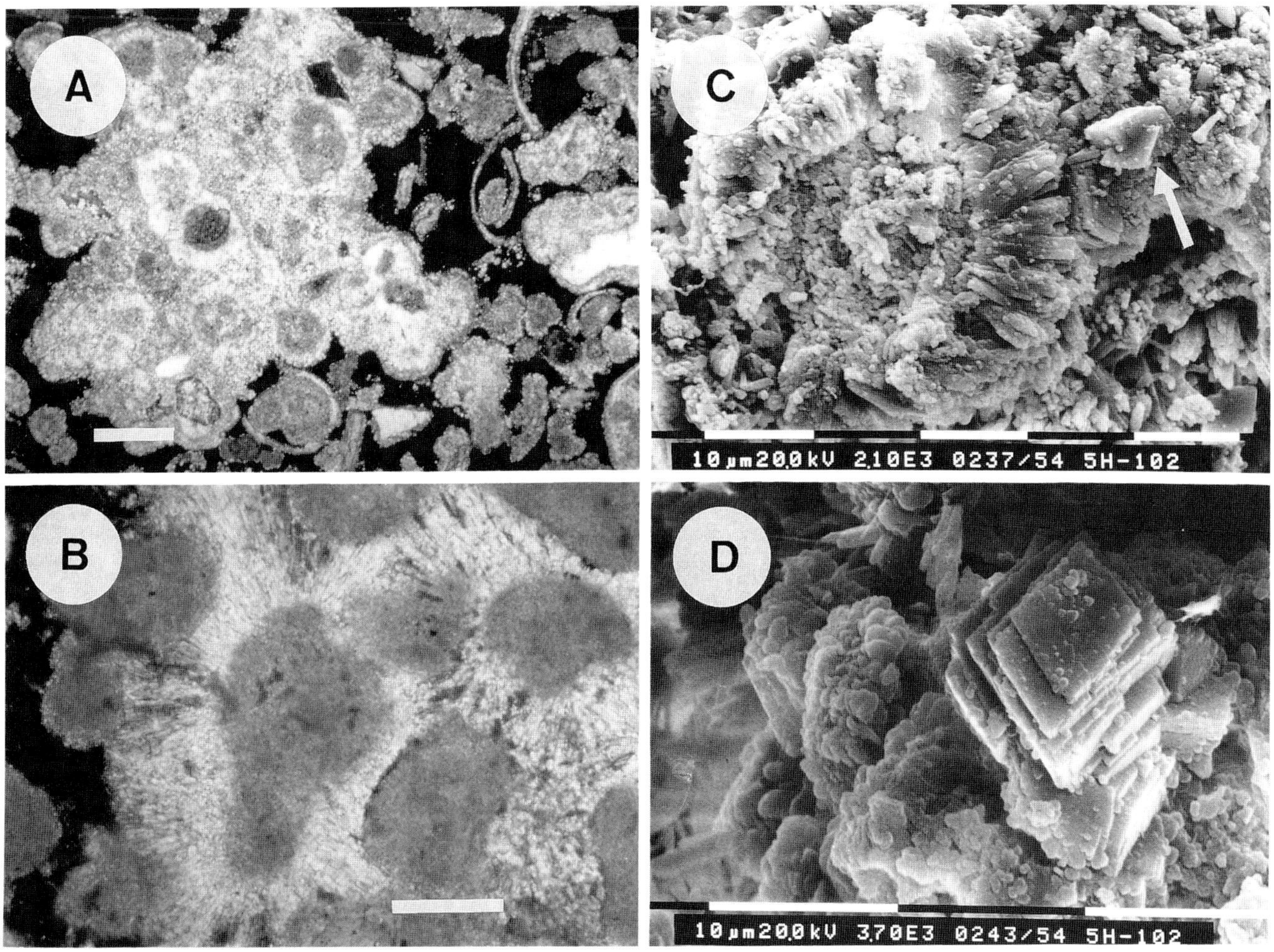

FIG. 11.—(A) Thin section photomicrograph of lake sediments showing cemented peloids, carbonate encrusted quartz grains, diagenetically altered and encrusted ostracods and isopachous aragonite cements. Scale bar is 0.2 mm. (B) Close-up of radial isopachous aragonite cementing peloids. Scale bar is 0.1 mm. (C) Backscatter scanning electron microscope image of isopachous aragonite cement around a peloid. Scattered dolomite rhombs (arrow) are present and small (<1 µm) spherical shapes of bacterial bodies occur throughout the image. Each division on the scale bar is 10 µm. (D) SEM image of diagenetic dolomite rhombs in the lake sediment. Bacterial bodies are also present. Each division on the scale bar is 10 µm.

graded "*Chromatium*" BMC in the surficial sediment resembles the yellowish brown organics of this subfacies.

Facies Da—carbonate crust package.—

The carbonate crust girdle is a wedge-shaped package of individual crust laminae interlayered with unlithified and weakly lithified sandy grainstone (Fig. 12). The package thins rapidly toward the tree-line. For example between cores 5H and 19H (25 m apart, Fig. 2), the crust package thins from approximately 1.6 m to 0.4 m. Individual crust laminae in 5H are more widely spaced and constitute much less of the interval than in 19H (Fig. 9). In the latter, individual crusts have coalesced to the point where correlation with other cores is impossible. Core 21H (Fig. 9) is an intermediate example of this facies. Fossils are the same as that recorded in laterally equivalent facies Db; however, reworked marine species, particularly *Katelysia scalarina*, are more abundant. Dissolution of early formed crust in the uppermost littoral zone has led to recycling of previously bound *K. scalarina*, which in places has formed a lag, overlying karstified crust.

LACUSTRINE FAUNA

Only one ostracod species in the marine inlet phase at Lake Hayward, *Cyprideis australiensis*, successfully made the transition to the lacustrine environment. At times, *C. australiensis* would have been prolific in paleo-Lake Hayward as shown by its abundance in the sediments. De Deckker (1983) notes that *C. australiensis* is one of only two halobiont ostracods of marine ancestry in Australia. This species is not found in ephemeral waters because its eggs cannot withstand periods of desiccation. This suggests that Lake Hayward has always been a permanent water body.

Yassini and Kendrick (1988) report that *C. australiensis* has a salinity tolerance of 2–35 g/l, a temperature range of 13–25°C, and a pH tolerance of 8.0–8.5. De Deckker (1983) also records a salinity tolerance of seawater or less. The relative abundance of *C. australiensis* tests in the sequence at Lake Hayward may indicate a low-salinity brine. However, *C. australiensis* could be more salt tolerant than previously reported.

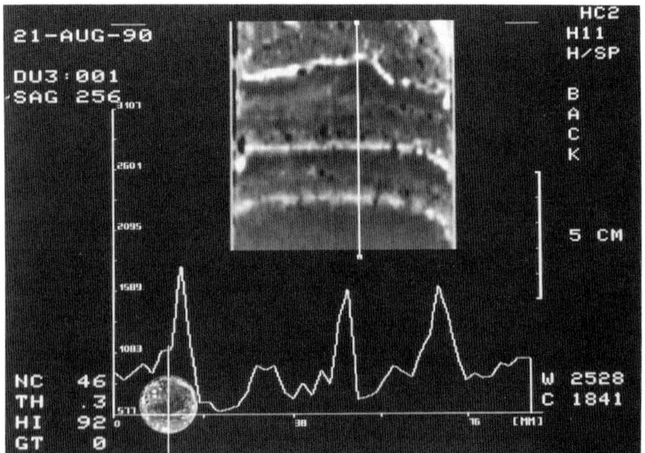

FIG. 12.—X-ray computed tomographic (CT) image of laminated lake sediments (Unit Da) showing crusts (white) (See Coshell and others, 1991).

The gastropods *Hydrococcus brazieri* and *Coxiella striata* also span both the marine inlet and lacustrine phases at Hayward. *C. striata* is, however, rarely preserved in the sequence. *H. brazieri* occurs at seemingly random intervals throughout the sequence and is abundant in places. Wells and Threlfall (1981) found *H. brazieri* to be active over a wide range of temperatures and salinities (at least up to 55 g/l) in nearby Peel Inlet.

In summary, the ostracod and molluscan fauna in the lake sequence indicate that Lake Hayward was a permanent water body in the past, like today. There is no faunal or sedimentological evidence to show that the lake has completely desiccated since formation. There is also evidence that Lake Hayward was at times considerably less saline than today.

RADIOCARBON DATES

Radiocarbon dates have been determined for three short (<10 cm) intervals of the stratigraphic sequence by analysis of the total organic matter fraction. Another date, also over a short interval, was obtained from *Katelysia scalarina* shells. The dates (Table 2) are presented uncorrected, although we recognize that some correction needs to be applied to marine carbonates because of oceanic reservoir effects. If it is assumed that there was negligible groundwater influence when *K. scalarina* were growing in the Hayward paleolagoon, the marine correction of Bowman (1985) for Southern Australia may be applicable. This would reduce the *K. scalarina* date to approximately 7,400 yr BP.

Aquatic and terrestrial organics are more problematic. In the marine environment, fully submerged plants, algae and BMC are also subject to oceanic reservoir effects. However, a fully marine origin for the organic matter tested in 3H (1.24–1.34 m) is unlikely. Marine vegetation in lagoonal environments is not necessarily fully submerged. Nonmarine organic matter would also have been sourced from onshore. Moreover, reworking of previously deposited freshwater peat is also likely. Samples from 3H (1.54–1.66 m) and 5H (2.0–2.09 cm) are considered to have been deposited in a freshwater swamp, and therefore have radiocarbon dates that reflect their different carbon reservoir—in particular, older carbon derived from groundwater that has passed through a Pleistocene carbonate aquifer. The presence of *Physastra georgiana*, a freshwater snail, indicates quiet water with partially submerged plants and reeds. Carbon would, therefore, have also been obtained from the atmosphere, thus further complicating any correction to these dates. An example of the possible effect of mixing organic matter derived from submerged and non-submerged sources is provided by Shottan (1972), who documented an apparently constant hardwater effect in a lake environment by comparing dates from fully submerged plant matter and allochthonous organics. Turner and others (1983) point out that totally submerged aquatic plants must obtain their carbon from that dissolved in the lake water (see their Fig. 1).

Thus, dates obtained for the marine and swamp phases at Lake Hayward can only at this stage be viewed as ap-

TABLE 2.—UNCORRECTED ^{14}C AGE DETERMINATIONS FOR LAKE HAYWARD SAMPLES

Hole No.	Interval (m)	Facies	Material dated	^{14}C age (years B.P.)	$\delta^{13}C$ (‰ PDB)
5H	2.00–2.09	Bc	Undifferentiated Organic matter	8060 ± 130	−27.8
3H	1.54–1.66	Bc	Undifferentiated Organic matter	7050 ± 160	−28.8
3H	1.24–1.34	Cb	Undifferentiated Organic matter	6960 ± 170	−29.0
5H	1.62–1.70	Cb/Ca	Bivalve shells *Katelysia scalarina*	7880 ± 300	−4.0

parent maxima. Ongoing geochemical, hydrological and paleontological studies may provide data for suitable corrections. The uncorrected dates suggest that the Lake Hayward depression experienced marine inundation at about 7,000 yr BP. Although imprecise, this is important in light of published sea-level history curves for coastal southwestern Australia.

DISCUSSION

Holocene History

The stratigraphic record of sediments below modern Lake Hayward reveals a detailed depositional history through the infilling of a coastal interdunal depression. This record begins at at least 7,000 yr BP (although probably as early as 8,000 yr BP). Holocene deposition began with a rise in sea level resulting from rapid global warming in the late Pleistocene and early Holocene. A rising water table caused a freshwater swamp to develop over the Pleistocene dune sand and calcretized calcarenite foundation of the depression. Lacustrine conditions gradually became predominant with progressively higher sea level and concomitant rise in water table. Ultimately, dune barriers were breached and the Lake Hayward depression became part of a complex marine inlet. This surface marine connection was apparently short-lived, probably less than 1,000 years. Gradual lowering of sea level, possibly aided by the accumulation of sand at the northern and southern end of the modern lake, isolated Lake Hayward from the rest of the system. Since that time, at approximately 7,000 yr BP, a saline lacustrine environment has prevailed.

Sea-Level Curves

There are significant differences in interpretations of the Holocene sea-level history along the coastal margin of southwestern Australia (Fig. 13). Indeed, no generally accepted correlations exist elsewhere in Australia (Thom and Chappell, 1975). Tectonic factors have been invoked to account for the lack of correlation (Playford, 1977, 1983). From a review of Swan Coastal Plain research, Playford (1988) concluded that although there is increasing acceptance of a tectonism hypothesis, more detailed studies are needed.

Sea-level history curves have been constructed from former beach evidence at Leschenault Inlet south of Lake Hayward, at Rockingham to the north (Semeniuk and Searle, 1986; Searle and Woods, 1986; Searle and others, 1988), and at Peel Inlet (Semeniuk and Semeniuk, 1991). The curves from Rockingham and Peel Inlet are similar (Fig. 13). Sea level rose above present level at approximately 7,500 yr BP, reaching a 2- to 2.5-m maximum at approximately 7,000 yr BP, followed by an essentially steady decrease. In contrast, the Leschenault curve shows sea level at -2 m until 5,500 yr BP, rising to a 3.5-m plateau from 5,000 to 3,500 yr BP, before falling to the present level at 2,000 yr BP. Data from Lake Hayward show the upper part of the marine inlet phase to have occurred at 7,000 yr BP, suggesting that the highest sea level in the area ended about this time. Thus, Lake Hayward results fit well with the sea-level history from Rockingham and Peel Inlet.

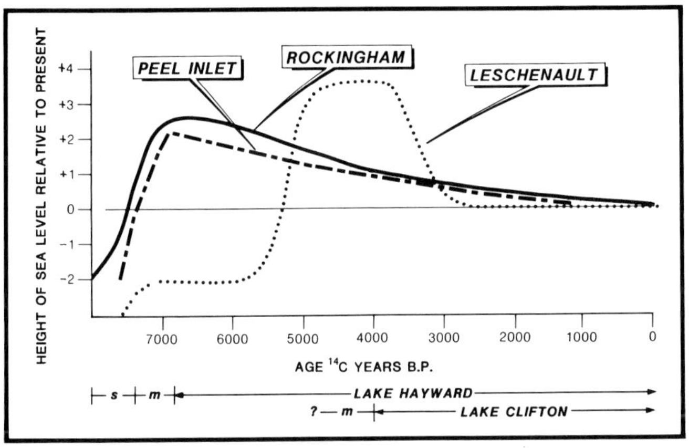

FIG. 13.—Sea-level history curves for the Rockingham, Peel Inlet and Leschenault areas (after Semeniuk and Semeniuk, 1991) compared with the lacustrine history of Lake Clifton (after Moore and Burne, undated) and Lake Hayward. Bar at the bottom of the diagram indicates the timing of the swamp phase(s) in Lake Hayward and the marine phase (m) in Lake Hayward and Lake Clifton. Dates are ^{14}C uncorrected except Lake Clifton, which was not stated in Moore and Burne (undated). Geographic locations are shown in Figure 1.

At Lake Clifton, the latest part of a pre-lacustrine marine-inlet phase was dated by Moore and Burne (undated) at between 4,670 and 3,890 yr BP (correction, if any, not stated). A marine connection, therefore, existed at least during this time but no date for the initiation of marine inundation or details of underlying stratigraphy are provided. They contend that this supports a Leschenault-type history for Lake Clifton and the adjacent lakes. The marine phase at Lake Clifton certainly appears to have ended later than at Lake Hayward, implying that Lake Clifton has a shorter lacustrine record.

A well-developed terrace approximately 2 m above lake level is associated with all the lakes in the Clifton-Preston System (McArthur and Bartle, 1980; Gozzard, 1987). This suggests that all the lakes were once connected, forming a complex marine inlet. A lower terrace at Lake Clifton has been dated at 4,400 yr BP (Moore and Burne, undated).

PART III
ANCIENT SALINE LAKES

Sedimentology and Geochemistry of Modern and Ancient Saline Lakes, SEPM Special Publication No. 50

MIOCENE GLAUBERITE DEPOSITS OF ALCANADRE, EBRO BASIN, SPAIN: SEDIMENTARY AND DIAGENETIC PROCESSES

JOSEP M. SALVANY

Departament Enginyeria del Terreny i Cartogràfica, Universitat Politècnica de Catalunya, 08034, Barcelona, Spain

AND

FEDERICO ORTÍ

Departament Geoquímica, Petrologia i Prospecció Geològica, Universitat de Barcelona, 08071 Barcelona, Spain

ABSTRACT: The Alcanadre glauberite deposits are interbedded within the Lerín Gypsum Formation (Oligocene to Lower Miocene) in the western part of the Tertiary Ebro Basin of Spain. Glauberite forms several beds associated with other evaporitic minerals such as halite, polyhalite, gypsum, anhydrite, magnesite and dolomite. This association formed by sedimentary and early diagenetic processes in a playa-lake complex. The precipitation sequence was: carbonate, gypsum-anhydrite, glauberite, halite-polyhalite.

Gypsum occurs as a laminated facies precipitated as fine crystals on the floor of a shallow saline lake. Anhydrite formed at the lake margin and grew as interstitial nodules and enterolithic layers, or replaced pre-existing gypsum. Glauberite occurs both in nodular and enterolithic facies, commonly with convolution-like structures. Massive and banded facies are also present. Glauberite has textures ranging from fine-grained to euhedral macrocrystalline aggregates (up to 3 cm) that formed either from interstitial primary growth or from gypsum (or anhydrite) replacement. Polyhalite replaced all former sulfate phases (mainly gypsum and glauberite) and displays laminated facies and spherulitic microstructures. The polyhalite is associated with halite, which precipitated as primary crystals on the floor of the saline lake or as interstitial cement. The high Mg^{2+} content of the brines probably favored early transformation of all original $CaCO_3$ phases into dolomite or magnesite, both of which show micritic textures.

Because of dissolution by less concentrated (meteoric) brines, some retrodiagenetic processes have also been observed; polyhalite was replaced by glauberite, and polyhalite and glauberite by anhydrite. Burial led to further transformation of gypsum to anhydrite, and during exposure all pre-existing sulfates (anhydrite, glauberite, polyhalite) were transformed into secondary gypsum near the surface.

INTRODUCTION

The Tertiary Ebro Basin of Spain contains an important sequence of non-marine glauberite deposits of Oligocene to Miocene age (Fig. 1). Glauberite and its associated evaporite minerals are usually confined to the subsurface. Near the surface they have been extensively dissolved or transformed into secondary gypsum. The Alcanadre and Cerezo del Río Tirón deposits have been known for several years (Riba, 1955, 1964; Menduiña and others, 1984; Ordóñez and others, 1982). Several glauberite layers crop out, and were formerly mined (Alcanadre, Arrúbal); some deposits are still mined today (Cerezo, Belorado). More recently, glauberitic beds have been discovered in the subsurface evaporites, by new mining and oil boreholes (Fernández-Nieto and Galán, 1979; Mandado, 1987; Salvany, 1989a; García-Veigas and others, 1991). Overviews of the glauberite occurrences in the Ebro Basin are presented by Ortí and others (1989), and Ortí and Salvany (1991). However, their distribution and characteristics are still poorly known because of the paucity of subsurface data and drill core. Each deposit has its own specific mineral paragenesis, origin and differences.

In this paper we describe the main sedimentary and diagenetic features of the Alcanadre glauberite deposits. These are interbedded within the evaporites of the Lerín Gypsum Formation (Oligocene to Miocene boundary) in the western extremity of the Ebro Basin (Fig. 2). Besides glauberite, significant amounts of halite, polyhalite, magnesite and dolomite are also present, with gypsum and anhydrite forming the major part of the evaporitic host sediments. Preliminary results were given by Ortí and others (1986), Salvany and Ortí (1987), and Salvany (1989a).

REGIONAL SETTING

The Tertiary Ebro Basin is a foreland basin in the northeastern Iberian Peninsula. It is bounded by three Alpine chains that underwent uplift concurrently with the sedimentary filling of the basin: the Pyrenees to the north, the Iberian Ridge to the southwest, and the Catalonian Coastal Ranges to the southeast (Fig. 1). Basin evolution was controlled mainly by the Pyrenean uplift, which acted as a highly active margin, although significant structural roles were also played by the two other chains.

Two broad sedimentary cycles have been recognized in the basin (Riba and others, 1983): (1) from the early Tertiary until the upper Eocene the northern and eastern parts of the basin were the site of major siliciclastic, carbonate and evaporite marine sedimentation. To the south these sediments passed laterally into non-marine deposits; (2) since the late Eocene the basin has been the site of a thick sequence of non-marine sediments, both detrital and chemical. These include alluvial fan deposits that developed along the basin margins, and lacustrine carbonates and evaporites in the more central parts.

Tectonically, the Tertiary deposits of the Ebro Basin lie in two structural domains: (1) a northern domain, in which deformed sediments were incorporated into the South-Pyrenean nappes, and (2) a southern domain, where the sediments are in place and have been only affected by gentle folding related to emplacement of the nappes. The northern domain includes marine Paleocene and Eocene sediments, while most sediments in the southern domain are non-marine and Oligocene to Miocene in age (Fig. 1).

Glauberite deposits are found only within the large non-marine evaporite formations of the southern (autochthonous) domain. There, they are widespread and distributed along the western and central part of the basin. The evaporitic deposits were precipitated in playa-lake complexes. They indicate the former position of the basin depocenter, which shifted laterally during the Tertiary under the tectonic controls. Ortí and Salvany (1990) summarized these large evaporite sequences.

Sedimentology and Geochemistry of Modern and Ancient Saline Lakes, SEPM Special Publication No. 50
 ISBN 1-56576-014-X

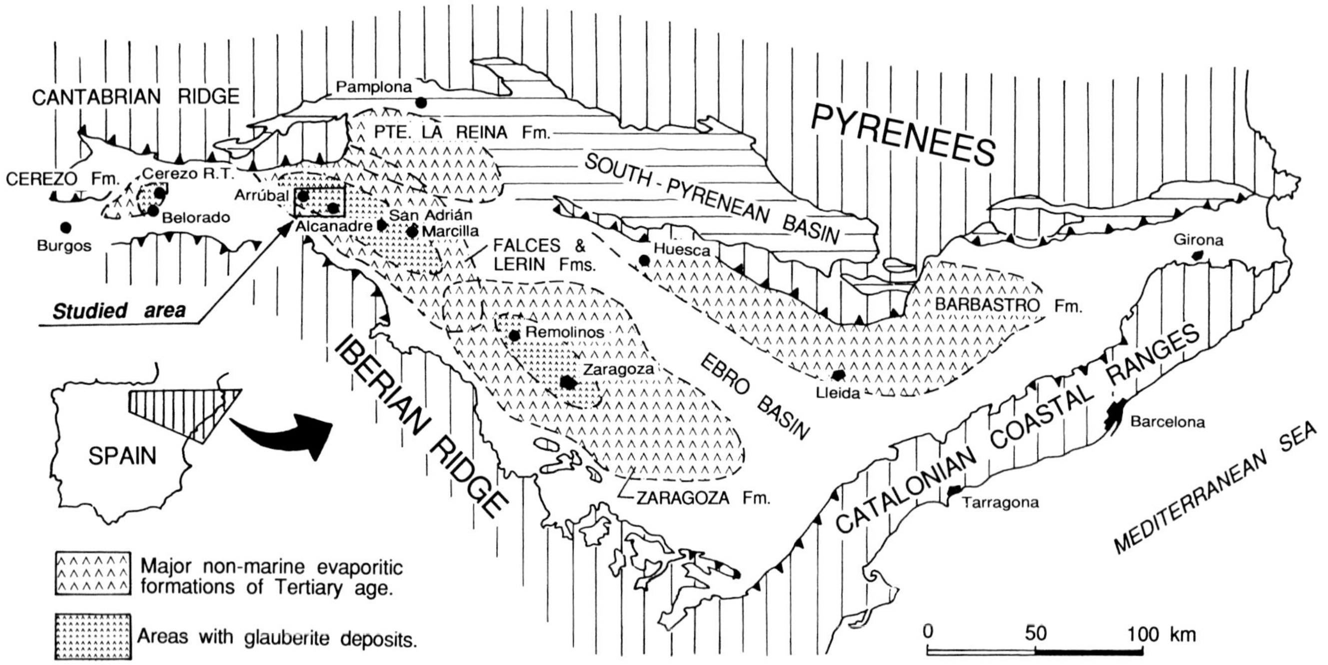

FIG. 1.—Geologic map of the Tertiary Ebro Basin showing distribution of major non-marine evaporitic formations and associated glauberite deposits.

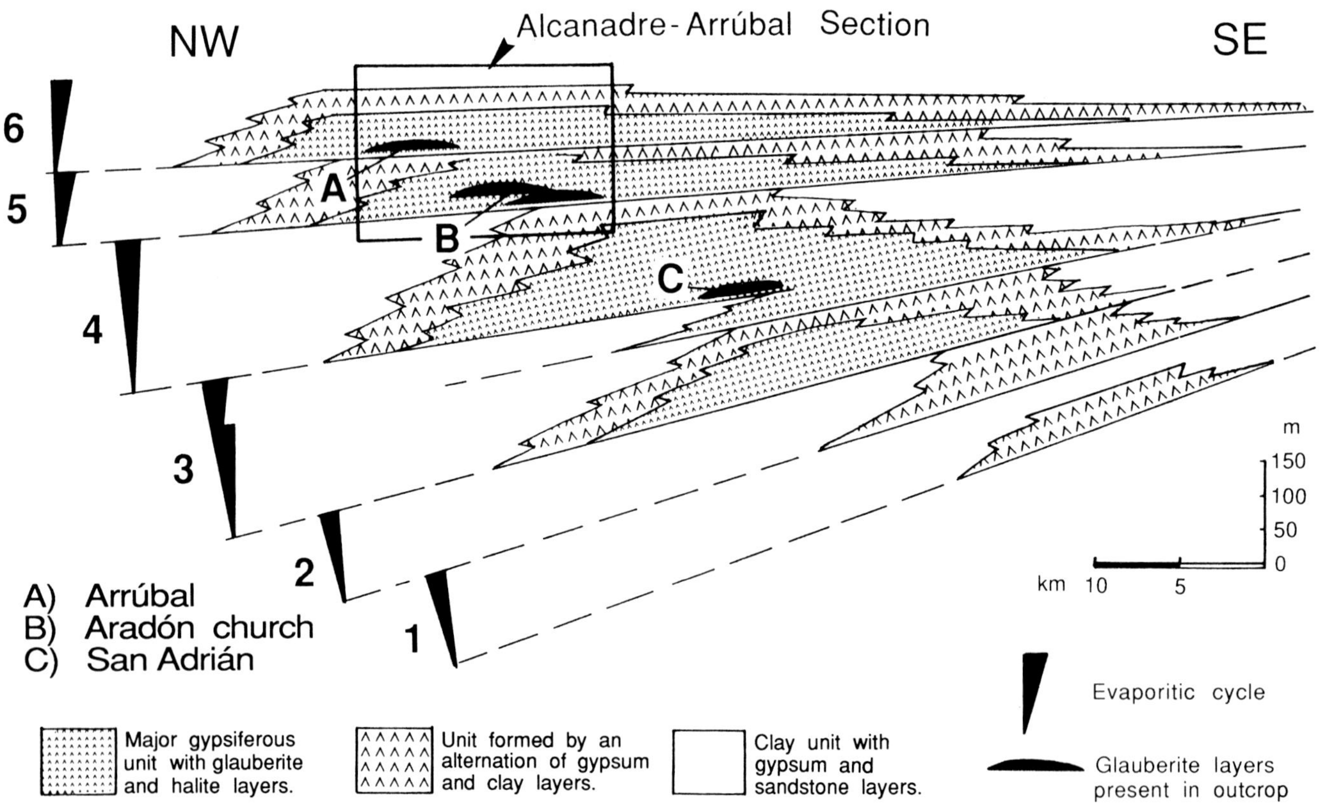

FIG. 2.—Stratigraphy of the Lerín Gypsum Formation subdivided into evaporite-detrital cycles. Location of glauberite deposit outcrops is also shown.

FIG. 5.—Anhydrite textural varieties: (A) elongated tabular crystals growing on (and replacing) polyhalite (scale bar= 300 μm), (B) equant to prismatic crystals growing on (and replacing) glauberite (scale bar= 300 μm).

eral changes also occurred during burial and final exposure (late diagenesis). Because of their long diagenetic history, primary evaporite minerals (i.e., any mineral precipitated and grown from brine, either in open water or interstitially) have only partly survived, but in places they can be identified as pseudomorphs, now composed of secondary minerals. Pseudomorphs may also have originated from secondary minerals, thus the final mineral may have had two or more precursor solid phases. The presence of partially replaced minerals and relict inclusions have allowed us to recognize the various mineralogical transformations that have taken place.

Gypsum

All the gypsum ($CaSO_4 \cdot 2H_2O$) in these deposits is secondary, and displays the well-known alabastrine, porphy-

roblastic and meg
1977). Gypsum fo
by rehydration of
also result from
glauberite and po
samples, replacen
it is limited to the
eral facies, the im
precursor was gla
recognizable even
mineral relicts are
gypsum. In conti
commonly shows
crystalline varietie
Secondary gyps
ular-enterolithic fa
by individual crys
sette shaped aggre
mary gypsum that
laminae formed in
lake. Microbial st
inae; wave ripples
and other sediment
the laminated and
gypsum that trans
genesis or burial. I
enterolithic layers
represent anhydrite
early diagenesis. S
tion of the various

Anhydrite (CaSO
tens of meters (Fig
for secondary gypsu
grained (30–150 μ
plays a granular or
bimodal trend may
Anhydrite forme
terstitial nodules or
Displacive nodular
sediments and as n
playa during perioc
hydrite caused def
nated gypsum was r
concurrently with g
Commonly, anhy
minerals (polyhalite
genesis. In polyhali
ular crystals up to
purest polyhalite zo
laths are randomly o
the host impurities (
inal polyhalite fabr
fractured, probably
replacing glauberite
and resemble isolate
5B). Glauberite rep
from other Tertiary

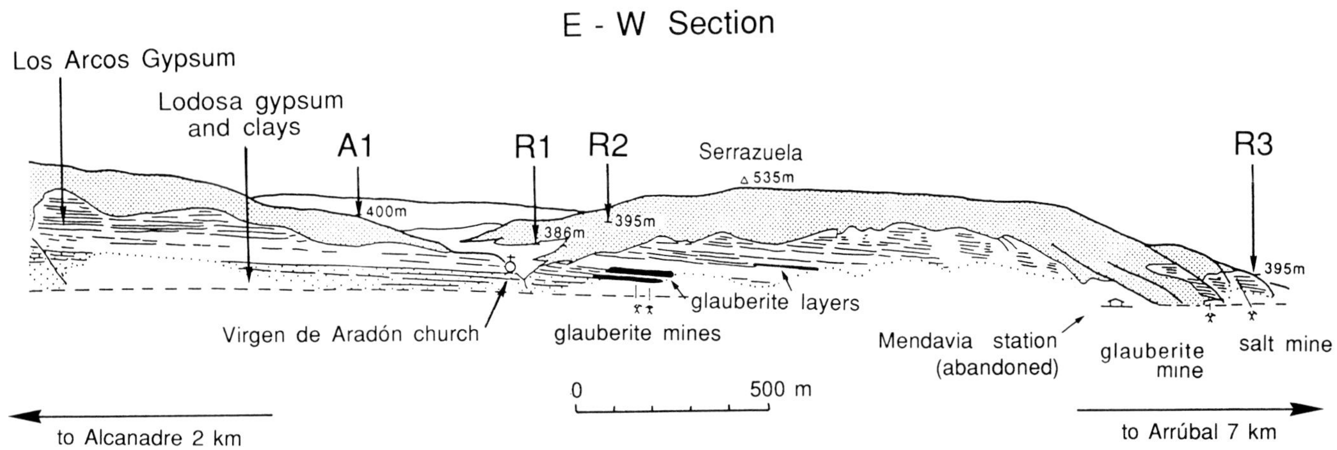

FIG. 3.—View of the Lerín Gypsum Formation at the Aradón church section near Alcanadre. Location of glauberite outcrops and boreholes is indicated (see Figure 4).

STRATIGRAPHIC SETTING AND EVAPORITE CYCLES

During Oligocene and Miocene time, non-marine sediments were deposited almost continuously in the western Ebro Basin, reaching 4 km in thickness. Three stratigraphic units with evaporites are recognized: the Puente La Reina Gypsum Formation (Lower Oligocene; 400 m thick), the Falces Gypsum Formation (Middle to Upper Oligocene; 800–1,000 m thick), and the Lerín Gypsum Formation (Oligocene to Lower Miocene; 400–600 m thick) (Castiella and others, 1978; Salvany, 1989a, b). These formations extend over an area about 100 km long by 30–40 km wide, and are separated by clastic units with subordinate evaporitic layers.

These three evaporitic formations are formed mainly by Ca-sulfates (gypsum, anhydrite) and halite. Anhydrite and halite are found in the subsurface, whereas gypsum occurs near the surface. In the Falces Formation and Lerín Formation, glauberite deposits are interbedded with Ca-sulfates. Glauberite deposits are unknown in the Puente La Reina Formation.

The Lerín Formation is composed of several evaporitic cycles about 100 m thick and more than 50 km in length (Fig. 2) (Salvany, 1989a, c). These cycles appear to have shifted progressively toward the northwest with time, with a concurrent shift of the basin depocenter. Maximum evaporite development occurs toward the top of the formation, the upper cycles becoming more gypsiferous than those below.

The cycles reflect the advance and retreat of the alluvial fan systems along the basin margins, which in turn affected the playa-lake complex in the central part of the basin. The base of an individual evaporite cycle represents the maximum expansion of lacustrine evaporitic conditions, while upward there is increasing influence of the progressive progradation of the alluvial fans. The tops of some cyclic units are formed by a thick (up to 10 m) layer of clays and fine-grained sandstones. This represents the maximum extension of the alluvial sediments that preceded the next cycle. The glauberite layers develop preferentially at the base of each cyclic unit and are confined toward the more central parts (Fig. 2).

THE GLAUBERITE DEPOSITS OF ALCANADRE

The uppermost two cycles that form the Lerín Formation are well exposed between the Alcanadre and Arrúbal (Fig. 3). Together these cycles form the most important evaporite level of the Lerín Formation, known as the "Los Arcos Gypsum" (Riba, 1964). In the eastern part of the Alcanadre-Arrúbal region the upper part of the fourth cycle of the Lerín Formation crops out and is composed of reddish detrital gypsiferous materials ("Lodosa Gypsum and Clay"; see Salvany, 1989a). Glauberite layers are found mainly at the base of the two evaporitic cycles that form the Los Arcos Gypsum. The lower one produces a steep escarpment near Aradón church. Two major glauberite layers, with thicknesses of 2 m and 4 m respectively, are exposed in the escarpment near sites of former mines. Near Aradón church the upper cycle has been eroded, but remains intact west of the old Mendavia railroad station. Overlying the upper cycle the clay and sandstones of the lower part of the Alfaro Formation crop out, representing the beginning of a new pattern of sedimentation, dominated by the fluvial environments.

The lower cycle was penetrated by the R1, R2 and A1 exploratory boreholes drilled by the Unión Salinera de España S. A. in 1983–84, near Aradón church (Figs. 3, 4). Glauberite layers that are thicker and more abundant than those exposed in the escarpment were cut by these boreholes. Borehole R3, located near Arrúbal, cut completely through the upper cycle. The glauberite layers are less common and less well developed than near Aradón church zone, but a thick halite bed is present, which was once exploited by a small mine near the top of the escarpment (Fig. 3).

EVAPORITE PETROGRAPHY

Petrographic analysis of the glauberite deposits has revealed extensive development of secondary minerals. These formed mainly during early diagenesis, but significant min-

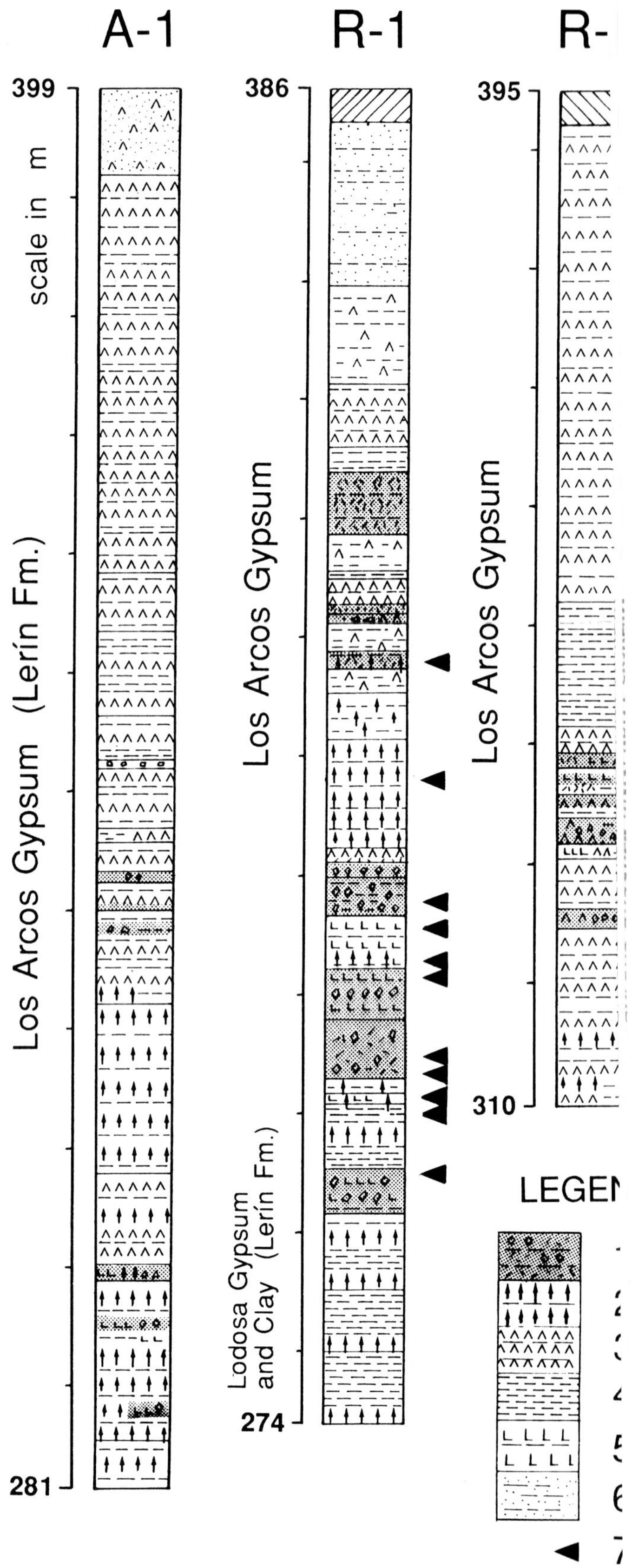

FIG. 4.—Simplified stratigraphy of boreholes drilled by Union Salinera de España, S.A., in th
(1) glauberite, (2) anhydrite, (3) secondary gypsum, (4) clay, (5) halite, (6) siltstone and sands

FIG. 10.—Polyhalite textural varieties: (A) glauberite crystals (g) partly replacing a polyhalite fabric (p) (scale bar= 200 μm), (B) polyhalite prisms growing along crystal boundaries of halite (dark) (scale bar= 500 μm), (C) spherulitic aggregates of polyhalite (scale bar= 500 μm), and (D) sheaf-shaped aggregates of polyhalite (scale bar= 500 μm).

Na^+, SO_4^{2-}, and Cl^-. Sulfate was the dominant anion, since sulfate minerals precipitated from the earliest stages of evaporation (gypsum, anhydrite) until the final most concentrated brine (glauberite and polyhalite paragenetic with halite). Carbonates are subordinate to sulfate, suggesting that HCO_3^- and CO_3^{2-} were less important that the other anions in the lake waters. Small amounts of Mg^{2+} and K^+ were also present, as shown by the occurrence of dolomite, magnesite and polyhalite.

After the initial precipitation of carbonate minerals (calcite or aragonite?), early evaporative concentration led to major precipitation of gypsum and anhydrite, causing significant loss of Ca^{2+} and SO_4^{2-} from the waters. Further brine evolution resulted in precipitation of glauberite and possibly Na-sulfates (thenardite, mirabilite) that were not preserved. In the final stages of evaporation halite precipitated, with small amounts of polyhalite. This partly depleted the residual brine in K^+ and Mg^{2+}, and completely removed remaining Ca^{2+}.

The increasing Mg^{2+} content of the brine changed all carbonate precursors into dolomite and magnesite. However, interstitial growth of magnesite in highly concentrated brines is also possible (Pueyo and Inglés, 1987). Although pure Na-sulfates (mainly thenardite) have not been found in the Alcanadre deposits, thenardite is present in other formations of the Ebro Basin (e.g., Zaragoza Formation: Mandado, 1987) and the Tajo Basin (Ortí and others, 1979). Although Ca^{2+} in polyhalite could be supplied mainly from the diagenetic replacement of former Ca-sulfates, the presence of beds of spherulitic polyhalite lacking pseudomorphs suggests direct precipitation from brines that retained some Ca^{2+}.

Thus, from the mineral paragenesis and diagenetic alteration observed in these deposits, the inferred hydrochem-

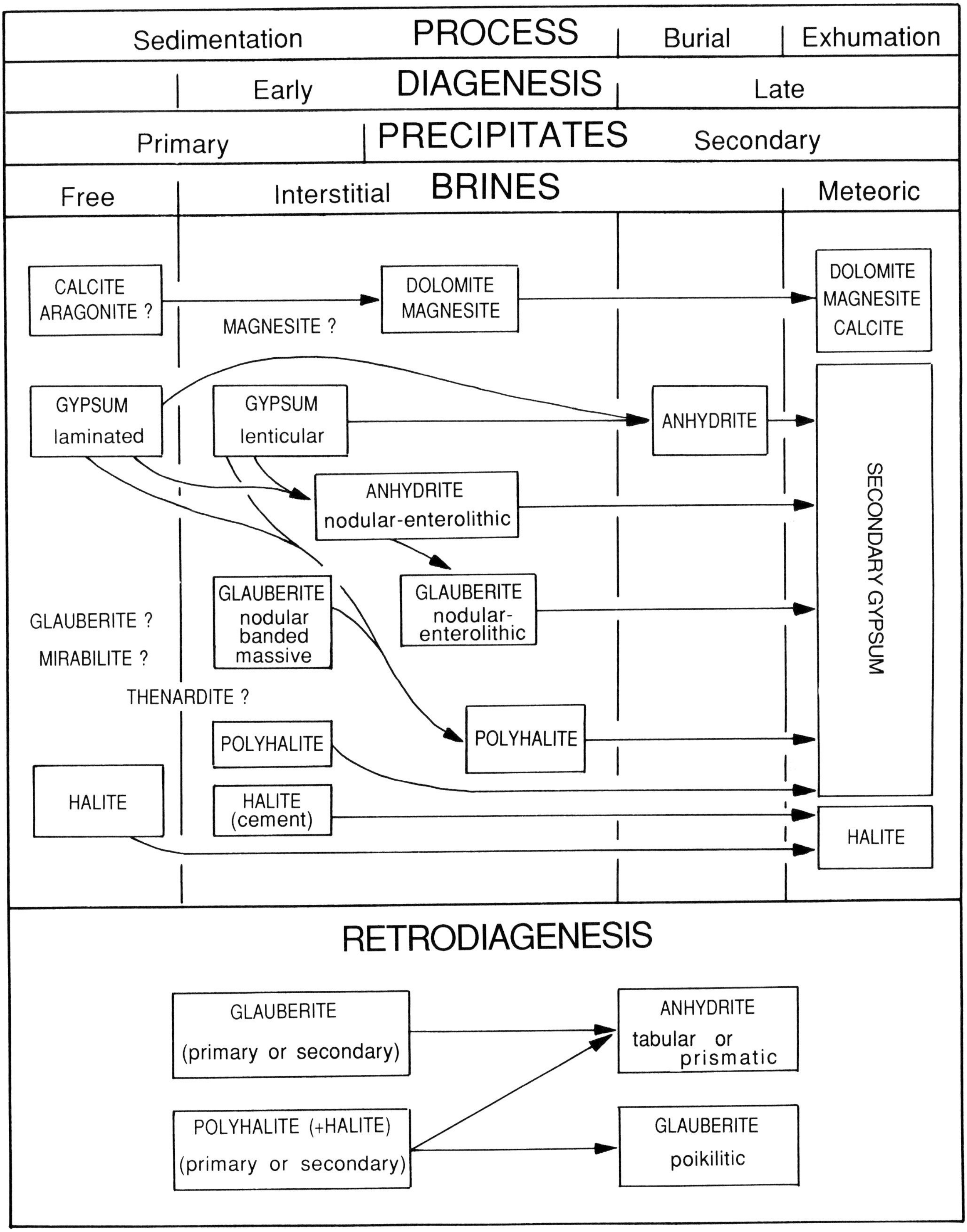

FIG. 11.—Summary of the diagenetic evolution of the glauberite deposits of Alcanadre.

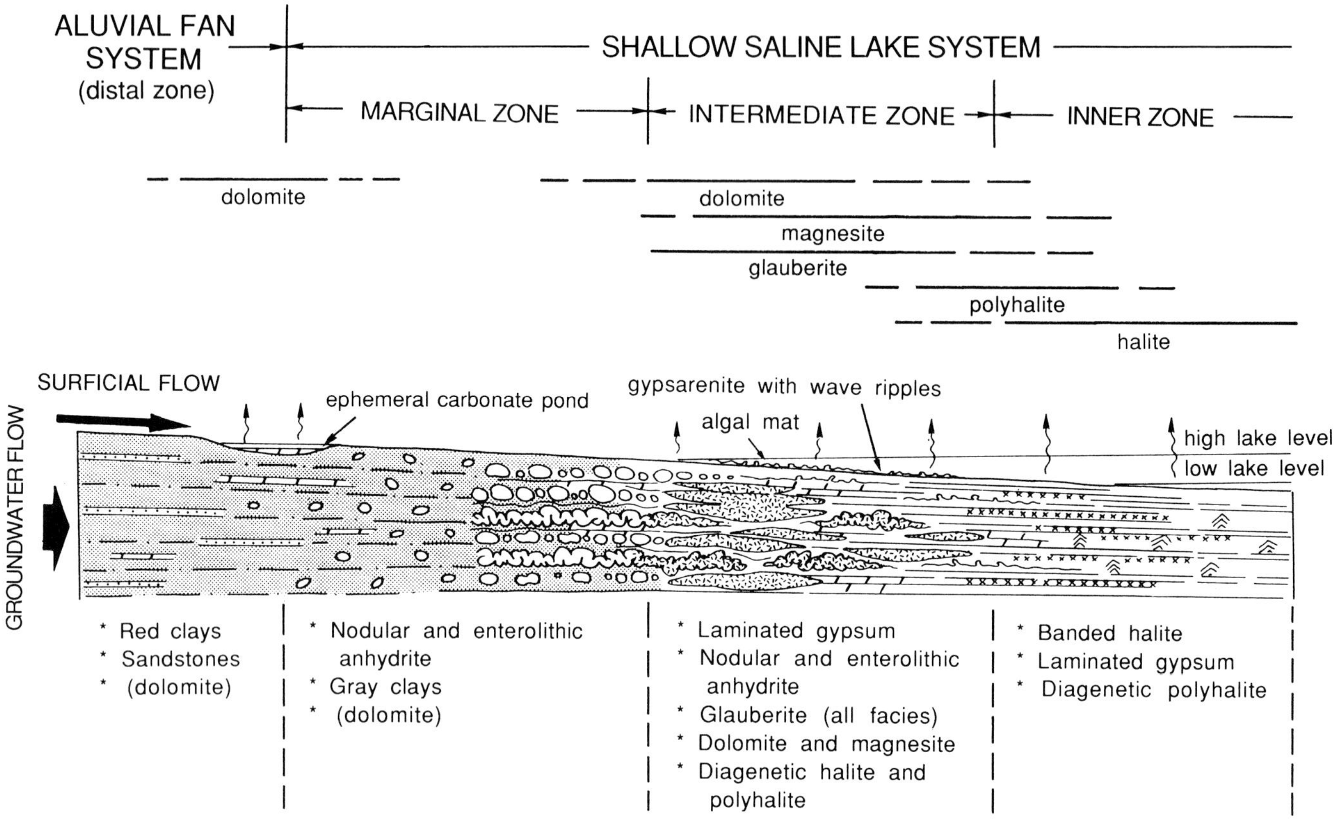

FIG. 12.—Sedimentological model for deposition of the Alcanadre glauberite deposits showing facies arrangement and evaporitic environments.

istry is in accord with the expected composition of major solutes that were drained from the marginal ridges of the Ebro Basin during the Tertiary. Significant Mesozoic evaporite deposits are present in the former source areas, mainly as anhydrite, gypsum and halite, which may explain the ionic trend of lake waters. This chemical recycling mechanism is also supported by stable isotopic evidence (Birnbaum and Coleman, 1979; Utrilla and others, 1991).

Glauberite has been reported from Holocene non-marine saline lakes with brines of similar composition to that inferred in this study in several countries (e.g., Australia, China, U. S. A., Canada, etc.). However, detailed petrographic descriptions of the mineral occurrence are usually lacking. Of particular relevance are the studies by Hardie (1968) of Saline Valley (California, U. S. A.) and more recently by Arakel and Cohen (1991) on the Karinga Creek playas (Amadeus Basin, Australia). In both examples, glauberite forms a significant part of the evaporite deposits of these playa-lake systems, having grown interstitially from the phreatic or vadose brine at the lake margins and is associated with other sulfates such as gypsum, thenardite, mirabilite or epsomite, as well as with halite.

Arakel and Cohen (1991) gave detailed descriptions of the glauberite textures in the Karinga Creek system, which developed either as primary or secondary phases. Primary crystals grow in the phreatic zone displacing the detrital host sediment, producing nodular aggregates and lenticular beds of glauberite. In the vadose zone, glauberite crystals grow in the same manner, but some are of secondary origin and display rim and poikilotopic textures on gypsum crystals that have been partially or totally replaced. The salinity of the interstitial brines attains 300 g/l TDS.

Polyhalite is less frequently cited in non-marine saline lakes than in ancient formations or recent coastal evaporitic lakes. Nevertheless, it is common in the continental Tertiary basins of Spain, where it always has the same textural features and a diagenetic origin (Ortí and Pueyo, 1980). Holser (1966) described similar diagenetic polyhalite in coastal Baja California that grew from interstitial brines with a salinity of almost 400 g/l TDS.

CONCLUSIONS

1. The glauberite deposits of Alcanadre and their associated minerals (gypsum, anhydrite, polyhalite, halite, dolomite and magnesite) precipitated in a shallow, ephemeral saline lake. The solutes were derived from dissolution of Mesozoic evaporites, mainly Ca-sulfates and halite, in the surrounding mountains (Pyrenees and Iberian Ridge).

2. With progressive evaporative concentration of the saline lake brine, the precipitation sequence was: carbonates, gypsum (+ anhydrite), glauberite, halite (+ polyhalite). During stages of low salinity and lake expansion, the less soluble minerals precipitated on the floor of the lake (primary gypsum and carbonates) and around the lake (diagenetic nodular anhydrite). During more saline stages with lake contraction or desiccation, the most soluble minerals grew mainly in saline mudflats surrounding the lake (dia-

genetic glauberite and polyhalite). Thin halite beds formed in the lake before desiccation.

3. The petrographic study shows that in many cases anhydrite, glauberite and polyhalite are secondary minerals. The alteration occurred during early diagenesis. Brines reacted with solid precursor phases by two main methods: (1) reaction of concentrated brines with solid phases precipitated during stages of low salinity (e.g., anhydrite altered to glauberite; glauberite altered to polyhalite) and (2) reaction of dilute brines with solid phases precipitated in stages of higher salinity (e.g., glauberite altered to anhydrite; polyhalite altered to glauberite or anhydrite).

4. Burial and exhumation produced further mineral transformations. With burial, gypsum altered to anhydrite. With exhumation, anhydrite, glauberite and polyhalite altered to secondary gypsum.

ACKNOWLEDGMENTS

We thank Union Salinera de España, S. A., for permission and facilities to study borehole samples, and for providing data from exploration surveys in the Alcanadre-Arrúbal zone.

REFERENCES

ARAKEL, A. V., AND COHEN, A., 1991, Deposition and early diagenesis of playa glauberite in the Karinga Creek drainage system, Northern Territory, Australia: Sedimentary Geology, v. 70, p. 41–59.

BIRNBAUM, S. J., AND COLEMAN, M., 1979, Source of sulphur in the Ebro Basin (northern Spain) Tertiary non-marine evaporite deposits as evidenced by sulphur isotopes: Chemical Geology, v. 25, p. 163–168.

CASTIELLA, J., SOLE, J., AND DEL VALLE, J., 1978, Mapa Geológico de Navarra, 1:200.000: Navarra, Diputación Foral de Navarra.

FERNANDEZ-NIETO, C, AND GALAN, E., 1979, Mineralogía de los depósitos de sales de Remolinos (Zaragoza): Sociedad Española Mineralogía, Volumen Extra 1, p. 51–65.

GARCÍA-VEIGAS, J., FERNANDEZ-NIETO, C., AND ORTÍ, F., 1991, Nota sobre la mineralogía y petrología de la Formación Zaragoza en el sondeo de Purasal: Boletin Sociedad Española Mineralogía, v. 14, p. 82–83.

HARDIE, L. A., 1968, The origin of the recent non-marine evaporite deposit of Saline Valley, Inyo Country, California: Geochimica et Cosmochimica Acta, v. 32, p. 1279–1301.

HOLSER, W. T., 1966, Diagenetic polyhalite in Recent salt from Baja California: American Mineralogist, v. 51, p. 99–109.

MANDADO, J., 1987, Litofacies Yesíferas del Sector Aragonés de la Cuenca Terciaria del Ebro. Petrogénesis y Geoquímica: Unpublished Ph.D. Thesis, Universidad de Zaragoza, Zaragoza, 442 p.

MENDUIÑA, J., ORDÓÑEZ, S., AND GARCÍA DEL CURA, M. A., 1984, Geología del yacimiento de glauberita de Cerezo del Río Tirón (provincia de Burgos): Boletin Geológico y Minero, v. 95, p. 33–51.

ORDÓÑEZ, S., MENDUIÑA, J. AND GARCÍA DEL CURA, M. A., 1982, El sulfato sódico natural en España: Tecniterrae, v. 46, p. 16–33.

ORTÍ, F., 1977, Aproximación al estudio petrográfico de las microestructuras de rocas de yeso secundario y a su origen: Revista del Instituto de Investigaciones Geológicas Diputación Provincial Universidad de Barcelona, v. 32, p. 87–152.

ORTÍ, F., AND PUEYO, J. J., 1977, Asociación halita bandeada-anhidrita nodular del yacimiento de Remolinos, Zaragoza (sector central de la Cuenca del Ebro). Nota petrogenética: Revista del Instituto de Investigaciones Geológicas Diputación Provincial Universidad de Barcelona, v. 32, p. 167–202.

ORTÍ, F., AND PUEYO, J. J., 1980, Polihalita diagenética en una secuencia evaporítica continental (Mioceno, Cuenca del Tajo, España): Revista del Instituto de Investigaciones Geológicas Diputación Provincial Universidad de Barcelona, v. 34, p. 209–222.

ORTÍ, F., PUEYO, J. J., AND SAN MIGUEL, A., 1979, Petrogénesis del yacimiento de sales sódicas de Villarrubia de Santiago, Toledo: Boletin Geológico y Minero, v. 90, p. 347–373.

ORTÍ, F., ROSELL, L., SALVANY, J. M., INGLÉS, M., PERMANYER, A., AND GARCÍA-VEIGAS, J., 1989, Sedimentología y diagénesis como criterios de prospección de recursos en el Terciario evaporítico de la Cuenca del Ebro: Simposios, 12 Congreso Español de Sedimentología, Leioa-Bilbao, p. 253–262.

ORTÍ, F., AND SALVANY, J. M., 1990, Formaciones Evaporíticas de la Cuenca del Ebro y Cadenas Periféricas, y de la Zona de Levante. Nuevas Aportaciones y Guía de Superficie: Barcelona, ENRESA-GPPG, Universidad de Barcelona, 307 p.

ORTÍ, F., AND SALVANY, J. M., 1991, Depósitos de glauberita en España: aspectos sedimentológicos y petrológicos generales, *in* Pueyo, J. J., ed., Génesis de formaciones evaporíticas: Barcelona, Modelos Andinos e Ibéricos: Publicacions de la Universitat de Barcelona, Estudi General, p. 193–230.

ORTÍ, F., SALVANY, J. M., QUADRAS, M., AND PUEYO, J. J., 1986, Glauberite growth in playa-lake environments (Tertiary of Ebro Basin, Alcanadre-Arrúbal Zone) (abs.): Granada, Abstracts, International Symposium on Crystal Growth Processes in Sedimentary Environments, p. 142–143.

PUEYO, J. J., AND INGLÉS, M., 1987, Magnesite formation in recent playa lakes, Los Monegros, Spain, *in* Marshall, J. D., ed., Diagenesis of Sedimentary Sequences: Oxford, Special Publication of the Geological Society of London 36, p. 119–122.

RIBA, O., 1955, Sur le type de sédimentation du Tertiaire continental de la partie Ouest du Bassin de l'Ebre: Geologische Rundschau, v. 43, p. 363–371.

RIBA, O., 1964, Estructura sedimentaria del Terciario continental de la Depresión del Ebro en su parte riojana y navarra: London, Proceedings of the 20th International Geological Congress, p. 127–138.

RIBA, O., REGUANT, S., AND VILLENA, J., 1983, Ensayo de síntesis estratigráfica y evolutiva de la Cuenca terciaria del Ebro, *in* Instituto Geológico y Minero de España, ed., Geología de España: Madrid, v. 2, p. 131–159.

SALVANY, J. M., 1989a, Las formaciones evaporíticas del Terciario continental de la cuenca del Ebro en Navarra y La Rioja. Litoestratigrafía, petrología y sedimentología: Unpublished Ph. D. Thesis, Universitat de Barcelona, Barcelona, 397 p.

SALVANY, J. M., 1989b, Los sistemas lacustres evaporíticos del sector navarro-riojano de la Cuenca del Ebro durante el Oligoceno y Mioceno inferior: Acta Geológica Hispánica, v. 24, p. 231–241.

SALVANY, J. M., 1989c, Ciclos y megaciclos evaporíticos en las formaciones Falces y Lerín, Oligoceno—Mioceno inferior de la Cuenca del Ebro (Navarra-La Rioja): Somarios, 12 Congreso Español de Sedimentología, Leioa-Bilbao, p. 83–86.

SALVANY, J. M., AND ORTÍ, F., 1987, La paragénesis de sulfatos de Ca y Na en el Mioceno continental de Alcanadre-Arrúbal (La Rioja) y San Adrián (Navarra): Boletin Sociedad Española Mineralogía, v. 10, p. 47–48.

UTRILLA, R., ORTÍ, F., PIERRE, C., AND PUEYO, J. J., 1991, Composición isotópica de las evaporitas terciarias continentales de la Cuenca del Ebro: relación con los ambientes deposicionale: Revista de la Sociedad Geológica de España, v. 4, p. 355–362.

TERTIARY DETRITAL GYPSUM IN THE MADRID BASIN, SPAIN: CRITERIA FOR INTERPRETING DETRITAL GYPSUM IN CONTINENTAL EVAPORITIC SEQUENCES

MARÍA ESTHER SANZ, JUAN PABLO RODRÍGUEZ-ARANDA, JOSÉ PEDRO CALVO, AND SALVADOR ORDÓÑEZ
Departamento de Petrología y Geoquímica, Facultad de Ciencias Geológicas, UCM, 28040 Madrid, Spain

ABSTRACT: The lower beds of the Miocene Intermediate Unit in the southern and southeastern parts of the Madrid Basin, central Spain, are formed mainly of mudstones, marls and gypsum. The mudstones contain euhedral gypsum crystals and are interpreted as the sediments of a saline mudflat environment. Two main types of gypsum deposit have been identified: (1) chemical precipitates, which include selenite and lenticular gypsum crystals, and (2) detrital gypsum grains. The latter form prominent deposits throughout the study area and display a wide variety of hydrodynamic structures, including scours, wavy bedding, parallel stratification, ripple lamination, parting lineation, graded bedding, and low-angle cross-stratification. The detrital gypsum beds are formed of sand-sized, hemipyramidal, lenticular or prismatic crystals that commonly exhibit abraded and/or partially dissolved edges.

We recognize four facies associations. The first, formed of mudstone and gypsum-bearing mudstone, is a typical saline mudflat association. The second consists of detrital gypsum deposits interbedded with gypsum-bearing mudstones. In this association, the detrital gypsum is interpreted as sheetflood deposits derived from the reworking of the mudflat facies during periods of heavy rainfall. The other two facies associations comprise different types of detrital gypsum deposits that are interbedded with marginal and/or more central, lacustrine evaporite facies. These facies associations allow us to characterize the different sedimentary subenvironments throughout a saline mudflat-saline lake complex.

INTRODUCTION

The reworking and redeposition of gypsum has been recognized as a significant process in many evaporite basins. Most of the literature deals with ancient marine evaporite successions, mainly from the Messinian of the Mediterranean region (Hardie and Eugster, 1971; Parea and Ricci-Luchi, 1972; Schreiber and others, 1976; Schlager and Bolz, 1977; Vai and Ricci-Luchi, 1977; Rouchy, 1982; Aigner and Bachmann, 1989). There are only a few detailed studies of detrital gypsum in continental (lacustrine) settings (e.g., Truc, 1978, 1980), although some sedimentological work has been carried out in the Neogene successions of the Iberian Peninsula (Megias and others, 1981,1982; Dabrio and others, 1982; Mediavilla, 1988; Mediavilla and Dabrio, 1988; Corrochano and Armenteros, 1989; Sanz and others, 1991).

The recent lacustrine and coastal environments of Australia have provided useful sedimentary models for interpreting detrital gypsum deposits in ancient evaporite sequences (Caldwell, 1976; Arakel, 1980; Warren, 1982; Logan, 1987; Magee, 1991). These models supplement those proposed by Hardie and others (1978) and Eugster and Hardie (1978) from other closed-lake systems.

The terminology in the literature used to describe reworked or redeposited gypsum is varied and includes: clastic gypsite (Caldwell, 1976; Arakel, 1980), clastic gypsum (Schreiber, 1978), detrital gypsum (Megías and others, 1981; Rouchy and others, 1986), crystal clasts (Logan, 1987; Lowenstein and Hardie, 1985), and detrital gypsarenite/gypsorudite (Arakel, 1980). In this paper, we use the general term "detrital gypsum." Usage is restricted to those gypsum facies that clearly show detrital texture and structure. We use the term "chemical gypsum" for precipitated crystalline gypsum fabrics that lack evidence of reworking. In this paper, we describe several occurrences of well-developed detrital gypsum deposits of Miocene age in the Madrid Basin, central Spain (Fig. 1). San José (1975) gave the first description of the deposits. Megías and others (1981, 1982) suggested that the detrital gypsum deposits resulted from erosion of the underlying evaporite units and proposed that the clastic gypsum was deposited by lacustrine turbidity flows. The geochemical characteristics of the gypsum deposits have been analyzed by Ordóñez and others (1987, 1989). More recently, Sanz and others (1991) made a preliminary study of their sedimentology in the southern part of the basin. They suggested that sheetflooding and reworking of lacustrine evaporites were the main depositional processes.

In this paper, we attempt to reconstruct the depositional environments of the gypsum deposits based on facies analysis and their sequential arrangement. The wide variety of detrital gypsum deposits present in continental settings can make it difficult to interpret their fossil counterparts. With this in mind, we present criteria that may be useful for the recognition and interpretation of detrital gypsum deposits in other ancient evaporites.

GEOLOGICAL SETTING

The Madrid Basin, which has an area of more than 10,000 km^2, is located in the center of the Iberian Peninsula (Fig. 1). The basin was filled by a sequence of Tertiary sediments that are up to 3,500 m thick in the western part of the basin, and nearly 2,000 m in the central and eastern parts.

Tectonically, the Madrid Basin is a complex intracratonic basin that resulted from differential strain within the Iberian plate during the Alpine orogeny (Vegas and Banda, 1982). The northern basin margin, termed the Central System (Guadarrama and Somosierra Ranges; Fig. 1), forms a major mountain uplift generated either by basement thrusting (Warburton and Alvarez, 1989) or by compressive deformation of rigid intraplate areas (Vegas and others, 1990). In the northeast, the basin margin is formed by the Iberian Range, composed of NW-SE folded, Mesozoic formations. The Altomira Range to the east of the basin resulted from the emplacement of nappes, thrust from E to W. Although the tectonic evolution of the Toledo Mountains, to the south of the basin, is less well known, this margin seems to have been fairly active during the Paleogene, becoming moderately stable during the Neogene.

Sedimentology and Geochemistry of Modern and Ancient Saline Lakes, SEPM Special Publication No. 50

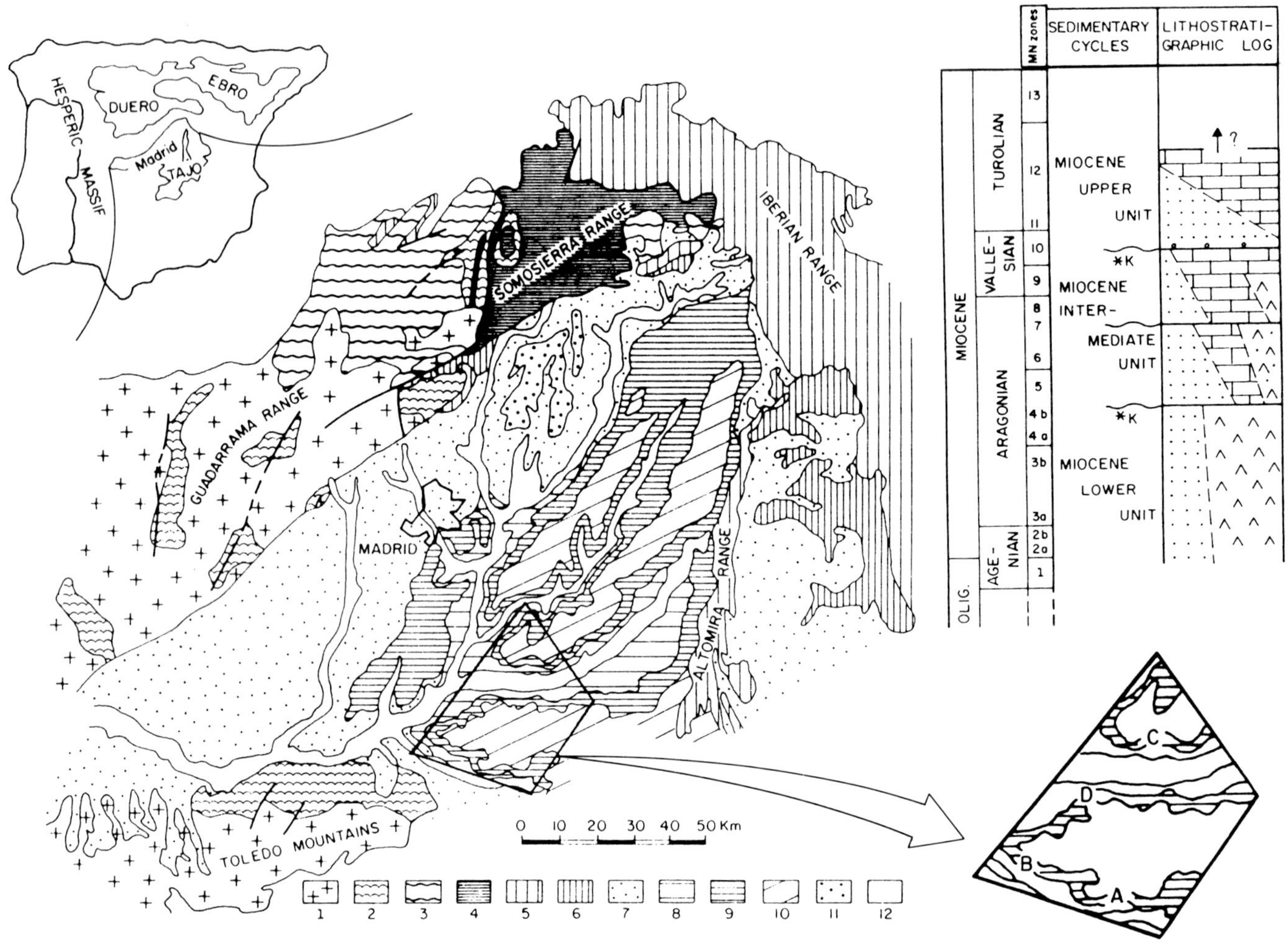

FIG. 1.—Schematic geologic map, and stratigraphic chart for the Miocene sedimentary record of the Madrid Basin (modified from Calvo and others, 1989). Study area is highlighted on the map. Locations of the stratigraphic logs (Fig. 2) are arrowed. Numbers on the map indicate the lithology of the basin margins and the main Tertiary units: 1- plutonic rocks; 2- shales, marbles, quartzites and gneisses; 3- shales and metagreywackes; 4- shales, quartzites and metavolcanics; 5- Mesozoic terranes (mainly carbonates); 6- Paleogene (terrigenous and carbonates); 7- lower to upper Miocene terrigenous sediments; 8- Lower Unit, Miocene; 9- Intermediate Unit, Miocene; 10- Upper Unit, Miocene; 11- Pliocene; 12- Quaternary. Symbols * and K in the chart refer to paleokarstification surfaces.

The Neogene continental strata range in total thickness from 1,200 to 800 m. The thickest exposed sections (up to 300 m) occur in the central parts of the basin. The lowermost part of the Neogene sequence is observed near the basin margins, where it unconformably overlies Paleogene deposits.

Three main Miocene sedimentary units have been distinguished in the Madrid Basin, termed Lower, Intermediate and Upper, respectively (Alberdi and others, 1983; Junco and Calvo, 1983; Hoyos and others, 1985). The boundaries between these units are major sedimentary discontinuities, represented by paleokarst and/or erosional surfaces (Fig. 1).

The Lower Unit, from 600–1,000 m thick, spans Agenian to middle Aragonian time. In the central parts of the basin, the unit is composed mainly of evaporite deposits, principally halite, anhydrite, thenardite and glauberite (Ordóñez and others, 1991). These pass laterally into alluvial sediments that become coarser-grained near the basin margins. The general sedimentary model for this unit is one with a concentric facies arrangement, characteristic of an arid to semi-arid, closed basin with an evaporitic saline lake.

The Miocene Intermediate Unit, from 50 to 200 m thick, spans the middle Aragonian to the early Vallesian, and is best developed in the northern part of the basin. A division into two subunits has been suggested based on changes in vertical succession (Fig. 1). The Intermediate Unit is formed of a variety of alluvial and lacustrine deposits. These facies are arranged in a complex concentric pattern (Calvo and others, 1989) characteristic of a lake basin that is both hydrologically and topographically closed.

The Upper Unit spans most of late Miocene (late Vallesian to late Turolian) time and has a maximum thickness of about 50 m in central and southern parts of the basin. This unit consists mainly of terrigenous fluvial deposits and shallow lacustrine carbonates.

The gypsum deposits described in this paper are located in the lower part of the Miocene Intermediate Unit. They lie either paraconformably on mudstones, or on karstified gypsum deposits of the Lower Unit. The thickness of the sequence containing detrital gypsum reaches up to 40 m. In the study area (Fig. 2), the individual detrital gypsum beds thicken from south to north. In the southern part, the detrital gypsum facies are overlain by mudstones with pedogenic features, whereas to the north, they are capped by monotonous successions of massive lenticular gypsum beds.

SEDIMENTOLOGY

Facies Description

We distinguish four main lithofacies groups based on lithology and characteristic sedimentary structures at outcrop scale. These lithofacies groups are: (1) detrital gypsum, (2) chemically precipitated gypsum, (3) mudstones and gypsiferous mudstones, and (4) carbonates (specifically, marls). Each group, in turn, is divided into several facies that represent different types of sedimentation and/or distinct locations within the closed lacustrine depositional system.

Detrital gypsum.—

This category includes sediments that have clastic textures, with detrital gypsum crystals (i.e., grains) 60 μm to 1 cm long, mixed in places with fine-grained mud-clasts. The detrital gypsum grains are arranged in grainstone or packstone fabrics (in the sense of Logan, 1987). The matrix of the packstones is mudstone or marlstone. Grain shape varies from euhedral to anhedral (Fig. 3). Several crystal geometries are recognized: tabular hemipyramidal, lenticular, prismatic, and rounded.

Monocrystalline gypsum grains are far more abundant than twinned crystals. Both the apices and edges of the detrital

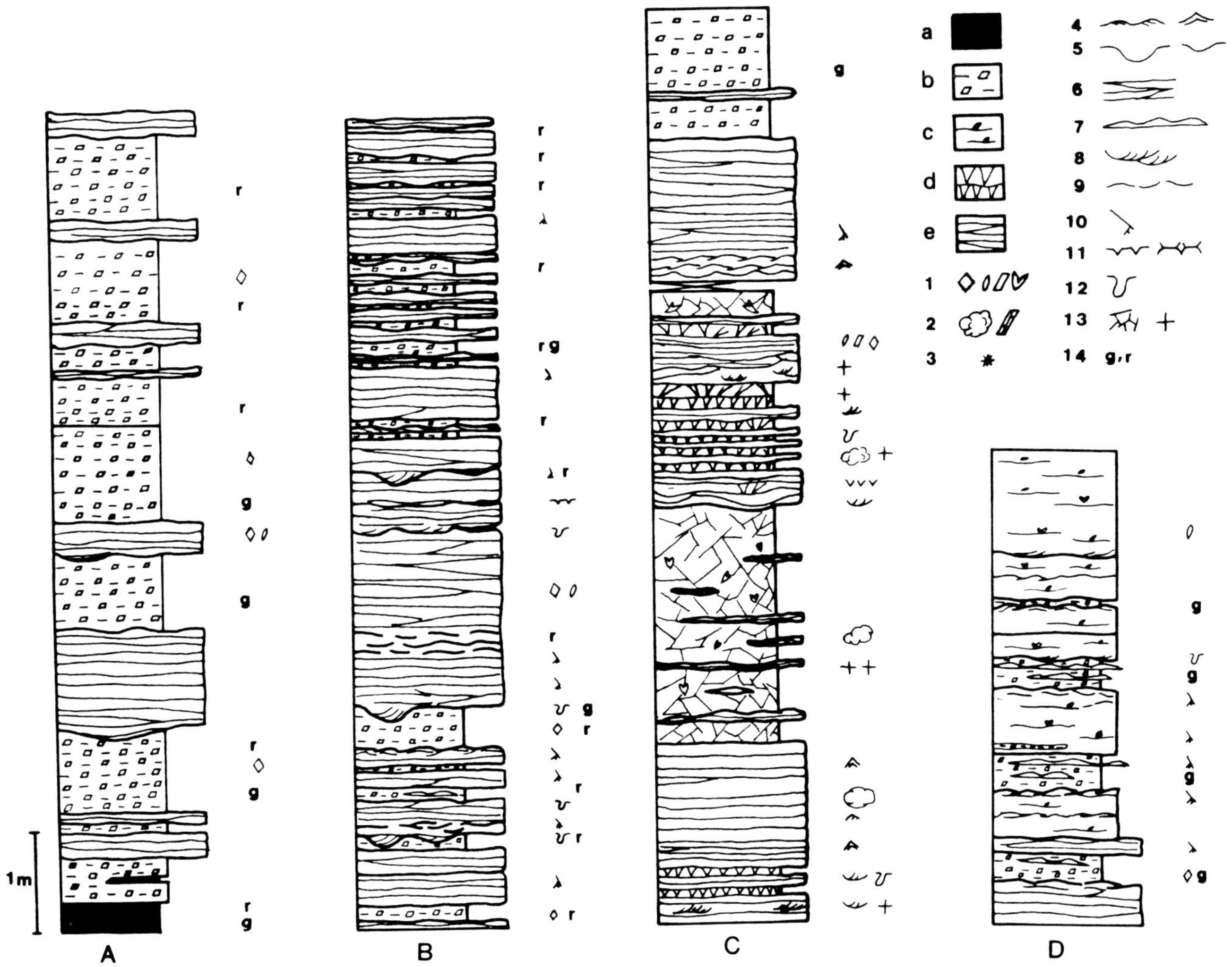

FIG. 2.—Stratigraphic logs of the lower levels of the Miocene Intermediate Unit. Letters below the logs refer to the location map in Figure 1. Lithology: a) mudstones, b) gypsum-rich mudstones, c) lenticular gypsum with carbonate matrix, d) selenitic gypsum, and e) detrital gypsum. Symbols represent: 1) gypsum crystal morphologies, from left to right: hemipyramidal, lenticular, prismatic, selenitic; 2) anhydritization; 3) soluble salts; 4) current and oscillation ripples; 5) scours, erosive surfaces; 6) parallel lamination and low-angle cross bedding; 7) wavy bedding; 8) trough-cross bedding; 9) flaser bedding; 10) bioturbation; 11) desiccation cracks; 12) hydroplastic deformation; 13) recrystallization; and 14) color of mudstones (g, green; r, red).

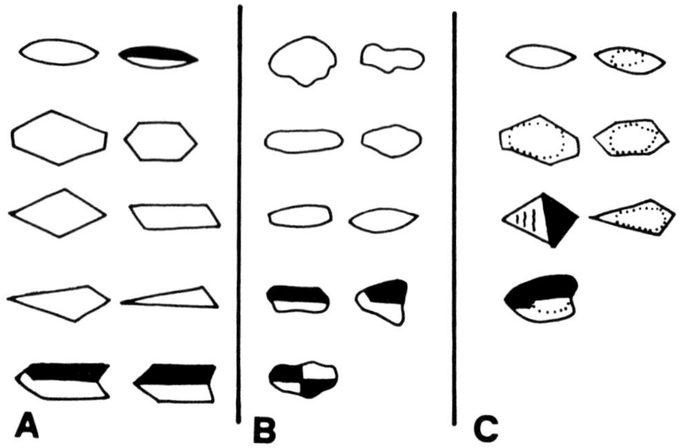

FIG. 3.—Microscopic sections of gypsum crystals. (A) Primary morphologies. (B) Crystals showing abrasion/dissolution features. (C) Abraded/dissolved crystals with epitaxial overgrowths.

crystals are commonly abraded and/or broken. These features are commonly masked by well-developed pyramidal overgrowths in optical continuity with the abraded detrital core. The distinction between the core and the overgrowths is clearly seen where the boundary is outlined by a zone of included impurities (e.g., clay, anhydrite pseudomorphs, iron oxides, fluid inclusions) (Fig. 4A). The greater abundance of inclusions within the core than in the overgrowth also allows recognition of the original gypsum crystals (Fig. 4A). Some crystals display overgrowths that, in turn, show evidence of reworking.

The long axes of the detrital gypsum grains commonly lie parallel to the bedding. Parting lineation is observed on some bedding planes. The detrital gypsum deposits display a variety of hydrodynamic structures. These structures, together with the fabric and textural characteristics of the reworked gypsum crystals, have led to the differentiation of three main detrital gypsum facies:

Scour-and-Fill (sf).—Scour-and-fill structures, 10 to 20 cm high and 20 to 50 cm wide (Fig. 5B), occur typically at the base of the detrital gypsum beds (Fig. 2B). The scour-and-fill structures are spaced from a few centimeters up to 80 cm apart along the same bed. They are concave or U-shaped in cross section. Commonly, these shapes have been modified by hydroplastic deformation.

The scour-and-fill structures are made up of discrete, centimeter-thick laminae that are either parallel or pinch out near the scour margins. The laminae are composed of poorly sorted, coarse-grained hemipyramidal gypsum crystals. These alternate with thin laminae of mudstones that contain rare hemipyramidal gypsum crystals. In some areas, especially to the NE (Fig. 2C), scour-and-fill facies occur as wider sedimentary bodies (up to 1 m in width, 15 cm in height) that contain a mixture of hemipyramidal and prismatic gypsum crystals.

Parallel-Laminated and Low-Angle Cross-Laminated Detrital Gypsum (pdg).—This facies comprises decimeter-thick (20–170 cm), sheet-like beds of detrital gypsum that are traceable laterally for several hundred meters (Fig. 4B). The beds are composed of layers of detrital gypsum, 2 mm to 5 cm thick, separated by thin (up to 5 mm) mudstones or gypsum-rich mudstone laminae (Fig. 4C). The layers are parallel to the bedding planes or show low-angle cross-stratification. The individual layers may be massive, but commonly they show normal grading and ripple-cross lamination (Fig. 4C). Current ripples, and wave and interference ripples, have been recognized in the detrital gypsum layers.

These sediments have grainstone and packstone fabrics (Arakel, 1980; Logan, 1987) made up of sand-sized hemipyramidal and lenticular gypsum crystals; twinned hemipyramidal and prismatic crystals have also been recognized. Sorting is generally moderate to poor. Bioturbation by roots and hydroplastic deformation are common.

Trough and Planar Cross-Stratified Detrital Gypsum (cdg).—This facies is represented by sheet-like beds, 10 to 20 cm thick, that locally have erosive bases. The beds are formed of moderately sorted, sand-sized gypsum grains showing trough cross-bedding or, less commonly, planar cross-bedding (Figs. 4D, 5A). Single troughs are a decimeter wide, and 5 to 10 cm high. This facies is usually interbedded with parallel-laminated and low angle cross-laminated detrital gypsum.

Chemically precipitated gypsum.—

Selenitic gypsum (sg).—This facies consists of laterally continuous beds of prismatic gypsum crystals that are arranged vertically or obliquely to the bedding planes (Fig. 5C). The thickness of the beds depends mainly on the length of the crystals, which usually ranges from 5 to 30 mm. Some crystals show (100) twins. Selenite crystals are commonly accompanied by minor amounts of detrital gypsum and marl between the prisms (Fig. 4E).

The tops of the selenite beds may be irregular due to truncation or partial erosion of the crystals, recrystallization and/or epitaxial growth from the original crystals, or even primary domal growth of the selenite. Hydroplastic deformation, deformation caused by pressure of crystallization, and the irregular geometry of the underlying gypsum (mainly ripple-bedding) also contribute to the irregular appearance of the selenite layers. The 2- to 3-cm long, prismatic crystals commonly show "double twinning," attributed to nucleation on gypsum grains (Warren, 1982). The primary selenite fabric has commonly been modified by tractive reworking, leading to individual selenite crystals with long axes parallel or oblique to bedding (Schreiber, 1978).

Lenticular Gypsum (lg).—This facies includes decimeter-thick, laterally continuous, sheet-like beds of lenticular gypsum with variable amounts of marly matrix (Fig. 4F). The fabric varies from grainstone to wackestone. On the basis of crystal size, these are divided into microcrystalline (60–300 μm long) and macrocrystalline (500 μm-5 mm). In the latter case, the crystals are usually twinned, and occur as aggregates. Most lenticular gypsum crystals are arranged parallel or sub-parallel to the bedding planes, although they are aligned vertically where aggregated. Root molds filled by mudstone are common. The lenticular gypsum beds are commonly capped by laminae (up 10 cm thick) of detrital lenticular gypsum displaying oscillation and interference ripples (Fig. 5D).

FIG. 4.—(A) Gypsum grains with included impurities and epitaxial overgrowths (Scale bar= 250 μm). (B) Outcrop view of parallel-laminated detrital gypsum. (C) Hand specimen showing parallel and ripple cross-laminated gypsum overlain by mudstone films (Scale in cm). (D) Hand specimen showing trough and planar-cross stratified gypsum (Scale in cm). (E) Prismatic gypsum crystals arranged vertically with respect to the bedding plane. Detrital gypsum grains overlying the top of the layer are observed (Scale bar= 1 mm). (F) Macro-and microlenses of gypsum with micrite matrix (Scale bar= 250 μm).

FIG. 5.—(A) Interbedded trough and planar cross-stratified gypsum with gypsum-rich mudstones. (B) Scours in the top of a gypsum-rich mudstone layer. Note hydroplastic deformation disrupting the scours. (C) Interbedded selenite layer with parallel laminated detrital gypsum. Note deformation in the selenite layer. (D) Bands of massive lenticular gypsum alternate with interbedded mudstone and gypsum with wavy bedding.

Mudstones and Gypsiferous Mudstones.—

This facies group comprises siliciclastic mudstones with variable amounts of silt grains, gypsum-rich mudstones, and beds showing alternation of both these facies. The composition of the mudstones is mainly illite with variable amounts of quartz. The carbonate content is typically less than 5 percent.

Mudstones (m).—The mudstones mainly occur as beds a few millimeters to 40 cm thick, averaging 10 cm. They are usually massive and show red, green and greyish colors, which suggest that hydromorphic processes took place (Retallack, 1988).

Gypsum-Rich Mudstones (grm).—This facies consists of greenish to reddish, massive to weakly laminated, mudstones that contain abundant intrasedimentary gypsum crystals (up to 50 percent by volume). The thickness of the mudstone beds ranges from 5 cm to 1 m. Crude lamination within the beds is caused by the variable amount of gypsum crystals in different horizons. Commonly, the layers show an undulating appearance due to the displacive growth of gypsum. Desiccation cracks are locally present. Most gypsum crystals have euhedral and hemipyramidal shapes and may be twinned, similar to those described by England (1982). Their average size ranges from 0.5 to 3 mm. Lenticular gypsum crystals are rare.

Interbedded Mudstone and Gypsum with Wavy Bedding (wmg).—This facies association consists of a centimeter-thick alternation of rippled detrital gypsum and the gypsum-rich mudstone facies (grm). Together, they form tabular beds 10 to 70 cm thick with poorly defined lower surfaces (usually transitional to mudstone or gypsum deposits) and sharp upper surfaces. The detrital gypsum crystals are lenticular. This facies commonly shows bioturbation by roots.

Carbonates (c).—

Only a few carbonate deposits are associated with the predominant mudstone and gypsiferous facies. Where present, they form thin, centimeter-thick layers made up of massive dolomitic marlstone, micrite with charophytes, or microbial mats.

Diagenetic Features

The sedimentary facies described above have in many cases been affected by diagenetic processes that partially destroy the primary features. The main diagenetic processes and/or products that have been recognized are:

Recrystallization.—

Recrystallization, resulting in anhedral macrocrystalline gypsum mosaics, is observed in both detrital and selenitic gypsum facies. In places, remnants of the original facies can be preserved within the recrystallized patches. Recrystallization may occur at specific levels or be found discontinuously.

Late Precipitation of Calcium Sulphate.—

Late precipitation of calcium sulphate took place in joints between mudstones and detrital gypsum beds, leading to the formation of 1 cm-thick aggregates of euhedral hemipyramidal gypsum crystals along the joints. These are interpreted as resulting from crystal growth in more permeable zones under phreatic conditions (Arakel, 1980).

Displacive Growth of Hemipyramidal Crystals.—

Some glauberite and gypsum crystals have grown displacively within the detrital gypsum deposits. These secondary crystals, up to 1 cm long, are oriented vertically. Their early diagenetic nature is demonstrated by their occurrence as lags at the base of some beds of detrital gypsum, suggesting that they were reworked rapidly after formation.

Anhydritization.—

Displacive-replacive growth of anhydrite nodules up to 30 cm in diameter, now altered to alabastrine gypsum, has also been observed, although rarely. Additional evidence for anhydritization is found in small prisms of anhydrite partially replacing the gypsum crystals.

Carbonate Replacements.—

Gypsum is locally replaced by carbonate, usually dolomite and/or calcite due to dedolomitization. This process is recognized by corrosion of the gypsum crystal margins or, in some cases, by sparite crystals replacing displacive gypsum. Where the replacement is well developed, it destroys the primary fabric.

Facies Associations and Depositional Environments

Representative logs of the Miocene gypsiferous series in the study area are shown in Figure 2. We distinguish four main facies associations (Fig. 6). Their sedimentological interpretation follows the general model proposed by Hardie and others (1978) for closed continental evaporite basins.

Association 1: Dry Mudflat-Saline Mudflat Environment.—

This association is composed of mudstones (*m*) and gypsum-rich mudstones (Fig. 6), exposed mainly in the southern part of the study area (Fig. 2A). Both the paleogeographic location and facies characteristics suggest deposition in marginal areas of the lake. The mudstones were likely deposited from suspension from low energy, viscous sheetfloods in low-relief areas and/or in ponds. They represent typical facies of dry mudflat-saline mudflat environments (Hardie and others, 1978). The sheetfloods were related to alluvial fan systems that developed at the southern basin margin (Toledo Mountains, Fig. 1) (Sanz and others, 1991). In this setting, gypsum-rich mudstones characterize the saline mudflat environment. The intrasedimentary displacive growth of gypsum crystals within the mudstones may have taken place at times when the mudflat surface became dry, but where the phreatic zone was near the surface. The most likely mechanism for displacive growth of gypsum under these conditions is evaporative pumping (Hsü and Siegenthaler, 1969). The distinction between massive and layered gypsum-rich mudstones may depend strongly on the position of the groundwater table. Massive gypsum-rich mudstones probably correspond to growth of gypsum under phreatic conditions, whereas layered mudstone beds may form in vadose settings.

Association 2: Inner Mudflat-Lake Margin Transition.—

This is the most common association in the southern part of the study area (Figs. 2A, 2B). It comprises both mudstones and gypsiferous mudstones (facies *gbm*, *wmg*), together with several facies composed of detrital gypsum (Figs. 5A, 5B, 6). Variations are related to the predominant facies type and the relative thickness of this facies in vertical sequences. The most common sequential arrangement consists of basal gypsum-rich mudstones with subordinate *wmg* lithofacies, that pass upward into detrital gypsum beds.

The lower mudstone facies is characteristic of a saline mudflat environment; the occurrence of facies *wmg* indicates episodic reworking by flowing water. The upper detrital gypsum facies is interpreted as deposition by sheetfloods in the more central zones of the saline mudflat. This is supported by the characteristic sedimentary structures (current ripples, graded bedding, parallel and cross lamination), which are similar to those described by Hubert and Hyde (1982) and Tunbridge (1984).

The development of "overland flows" or sheetfloods in periods of sporadic heavy rainfall is characteristic of arid to semiarid regions. This process has been described in recent environments (Hardie and others, 1978; Handford, 1982; Chen and others, 1991). Water supplied by sheetfloods

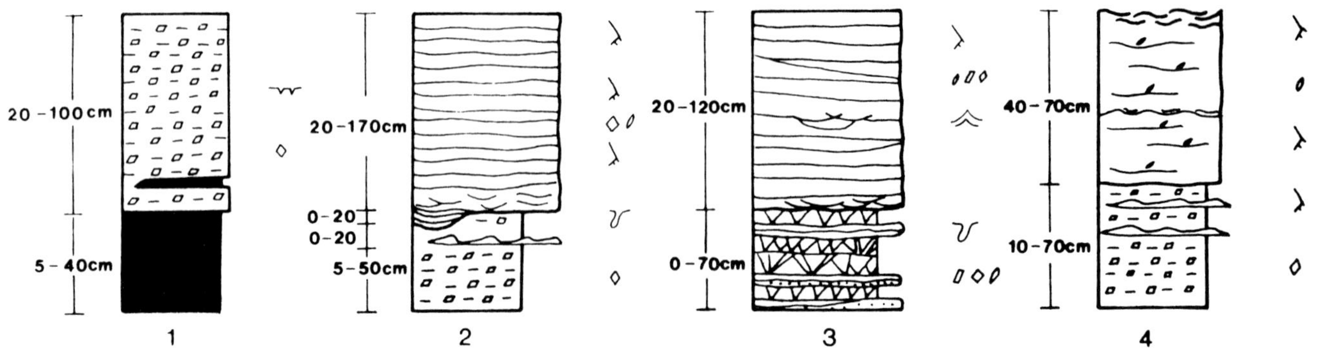

FIG. 6.—Sketch showing the facies associations: (1) Dry mudflat-saline mudflat association; (2) Inner mudflat-lake margin association; (3) Lake marginal association; and (4) Lake margin-open lake association (Legend as Fig. 2).

contributes significantly to lake expansion across the mudflats (Eugster and Hardie, 1978).

In the deposits studied, we believe that sheetfloods were the most important mechanism for reworking and redepositing gypsum crystals and grains in the saline mudflat and/or lake marginal environments. The close facies relationship between gypsum-rich mudstones and detrital gypsum deposits in vertical sequences supports this interpretation. The reworking of gypsum-rich mudstones may be favored by the disruption of the saline mudflat surface by mudcracks, surficial precipitation of efflorescent salts, and deflation (e.g., Hardie and others, 1978; Magee, 1991).

Figure 7 shows the successive stages in the development of this facies association. The presence of scour-and-fill structures at the base of the detrital gypsum beds corresponds to the first stages of sheetflow in proximal areas of the mudflat. Where scour-and-fill is absent, the detrital gypsum beds were deposited from unconfined tractional currents with waning power. It is probable that each lamina (ranging from 2–50 mm in thickness) was deposited by discrete sheetflow episodes, as a result of flow deceleration. Flow deceleration could have occurred when sheetfloods either reached ponded water on the mudflats or, more likely, the mudflat surface during periods of lake expansion.

Association 3: Lake Marginal Environment.—

Both detrital gypsum and selenite gypsum facies are grouped in this association (Figs. 5C, 6), which is mainly represented in the log of Figure 2C. This facies association is well developed in the eastern part of the study area.

Growth of selenitic gypsum usually takes place in shallow water bodies, the crystals forming at the sediment-brine interface (Schreiber, 1978; Rouchy and others, 1986). In our example, we infer a shallow depth for gypsum formation from the lateral reduction in width of the selenite layers and from their irregular geometry, which was mainly controlled by the rippled surface of the lake floor.

Growth of selenite crystals requires conditions that remain relatively constant, and close to gypsum saturation for a long time (Magee, 1991). These conditions could have developed in restricted subenvironments, such as ponds at the lake margins. Relatively long periods of selenite growth might be followed, however, by phases of stronger evaporation, causing incipient anhydritization and/or recrystallization of the previously formed selenite.

The presence of detached and horizontally oriented crystals at the upper surface of the selenite layers provides evidence for reworking of the selenite facies. The selenite crystals are thought to have been the main contributors to the detrital gypsum deposits recognized in this facies association (Fig. 6C). The clastic gypsum is formed mainly of prismatic crystals, although hemipyramidal crystals have also been recognized. The origin of the hemipyramidal crystals in this facies is problematic. They may have resulted from diagenetic overgrowth of former prismatic cores (cf. Magee, 1991).

The detrital gypsum beds of this association were likely produced by the reworking of selenitic gypsum by waves in the littoral zone. This is supported by the widespread occurrence of oscillation and interference ripples, parallel and low angle cross-lamination, and parting lineation in the gypsum beds, which are typical of beaches (e.g., Elliot, 1986).

Figure 8 shows the proposed model for the formation of the selenitic and ripple-bedded detrital gypsum in this facies association. Initially, selenite crystals grow on an irregular, rippled surface. Reworking by waves and/or episodic sheetfloods across the lake margin leads to accumulation of selenite fragments or finer crystalline grains. Thin marly laminae, which are common between the gypsum layers, represent periods of relative dilution of the brine, probably coincident with higher water supply to the lake. Further brine evaporation might induce supersaturation with respect to gypsum, leading to the precipitation of "selenite grass," a facies typically formed of crystals less than 1 cm in size. The "selenite grass" could be reworked easily by waves and the small crystals could then undergo epitaxial overgrowth finally forming "crystal-clasts" (see Logan, 1987). This may explain the presence of hemipyramidal gypsum crystals within the detrital gypsum beds.

Association 4: Lake Margin-Open Lake Facies.—

This facies association, which is most common in the central and northern parts of the study area, comprises tabular beds of lenticular gypsum and beds of interbedded mudstone and gypsum with wavy structure (Figs. 2D, 6).

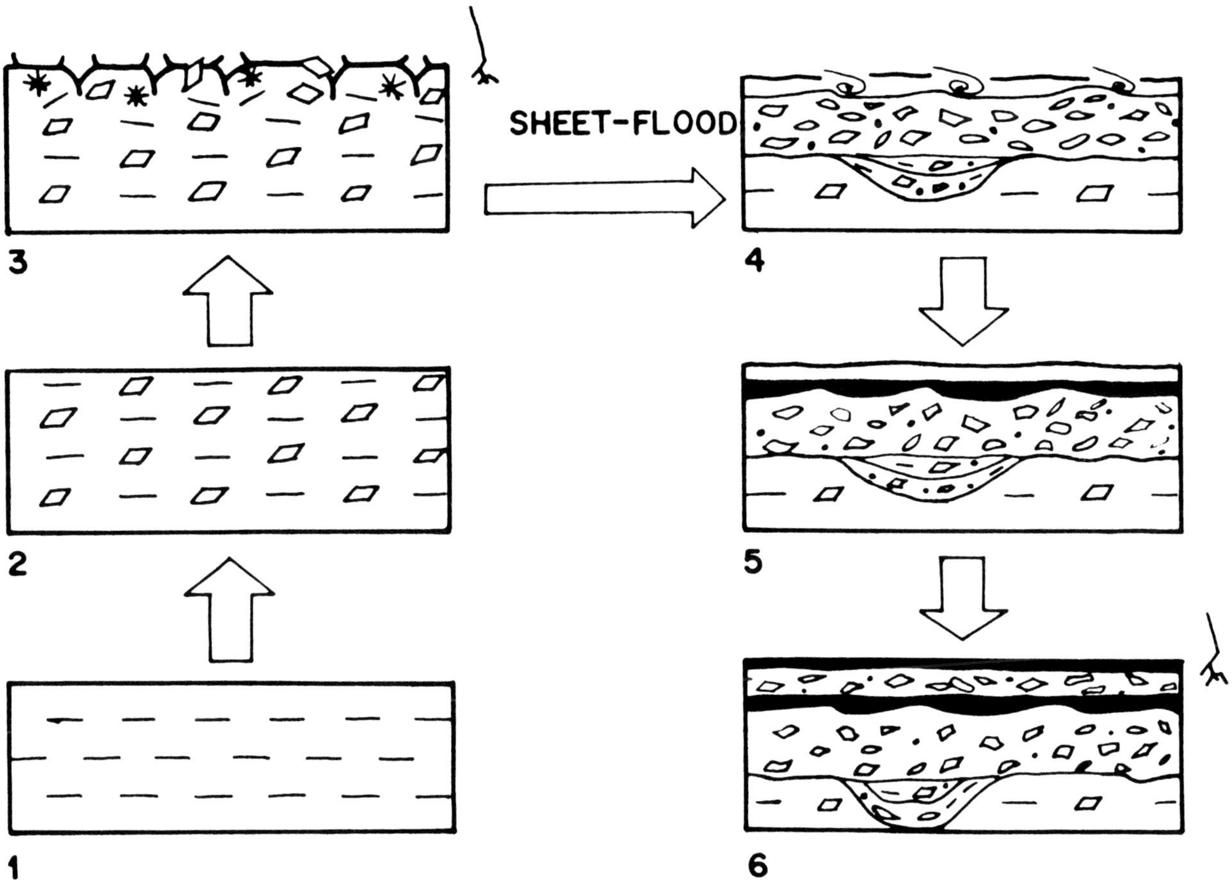

FIG. 7.—Proposed model for the formation of the inner mudflat-lake margin association. For explanation, see text (Legend as Fig. 2).

The lenticular gypsum facies was probably deposited in a shallow lake with a brine that did not reach a very high salinity. Gypsum and subordinate anhydrite are the only saline phases identified in this area and elsewhere in the basin.

A different origin is postulated for the presence of microlenticular and macrolenticular gypsum in these deposits (cf. Schreiber, 1978; Truc, 1980; Fort and others, 1982). Microscale lenses of gypsum probably formed at the air-brine interface or in the brine, and later sank to the lake floor. Macroscale lenses of gypsum grew at the sediment-brine interface or in the sediment. Periods of brine dilution might have led to precipitation of a carbonate (mainly dolomite) matrix between the gypsum crystals. Early diagenetic overgrowth of gypsum lenses, which are commonly recognized in this facies, indicates supersaturation of calcium sulphate in the residual brine. This may have been coincident with periods of low lake level.

Gypsum-rich mudstones at the base of the sequences (Figs. 5D, 6) are usually intercalated with interbedded mudstone-gypsum with wavy structure. This grouping seems to characterize the lower part of the saline mudflat. These facies include some layers of detrital gypsum produced by wave reworking during periods of lake expansion. Similar features are recognized at the top of the lenticular gypsum beds (Fig. 6D).

In summary, facies association 4 is formed by the alternation of true subaqueous lacustrine deposits with sediments that characterize the distal zones of saline mudflats. These deposits, in turn, record long-term fluctuations of lake level.

Sedimentary Model

The main paleogeographic characteristics of the Madrid Basin during middle Miocene deposition have been described by Calvo and others (1989,1991), and Ordóñez and others (1991). The general scenario is that of a topographically and hydrologically closed basin surrounded by high mountains that formed the catchment areas for huge alluvial fans. These fans sloped down to reach a large, shallow perennial saline lake. Figure 9 shows a sedimentary model depicting the lacustrine and peripheral alluvial systems in which sedimentation of gypsum took place. The location of the main sedimentary sequences shown in Figure 6 is also indicated.

Alluvial fans supplied fine-grained sediment to the fringing mudflats that became progressively saturated with brine closer to the saline lake. This brine-saturated subenvironment corresponds to the "saline mudflat" of Hardie and others (1978), which is characterized by intensive intrasediment

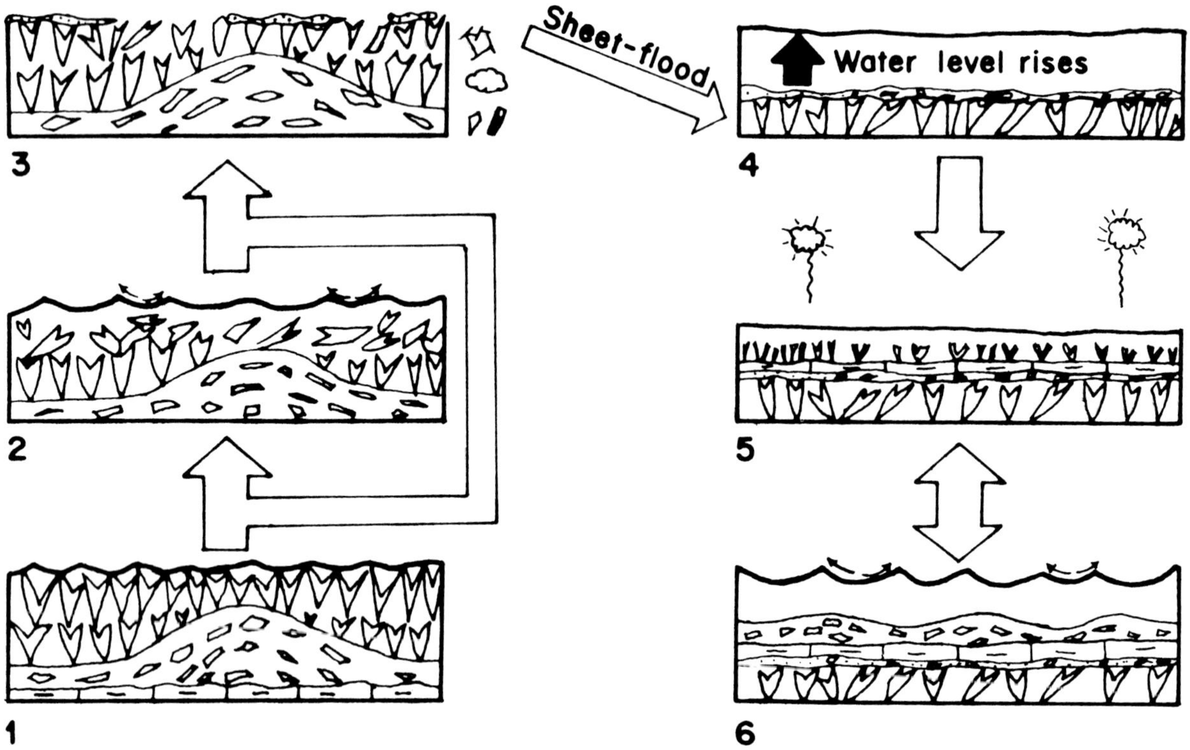

FIG. 8.—Successive stages of development of the lake margin association. For explanation, see text (Legend as Fig. 2).

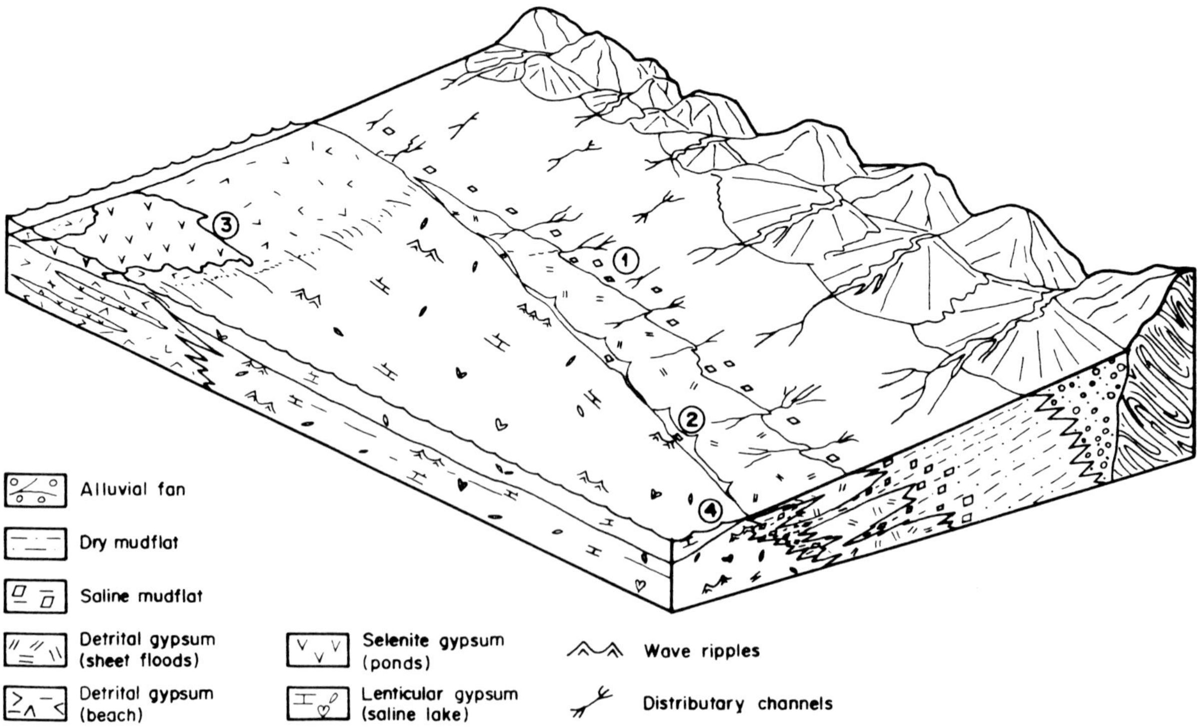

FIG. 9.—Inferred sedimentary model for the gypsiferous sequence. Numbers refer to the facies associations.

growth of salt crystals (in our case only gypsum). Growth of the gypsum crystals, which consist typically of euhedral hemipyramidal crystals within the mudstone (England, 1982), may have been caused by evaporative pumping.

Water supply to the mudflat occurred through episodic surficial, sediment-charged storm waters. Sheetfloods across the mudflats could account for the entrainment of gypsum crystals and mud fragments from the gypsum-rich mudstones previously deposited in the saline mudflat. The unchanelled flows transported the gypsum grains across the mudflat and redeposited them as graded beds as the flow waned. This process takes place in periods of storm flooding and subsequent lake expansion (Hardie and others, 1978).

This suggests that alternating drier and wetter periods developed, resulting in periodic exposure and flooding of the mudflats. These are also recorded in the saline lacustrine facies, accounting both for periodic concentration and dilution of the lake brine. An increase in brine concentration is recorded by crystal overgrowths and other processes such as the minor anhydritization and recrystallization. Dilution of the brine could explain the formation of marls interbedded with the gypsum deposits.

In the saline lake environment, two different subenvironments have been recognized:

1. A shoreline subenvironment characterized by the stacking of thin detrital gypsum beds with oscillation ripples and interbedded, discontinuous selenite layers. These facies formed at shallow depths in ponds where the growth of selenite crystals was favored by relatively stable brines. Further flooding of these ponds (during lake expansion?) would cause extensive reworking by waves and the formation of beaches, and shallow platform detrital gypsum deposits.
2. An open lake subenvironment characterized by massive deposition of lenticular gypsum. Alternating periods of concentration and dilution of the lake brine are recorded by the formation of lenticular gypsum crystals and thin carbonate laminae, respectively. A more drastic fall in lake level is suggested by the reworking of the lenticular gypsum deposits by waves.

CONCLUSIONS

1. Well-developed detrital gypsum deposits crop out in the lower part of the Miocene Intermediate Unit of the Madrid Basin. The detrital gypsum deposits are interbedded with mudstones and/or gypsum-rich mudstones that are considered to be the main source for the gypsum grains.
2. Deposition of the detrital gypsum and associated lithofacies took place in an evaporitic closed-lake basin. Several environments, namely saline mudflat, lake margin and open lake, have been distinguished, each of which is characterized by different facies associations and sequences.
3. From detailed sedimentological analysis of these facies associations, several mechanisms for the formation of detrital gypsum deposits are recognized: (1) reworking of gypsum-rich mudstones (saline mudflat deposits) by sheetfloods in periods of heavy rainfalls, (2) reworking of selenite deposits in ponds by waves during periods of flooding across shoreline areas, and (3) reworking by waves of open lake facies (lenticular gypsum deposits) during periods of low lake level.
4. All these processes resulted in the formation of a varied spectrum of detrital gypsum facies, each characterized by specific sedimentary structures. The morphology of the gypsum crystals (lenticular, hemipyramidal tabular, prismatic) within the beds depends both on the source sediments and early diagenetic processes. Overgrowth of primary gypsum cores commonly results in hemipyramidal gypsum crystals that can be confused with precipitated gypsum.

ACKNOWLEDGMENTS

We thank Drs. A. M. Alonso and M. A. Garcia del Cura for critical comments. We also thank Dr. J. M. Rouchy for discussions in the field and Robin Renaut and Bill Last, convenors of the conference where this paper was presented and discussed. P. Silva helped with preparation of the figures. This research was supported by the CSIC-CAYCIT, through project PB 89–0032.

REFERENCES

AIGNER, T., AND BACHMANN, G. H., 1989, Dynamic stratigraphy of an evaporite-to-red bed sequence Gipskeuper (Triassic), southwest German Basin: Sedimentary Geology, v. 62, p. 5–25.

ALBERDI, M. T., HOYOS, M., JUNCO, F., LOPEZ MARTINEZ, J., MORALES, J., SESE, C., AND SORIA, D., 1983, Biostratigraphie et évolution sédimentaire du Néogene continental de l'aire de Madrid (abs.): Montpellier, Abstracts of the International Colloquium on the Mediterranean Continental Neogene and its Paleoclimatic Evolution, p. 15–23.

ARAKEL, A. V., 1980, Genesis and diagenesis of Holocene evaporitic sediments in Hutt and Leeman lagoons, western Australia: Journal of Sedimentary Petrology, v. 50, p. 1305–1326.

CALDWELL, R. A., 1976, Holocene gypsum deposits of the Bullara Sunkland, Carnavau basin, western Australia: Unpublished Ph. D. Dissertation, University of Western Australia, Perth, 123 p.

CALVO, J. P., ALONSO ZARZA, A. M., AND GARCÍA DEL CURA, M. A., 1989, Models of Miocene marginal lacustrine sedimentation in response to varied depositional regimes and source areas in the Madrid Basin (Central Spain): Palæogeography, Palæoclimatolology, Palæoecology, v. 79, p. 199–214.

CALVO, J. P., HOYOS, M., MORALES, J., AND ORDÓÑEZ, S., 1991, Neogene stratigraphy, sedimentology and raw materials of the Madrid Basin, *in* Augustí, J., Domènech, R., Juliá, R., and Martinell, J., eds., Iberian Neogene Basins: Sabadell, Memoria Especial, Paleontologia i Evolució, v. 2. p. 61–95.

CHEN, X. Y., BOWLER, J. M., AND MAGEE, J. W., 1991, Aeolian landscapes in central Australia: gypsiferous and quartz dune environments from Lake Amadeus: Sedimentology, v. 38, p. 519–538.

CORROCHANO, A., AND ARMENTEROS, I., 1989, Los sistemas lacustres de la Cuenca Terciaria del Duero: Acta Geológica Hispánica, v. 24, p. 259–279.

DABRIO, C. J., MARTIN, J. M., AND MEGÍAS, A. G., 1982, Signification sédimentaire des évaporites de la dépression de Granade (Espagne): Bulletin de la Societé Géologique de France, v. 24, p. 705–710.

ELLIOT, T., 1986, Siliciclastic shorelines, *in* Reading, H. G., ed., Sedimentary Environments and Facies: Oxford, Blackwell, p. 155–188.

ENGLAND, B. M., 1982, Crystallized gypsum from the playa lake clays of Lake Gilles: Mineralogical Record, v. 5, p. 187–191.

EUGSTER, H. P., AND HARDIE, L. A., 1978, Saline lakes, *in* Lerman, A., ed., Lakes, Chemistry, Geology, Physics: New York, Springer, p. 237–293.

FORT, R., CALVO, J. P., AND ORDÓÑEZ, S., 1982, Contribución al conocimiento de la geología del borde oriental de la Cuenca de Madrid (sector Pastrana-Illana): Tecniterrae, v. 49, p. 23–40.

HANDFORD, C. R, 1982, Sedimentology and evaporite genesis in a Holocene continental sabkha playa basin, Bristol Dry Lake, California: Sedimentology, v. 29, p. 239–253.

HARDIE, L. A., AND EUGSTER, H. P., 1971, The depositional environment of marine evaporites a case for shallow, clastic accumulation: Sedimentology, v. 16, p. 187–220.

HARDIE, L. A., SMOOT, J. P., AND EUGSTER, H. P., 1978, Saline lakes and their deposits, *in* Matter, A. and Tucker, M. E., ed., Modern and Ancient Lake Sediments: Oxford, Special Publication of the International Association of Sedimentologists 2, p. 7–41.

HOYOS, M., JUNCO, F., PLAZA, J. M. RAMIREZ, A., AND RUIZ, J., 1985, El Mioceno de Madrid, *in* Alberdi, M. T., ed., Geología y Paleontología del Terciario Continental de la Provincia de Madrid: Madrid, Consejo Superior de Investigaciones Cientificas, p. 9–16.

HSÜ, K. J., AND SIEGENTHALER, C., 1969, Preliminary experiments on hydrodynamic movement induced by evaporation and their bearing on the dolomite problem: Sedimentology, v. 12, p. 11–25.

HUBERT, J. F., AND HYDE, M. G., 1982, Sheet-flow deposits of graded beds and mudstones on an alluvial sandflat-playa system: Upper Triassic Blomidon redbeds, St. Mary's Bay, Nova Scotia: Sedimentology, v. 29, p. 457–474.

JUNCO, F., AND CALVO, J. P., 1983, Cuenca de Madrid: Libro Jubilar J. M. Rios: Instituto Geológico y Minero de España, v. II, p. 534–543.

LOGAN, B. W., 1987, The Macleod Evaporite Basin, Western Australia: Tulsa, American Association of Petroleum Geologists Memoir 44, 140 p.

LOWENSTEIN, T. K., AND HARDIE, L. A., 1985, Criteria for the recognition of salt-pan evaporites: Sedimentology, v. 32, p. 627–644.

MAGEE, J. W., 1991, Late Quaternary lacustrine, groundwater, aeolian and pedogenic gypsum in the Prungle Lake, southeastern Australia: Palæogeography, Palæoclimatolology, Palæoecology, v. 84, p. 3–42.

MEDIAVILLA, R. M., 1987, Sedimentología de los yesos del sector central de la depresión del Duero: Acta Geológica Hispánica, v. 21–22, p. 35–44.

MEDIAVILLA, R. M., AND DABRIO, C. J., 1987, Controles sedimentarios neógenos en la depresión del Duero (sector central): Revista Sociedad Geológica de'España, v. 1, p. 187–195.

MEGÍAS, A. G., ORDÓÑEZ, S., AND CALVO, J. P., 1981, Tertiary clastic gypsum deposits in the Madrid Basin (abs.): Bologna, Abstracts 2nd European Meeting of the International Association of Sedimentologists, p. 109–112.

MEGÍAS, A. G., ORDÓÑEZ, S., CALVO, J. P., AND GARCÍA DEL CURA, M. A., 1982, Sedimentos de flujo gravitacional yesíferos y facies asociadas en la Cuenca Neógena de Madrid, España: Buenos Aires, Actas V Congreso Latinoamericano de Geología, v. II, p. 311–328.

ORDÓÑEZ, S., CALVO, J. P., GARCÍA DEL CURA, M. A., ALONSO ZARZA, A. M., AND HOYOS, M., 1991, Sedimentology of sodium sulphate and special clays from the Tertiary of the Madrid Basin (Spain), *in* Anadón, P., Cabrera, L., and Kelts, K., eds., Lacustrine Facies Analysis: Oxford, Special Publication of the International Association of Sedimentologists 13, p. 37–53.

ORDÓÑEZ, S., FONTES, J. C., AND GARCÍA DEL CURA, M. A., 1987, Estudio isotópico de la paragénesis sulfatada sódica, calcosódica y cálcica de la Unidad Salina de la Cuenca de Madrid: Soria, II Congreso Geoquímica, p. 95–98.

ORDÓÑEZ, S., FORT, R., AND BUSTILLO, M., 1989, Caracterización geoquímica de la ruptura sedimentaria entre los yesos de la Unidad Salina e Intermedia de la Cuenca de Madrid: Estudios Geológicos, v. 45, p. 45–53.

PAREA, G. C., AND RICCI-LUCHI, F., 1972, Resedimented evaporites in the Periadriatic trough: Israel Journal of Earth Sciences, v. 21, p. 125–141.

RETALLACK, G., 1988, Field recognition of paleosols, *in* Reinhart, J., and Sigleo, W. R., eds., Paleosols and Weathering through Geologic Time: Boulder, Special Paper of the Geological Society of America 216, p. 1–20.

ROUCHY, J. M., 1982, La Genése des Évaporites Messiniennes de Mediterrane: Paris, Memoire de Muséum National d'Histoire Naturelle, Série C, Tome L., 306 p.

ROUCHY, J. M., BERNET-ROLLANDE, M. C., AND MAURIN, A. F., 1986, Pétrographie descriptive des évaporites. Applications sur le terrain, en subsurface et au laboratoire, *in* Les Séries á Evaporites en Exploration Petrolière, 1: Méthodes géologiques: Paris, Editions Technip, p. 73–122.

SAN JOSÉ, M. A., 1975, Mapa y memoria explicativa de la Hoja 606 (Chinchón). MAGNA. E. 1/50.000: Instituto Geológico y Minero de España.

SANZ, M. E., RODRÍGUEZ ARANDA, J. P., CALVO, J. P., AND ORDÓÑEZ, S., 1991, Sedimentología de las facies de yesos detríticos de la Unidad Intermedia del Mioceno (sector sur de la Cuenca de Madrid): Vic, I Congreso del Grupo Español de Terciario, p. 316–319.

SCHLAGER, W., AND BOLZ, H., 1977, Clastic accumulation of sulfate evaporites in deep water: Journal of Sedimentary Petrology, v. 47, p. 600–609.

SCHREIBER, B. C., 1978, Environments of subaqueous gypsum deposition, *in* Dean, W., and Schreiber, B. C., eds., Marine Evaporites: Tulsa, Society of Economic Paleontologists and Mineralogists Short Course Notes 4, p. 43–73.

SCHREIBER, B. C., FRIEDMAN, G. M., DECIMA, A., AND SCHREIBER, E., 1976, Depositional environments of Upper Miocene (Messinian) evaporite deposits of the Silician Basin: Sedimentology, v. 23, p. 729–760.

TRUC, G., 1978, Lacustrine sedimentation in an evaporitic environment: the Ludian (Palaeogene) of the Mormorion basin, southeastern France, *in* Matter, A., and Tucker, M. E., eds., Modern and Ancient Lake Sediments: Oxford, Special Publication of the International Association of Sedimentologists 2, p. 189–203.

TRUC, G., 1980, Evaporites in a subsident continental basin (Ludian and Stampian of Mormorion-Pernes in southeastern France): Sequential aspects of deposition primary facies and their diagenetic evolution, *in* Evaporite Deposits: Paris, Technip, p. 61–71 and p. 234–259.

TUNBRIDGE, I. P., 1984, Facies model for sandy ephemeral stream and clay playa complex; the Middle Devonian Trentihoe Formation of North Devon, U. K.: Sedimentology, v. 31, p. 697–716.

VAI, G. B., AND RICCI-LUCHY, F., 1977, Algal crusts, autochthonous and clastic gypsum in a cannibalistic evaporite basin: a case history from the Messinian of Northern Apennines: Sedimentology, v. 24, p. 211–244.

VEGAS, R., AND BANDA, E., 1982, Tectonic framework and Alpine evolution of the Iberian Peninsula: Earth Sciences Evolution, v. 4, p. 320–343.

VEGAS, R., VAZQUEZ, J. T., SURIACH, E., AND MARCOS, A., 1990, Model of distributed deformation, block rotation and crustal thickening for deformation of the Spanish Central System: Tectonophysics, v. 184, p. 367–378.

WARBURTON, J., AND ALVAREZ, C., 1989, A thrust tectonic interpretation of the Guadarrama Mountains, Spanish Central System, *in* Asociación de Geólogos y Geofísicos Españoles del Petróleo, ed., Libro Homenaje a Rafael Soler: Madrid, p. 147–155.

WARREN, J. W., 1982, The hydrological setting, occurrence and significance of gypsum in late Quaternary salt lakes in South Australia: Sedimentology, v. 29, p. 609–638.

DEPOSITION AND DIAGENESIS OF SODIUM-CALCIUM SULFATE SALTS IN THE TERTIARY SALINE LAKES OF THE MADRID BASIN, SPAIN

SALVADOR ORDÓÑEZ[1] AND MARIA DE LOS ANGEL ES GARCÍA DEL CURA[2]

[1]*Departamento de Petrología y Geoquímica, Facultad de Ciencias Geologícas, Universidad Complutense, 28040 Madrid, Spain*

[2]*Consejo Superior Investigaciones Científicas, Spain*

ABSTRACT: The Tertiary Madrid Basin is located in the center of Spain and covers an area of more than 10,000 km^2. The basin is filled by Tertiary sediments that are up to 3.5 km thick on its northwestern side. Neogene sediments account for the upper third of the Tertiary record and are divided into four main stratigraphic units. The lowermost unit, called the "Lower Unit" or "Saline Unit," is of Miocene age and consists mainly of evaporite sediments. It is divided into two subunits, the Lower Saline Subunit and the Upper Saline Subunit, based on sedimentary criteria.

In the inner part of the basin, the Lower Saline Subunit is crudely laminated with alternating laminae of anhydrite-magnesite, glauberite-anhydrite, and polyhalite-halite. In the outer parts of basin, this subunit is rich in early diagenetic glauberite with many textural varieties. In the Upper Saline Subunit, on the south side of the basin, massive thenardite beds with minor glauberite crystals overlie homogeneous reddish muds with halite and/or glauberite.

The Lower Saline Subunit is interpreted to represent deposition in an elongate, arid closed basin, oriented north to south. A perennial saline lake surrounded by a broad mudflats probably occupied the central part of the basin. In these saline mudflats abundant glauberite formed, both by replacement of other salts and displacively. The mudflat deposits merge laterally into a coalescent alluvial–fan system. The alluvial fans along the eastern margin of the basin may be associated with emplacement of nappes from the east. The erosion and weathering of Upper Cretaceous-Paleocene marine evaporites that crop out along the thrust zone appear to have been the main source of solutes for the lake brine. A paleotopographic change resulting from tectonic movements may explain the different character of the brines and stratigraphic record of the Upper Saline Subunit.

INTRODUCTION AND PREVIOUS WORK

The non-marine sodium-calcium sulfate salts of the Miocene Saline Unit of the Madrid Basin have been known for 200 years. In this paper, we synthesize all the available data from drill holes, mines, springs, and the main outcrops of the Saline Unit. Supported by sedimentological and petrologic examination of more than a thousand drill hole samples, we propose a depositional model to explain the origin of the salts based on sedimentation in a closed saline lake basin.

Halite has been extracted for centuries in a mine near Villarrubia de Santiago and it was in the same mine, which is still active, that Dumeril first discovered and identified the mineral glauberite in 1807. Brogniart determined the chemical composition of glauberite ($CaSO_4 \cdot Na_2SO_4$) from Villarrubia de Santiago in 1807 (García del Cura, 1979).

The origin of the sodium-calcium sulfate minerals in the Madrid basin received the attention of many geologists during the last century. Sullivan and O'Reilly (1863) described the sodium sulfate deposits along the Jarama River valley and gave a paleogeographic interpretation based on the Kara Bogaz (Caspian Sea) model. Later, Calderon (1895) suggested that the Tertiary salts originated from leaching of older, mainly Triassic, evaporites.

Rios (1963, 1968) gave a general account of Spanish evaporite deposits. During the 1970's, many authors, including San Miguel and Montoto (1971), Ordóñez and others (1977), García del Cura and others (1978, 1979), García del Cura (1979), and Ortí and others (1979), described the main sedimentological, petrologic and geochemical features of the sodium-calcium sulfate salts and associated sediments, and proposed various paleogeographic models. Ordóñez and others (1977) showed that the carbonate mineral associated with the evaporites of the Madrid basin is always magnesite. García del Cura and others (1978) identified polyhalite in drill hole samples.

Ordóñez and others (1982) outlined the economic importance of sodium sulfate minerals in Spain. During the 1980's, research focused on the stable isotopic composition of the sulfates and carbonates (Ordóñez and others, 1983, 1987; Utrilla and others, 1987). These results support the old ideas of recycling of older evaporite deposits as the most likely origin for the sodium sulfate deposits.

Today the sodium sulfate deposits of the Madrid basin supply 50 percent of the natural sodium sulfate production in Spain (Table 1). For many years, Spain has been the principal European producer of natural sodium sulfate. In 1989, the total production of natural sodium sulfate (by Na_2SO_4 content) was 536,332 tonnes, 49.7 percent coming from Madrid basin (29.1 percent as thenardite; 20.6 percent as glauberite) (I.T.G.E, 1991).

There are two active mines in the Madrid basin (Figs. 1, 2), one near Villaconejos, the other near an eighteenth century rock salt mine in Villarrubia de Santiago. The mine near Villaconejos, is an open-pit solution mine. Glauberite is dissolved incongruently in a pool that overlies a thick bed (27 m) of glauberite-anhydrite. The brine is pumped to an evaporation plant to precipitate high quality sodium sulfate. The underground mine near Villarrubia de Santiago works a thenardite bed with minor glauberite, 5 to 8 m thick (Ordóñez and others, 1982).

Many small inactive mines exist in the Madrid basin, but historically the most famous method for recovering salts was from saline springs and wells, as for example at Espartinas, Carabaña, Carcaballana and Loeches (Fig. 2). Some springs and wells were exploited as early as Roman times. Sodium sulfate was produced mainly by freeze-out in winter.

GEOLOGICAL SETTING

The Madrid Basin (Fig. 1) is an intraplate continental basin, more than 10,000 km^2 in extent, that originated by tectonic events during the Alpine orogeny (Vegas and Banda,

Sedimentology and Geochemistry of Modern and Ancient Saline Lakes, SEPM Special Publication No. 50

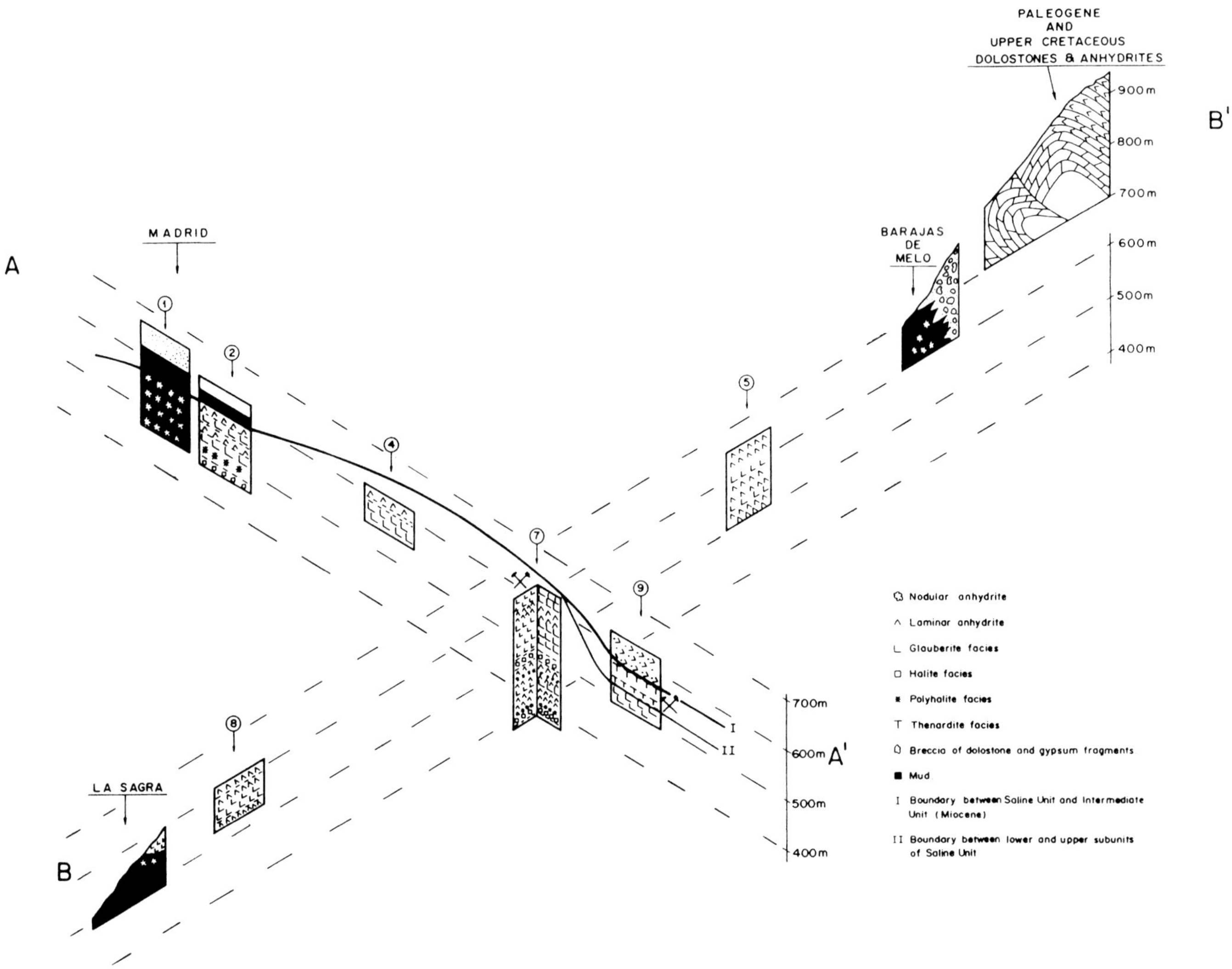

FIG. 3.—Schematic cross-section of the Saline Unit from drill hole data. The numbers of drill holes and location of sections are shown on Fig. 2. Elevations are given in meters above sea level.

ondary gypsum occur in the upper portion of the drill hole, of which as much as 30 m is replacing glauberite-anhydrite laminar facies. Some deeper beds have alternating millimeter-scale laminae of halite, polyhalite, glauberite, anhydrite and magnesite. Terrigenous sediments are rare and are laminated. Nodular and enterolithic features are scarce, but some poikilotopic features have been recognized.

Several kilometers from drill hole No. 7, an open pit, with a solution pond at the bottom, was probably excavated over the same massive gypsum beds cut by the drill hole. The face of the open pit mine (Fig. 5) shows a weathering layer of secondary gypsum up to 18 m thick. The upper part of weathering layer is composed of coarsely crystalline gypsum with anhydrite and calcite inclusions; the lower part is poikilotopic gypsum and gypsum pseudomorphing anhydrite and glauberite. The boundary between the weathering layer and unweathered section has a discontinuous layer of fibrous mirabilite a few centimeters thick. The unweathered zone cropping out in the face of open pit shows the following features, from top to bottom:

1. A layer about 5 m thick of coarse nodular anhydrite and glauberite. The diameter of individual anhydrite nodules may reach 60 cm. Some nodules are partially altered to glauberite, mainly along shrinkage cracks.
2. A massive glauberite layer, up to 5 m thick. This layer is formed by lenses of glauberite and magnesite overprinted by poikilotopic glauberite. Fibrous glauberite is common and associated with brecciated structures.
3. Bedded glauberite layers, with bedsets up to 3 m thick. These layers are formed by laminae of meso- to macrocrystalline euhedral discoid glauberite, alternating with micritic magnesite and/or magnesite marl laminae. Poikilotopic, branching and fibrous glauberite textures overprint and erase original laminar structures.
4. Laminated glauberite layers in which couplets of euhedral discoid glauberite and magnesite marls are the dominant feature.

Drill hole No. 6, located 4 km to the west, displays similar features to those described at No. 7. The zone around

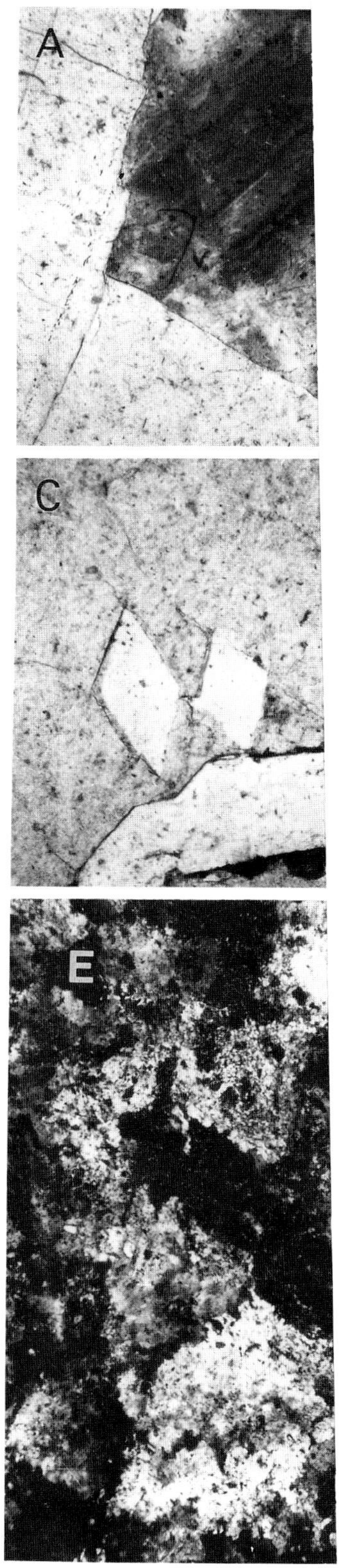

FIG. 6.—Thin-section photomicrographs of repres
(B) Fascicular polyhalite with replacive anhydrite. Dril
Plane polarized light. (D) Glauberite-magnesite coup
polarized light. (E) "Piedra Negra" facies showing anh
nicols. (F) Fibrous glauberite in magnesite-glauberite

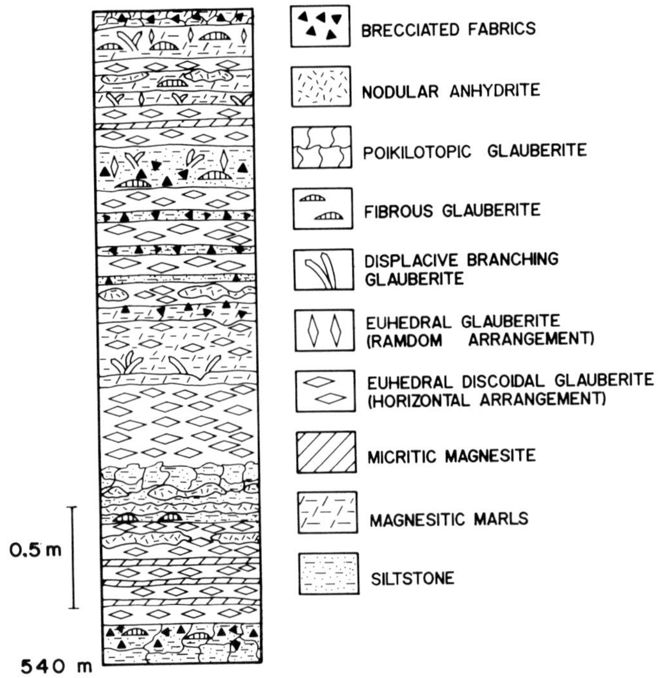

FIG. 4.—Detailed section of Saline Unit exposed in Consuelo Mine.

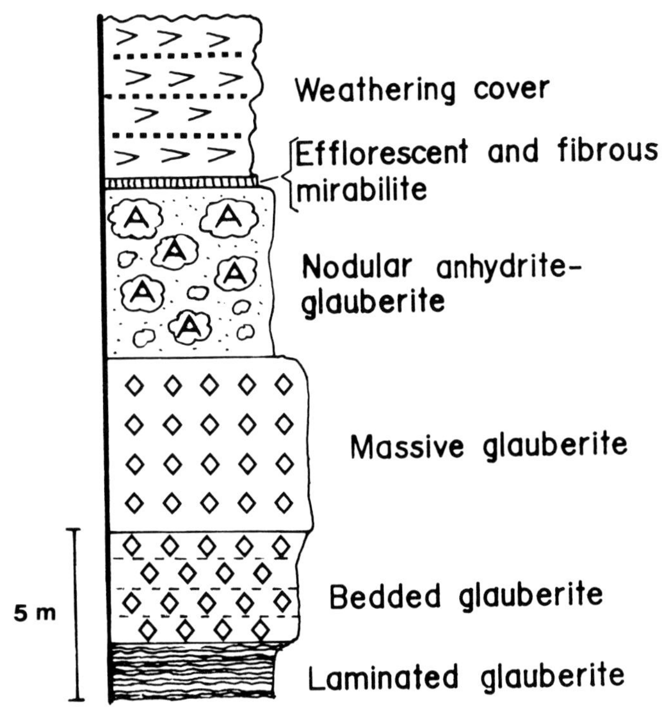

FIG. 5.—Section of Saline Unit exposed in Villaconejos Mine.

drill hole No. 9 is the most extensively studied, because it is near the underground mine of Villarrubia de Santiago. We have combined data from more than seven drill holes, as well as from observations in the mine, to describe the geological setting of this unique deposit of thenardite in the Madrid basin. The thenardite beds, which contain approximately 10 percent glauberite, have a thickness of up to 8 m. They overlie reddish saline muds containing crystals of halite, platy lensoid glauberite, and coarsely crystalline, euhedral discoid glauberite. In places, the boundary between the massive thenardite beds and the reddish saline muds is occupied by irregular beds of microcrystalline glauberite, known locally as "Piedra Negra" (Black Stone).

Drill hole No. 9, located 4-km SW of the underground mine (Fig. 3), has up to 20 m of detrital gypsum from the Miocene Intermediate Unit (see Sanz and others, this volume). Beneath these sediments is a thick bed of "mirror-like" gypsum at the top of the Saline Unit. Below this bed (between 575 and 520 m), we identified at least six repetitions of the following sedimentary cycle: (1) reddish mud containing halite and glauberite, and (2) massive thenardite. The thenardite beds have been hydrated to mirabilite and are partially dissolved. The reddish muds display secondary fibrous gypsum veins. Below 520 m, the Saline Unit has a similar stratigraphy to that in the drill hole No. 7 (Fig. 3). However, at the boundary between the sedimentary cycles of thenardite and the underlying laminar saline facies, we found some detrital grains of glauberite and anhydrite, both replaced by gypsum, that clearly have been reworked.

The distribution of the thenardite facies, and its location and sedimentological features, allow us to divide the Saline Unit in two subunits: (1) a Lower Saline Subunit and (2) an Upper Saline Subunit. The Lower Saline Subunit is common and extensive throughout the Madrid basin. The Upper Saline Subunit containing the thenardite-bearing sediments is restricted to the southern Tajo River Valley.

In Barajas de Melo (Fig. 3), breccias with fragments of dolostone and gypsum interfinger with, and onlap, saline-mudflat deposits containing nodular anhydrite. The gypsiferous dolomitic breccia is clearly the result of weathering and erosion of 700 m or more of Upper Cretaceous-Paleocene dolostones and anhydrite beds. These dolostones and anhydrite beds crop out along the Altomira Mountain Range (Fig. 1; Rodriguez Aranda and others, 1991).

In drill hole No. 5, a shallow exploration well for gypsum, we found glauberite beds replaced by gypsum, a few meters thick, interbedded with anhydrite. Drill hole No. 8, in the west of the basin contains a thick bed (>10 m) of coarsely crystalline glauberite with some mudstone. This bed onlaps a red saline mudflat deposit of the La Sagra region (Fig. 3).

PETROLOGICAL AND MINERALOGICAL FEATURES OF THE SALINE UNIT

The upper part of the weathering zone on the Saline Unit is commonly formed by coarsely crystalline *gypsum*, that locally displays pseudo-selenitic structures. Secondary calcite, in places forming radiating aggregates, is found intergrown with the macrocrystalline gypsum. In the deepest part of the weathering zone, poikilotopic gypsum crystals, containing inclusions of mud and primary evaporite minerals, are the principal features. Gypsum, pseudomorphing glauberite and anhydrite, is also common. Fibrous gypsum veins and fascicular fibrous gypsum masses are found in some outcrops of beds rich in soluble salts.

Anhydrite is a common unweather line Unit. It has two main fabrics in c even enterolithic, and (2) laminar. T habits including fascicular crystalline and, in the deepest part of some drill tures. Alternating laminae of anhydrit mon in some beds of the Lower Sub

Magnesite is the only unweathered the Saline Unit, occurring in continu or as an intercrystalline mineral. Th talline with crystals less than 2 μm. C per Saline Subunit, but also in the m Lower Saline Subunit, magnesite is m sediments (Fig. 7F). The latter are cc erals—mainly illite, kaolinite, minor (only in the northern basin)—togeth spars, and other silicates.

Halite commonly occurs in rhythm timeters thick. Two types of prima guished, both with bottom-nucleated cr faces parallel to the substrate, the ot ented chevron structures. Both types clusions. Poikilotopic clear halite and are also common textures. Reworked as inclusions in sodium sulfate miner

Polyhalite is uncommon in the M been identified throughout the Saline U low 520 m. Thin (mm) green layers halite with minor magnesite alternate nesite with minor polyhalite. Spher occurs as an intercrystalline mineral in ers. Felty masses of polyhalite local margins of glauberite crystals, and in morphic after glauberite. Fascicular p cive anhydrite crystals is locally prese

Glauberite is the most common salt Subunit, showing a wide variety of tex ble 2; Figs. 6, 7). Massive beds of glau minor magnesite and/or magnesite marl 7E). In the deepest part of some drill h occurs in laminae alternating with anhy Most glauberite occurs as euhedral disc parallel to the substrate (Fig. 7A). Fibr have been identified in these crysta glauberite crystals have inclusion-rich o syntaxial cements) (Fig. 7C). In place or magnesite marl laminae, glauberite prismatic habit, with the long axis p substrate.

In the Villaconejos and Consuelo Mi soid, contorted and brecciated structu both localities, fibrous glauberite vein- ondary euhedral discoid glauberite (Fi cive branching glauberite (Fig. 7B) are tative secondary textures. Non-glauberiti magnesite-rich, and even nodular anhy by poikilotopic anhedral glauberite that gree of induration.

In the lower part of some thenardite Saline Subunit, a hard nodular bed of g

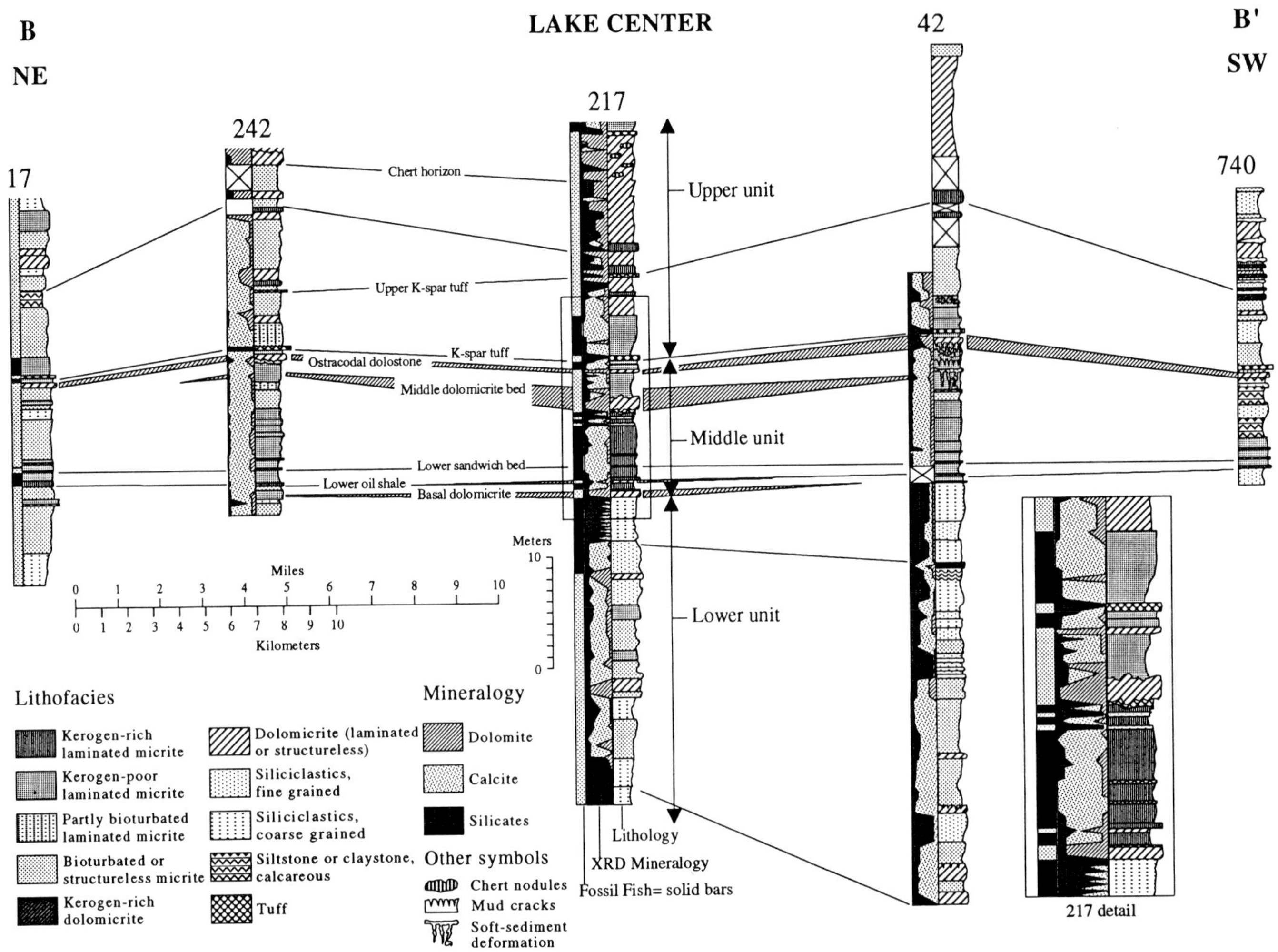

FIG. 2.—Stratigraphic sections from the northwestern lake margin, through lake center (about site 217), to the southern lake margin (site 740). Detailed mineralogic data from X-ray diffraction peak heights are plotted left of three sections. Note that most dolomite units of the middle unit grade laterally into calcite-dominated rocks, as do some upper unit dolomite beds. Occurrences of fossil fish occur with calcite and are absent from dolomitic rocks. Sections are hung on the lower sandwich bed, a set of tuffs that can be correlated basinwide. Detail of boxed area of section 217 is shown far right.

LITHOFACIES AND STRATIGRAPHIC RELATIONS

Table 1 summarizes the lithofacies in the Fossil Butte and Angelo members of the Green River Formation. Within the middle unit, any single sedimentary unit can be traced from basin center to margin through a predictable suite of lithofacies. This lithofacies assemblage can be recognized in vertical and lateral transitions, and as such follows Walther's Law (cf. Middleton, 1973). In the middle and lower units, the lithofacies assemblage from bottom to top and from basin center (near site 270) to margin includes kerogen-rich laminated micrite, kerogen-poor laminated micrite, partly burrowed laminated micrite, bioturbated micrite and, at more shoreward sites, siliciclastic siltstone or sandstone. The entire sequence occurs in vertical succession most commonly mid-way between basin center and margin. The lower lithofacies dominate basin center assemblages while the upper lithofacies are dominant near the basin margin. Lithofacies packages range from 1 to 7 m thick.

Lateral changes of most lithofacies reflect changes in Total Organic Carbon (TOC, primarily kerogen), laminae thickness, and degree of bioturbation. The partly burrowed laminated micrite lithofacies is transitional between kerogen-rich laminated micrite and bioturbated micrite. Partly burrowed and bioturbated micrite lithofacies have not previously been reported from the Green River Formation. These facies appear unique to Fossil Basin, where they account for up to 30 percent of the total thickness in measured sections. Facies of the lower unit are similar to the middle unit, except that bioturbated micrite dominates the section.

In detail, the facies arrangement is more complex. The lithofacies assemblage of the middle unit is capped by a dolomicrite that does *not* follow Walther's Law because the lateral facies transition does not grade into dolomicrite. Several dolomicrite units occur within the middle unit (Fig. 2). They are massive, chalky (Fig. 5E), and contain numerous soft-sediment deformation structures, including ball and pillow and hydroplastic intrastratal flow structures. The

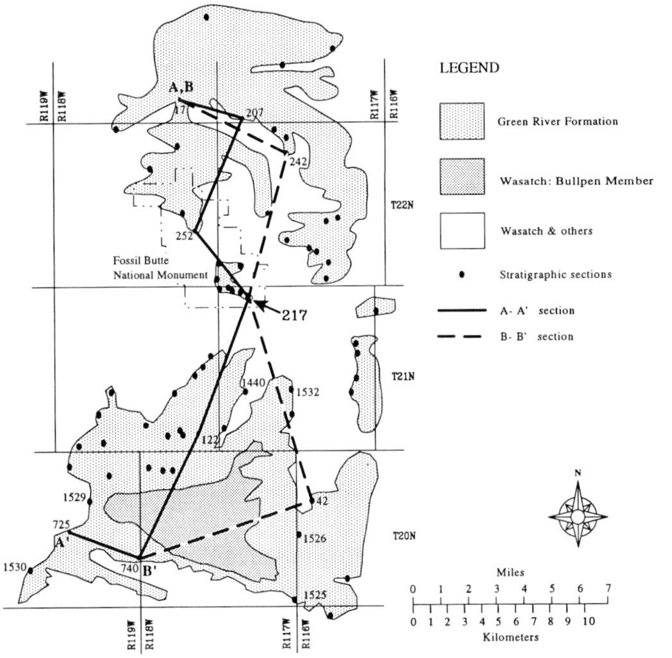

FIG. 3.—Map showing extent of the Fossil Butte Member and site of measured sections and study sites. See Figure 6 for facies cross section indicated by cross section line A-A' and for correlated stratigraphic sections indicated by cross section line B-B'.

top dolomicrite bed, a meter below the K-feldspar tuff, contains abundant ostracods, almost forming a coquina. The top and bottom contacts of these units are abrupt and non-gradational. All except the top unit either pinch out toward the margins or grade into bioturbated or massive calcimicrite.

The facies of the upper unit have not been organized into a specific facies assemblage, although repetitive sequences occur, such as laminated dolomicrite that grades upward into laminated dolomicrite filled with calcite casts of saline minerals such as trona. The facies across section (Fig. 6) suggest that a similar pattern of lithofacies does occur, except that dolomite is dominant in the laminated facies; these are well exposed at site 252 (Fig. 3) about 3-km north of Fossil Butte. Several kerogen-rich laminated dolomicrites in the upper unit contain sand to granule-sized calcite pseudomorphs after halite. They are so thick in some beds that they deform and in some cases, obscure the original laminated structures (Fig. 5G).

Unlike the lower unit, the dolomicrite of the upper unit is generally laminated and, in some cases, kerogen-rich. Correlation of stratigraphic sections (Fig. 2) and the facies diagram (Fig. 6) show that the upper unit grades laterally into bioturbated or massive calcimicrites and siliciclastics nearer the margins.

Several meters of laminated dolomicrite with abundant chert nodules occur near the top of the upper unit and are traceable basinwide. This unit is overlain by bed of laminated micrite with abundant fish, ostracodes and other fossils, followed by a siliciclastic sequence. This may represent a significant depositional change, which is discussed below.

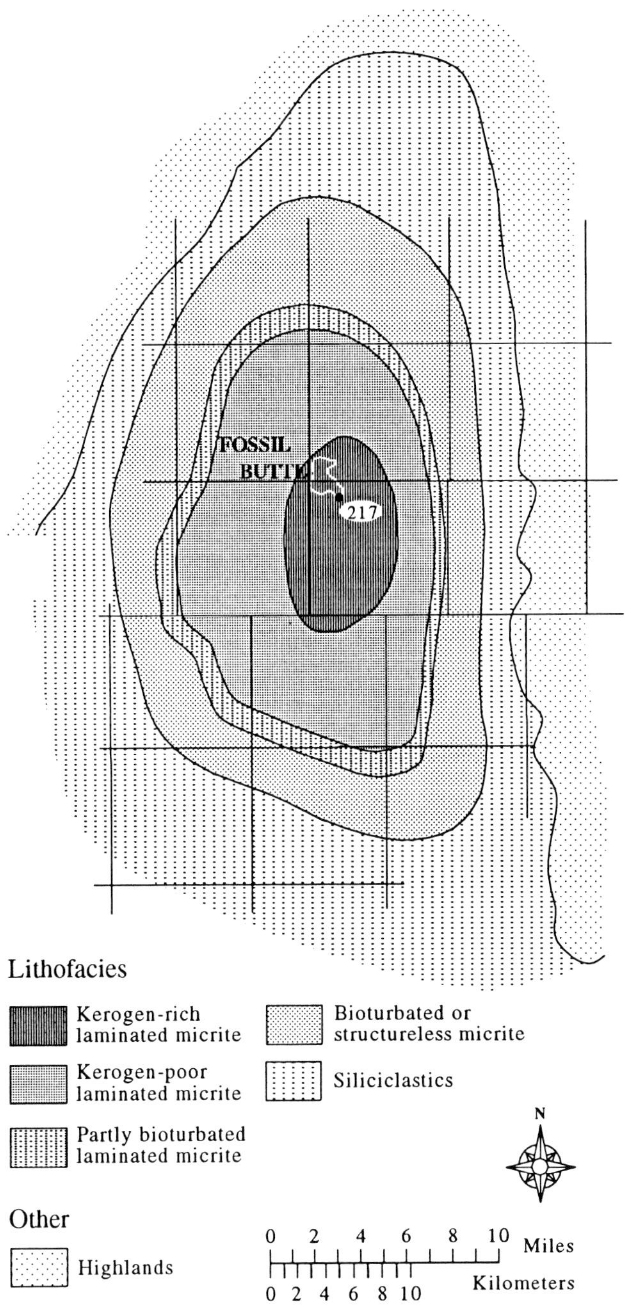

FIG. 4.—Fossil Basin lithofacies map showing major lithofacies and geographic features during middle unit time. The center of Fossil Lake can be from determined by these features as well as organic carbon distribution, shallow-water features increasing shoreward such as bioturbation, fossils, and shallow-water and mudflat sedimentary structures not shown on this map.

Kerogen-rich laminated dolomicrite beds in the upper unit are extensive and form marker beds nearly basinwide. Nearer the basin margins (e.g., site 740), kerogen-rich laminated dolomicrites are interbedded with thin, lenticular sand beds that pinch out within a few meters (Fig. 7). These sandstones fill mudcracks at their bases, are ripple-cross laminated, contain flat-pebble intraclasts and gastropods, and have upper current-rippled surfaces with clay mud-drapes

TABLE 1.—LITHOFACIES DESCRIPTIONS (see also Figure 5)

	Kerogen-rich laminated micrite	Kerogen-poor laminated micrite	Partly burrowed laminated micrite	Bioturbated micrite	Dolomicrite	Kerogen-rich laminated dolomicrite	Kerogen-poor laminated dolomicrite	Sandstone & siltstone
T.O.C. & Color	2–14 percent black to brown	>2 percent brown to white	>2 percent white	>2 percent white	>2 percent white	2–14 percent brown to black	>2 percent yellow buff	no data, brown
Sedimentary structures	laminated, laminae average 0.07 mm thick; composed of alternating calcite and kerogen laminae; ripple x-laminae common in interbedded k-spar tuff at loc. 224 & 1306	same as kerogen-rich laminated micrite except laminae are thicker (average 0.14 mm thick); kerogen laminae much less distinctive	same structures as kerogen-poor laminated micrite; in addition contains macroburrows 0.5 to 2 cm in diameter; both vertical & horizontal burrows; burrows contain meniscus fillings	structureless micrite to chaotic mixture of micrite supported clasts; abundant macroburrows; may contain microtubles (burrows 1–2 mm in diameter); bioturbation increases towards margins of basin	term here used to describe a massive and chalky, usually structureless sediment; one bed grades laterally into ostracod-rich dolomicrite with scour & fill structures	laminated; laminae even thickness to lenticular; soft-sediment deformation structures common; contain salt casts; laminae may be nearly completely disrupted by salt casts	laminated; laminae even thickness to lenticular; contain salt casts; laminae may be nearly completely disrupted by salt casts	trough & ripple x-bedding, current lineation; sandstone tongue contains 4-m thick unit with topset, foreset, bottom set beds & laminae; slump & other loading structures
Grain size	micritic (clay-size)	micritic (clay-size), calcite grains normal graded (10-6 microns)	same as kerogen-poor laminated micrite	clay-size matrix supporting sand to pebble size angular-carbonate clasts	clay-size	clay-size	clay-size; some silt to sand size dominated units	fine to course grained; some carbonate intra-clasts
Mineralogy	calcite with minor amounts of dolomite, quartz, feldspar, & clay	same as kerogen-rich laminated micrite, except the calcite content is higher	same as kerogen-poor laminated micrite	same as partly-burrowed laminated micrite	dolomite; some units may contain <20 percent quartz, feldspar, clay, & calcite	dolomite; most units contain >20 percent quartz, feldspar, clay, & calcite	dolomite; most units contain >20 percent quartz, feldspar, clay, & calcite	primarily quartz & feldspar; clays in fines
Paleontology (common fossils)	fossil fish, leaves, insects, ostracods, gastropods	fossil fish, leaves, insects, ostracods, gastropods	same as kerogen-poor laminated micrite	gastropods, pelecypods, ostracods, & fish bones	**No fish**; some units contain abundant ostracods	**No fish**, some units contain abundant ostracods	**No fish**, some units contain abundant ostracods	gastropods, pelecypods & burrows
Other notes	this facies grades imperceptively laterally & vertically into kerogen-poor laminated micrite	at loc. 217, a 4-m sequence of laminated micrites composed of alternating micrite and claystone laminae with interbedded siltstones grades southward into the sandstone tongue	this facies is gradational between laminated and bioturbated micrite	most common lithofacies, overall, makes up 30 percent of total sections measured	form individual beds 0.30–2 m thick within laminated micrites of the middle unit; some upper unit dolomicrites beds contain chert nodules & tufa-coated logs	associated siltstone & sandstone beds contain current structures including current lineation & small ripples (loc. 740)		foreset x-beds contain some calcimicrite laminae and abundant calcimicrite pebble-size intraclasts

and bird tracks. Some dolomite beds are replaced by silty calcimicrite.

Fossil fish are restricted to calcite-dominated units (Fig. 2). This relationship was first recognized in the Laney member of the Green River Formation (Buchheim and Surdam, 1981). In Fossil Basin, ostracodes and plant fragments are the only abundant fossils in dolomicrites, and only in specific units. Bioturbation is abundant in calcite-dominated units, but lacking in dolomicrites.

The common tuff beds that interrupt the lithofacies at random intervals (Fig. 2) are not a major facies, but are important in geochemical considerations. They are generally only a few centimeters thick, but range from a few millimeters to 25 cm thick. Near the basin center they consist of K-feldspar or analcime and grade laterally into clay minerals near basin margins. This mineralogic trend has been recognized elsewhere in the Green River Formation (Surdam and Sheppard, 1978).

DISCUSSION AND INTERPRETATIONS

Vertical and lateral lithofacies trends can be interpreted as representing a freshwater lake that experienced hypersaline episodes, returning each time to a mainly fresh, but alkaline lake. However, Fossil Lake finally became predominantly hypersaline during upper unit time. The vertical mineralogy and lithology summarized in the stratigraphic columns (Fig. 2) provide evidence of these trends. Calcite precipitated from dilute to saline, alkaline lake waters, whereas dolomite is either a replacement product of originally precipitated high magnesium calcite or a primary precipitate from hypersaline waters with high Mg^{2+}/Ca^{2+} ratios (cf. Last, 1990). This relationship between hydrochemistry and mineral precipitate has been demonstrated previously for the Green River Formation (e.g., Eugster and Surdam, 1973; Eugster and Hardie, 1975; Surdam and Wolfbauer, 1975; Surdam and Stanley, 1979). Other evidence supporting this interpretation includes the paleontology. Fossil fish only occur in units where the carbonates are dominated by calcite. Calcite pseudomorphs after saline minerals (salt casts) occur only in the upper unit where dolomite predominates, providing direct evidence of hypersalinity. Buchheim and Surdam (1981) recognized the same relationships in the Laney Member of the Green River Formation.

Abundant chert nodules near the top of the upper unit provide further evidence for hypersalinity. The most probable interpretation, consistent with a closed saline, alkaline lake model, is that of freshwater inflow over hypersaline, highly alkaline waters. This resulted in the precipitation of silica that eventually formed chert nodules, as described by

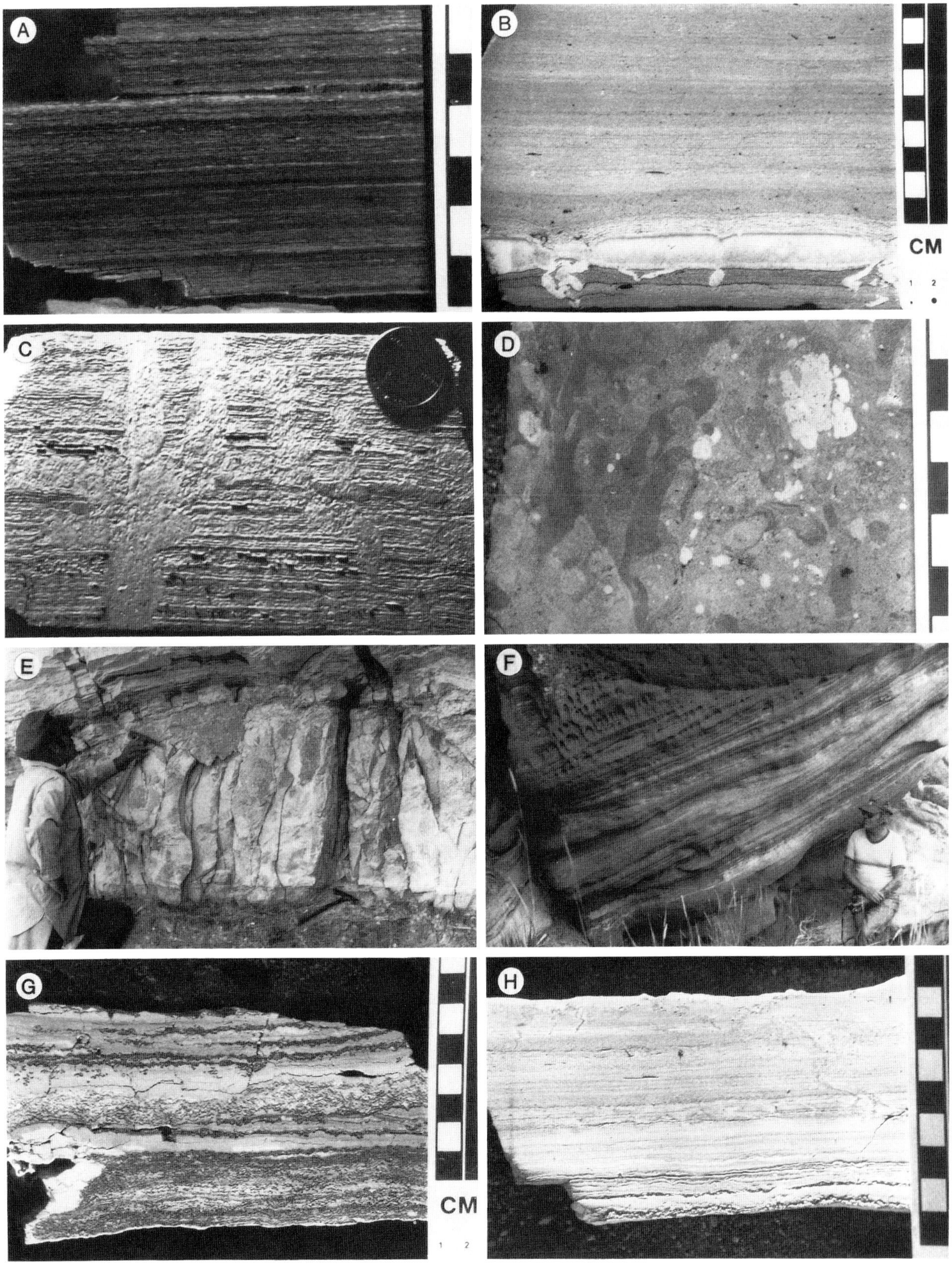

FIG. 5.—Common lithofacies include: (A) Kerogen-rich laminated micrite; (B) kerogen-poor laminated micrite; (C) partly burrowed laminated micrite; (D) bioturbated micrite; (E) dolomicrite; (F) claystone, siltstone, and sandstone; (G) kerogen-rich laminated dolomicrite; and (H) kerogen-poor laminated dolomicrite.

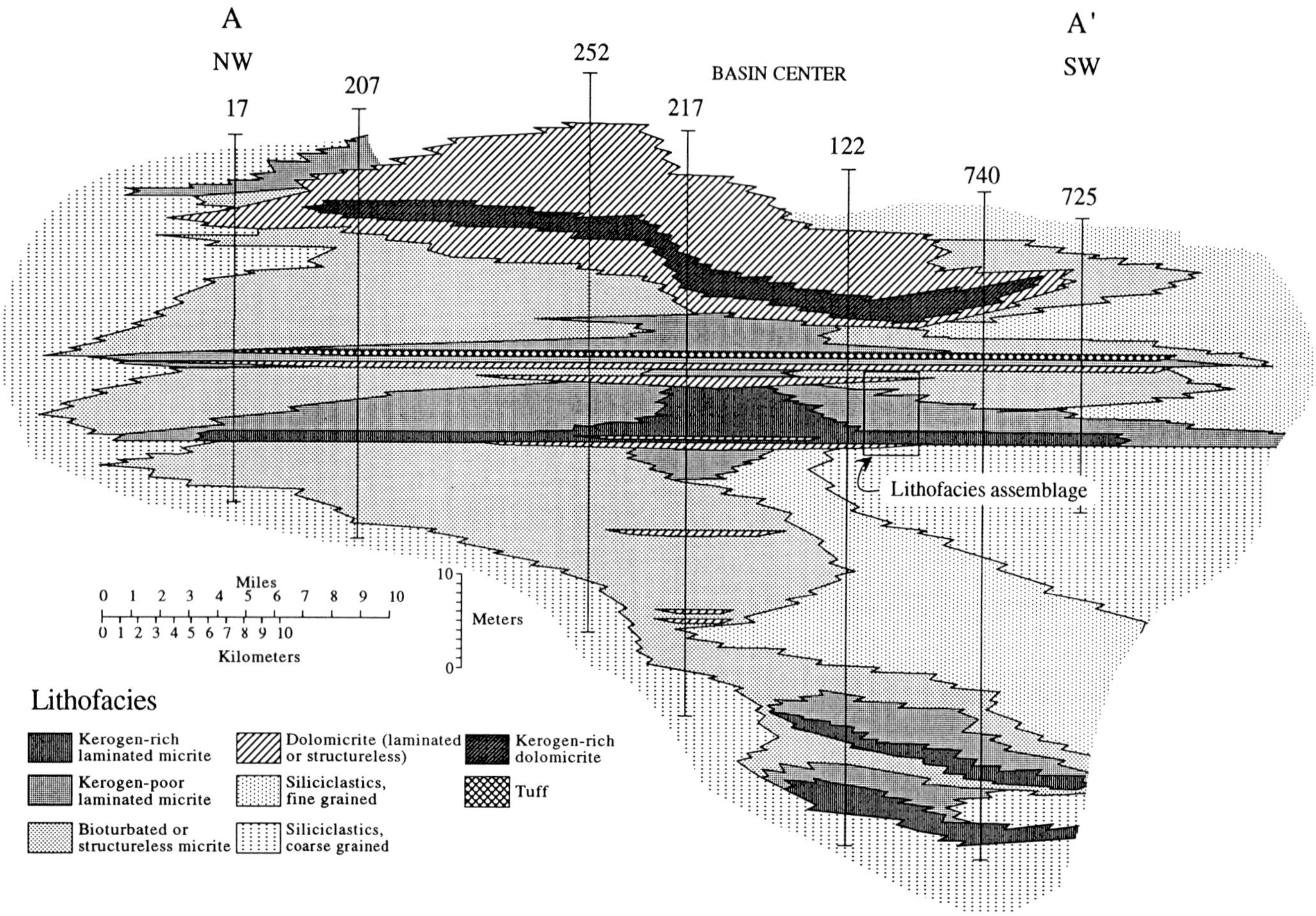

FIG. 6.—Facies cross-section of the Green River Formation, with measured sections indicated by bars headed with locality numbers. Datum is the lower sandwich tuff beds. This figure shows that the most common lithofacies assemblage that occurs within the lower and middle units can be seen in vertical section within the lithofacies box to the right of center. Note the vertical assemblage of kerogen-rich laminated micrite, kerogen-poor laminated micrite, bioturbated micrite, and dolomicrite. This transition is also well developed in the base of the lower unit. Note that these facies generally follow Walther's Law, except that they do *not* grade laterally into dolomicrite. Claystone, siltstone, and sandstone may form the top of the assemblage in basin marginal localities.

Eugster (1980) in Quaternary sediments of Lake Magadi in the Kenya Rift Valley. The large amount of dissolved silica in the hypersaline, high pH waters was forced to precipitate when the pH was abruptly lowered by overflow of dilute, low pH waters. The transition at the top of the upper unit into a meter thick bed of laminated calcimicrite with fossil fish, followed by fluvial siltstone of the mudstone tongue of the Wasatch Formation, may represent a continuation of this freshening trend.

The dolomitic beds that interrupt the calcite-dominated lower and middle units are interpreted as abrupt hypersaline episodes for reasons discussed above and because of their sharp top and bottom contacts with the laminated micrite. Dolomicrite in the Green River Formation has been interpreted as syndepositional rather than post depositional or late diagenetic dolomite (Eugster and Surdam, 1973; Eugster and Hardie, 1975; Surdam and Wolfbauer, 1975; Surdam and Stanley, 1979). Evidence for syndepositional dolomite includes the interbedding of thin (some on a centimeter scale) dolomicrite and micrite beds, the association of salt casts and chert nodules only with the dolomite, and the association of fossil fish only with micrites. The massive or non-laminated dolomicrites are interpreted as being directly precipitated from the water column. Evidence for this interpretation includes the massive "yoghurt"-like characteristic of these units (they are neither laminated nor peloidal unlike most upper unit dolomicrites). They are commonly nearly pure dolomite and show evidence for rapid precipitation. The latter includes their massive unstratified and commonly deformed character. The lateral variation in thickness over a few meters, ball–and–pillow and other soft-sediment deformation structures, and evidence for intrastratal flow (i.e., clasts of roof material incorporated and transported laterally centimeters to more than a meter) all indicate rapid deposition, and demonstrate the fluidity (high water content), and lack of shear strength even after deposition.

The massive dolomicrite beds represent low stands of Fossil Lake during lower and middle unit time. This is supported by the observation that the dolomicrite is at the top of the lithofacies assemblage that represents gradual regres-

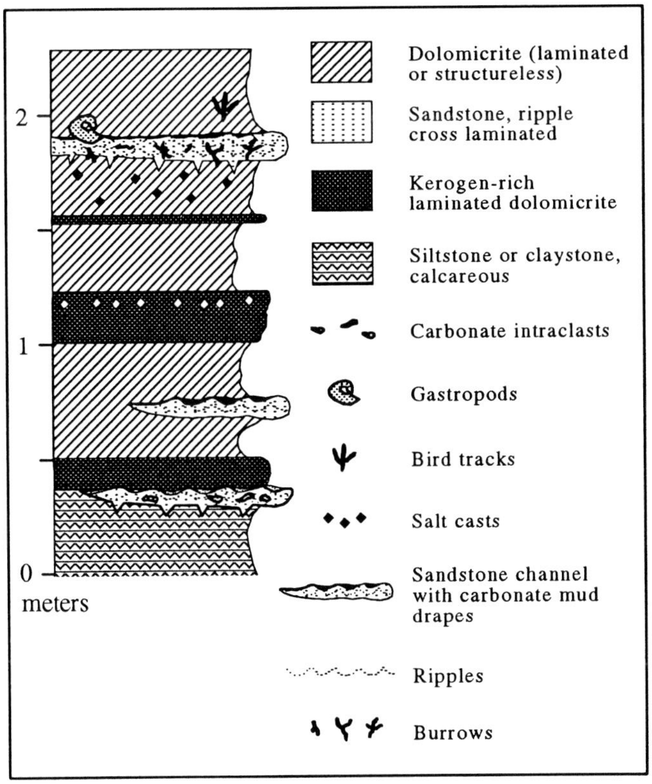

FIG. 7.—Thin, 5 to 15 cm, lenticular sandstones with numerous shallow water structures, burrows, gastropods and bird tracks are interbedded with kerogen-rich laminated dolomicrite of the upper unit at site 740. Note also the silty calcimicrite unit. Both gastropods and calcimicrite indicate freshwater conditions.

sion of the lake. Kerogen-rich laminated micrite, the lowermost unit, was deposited over a wide area in response to a rapid transgression during a highstand of the lake. Organic productivity was high and the lake was relatively deeper over a wider area, resulting in deposition and preservation of laminated organic-rich sediment. As the lake shrank in size, laterally adjacent facies prograded over the previously deposited organic-rich facies of the central basin (Fig. 6). Eventually, organic-poor, but still laminated, facies gave way to shallow water bioturbated micrites, and finally dolomicrites during the most shallow and hypersaline stage. Deposition of dolomicrite began and ended abruptly, as indicated by sharp basal and top contacts with laminated micrite.

The laminated dolomicrites with salt casts of the upper unit are distinctly different. They are associated with large amounts of siliciclastic input, probably volcanogenic. The lenticular nature of the laminae, and the association of kerogen-rich laminated dolomicrites with mudcracks and other shallow water features such as rippled sandstones with rip-up clasts and bird tracks (Fig. 7), all indicate deposition during upper unit time in a shallow lake, perhaps only one to a few meters deep. In addition, the "K-spar tuff" (unit 139) near the top of the middle unit, contains small, ripple cross-laminae in basin marginal areas indicating shallow lake water preceding the onset of hypersaline conditions. Some upper unit beds are peloidal or sandy, which indicates higher energy conditions. Unlike the massive dolomicrites of the lower and middle units, upper unit dolomicrites can be very kerogen-rich, with up to 14 percent TOC. Although Fossil Lake was very shallow during upper unit time, hypersalinity may have contributed to preservation of abundant organic matter.

Lateral facies gradations observed in both carbonates and tuff beds indicate a relatively fresher hydrochemistry, but possibly still saline, near the lake margins. Dolomicrites grade laterally into bioturbated micrite. This indicates that either the lake was fresher nearer the margins or that the bioturbated micrites are mudflat deposits affected by dilute groundwater flow. The lack of mudcracks, scour features, soil zones, etc., does not favor a mudflat interpretation for the bioturbated micrites. One would also expect development of higher salinities on exposed mudflats. Well-defined mudflat-type deposits associated with dolomite have been observed in the lower unit at site 42, and in the upper unit at site 740, discussed above. At site 42, dolomicrites are mudcracked at the base, lenticular laminated, have scour and fill structures, contain abundant ostracodes, which are common in shallow nearshore zones (Buchheim and Surdam, 1981), and are interbedded with fluvial and/or floodplain deposits.

Tuff beds also provide evidence that Fossil Lake was fresher nearer the margins. From center to margins, the mineralogy of the tuffs grades laterally from K- feldspar, to analcime and finally to clay minerals. This type of lateral facies transition is well documented elsewhere (Sheppard and Gude, 1968; Surdam and Sheppard, 1978) and indicates less alkaline, saline waters toward the marginal clay facies.

The occurrence only of analcimic tuff beds in the lower unit, interpreted as the freshest, initial shallow stages of Fossil Lake, and analcime and K-feldspar tuffs in the middle and upper units, is consistent with the latter having more saline waters. However, the occurrence of feldspathic tuffs in the middle unit is somewhat puzzling, especially with its calcimicrite composition and abundance of freshwater fish.

The question arises, can a hypersaline to saline or fresh water lateral chemistry be maintained for any length of time? Are there any modern analogs? The observations discussed above are evidence of such a hydrochemical gradient. The lake marginal water need not have been fresh for calcite rather than dolomite to have precipitated; calcite is precipitated from both dilute (fresh) and saline water.

The low topographic gradient on the lake floor may have been a contributing factor. Bottom gradients were probably less than 40 cm per kilometer (established from work of Bradley, 1929, 1964; Surdam and Wolfbauer, 1975; and Lundel and Surdam, 1975). Bradley (1929) used fluvial gradients of a modern river system in northern California for his estimates, concluding that once the lake was established for a period of time these gradients decreased to nearly horizontal. Gradients that I have measured in the Bear River Bay of Great Salt Lake are in centimeters per kilometer. The bay stands between the major inflow to Great Salt Lake and the main body of the lake itself. Inflow from the river freshens the Bay and flows out as an overflow over Great

Salt Lake during spring. Freshwater fish are abundant in the northern part of the bay and are even washed up on the shores of Great Salt Lake several kilometers away. They probably enter the main lake during spring overflow. The larger portion of this inflow is diked off and prevented from entering the lake, so a true picture of the total effect of such inflow on a very low gradient saline lake is unknown.

Lake Natron in the East African Rift of Tanzania has large 'lagoonal' areas several kilometers in extent on its margins (R. Surdam, pers. commun.; Willock, 1974). Lake Turkana in Kenya is alkaline and saline yet maintains relatively fresh waters in its northern portion that receives inflow from the Omo River (Yuretich and Cerling, 1983). Fossil Lake may have also maintained a broad (up to 10 km), shallow, saline to fresh margin, particularly near inflow areas. It is possible that conditions of hypersalinity with fresh inflow would result in meromixis, or a chemically stratified lake initiated by dilute overflow (cf. Renaut and Tiercelin, this volume).

The massive dolomite beds of the middle unit that pinch toward the basin margins may represent deposits that accumulated in remanent, shallow hypersaline lakes formed at the end of periods of lake regression, with non-deposition on the surrounding, contemporary mudflats. There is no evidence laterally for highly desiccated mudflats that would be expected. There is no question that the dolomicrites of the upper unit grade laterally into calcimicrites (Figs. 2, 6). The default explanation has to be nearshore freshening, as calcite is not normally precipitated from hypersaline solutions required for dolomite deposition.

Despite the hypersaline intervals, Fossil Lake continued to have a varied, prolific fish population. Evidence discussed indicates that lake marginal areas were relatively fresh during periods of hypersalinity. At least some fish species could have survived in these areas and in adjacent floodplain rivers and ponds.

CONCLUSIONS

The deposits of Eocene Fossil Lake represent one of the best preserved and most complete records of a freshwater lake that was interrupted by episodes of hypersalinity before finally succumbing to an extended period of hypersalinity. Less well understood, however, is the conflicting evidence of large numbers of well-preserved fossil fish with tuff beds in the middle unit that are altered to authigenic K-feldspar and analcime (indicating high salinity and alkalinity), and minor amounts of associated or bedded dolomite, respectively. The fact that the fossil fish are nearly always found in calcimicrites and rarely in dolomicrites is consistent with an interpretation that calcite is precipitated from mildly saline, alkaline waters, whereas dolomite is either precipitated or formed syndepositionally in hypersaline lake or sediment-pore waters. The massive dolomite beds interbedded in the lower and middle calcimicrite units could have been precipitated in the water column. Either way, they provide evidence for hypersalinity. Although the evaporite facies (dolomite) contain disseminated calcite pseudomorphs after saline minerals, chert, and mudcracks, bedded salts were never preserved.

In summary, the Green River Formation in Fossil Basin provides a unique set of sedimentary facies not found in similar deposits elsewhere in Wyoming, Utah, or Colorado. The vertical and lateral relations observed provide new insights into the history of saline, alkaline lakes in the geologic record, and provide a framework for further studies to clarify our understanding of ancient saline, alkaline lakes.

ACKNOWLEDGMENTS

This research was funded grants from the University of Wyoming—National Park Research Center, and Loma Linda University. The field, laboratory, drafting, and other research assistance of Ted Fremd, Robert Biaggi, Carole Buchheim, Greg Morgan, and Roberto Cushman is greatly appreciated. The cooperation and support of Fossil Butte National Monument contributed to the successful completion of this research. The helpful direction and inspiration of Hans Eugster, who was instrumental in the early stages of this work, is gratefully acknowledged.

REFERENCES

BRADLEY, W. H., 1929, The varves and climate of the Green River Epoch: Washington, D. C., United States Geological Survey Professional Paper 158, p. 87–110.

BRADLEY, W. H., 1964, Geology of the Green River Formation and associated Eocene rocks in southwestern Wyoming and adjacent parts of Colorado and Utah: Washington, D.C., United States Geological Survey Professional Paper 496-A, 57 p.

BUCHHEIM, H. P., AND SURDAM, R. C., 1981, Paleoenvironments and fossil fish of the Laney Member, Green River Formation, Wyoming, *in* Gray, J., Boucot, A. J., and Berry, B. N., eds., Communities of the Past: Stroudsburg, Hutchinson Ross, p. 415–452.

EUGSTER, H. P., 1980, Lake Magadi, Kenya and its precursors, *in* Nissenbaum, A., ed., Hypersaline Brines and Evaporitic Environments (Developments in Sedimentology 28): Amsterdam, Elsevier, p. 195–232.

EUGSTER, H. P., AND HARDIE, L. A., 1975, Sedimentation in an ancient playa-lake complex:- the Wilkins Peak Member of the Green River Formation of Wyoming: Geological Society of America Bulletin, v. 86, p. 319–334.

EUGSTER, H. P., AND SURDAM, R. C., 1973, Depositional environment of the Green River Formation of Wyoming: A preliminary report: Geological Society of America Bulletin, v. 84, p. 1115–1120.

GRANDE, L., 1984, Paleontology of the Green River Formation, with a review of the fish fauna: Cheyenne, Bulletin of the Wyoming Geological Survey 63, 333 p.

LAST, W. M., 1990, Lacustrine dolomite — an overview of modern, Holocene, and Pleistocene occurrences: Earth Science Reviews, v. 27, p. 221–263.

LUNDELL, L. L., AND SURDAM, R. C., 1975, Playa-lake deposition, Green River Formation, Piceance Creek Basin, Colorado: Geology, v. 3, p. 493–497.

MIDDLETON, G. V., 1973, Johannes Walther's Law of the correlation of facies: Geological Society of America Bulletin, v. 84, p. 979–988.

ORIEL, S. S., AND TRACY, J. I., 1970, Uppermost Cretaceous and Tertiary stratigraphy of Fossil Basin, southwestern Wyoming: Washington, D.C., United States Geological Survey Professional Paper 635, 53 p.

SHEPPARD, R. A., AND GUDE, A. J., 1968, Distribution and genesis of authigenic silicate minerals in tuffs of Pleistocene Lake Tecopa, Inyo County, California: Washington, D. C., United States Geological Survey Professional Paper 597, 38 p.

SURDAM, R. C., AND SHEPPARD, R. A., 1978, Zeolites in saline, alkaline-lake deposits, *in* Sand, L., and Mumpton, F., eds., Natural Zeolites: Occurrence, Properties, Use: New York, Pergamon, p. 145–174.

SURDAM, R. C., AND STANLEY, K. O., 1979, Lacustrine sedimentation during the culminating phase of Eocene Lake Gosiute, Wyoming (Green

River Formation): Geological Society of America Bulletin, v. 84, p. 1115–1120.

SURDAM, R. C., AND WOLFBAUER, C. A., 1975, Green River Formation, Wyoming: A Playa-Lake Complex: Geological Society of America Bulletin, v. 86, p. 335–345.

WILLOCK, C., 1974, Africa's Rift Valley: Amsterdam, Time Life International, 184 p.

YURETICH, R. F., AND CERLING, T. E., 1983, Hydrogeochemistry of Lake Turkana, Kenya: Mass balance and mineral reactions in an alkaline lake: Geochimica et Cosmochimica Acta, v. 47, p. 1099–1109.

THE JURASSIC EAST BERLIN FORMATION, HARTFORD BASIN, NEWARK SUPERGROUP (CONNECTICUT AND MASSACHUSETTS): A SALINE LAKE–PLAYA–ALLUVIAL PLAIN SYSTEM

ELIZABETH GIERLOWSKI-KORDESCH
Department of Geological Sciences, Ohio University, Athens, Ohio, 45701–2979
AND
BRIAN R. RUST†
Department of Geology, University of Ottawa and Ottawa-Carleton Geoscience Centre, Ottawa, Ontario K1N 6N5 Canada

ABSTRACT: The Lower Jurassic East Berlin Formation of the Hartford Group (Newark Supergroup), located in the Hartford Basin (Connecticut and Massachusetts), is an alluvial and lacustrine half-graben succession deposited in a semi-arid climate. The formation mainly comprises massive red mudstones, reddish rippled siltstones and shales with minor gray-green mudrocks and shales, black shales, and sheet sandstones. The red mudstones, which contain desiccation cracks up to 1m deep and pedogenic features, are interpreted as fossil vertisols. Like modern analogs in central Australia, these paleosols produced abundant sand-sized mud aggregates that were transported as bedload, and are still recognizable within sandy facies, but were obscured by compaction in the mudstones. Thus, although overwhelmingly muddy, the alluvial system is interpreted as a bedload-dominated distal alluvial plain to sandflat. Sheet sandstones comprise trough cross-stratified sandstone overlain by lineated planar-laminated sandstone or rippled siltstone; all are attributed to sheetfloods, some of which entered the lake.

Lacustrine deposits are represented by stratified gray-green mudrocks and black shales. Sedimentologic, petrologic, and ichnologic evidence all indicate a saline lacustrine paleoenvironment. The lake deposits are associated with disrupted shales that are interpreted as playa sediments. Disrupted shales, interbedded with massive mudstones, show similar vertisol development to the playa muds. The lack of prominent shoreline deposits and large-scale deltaic features is attributed to frequent lake-level fluctuations, which caused extensive migration of the lake shore across distal parts of the uniformly sloping alluvial plain. Major changes in lake level through time in the Hartford Basin are attributed to climatic influences, whereas changes in hydrologic drainage appear to be related to tectonic activity.

INTRODUCTION

The East Berlin Formation (EBF) lies within the Hartford Basin (Fig. 1), an eastward-tilted half-graben, representing one of several basins in which the Mesozoic Newark Supergroup accumulated (Hubert and others, 1976; Lorenz, 1987; Phillips, 1988; Olsen and others, 1989; Hubert and others, 1992). The EBF is of Early Jurassic (Hettangian) age (Table 1) based on radiometric dating of associated basalts (Seidemann and others, 1984) and palynology by Cornet and Traverse (1975). The EBF is approximately 150 m thick in central Connecticut (Olsen and others, 1989).

The upper 107 m of the East Berlin Formation is well exposed near East Berlin, Connecticut (Fig. 1), where it mainly comprises red mudstones and shales with minor sandstones, gray-green mudrocks and shales, and black shales. Sanders (1968) and Hubert and others (1976) interpreted the red mudstones and shales with interbedded sandstones as deposits of meandering rivers. The remaining non-red fine-grained rocks were divided into five cycles cored by black shales and interpreted as cycles recording the expansion and contraction of oligomictic lakes. Demicco and Gierlowski-Kordesch (1986) and Hubert and others (1982, 1992) reinterpreted the red sandstones, mudstones and shales as playa deposits interbedded with sheetflood sandstones of an ephemeral floodplain. Demicco and Gierlowski-Kordesch (1986) identified 15 cycles, both with and without black shale units, that they interpreted as sediments resulting from the expansion and contraction of lakes of variable depths. Olsen and others (1989) attributed the red mudstones and shales of the EBF to shallow lacustrine sedi-

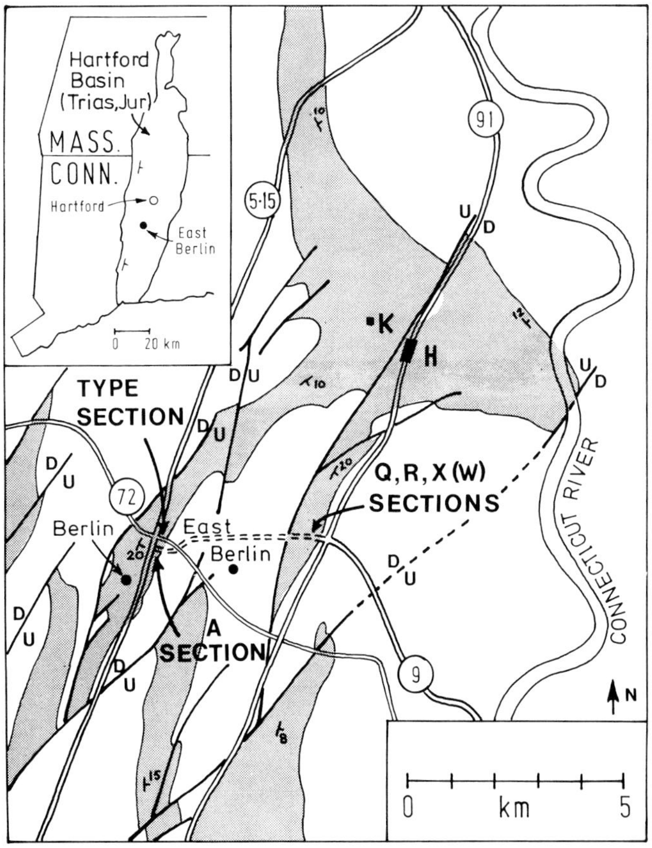

FIG. 1.—Distribution of the East Berlin Formation in central Connecticut (stippled) with locations of measured sections; inset map shows position of East Berlin study area in relation to the Hartford Basin. U,D: upthrown, downthrown sides of faults.

†Deceased

Sedimentology and Geochemistry of Modern and Ancient Saline Lakes, SEPM Special Publication No. 50

TABLE 1.—GENERALIZED STRATIGRAPHY OF THE HARTFORD BASIN (TALCOTT BASALT DOES NOT COVER THE ENTIRE BASIN). DATA FROM LEHMAN (1959), LETOURNEAU (1985), OLSEN AND OTHERS (1989), SMOOT (1991), AND VARIOUS OTHER SOURCES.

Series	Stage	Thickness (m)	Stratigraphic Unit	Lithology	Environmental Interpretation	Hydrologic Drainage	Tectonics
	Sinemurian-Hettangian	~2000+	Portland Fm.	black shales mud/siltstones sandstones conglomerates minor limestones	lacustrine & alluvial	open ↑ closed	↑ Subsidence slows
		30–65	Hampden Fm.	tholeiitic basalt	volcanic flow	—	↑
		150–300	East Berlin Fm.	black shales mud/slit/sandstones conglomerates minor limestones	lacustrine & alluvial	closed	
		80–150	Holyoke Fm.	tholeiitic basalt	volcanic flow	—	
		17–140	Shuttle Meadow Fm.	black shales mud/siltstones sandstones minor limestones	lacustrine & alluvial	closed	
		0–80	Talcott Fm.	tholeiitic basalt	volcanic flow	—	201 Ma—Extension & Igneous Activity
Upper Triassic	? Early Norian-basal Hettangian	~2400	New Haven Arkose	mud/siltstones sandstones conglomerates	alluvial	open	↑ 220 Ma—Extension

mentation with minor fluvial contribution, but gave no firm criteria to distinguish fluvial from lacustrine redbeds.

The aims of this paper are twofold: (1) to describe the facies of the EBF and to reinterpret the interrelationships between playa-lake and alluvial plain-sandflat sediments, and (2) to review existing sedimentologic, ichnologic, petrologic, and isotopic data to interpret the salinity of the East Berlin paleolakes.

FACIES DESCRIPTIONS

Introduction

The East Berlin Formation is well exposed at a new road cut (Section A) adjoining the Type Section (Fig. 1). Although road construction destroyed a significant portion of the Type Section, Section A (Fig. 2) provides a supplementary reference section for the EBF. Eight facies are recognized in the EBF (Table 2). Facies summary below includes data from Gierlowski-Kordesch (1985, 1991, and unpublished) and Demicco and Gierlowski-Kordesch (1986).

Sources for additional geologic data on the EBF include work on organic geochemical analyses (Pratt and others, 1985, 1986b; Spiker, 1985; Kotra and others, 1988; Spiker and others, 1988; Kruge and others, 1990, 1991), clay mineralogy (April, 1980, 1981a, b; Gottfried and Kotra, 1988), petroleum prospects (Ziegler, 1983; Pratt and others, 1986a, 1988; Pratt and Burruss, 1988; Hubert and others, 1992), inorganic geochemical analyses (Dean, 1988; Smoot and Robinson, 1988), and stable isotopes of carbonates (Pratt and others, 1985, 1986b; Suchecki and others, 1988).

Trough Cross-Stratified Sandstones (St)

Fine- to medium-grained sandstones are medium- to well-sorted. Sets are 10 to 20 cm thick within cosets 10 to 100 cm in thickness. Sets contain mostly small-scale cross-stratification, which commonly fine upwards. Cosets commonly have erosional bases with gradational upper contacts into overlying mudrocks. Color is mostly light red (a few units are gray or mottled). Grains comprise quartz, feldspar, and sand-sized mud aggregates (Fig. 8A in Rust and Nanson, 1989). Also, mudstone intraclasts up to 10 cm long occur. This facies appears to be transitional between facies F2 and F3 of Smoot (1991) and is similar to the large-scale cross-stratified sandstones of Demicco and Gierlowski-Kordesch (1986).

Horizontally Stratified Sandstones (Sh)

These sandstones are similar to St in color, sorting, and composition, but are finer in grain size and show better segregation into mud aggregate and arkosic sand layers. Parting lineations are present. They are commonly interbedded with St, as sets up to 20 cm thick within cosets up to 50 cm thick. Many sets have erosional bases, fine upward, and are overlain by a cracked mud drape. Trace fossils (*Planolites montanus* and spherical pellets) are common in the mud drapes capping sedimentation units of Facies St and Sh.

The main sandstone bodies (St, Sh) are sheets extending to limits of outcrop (tens to hundreds of meters) without significant variation in thickness, commonly 50 to 100 cm (Fig. 3). Only one channel is present in the study area (Fig. 1- Section K), and it contains large burrows. More channels were observed to the north in Massachusetts. Sheets exhibit erosional lower contacts and gradational upper contacts into overlying mudrocks. Sheet sandstones (St, Sh) interbed mainly with red mudrock sequences (Fig. 2), but also occur within gray-green sequences (e.g., Section A at 70.1–70.5 m; Fig. 2) as a massive, mottled gray/red sandstone. In association with ripple cross-laminated siltstones (Fr), sheet sandstones occur within fining-upward sequences grading

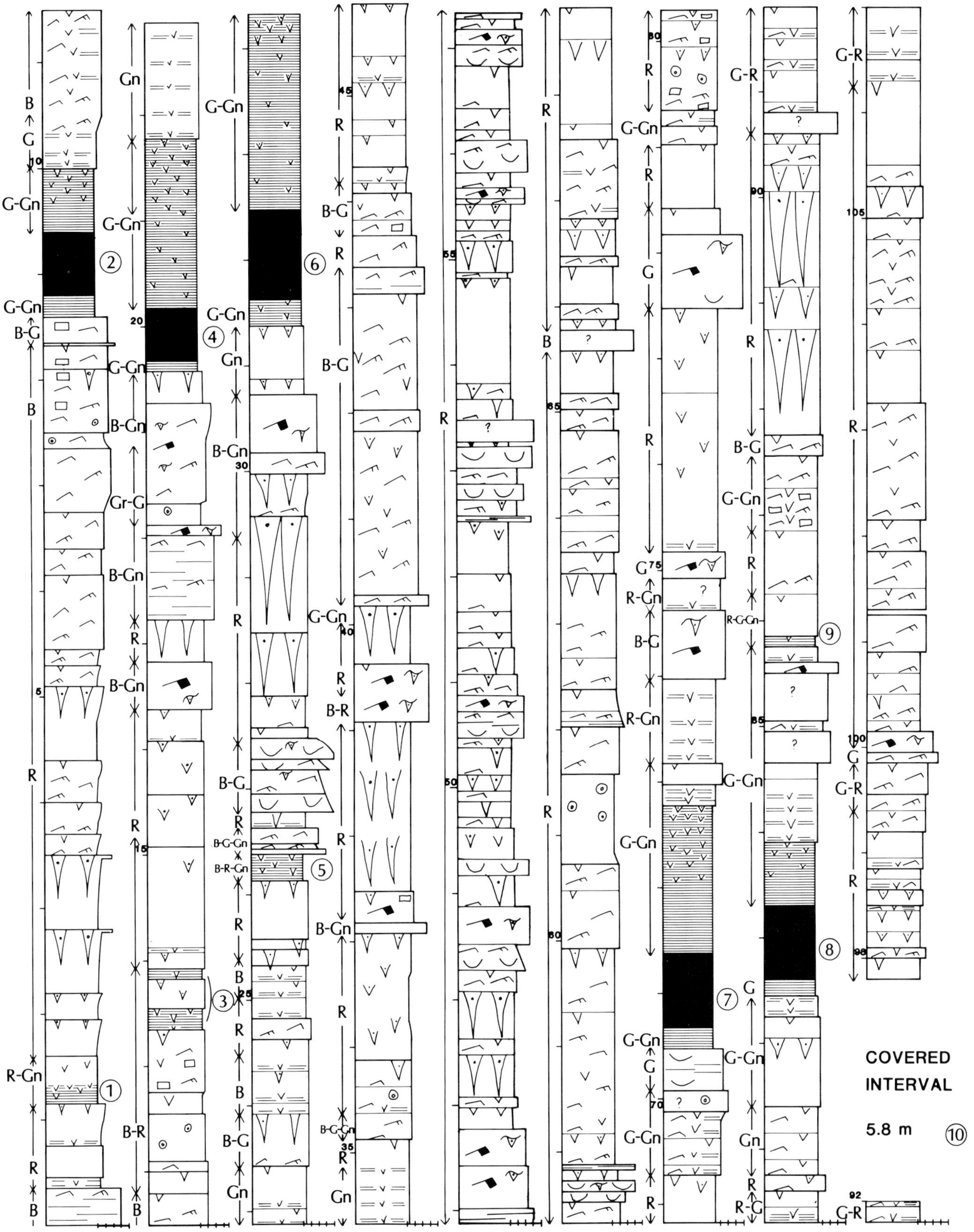

FIG. 2.—Section A of the East Berlin Formation (see Fig. 1 for location). Tick marks and numbers on vertical scale indicate thickness in meters. Horizontal scale at bottom shows grain size scale.

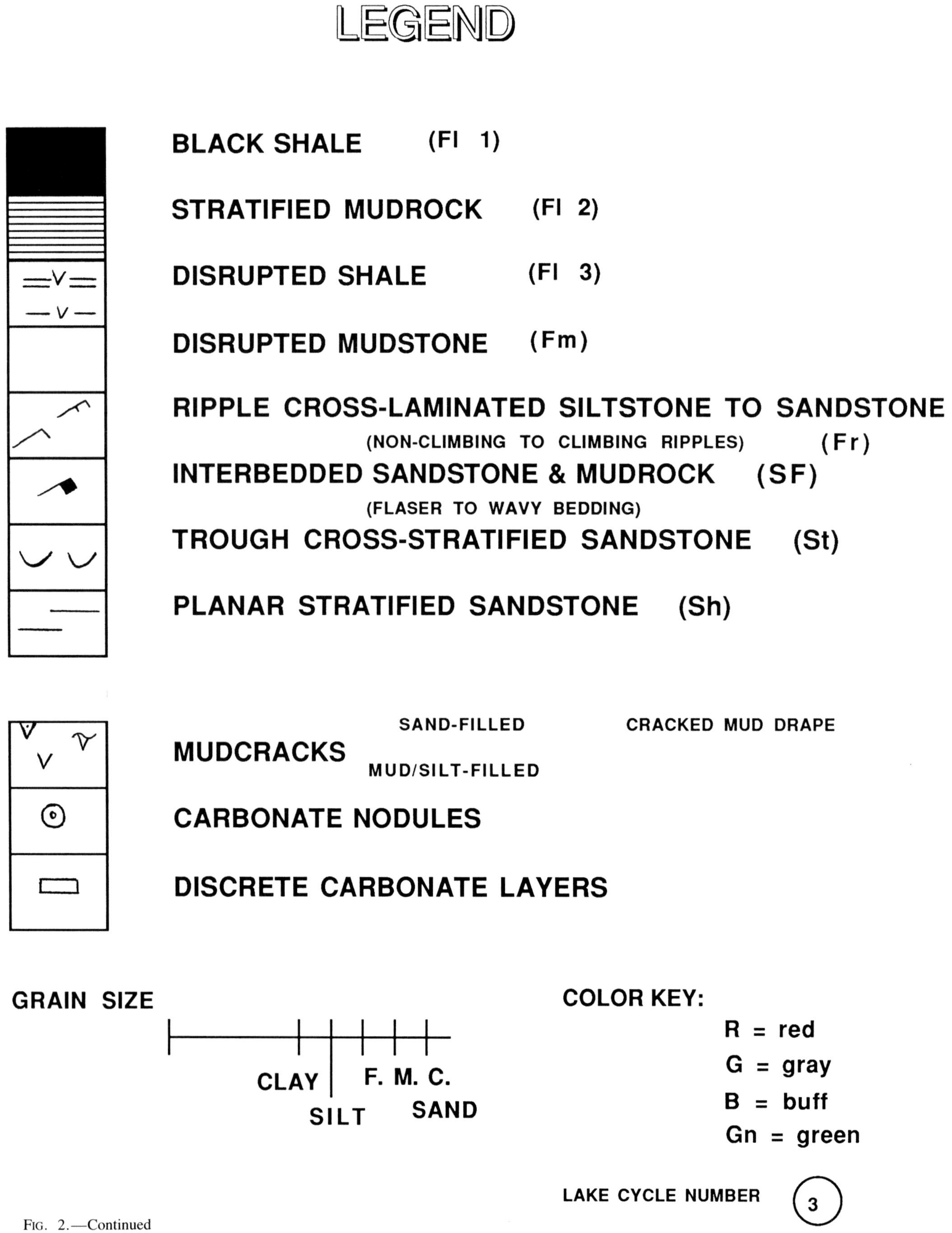

FIG. 2.—Continued

TABLE 2.—SUMMARY LIST OF FACIES AND THEIR INTERPRETATION FOR THE EAST BERLIN FORMATION OF THE NEWARK SUPERGROUP

FACIES	DESIGNATION	INTERPRETATION
Trough cross-stratified sandstones	**(St)**	Alluvial plain/sandflat (ephemeral)
Horizontally stratified sandstones	**(Sh)**	Alluvial plain/sandflat (ephemeral)
Interbedded sandstones and mudrocks	**(SF)**	Alluvial plain/sandflat (ephemeral)
Ripple cross-laminated siltstones	**(Fr)**	Alluvial plain/sandflat (ephemeral)
Black shales	**(Fl_1)**	Offshore lacustrine (saline)
Stratified mudrocks	**(Fl_2)**	Shallow lacustrine (saline)
Disrupted shales	**(Fl_3)**	Playa mudflat (aggrading)
Disrupted mudstones	**(Fm)**	Paleosols (vertisols)

FIG. 3.—Middle portion of Section X. Meter-thick layer of sandstone extends full length of the outcrop without variation in thickness. Senior author (arrowed) for scale.

from medium- to fine-grained sandstone to siltstone: Sh → SF → Fr, St → SF → Fr, St → Fr, Sh → Fr, and SF → Fr. Examples from section A include 26.3–27.2 m, 43.1–43.3 m, and 55.8–56.1 m (Fig. 2). Paleocurrents for both sheet sandstones and rippled siltstones in all sections (Fig. 1) show unidirectional eastward trends (Fig. 4). This facies is similar to a single component of facies F1 of Smoot (1991) and is similar to planar cross-stratified sandstones of Demicco and Gierlowski-Kordesch (1986).

Interbedded Sandstones and Mudrocks (SF)

Fine- to medium-grained sandstone laminae or thin beds up to 5 cm thick intercalate with continuous to discontinuous laminae of mudstone up to 1 mm thick. SF units are 0.25 to 1.5 m thick with erosional bases. Colors include red to pink and gray. Sandstone laminae/beds contain asymmetric and symmetric ripple cross-lamination or horizontal laminae. Mudstones comprise flasers or continuous laminae that are commonly mudcracked. The entire spectrum from flaser to wavy bedding occurs in SF units. Trace fossils are mostly restricted to the upper portions of depositional units; these include *Scoyenia gracilis*, *Skolithos* ichnosp., *Fuersichnus* ichnosp., and spherical pellets. This facies is similar to some components of facies F1 of Smoot (1991) and represents the interbedded sandstones and mudstones of Demicco and Gierlowski-Kordesch (1986).

Ripple Cross-Laminated Siltstones (Fr)

This facies is composed of muddy to fine siltstones and very fine-grained sandstones containing claystone aggregates (Fig. 6 in Demicco and Gierlowski-Kordesch, 1986). Fr units commonly occur fining-upward transitions with St. Colors are mostly red with a few gray units. Cosets are 0.25 to 1.5 m thick with multiple sets separated by erosional surfaces. The upper few cm of cosets commonly contain irregular patches of disrupted massive mudstone or a complex system of mudcracks. Ripple forms include mostly asymmetric current ripples, both climbing and erosional. Uniformity of paleoflow within cosets is notable. Paleocurrents for both rippled siltstones and sheet sandstones in all sections (Fig. 1) show unidirectional eastward trends (Fig. 4).

Yellow, carbonate-cemented, well-sorted, fine siltstone beds (up to 5 cm thick) are present locally as discrete layers within a red siltstone sequence (e.g., Section A at 12.8 m,

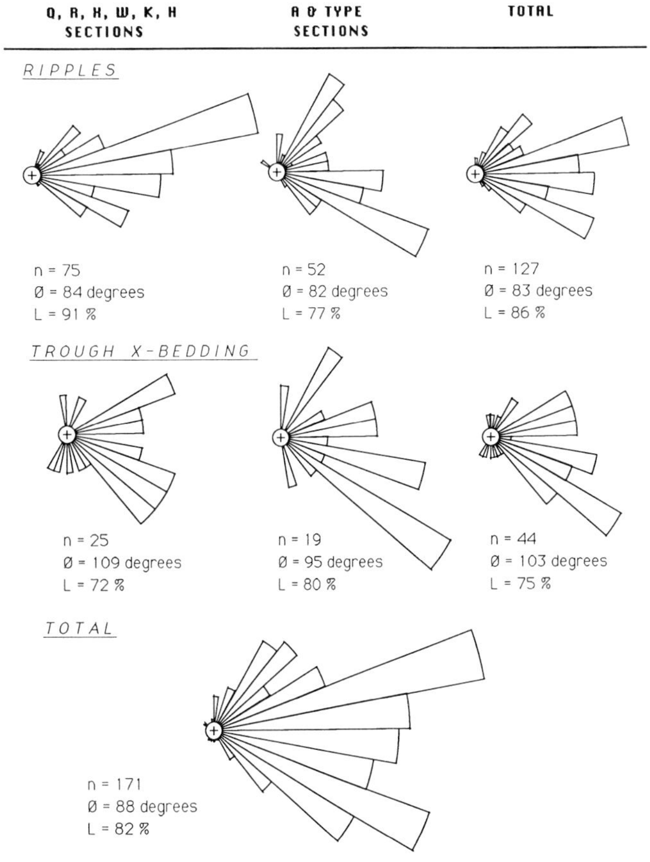

FIG. 4.—Paleocurrent rose diagrams for measured sections of East Berlin Formation (Type, A, X, R, Q, W, H, and K Sections; see Fig. 1 for localities). Scale factor=9 for the six upper rose diagrams; rose diagram for all paleocurrents has scale factor of 4.5. n=sample number, Ø=azimuth of resultant vector, and L=vector magnitude. Method after Curray (1956).

79.4 m); primary ripple lamination can be well-preserved in these apparently little compacted beds. Star-shaped "balls" of calcite about 5 mm in diameter (Fig. 5) are present in the uppermost portions of Section X (Fig. 6). Erosive contacts show that some "balls" were eroded, transported, and sorted. Secondary calcite is suspected due to unusual habit and calcite-filled fusiform burrows. Also, gypsum molds and calcite pseudomorphs after gypsum occur. Trace fossils are restricted to upper portions of depositional units; they include *Scoyenia gracilis*, *Planolites montanus*, fusiform burrows, spherical pellets, and an escape structure. Fr is similar to facies FM3 of Smoot (1991).

Black Shales (Fl_l)

Three types of black shales occur (Figs. 7, 8): shale type 1 (irregularly laminated shales), shale type 2 (shale with magnesite microspar nodules), and shale type 3 (finely laminated shale). All shale types are transitional to (i.e., gradual contacts) and enveloped by Fl_2. This facies is the laminated black mudstone of Demicco and Gierlowski-Kordesch (1986).

1. Irregularly laminated black shale is most common in complete lake cycles 4, 6 (upper portion only), 7, 8, and 10 (Fig. 2). Fine irregular (discontinuous) laminae (mm to submm scale) are composed of carbonate to arkosic siltstone, up to 20 mm thick. True laminae have parallel lamination (Fig. 2A in Demicco and Gierlowski-Kordesch, 1986). Other laminae appear to be displaced clasts of siltstone laminations; evidence includes laminae oblique to bedding (at all angles) within these clasts (Fig. 7- lake cycle 6, at 32 m; Fig. 8A). Some very rare chaotic bedding is present. Rare radiating crystal impressions (of an evaporitic mineral?) occur on bedding surfaces (Fig. 2B in Demicco and Gierlowski-Kordesch, 1986). Organic matter content of Fl_1 and Fl_2 varies from 1.2–3.8 weight percent (Pratt and others, 1986b). Poorly preserved fish fossils and rare conchostracans (*Cornia* sp.) are present (McDonald, 1982, 1992).

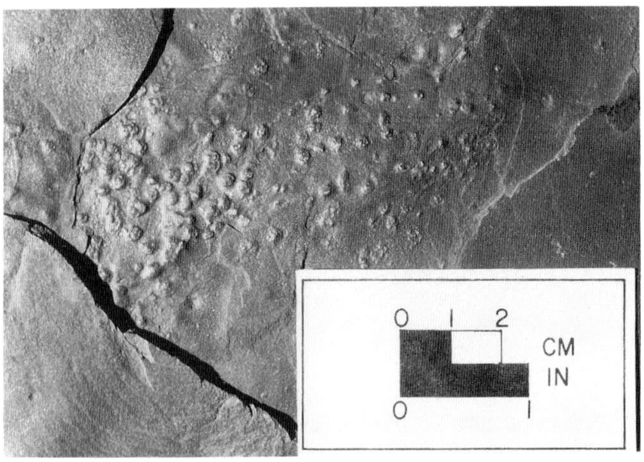

FIG. 5.—Ripple cross-laminated siltstone (Fr) from upper portion of Section X, showing the star-shaped calcite "balls" exposed on a weathered bedding plane.

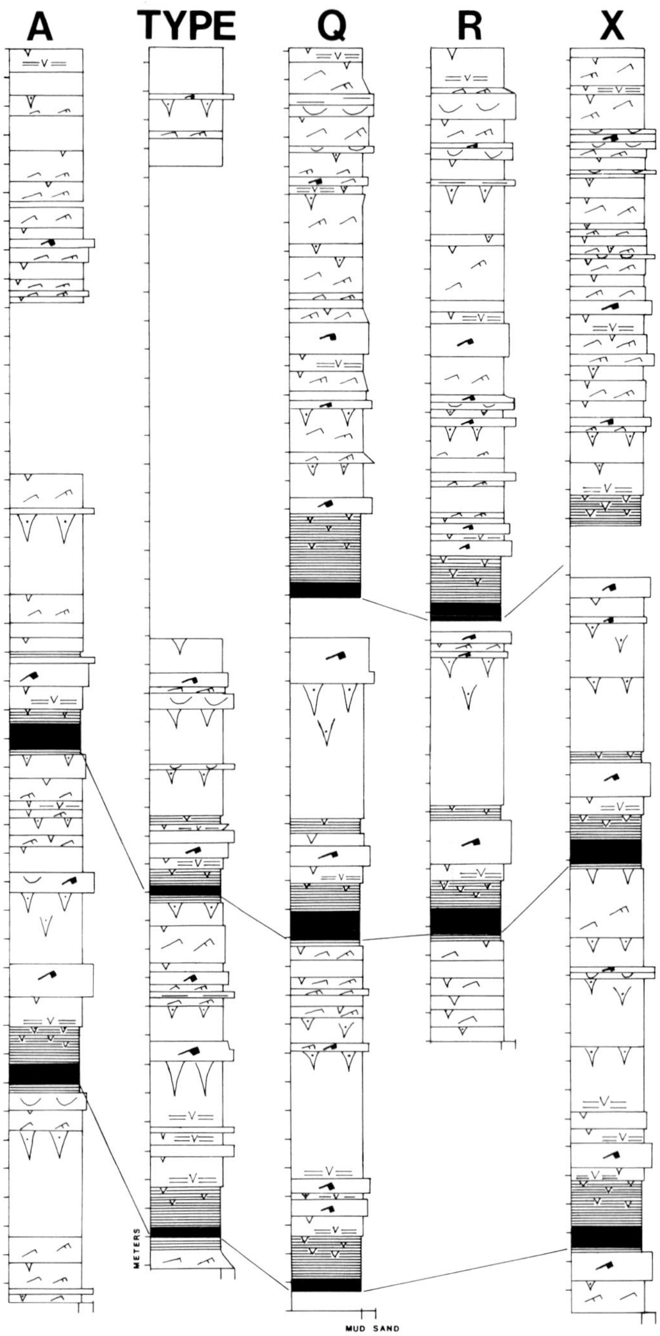

FIG. 6.—Measured sections of the upper 44 m of Sections A, Type, Q, R, and X. Locations given in Figure 1. Top of each section is defined by the Hampden Basalt. Legend given in Figure 2.

2. Magnesite-bearing shale contains magnesite microspar nodules 2–3 mm in diameter (as well as larger crystal masses to lenses) disseminated in varying densities (Fig. 8B). One example of unit is 60 cm thick (Section A at 8.8 m, Fig. 2- lake cycle 2; Fig. 7).

3. Finely laminated shale comprises continuous, alternating clay and carbonate siltstone lamination (Fig. 8C). Single example of unit is 25 cm thick (Section A at 31.7 m, Fig. 2- lake cycle 6; Fig. 7—middle portion above a

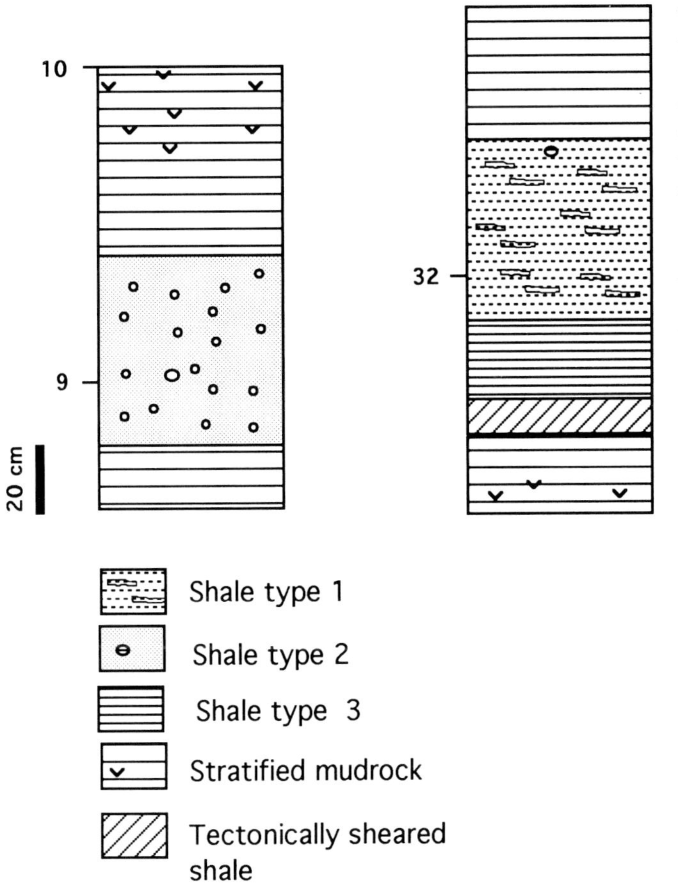

FIG. 7.—Detailed sections of the first and third black shale sequences (lake cycles 2 and 6; see Fig. 2; meter position in sequence at left), illustrating the three black shale types: shale type 1- irregularly laminated; shale type 2- magnesite-bearing; and shale type 3- finely laminated.

tectonically sheared shale unit). Fossils include well-preserved, articulated fish remains (*Semionotus* up to 40 cm in length, *Redfieldius* avg. 20 cm in length, *Diplurus* 80 cm maximum length), conchostracans (*Cornia* sp.), rare ostracodes (*Darwinula* sp.) and coprolites (Westfield fish bed; McDonald, 1982; 1992; Olsen and others, 1989). This shale type is similar to some features in facies L1 of Smoot (1991).

Stratified Mudrocks (Fl_2)

Gray-green mudrocks are gray to black when fresh (Fig. 9A). Some Fl_2 units contain minor red laminae and yellowish calcitic or dolomitic laminae. Stratification is continuous and planar with mostly non-erosive boundaries. Contacts with facies Fl_1 are gradual and non-erosive. In upper portions of Fl_2 units above black shales (Fl_1), arkosic, fine-grained sandstone to siltstone laminae and lenses intercalate and contain ripple cross-lamination or parallel lamination. Individual laminae are a few mm to 1.5 cm thick. Mudcracks are generally a few centimeters long. Simple to multiple mudcracks contain sandstone and/or mudstone fillings. Mudcrack density increases upward in most units. Fl_2 units above black shales (Fl_1) are up to 2 m thick, whereas those beneath are a few decimeters thick and rarely contain mudcracks (Figs. 2, 6, 7).

Fossils include carbonized macrophytic plant fragments (conifers- *Brachyphyllum* and *Pagiophyllum*; cycadophyte- *Otozamites*; Olsen and others, 1989), dinosaur trackways, very rare burrows (*Planolites montanus*), and rare conchostracan shells replaced by calcite (McDonald, 1982, 1992). Pyrite is dispersed randomly along layers. Traces of analcime and gypsum and very rare molds of halite, glauberite? (Hubert and others, 1976, p. 1205) and gypsum (Kruge and others, 1990, 1991) are present. This facies is similar to facies L2 of Smoot (1991).

Disrupted Shales (Fl_3)

Red and green shales contain calcitic and dolomitic laminations. Green shales occur directly above stratified mudrocks (Fl_2), whereas red shales are interbedded with Fm and Sr. Discontinuous siliciclastic laminations (2–5 mm thick) consist of "thick" fine silt laminae alternating with "thin" clay laminae (Fig. 9B). Discontinuous carbonate laminations are composed of mud to silt-sized particles. Fl_3 units are termed "disrupted" because layering is rarely preserved across an entire exposed vertical surface. Thin beds (2–15 cm thick) commonly occur as patches within disrupted mudstones (Fm) (Fig. 4A in Demicco and Gierlowski-Kordesch, 1986). Complex mudcracks, both sandstone- and mudstone-filled, vary from regularly-spaced cracks that penetrate through entire beds to separate mm-scale cracks within individual laminae. Adjacent displaced blocks contain distorted, concave-upward "mud curls" (a few centimeters in width), that rarely fold up into roll-up structures with small, discrete, laminated intraclasts floating in adjacent mudstone (Fm). Rare ripple structures occur, seemingly transitional to totally rippled siltstones (Fr).

Gypsum molds and calcite psuedomorphs after gypsum are present. Fossils include rare carbonized macrophytic plant fragments (conifers- *Brachyphyllum* and *Pagiophyllum*; cycadophyte- *Otozamites*; Olsen and others, 1989), very rare fish scales, and trace fossils (*Skolithos* ichnosp., *Planolites montanus*, and spherical pellets) which occur only toward the tops of laminae sets. This facies is similar to some features in L4 facies of Smoot (1991) and is included as crinkly stromatolitic laminae in disrupted mudstone facies of Demicco and Gierlowski-Kordesch (1986).

Disrupted Mudstones (*Fm*)

Massive mustones, composed of units up to several meters thick, are interbedded with both facies Fr and Fl_3. Colors are red with rare green or red/green mottled units. Textures are patchy with numerous mudcracks in polygonal networks (Fig. 4B in Demicco and Gierlowski-Kordesch, 1986), with curved clay slickensides and small fenestrae or subspherical cavities (clay-lined or filled with dolomite or calcite, Fig. 9C). Large mudcracks are up to 1 m deep, with sandstone, mudstone, or mixed fills (Fig. 9D). Smaller mudcracks can be up to 10–15 cm in length, also with three types of fill. A few sandstone-filled cracks contain vague internal lamination parallel to mudcrack walls, suggesting that filling occurred in several stages. Most sandstone-filled

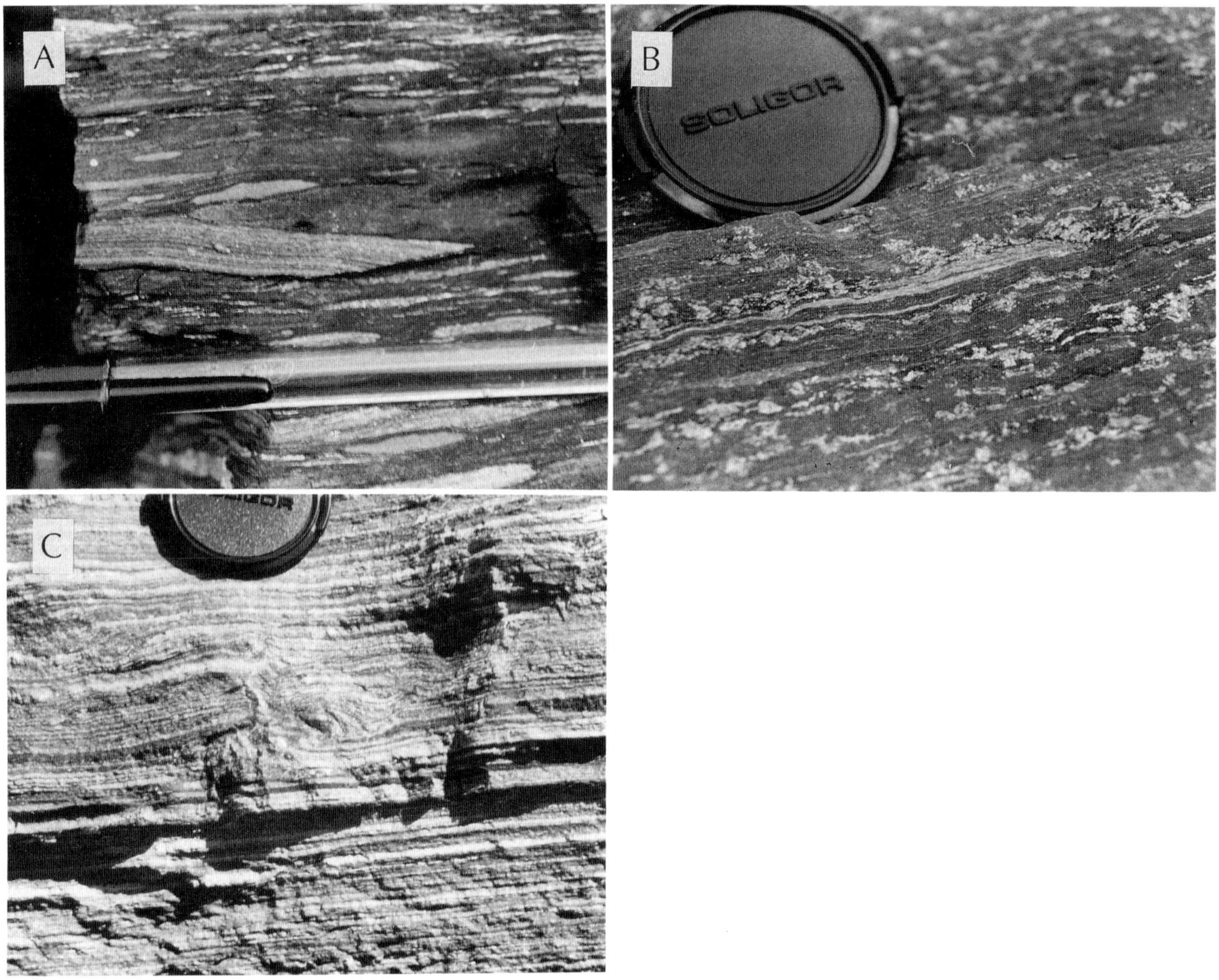

FIG. 8.—Features of the three black shale types: (A) irregularly laminated shale (pen for scale), (B) magnesite-bearing shale, and (C) finely laminated shale. Lens cap is 5 cm in diameter.

cracks have no overlying source bed of sandstone, although they are fully filled with sandstone. In the terminology of Allen (1987), these mudcracks are "filled but unbridged cracks." The upper portion of mudstone units commonly coarsens upward because of sand introduced into thoroughly disrupted crack networks. Closely-spaced subspherical holes up to 2 mm in diameter are also present within sandstone-filled mudcracks, particularly where primary features are accentuated by weathering (Fig. 9C). Mudcracks filled with mudstone are discernible on fresh outcrops or polished slabs due to faint color contrast between fill and surrounding mudstone. Dispersed calcitic/dolomitic nodules occur locally and can be up to 8 cm in diameter. Gypsum molds and calcite psuedomorphs after gypsum are present. A single specimen of *Palaeophycus* ichno sp. is present at Section K (Fig. 1). This facies is similar to some features in L4 facies of Smoot (1991).

Petrographically, this mudstone contains skeleton grains of quartz, feldspar, and mica floating in a groundmass of clay and iron oxides. Heterogenous, agglomeratic to porphyritic texture indicates a badly sorted structure. Other salient features include corroded quartz grains, ubiquitous ferruginous cementation, and distinct clay coatings and concentrations (Fig. 4C in Demicco and Gierlowski-Kordesch, 1986).

FACIES INTERPRETATION

Introduction

The East Berlin Formation in central Connecticut represents sedimentation in three continental environments: (1) alluvial plain to sandflat, (2) playa, and (3) perennial saline lake (Fig. 10). Red sandstone/siltstone facies (St, Sh, SF, and Fr) are attributed to sheetfloods on a distal braidplain to sandflat. These deposits are interstratified with disrupted mudstones (Fm) that represent paleosols on alluvial and playa muds. Playa mudflat (Fl_3) to ephemeral, shallow lacustrine facies (Fl_2) are distinguished by their lamination. Fl_3 and

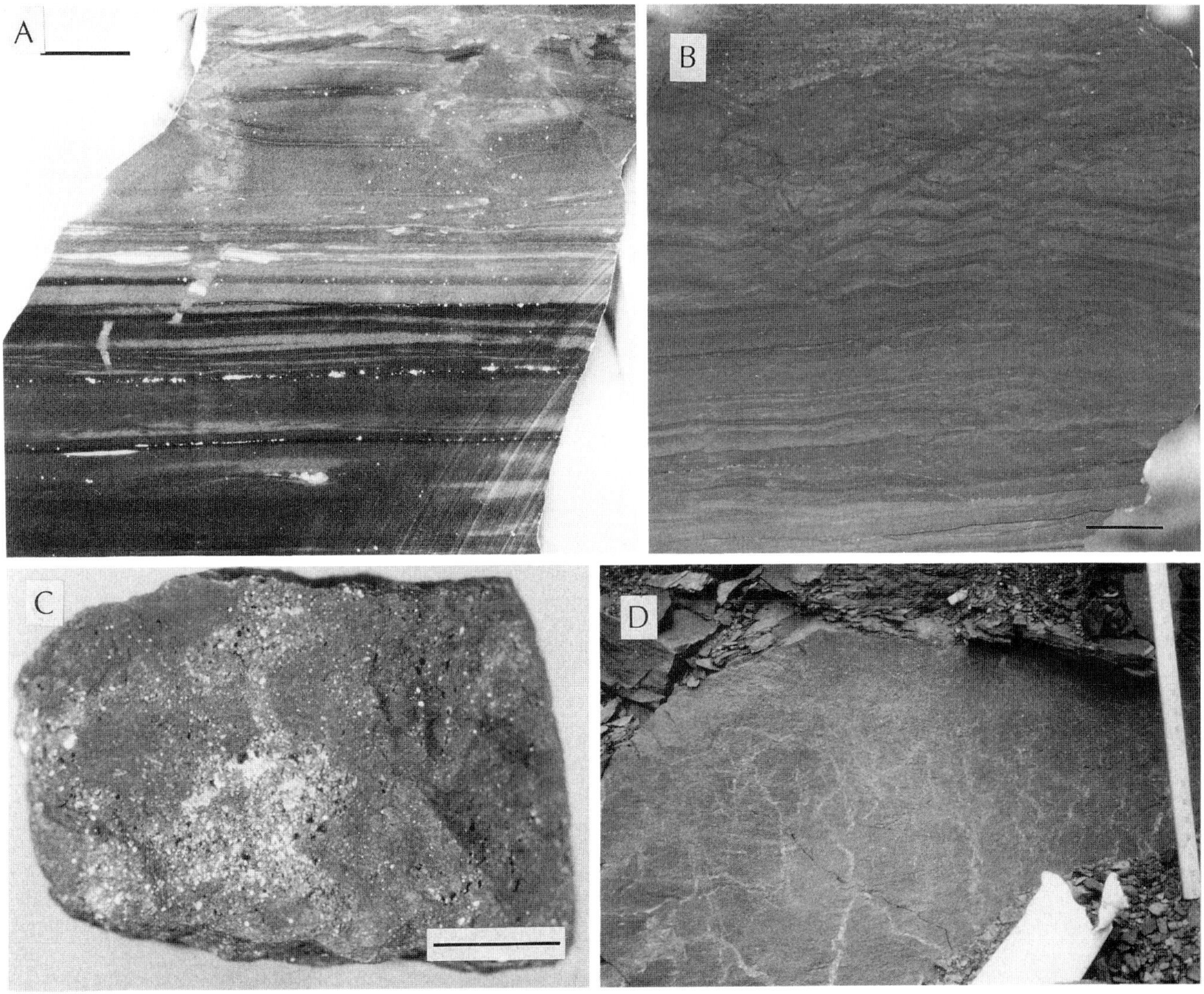

FIG. 9.—Features of mudrocks and shales. (A) Slab of stratified mudrock (Fl_2), showing lamination style and upward increase of mudcracks. White spots are pyrite crystals. Scale bar=2 cm. (B) Slab of large patch of disrupted shale (Fl_3) capped by trough cross-stratified sandstone (St), from the Type Section. Note siltstone/mudstone laminae alternations and upward disruption of lamination in Fl_3. Carbonate laminae are white. Scale bar=1 cm. (C) Hand sample of red disrupted mudstone (Fm) from Section A. Subspherical pores about 2 mm in diameter trace outline of mudcracks, in both sandstone-filled and mudstone-filled types. Some pores are lined with clay cutans. Scale bar=1 cm. (D) Disrupted mudstone facies (Fm) with sandstone-filled mudcracks up to a meter deep in Section A, interpreted as a vertisol profile. Meter stick for scale.

Fm are interstratified with the alluvial to sandflat facies. The offshore perennial saline lake facies (Fl_1) is conformably interstratified only with its surrounding shallow lake deposits (Fl_2), indicating contemporaneous deposition. Playa conditions alternated with perennial saline lake conditions throughout East Berlin time.

Alluvial Plain/Sandflat Facies Assemblage (St, Sh, SF, Fr, Fm)

Disrupted Mudstone Facies (Fm).—

Units of this facies are interpreted as alluvial, sandflat, and playa muds (Fr, SF, and Fl_3) that have undergone pedogenesis. Micromorphological features suggesting paleosol development include: (1) poor sorting, (2) corrosion of quartz grains, and (3) clay concentrations coating grains and filling small voids or cavities (cutans or clays skins of pedologists; e.g., Brewer, 1976). The paleosols of this facies are interpreted as fossil vertisols.

Vertisols are gray to brown clay-rich soils containing swelling clays that are deeply cracked and commonly display gilgai (FitzPatrick, 1983; Buol and others, 1989) They are characteristic of semi-arid to arid climates. Alternations between saturation and desiccation in vertisols promote churning (horizon displacements), as shown by lack of stratification or soil horizon development, curved and intersecting clay slickensides, wide and deep mudcracks (Yaalon and Kalmar, 1978; Wilding and Tessier, 1988; Re-

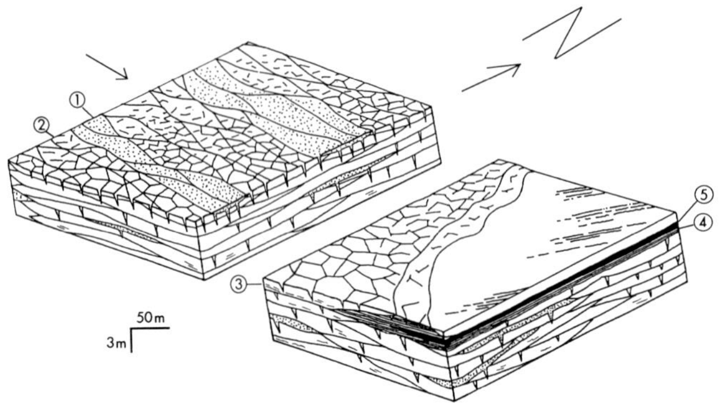

FIG. 10.—Schematic reconstruction of the alluvial plain/sandflat to lacustrine depositional environments of the East Berlin Formation. Alluvial plain/sandflat to the west with lake in the basin axis near the eastern border fault; transition between two environments is gradual and migrates back and forth depending on climatic and tectonic influences. The ephemeral alluvial plain/sandflat is represented by sheets of quartz sand (1) and mud aggregate sand (2), the latter deeply cracked by pedogenesis, forming vertisols. For clarity, bedforms are omitted from both types of sheets, braided channel network is indicated. Playa muds are represented by relict lamination and mudcracked to massive mudstones (3) which mixed with alluvial plain/sandflat deposits. Shallow to offshore lacustrine deposition is represented by the stratifed mudrock (4) and black shales (5). Top arrow indicates paleocurrent direction.

tallack, 1988; Buol and others, 1989), and granular ped development (Martins and Pfefferkorn, 1988). Carbonate concretions are common (FitzPatrick, 1983; Nettleton and Sleeman, 1985; Buol and others, 1989). Coarsening-upward sequences may result from sand introduced into the crack network from above (sandy clastic dikes in pedologic terminology; see Retallack, 1988). Similar pedogenic processes occur on playas due to repeated wetting and drying of clays (Smoot, 1981). The extent of paleosol development depends on the local sedimentation rate within the alluvial plain, sandflat, and playa areas (cf. Bown and Kraus, 1987; Kraus, 1987; Kraus and Aslan, 1993; Smith,1990).

A possible modern analog are the vertisols on the muddy floodplain of Cooper Creek and other rivers of the Lake Eyre Basin, central Australia. These rivers are mostly braided and transport mainly mud as bedload in the form of sand-sized pedogenic aggregates. The vertisols are deeply cracked, producing abundant sand-sized clay aggregates that are durable enough to withstand bedload transport during periodic floods (Rust and Nanson, 1989). Transport of sand-sized mud aggregates generates bedforms, but the aggregates are subsequently obscured by compaction and pedoturbation between floods and are barely visible at burial depths of 2 m (Rust and Nanson, 1991).

Vertisol development in the EBF is demonstrated by sandstone to mudstone-filled cracks up to 1m deep (Fig. 9D, and Fig. 7 in Rust and Nanson, 1989), cutans (Fig. 9C; and Fig. 4C in Demicco and Gierlowski-Kordesch, 1986), pedogenic textures, and carbonate nodules and layers.

The development of vertisols also depends on swelling clays (Hubble, 1984). Bulk analysis of the "red mudstones" of the EBF by April (1981a) confirmed illite and chlorite with vermiculite, mixed layer illite/smectite, and kaolinite. The predominate clays in lacustrine mudstones and shales are smectites and mixed-layer clays (e.g., illite/smectite and chlorite/smectite) with illite and some kaolinite (April, 1981b; Gottfried and Kotra, 1988). Gottfried and Kotra (1988) also noted that the clay mineralogy of the EBF is unique (i.e., its swelling clay content) compared with mudstones of three other basins of the Newark Supergroup. Wright and Robinson (1988) attributed mixed layer illite/smectite in ancient vertisols to K-fixation during wetting and drying cycles.

The abundance of cracks toward the top of paleosols resulted in total homogenization of sand and mud. Closely-spaced subspherical holes in the mudstone and within mudcracks (Fig. 9C) represent soil pores formed during rapid crack filling, shrinkage pores (Wilding and Tessier, 1988), or perhaps filled rootlets (Nettleton and Sleeman, 1985). In some examples, they are lined with clay or filled with micrite (interpreted as clay cutans (argillans) or calcans). The abundant unbridged cracks imply that either the original source bed of arkosic sand was subsequently removed, was diluted by subsequent transport of mud-aggregate sand (see Rust and Nanson, 1989, 1991 for details), or never fully developed. Arkosic sand could have been squeezed from the cracks as the surrounding clays swelled when wetted (cf. Yaalon and Kalmar, 1978; Blokhuis and others, 1990). A possible additional mechanism to produce sand-filled but unbridged cracks in pedogenic sequences is eolian transport of sand into desiccation cracks (e.g., Smith, 1990).

We interpret all these features to have developed on an alluvial plain to sandflat environment in which most of the mud was transported as bedload in sand-sized mud aggregates. Mud aggregate transport probably developed bedforms, as shown experimentally by Nanson and others (1986, their Fig. 5) and observed after a flood in the Lake Eyre Basin (Rust and Nanson, 1991). Because of density differences, flow sorted bedload into arkosic sand and mud-aggregate sand in the EBF (Fig. 10), depositing them separately or in alternate strata (Fig. 6B in Demicco and Gierlowski-Kordesch, 1986), as reported from the ephemeral floodplain of Cooper Creek (Rust and Nanson, 1989). Evidence for bedload mud aggregates in the EBF is their occurrence within sandstone-filled mudcracks and in sandstone facies (cf. Rust and Nanson, 1989), and the existence of mud-cracked sets of ripple cross-laminated siltstones.

Sandstone and Rippled Siltstone Facies (St, Sh, SF, Fr).—

These facies are interpreted as deposits of sheetfloods with dunes, plane beds, and ripples on an ephemeral alluvial plain to sandflat (Fig. 10). Rare channeling occurs in association with sheet sands or sheet mud-aggregates. Modern deposits from sheetfloods have been described by McKee and others (1967), Williams (1971), Karcz (1972), Handford (1982), and Sneh (1983). Ancient examples in a playa system include those by Tunbridge (1984), Cheadle (1986), Southgate and others (1989), and Mazullo and others (1991).

The sandstones of the EBF were originally interpreted by Hubert et al. (1976) as channel deposits of meandering streams, based on fining-upward trends. However, in a meandering system, lateral accretion deposits characterized by inclined heterolithic stratification (IHS, Thomas and others, 1987) should be common, but have not been described from the EBF. According to Collinson (1978), meanderbelt width for meandering channel sandstones 0.5–1 m thick should range from 10–30 m; thus, lateral terminations of meanderbelt sandstones should be common in the extensive roadcut exposures of the EBF, but they are absent. The facies patterns are better explained by sheetflood deposition with rare channels.

Paleocurrents show consistent unidirectional eastward trends (Fig. 4) for both sheet sandstones and rippled siltstones. This trend is attributed to a consistent eastward paleoslope, which fits the tectonic model of an eastward-tilted half-graben (Phillips, 1988; McDonald and LeTourneau, 1990; Schlische, 1993). Eastward paleoflow was also deduced from the alluvial redbeds of the EBF by Hubert and others (1976), but they interpreted the paleoslope as southeastward, based on "slump folds" in the lacustrine facies. More recently, Hubert and others (1992) have re-interpreted these paleocurrent data as eastward flowing. Other evidence also supports an eastward-dipping lacustrine paleoslope—for example, black shales of the EBF thicken eastward toward the border fault (LeTourneau, 1985; McDonald and LeTourneau, 1988).

Sheetflood sandstones occur mainly within red alluvial plain to sandflat sequences containing paleosols. Some sandstones, however, are present within gray-green sequences (e.g., 70.1–70.5 m; Fig. 2) where they commonly have a mottled gray/red coloration. These sandstones are attributed to sheetfloods that spread across the playa surface and were subsequently inundated by a rise in lake level, or which entered marginal parts of the lake. They are synonomous with facies LM3 of Smoot (1991) and analogous to the deltaic sheets or sheet deltas of Smoot (1985) and Smoot and Lowenstein (1991). The main difference between our interpretation of these East Berlin rippled siltstones and that of Smoot (1991) lies in the exact distribution of this facies. We believe that the rippled siltstones in gray-green sequences can be defined as deltaic sheets associated with lake margins, whereas those in red sequences are sheetflood deposits within an alluvial-sandflat setting (similar to Cooper Creek). Both types of rippled siltstones are sheetflood deposits resulting from flow deceleration due to high water levels, but we differentiate between deposition on a flooded lake margin (which is changing position with perhaps each flood) and on a flooded alluvial plain.

The ephemeral nature of the alluvial-sandflat system is shown by abundant desiccation features. A mudcracked upper surface is common in sequences with stacks of Facies Fr units (meter 44.8, 59.4, 105.0; Fig.2). Polygonally-cracked mud drapes occur at the tops of sandstone units and cracked mud flasers and layers within the SF facies. Some flasers may have been deposited as lenses of mud-aggregate sand, rather than as drapes that settled from suspension.

Carbonate nodules and discrete carbonate beds in Fm and Fr, respectively, contain relict ripple structures of silt and mud aggregate grains. This suggests synsedimentary to very early diagenesis before compaction. The carbonates are attributed to pedogenesis in alluvial deposits rather than formation in lake deposits, as suggested by Demicco and Gierlowski-Kordesch (1986), reducing the number of interpreted lake cycles from 15 to 10. Carbonate cementation can develop on alluvial plains and in soils during periods of low sedimentation and/or fluctuating water tables (Reeves, 1976; Goudie, 1983) and is characteristic of vertisols (FitzPatrick, 1983). Hubert and others (1992) discussed the origin of dolomitic and calcitic cement in the East Berlin sandstones, concluding that green-colored sandstones associated with lake deposits (delta sheets) underwent early cementation and red-colored alluvial sandstones underwent post-compaction, late cementation. However, relict mud to silt ripple structures preserved in neighboring red mudstones suggest that the timing of cementation needs re-evaluation.

Lacustrine Facies Assemblage (Fl_1, Fl_2, Fm)

Disrupted shale and massive mudstone facies (Fl_3, Fm).—

The green and red shales are interpreted as deposits of a dry playa mudflat (cf. Hardie and others, 1978); the interbedded massive mudstones represent pedoturbated muds (vertisols), a product of aggradational processes (Fig. 10). The disrupted shales contain thin carbonate crusts, discontinuous lamination, fine alternations of siltstone and claystone (Fig. 9B), desiccation cracks, and possible microbial lamination (similar modern features are described by Hardie and others, 1978; Bauld, 1981, 1986; Cornée and others, 1992; Renaut, 1993). Supporting evidence for lacustrine microbes in the EBF was found in the geochemical analysis of kerogen (Spiker and others, 1988; Kotra and others, 1988; Kruge and others, 1990, 1991).

These East Berlin playa deposits contain rare ripple cross-lamination and appear to be laterally transitional in form to the associated rippled siltstone facies (Fr). The gradational features between the Fr and Fl_3 facies indicate that the playa to alluvial plain/sandflat transition was gradual, and occurred on a uniformly sloping surface.

The gradual transition from alluvial plain/sandflat to lake in the EBF contains no well-defined shoreline deposits, reflecting the gently, uniform slope of the basin floor. Changes in water level caused significant shoreline migration, but lake level was never in equilibrium long enough to develop beach deposits—a typical feature of playa lakes (Smoot, 1985). The alluvial deposits that interbed with lake sediments contain the same rippled structure as those deposited exclusively on the alluvial plain/sandflat, signaling similar sheetflood deposition.

Nearly constant pedoturbation of the muds in playa, sandflat, and alluvial plain paleoenvironments also contributed to the gradual nature of the alluvial/lake transition in the EBF (cf. Bowler, 1986; Smoot and Lowenstein, 1991). The EBF is unusual because its sandflat/alluvial plain paleoenvironment contained mostly muddy sediment. Vertisol features can develop in 10 to 10^3 years (Allen and Wright, 1989). Similar soil processes on the playa mudflats and alluvial plain/sandflat could have rapidly eradicated all original structures.

Stratified mudrock facies (Fl_2).—

This facies is interpreted as ephemeral shallow lake deposits based on its continuous planar stratification, a few mm to 1.5 cm thick, that is commonly disrupted in upper portions by desiccation cracks (Fig. 9A). A Pleistocene sequence of laminated lake clays grading up to disrupted playa muds is illustrated for comparison in Demicco and Gierlowski-Kordesch (1986, their Fig. 12). The gray-green mudrock units, both in complete and incomplete lake sequences or cycles (Muddy Facies sequences A and B of Fig. 8, Demicco and Gierlowski-Kordesch, 1986) of the EBF, record the farthest extent of lacustrine influence as fringing ephemeral to shallow lake deposits. Intensive bioturbation, which would be expected in such a shallow lake environment, is absent in this facies. The upward increase in mudcrack density is attributed to lake desiccation, as suggested by rare evaporites (Hubert and others, 1976, 1992; Kruge and others, 1990, 1991).

Black shale facies (Fl_1).—

These black shales (Figs. 7, 8) represent offshore sediments in perennial lakes (Fig. 10). A saline lake interpretation is consistent with all observations. The most common black shale (Fig. 2A in Demicco and Gierlowski-Kordesch, 1986) with irregular lamination (shale type 1 in Fig. 7) is similar to "offshore" sediments of only a few meters depth in modern shallow saline lakes cored in Australia and Canada (Last, 1990; Last and De Deckker, 1990, 1992). Discontinuous carbonate laminae (calcareous or dolomitic) commonly alternate with clay laminae. Some of the carbonate laminae are cemented syndepositionally into thin hardground crusts. This style of lamination dominates the black shales of the EBF. Rare chaotic bedding and isolated groups of laminated intraclasts containing laminae discordant to stratification may represent intrasediment crusts that were broken and subsequently settled out through syndepositional gravity sliding (cf. Jackson and Galloway, 1984) in plastic, low viscosity material (e.g., overpressured muds) on slopes of one degree or more.

The lowest black shale in the East Berlin sequence (Section A at 8.8 m, Fig. 2; shale type 2 in Fig. 7) contains magnesite (Fig. 8B). Magnesite or a precursor alkaline earth carbonate may have grown interstitially in brine-saturated muds (cf. Last, 1989; Renaut, 1990; Renaut and Long, 1989; Smoot and Lowenstein, 1991). Perhaps this lake phase was relatively more saline than later lake episodes or the subsurface hydrology was changing due to active faulting. More work is needed to determine the origin of the magnesite. Modern saline lakes precipitating magnesite include Lake Beeac in Australia (Teller and Last, 1990; Last and De Deckker, 1992), Milk Lake, British Columbia (Renaut and Stead, 1991) and Freefight Lake, Saskatchewan (Last, 1993).

The finely laminated shale unit in the third black shale (Section A at 31.7 m, Fig. 2; shale type 3 in Fig. 7; Fig. 8C) resembles freshwater laminites (cf. Platt and Wright, 1991) and is interpreted as a "fresher" lake episode than all others (Kruge and others, 1990). Fossil fish in this shale are abundant, and associated coprolites show that they inhabited the lake (McDonald, 1975, 1982) and were not simply swept in by sheetfloods. The upper salinity tolerance of most fish from modern saline lakes is around 20‰, with a maximum of 145‰ (Table 10.2 in Hammer, 1986). Conchostracans, also found in this facies, today inhabit mostly freshwater; however, some varieties are subsaline to stenohaline (Hammer, 1986).

Additional Evidence for Salinity

Ichnologic evidence also supports the saline lake interpretation for the East Berlin lakes. The lack of intensive bioturbation in Facies Fl_1, Fl_2 and Fl_3 shows that conditions may not have been conducive for burrowing organisms (Gierlowski-Kordesch, 1991). Because density stratification can occur in perennial saline lakes only a few meters deep (Smoot and Lowenstein, 1991), oxygen-poor conditions are possible in shallow lake sediments. This, with the high salinity of lake waters, can exclude a burrowing invertebrate fauna. Olsen (1991) postulated that the lack of burrowers in the EBF was due either to "unknown stratification processes in freshwater lakes in the Mesozoic" or to the absence of key burrowers because they had not yet evolved, based on the body fossil record. Examples of bioturbated lacustrine sediments from early Mesozoic sequences (e.g., Van Dijk and others, 1978; Bromley and Asgaard, 1979; Pollard, 1981) as well as Late Paleozoic sequences (Aceñolaza and Buatois, 1993) suggest that burrowers had certainly evolved by then (Miller, 1984; Buatois and Mángano, 1993). The absence of burrowers in the East Berlin lakes is best explained by high salinity and anoxia in bottom sediments due to chemical stratification.

Preserved evaporitic minerals provide the best evidence for high paleosalinities. However, in some cases, the preservation potential of evaporitic minerals is low, despite closed drainage, due to subsurface dilution by groundwater (e.g., Bowler, 1986)—Lake Tyrrell is a good modern example (Teller and others, 1982; Bowler and Teller, 1986). Also, some perennial closed lakes do not precipitate evaporites over a long period of time (Smoot and Lowenstein, 1991).

The evidence for evaporitic minerals in the EBF is sparse. Facies Fl_1, Fl_2, Fl_3, Fr, and Fm contain crystal impressions, molds of crystals, and calcitic or (possible) magnesitic pseudomorphs of gypsum and some unidentifiable minerals (see also Gierlowski-Kordesch, 1985; Demicco and Gierlowski-Kordesch, 1986). Analyses of Fl_2 showed traces of analcime and gypsum and, in addition, preserved gypsum crystals, rare laminae of gypsum, and casts of halite, gypsum, and possibly glauberite crystals have been identified (Hubert and others, 1976, 1992; Kruge and others, 1990). Smoot (1991) argued that some of these minerals could be diagenetic replacements or cements. Demicco and Gierlowski-Kordesch (1986) observed impressions of radiating crystals in Fl_1 which Smoot (1991) postulated to be cumulate trona crystals (crystals that formed at the air-water interface, settled to the lake bottom, and dissolved). The magnesite-bearing shale unit of Fl_1 has the characteristic texture of post-depositional intrasediment growth. Molds of gypsum crystals and calcite pseudomorphs after gypsum are present in Fr, Fm, and Fl_3. Evidence for synsedimentary to early post-depositional evaporitic growth is shown by reworking of the crystal "balls" of calcite pseudomorphs in Fr. Carbonate replacement of gypsum is a phenomenon known from ancient and modern lacustrine deposits (see Anadón and others, 1992). Either the subsurface hydrology of the East Berlin paleoenvironment may have prevented preservation of evaporites, or the saline lake waters were relatively dilute. Such dilute saline lakes can precipitate alkaline earth carbonates such as dolomite, high-Mg calcite, protodolomite, hydromagnesite, and magnesite (Renaut, 1990; Renaut and Stead, 1991; Smoot and Lowenstein, 1991; Last and De Deckker, 1990, 1992). Robinson and Woodruff (1988) postulated saline brines to explain barite and base-metal vein deposits found in various Newark Supergroup basins.

Carbon and oxygen isotopic variations in primary lacustrine carbonates can help to determine whether the carbonates were produced in hydrologically open or closed basins (Talbot, 1990; Talbot and Kelts, 1991). The isotopic data on samples from the EBF (Pratt and others, 1986; Suchecki and others, 1988) give no clear trend on a covariance plot. The problem may lie in the measurement of both primary and secondary lacustrine carbonates in the samples; it is not clear whether the carbonates in the red mudstone samples were separated by facies, e.g. soils, playa muds, or alluvial sheets, since the isotopic signature would surely be different. Carbonates are very reactive at surface temperature and pressure and easily undergo syndepositional to very early diagenesis in shallow lacustrine and alluvial environments (Arríbas, 1986; Gierlowski-Kordesch and others, 1991). Mixed samples for isotopic analyses would give mixed signals. Detailed work on the origins of the East Berlin lake carbonates is required before testing for isotopic variations can proceed.

Authigenic corrensite within the sandstones and mudstones of the EBF (April, 1980, 1981b; Hubert and others, 1992) can be interpreted as an early diagenetic clay mineral which forms under alkaline saline conditions containing abundant magnesium (Andreason, 1992). The presence of abundant dolomitic cement and nodules of an early diagenetic to synsedimentary nature in the EBF also indicates high concentrations of Mg^{2+} in the East Berlin paleolakes, supporting a playa/saline lake complex interpretation. Kruge and others (1990, 1991) concluded that the East Berlin lakes were anoxic and moderately saline from geochemical analyses of the lacustrine organic matter.

TECTONIC AND CLIMATIC IMPLICATIONS

Figure 4 shows facies sequences for only the upper 44 m of the East Berlin Formation in other measured sections (including the old type section) in central Connecticut (locations in Fig. 1). Correlating all five measured sections (Figs. 2, 4), six complete lake sequences or cycles (muddy facies sequence A of Demicco and Gierlowski-Kordesch, 1986) are present in the upper 107 m of the EBF (lake cycle numbers 2, 4, 6, 7, 8, and 10 in Fig. 2). Olsen and others (1989) also identified six complete lake cycles in the EBF. Additionally, there are four "incomplete" lake cycles that lack black shale (muddy facies sequence B of Demicco and Gierlowski-Kordesch, 1986); these sequences (lake cycle numbers 1, 3, 5, and 9 in Fig. 2) presumably indicate shallower lakes of shorter duration, the edges of larger lakes, or subsequent deflation of the upper part of a complete lake cycle (see Bowler and Teller, 1986, for similar conclusions). This gives ten lake cycles in the upper portion of the EBF in central Connecticut. Olsen and others (1989) recorded only eight lake cycles in central Connecticut (missing two of the incomplete lake cycles). However, ten lake cycles were recorded for the Mt. Tom Section (Olsen, 1988) near Springfield, Massachusetts, in the northern part of the Hartford Basin (see Fig. 1 in Demicco and Gierlowski-Kordesch, 1986, for exact location). In the southern portion of the basin, the EBF is twice as thick, with black shales interbedded with basin-edge conglomerates (Thorpe, 1929; Olsen, 1988). The East Berlin lake records in this region are postulated to be the deepest and of longest duration due to their location near the border fault (LeTourneau, 1985; Schlische, 1993). The exact number of lake cycles is unknown due to poor exposure, but Olsen and others (1989) postulated 8–10 cycles.

Lithologic correlation over long distances in a continental rift basin is mostly speculative. The exact thicknesses of some stratigraphic units in the Hartford Basin are still imprecisely known (Table 1). Outcrop exposures across the basin are incomplete and difficult to correlate to basin-edge sedimentary sequences. Whether each East Berlin lake cycle records one large lake or several smaller ones remains to be definitively proven, but the hydrologic history of lake levels during East Berlin time appears to have been consistent across the Hartford Basin. The repetitive nature of the East Berlin lake cycles was probably caused by oscillations in surface and groundwater levels due to changes in precipitation and evaporation rates (see Bowler and Teller, 1986). Lake cycles associated with alluvial fan deposits in the lower Portland Formation of the Hartford Group have also been attributed to climatic influences (LeTourneau, 1985; LeTourneau and McDonald, 1985; Smoot, 1991).

Many authors have attributed lake cycles in all the continental basins of the Newark Supergroup to climatic influence (Van Houten, 1962; Wheeler and Textoris, 1978; Olsen, 1986; Smoot, 1991; Glenn and Kelts, 1991). In general, such small-scale sequence changes as water-level fluctuations in lake basins appear to be mostly climatically controlled, whereas large-scale sequence changes affecting drainage patterns (such as open or closed drainage) appear to be tectonically controlled (see conclusions by IGCP 219 Spanish Group on Tertiary Basins, 1990). Many observations by Olsen and others (1989), Schlische and Olsen (1990), Smoot (1991), and Hubert and others (1992) on sedimentation patterns from the Newark Supergroup basins support this theory.

Table 1 outlines the stratigraphic units and their characteristics in the Hartford Basin, as well as the paleoenvironmental interpretation of each unit and postulated paleodrainage. Border faults were active throughout the history of the basin (Krynine, 1950; Schlische, 1993) and would have profoundly influenced sedimentation patterns (LeTourneau, 1985; Schlische and Olsen, 1990; Hubert and others, 1992). Arguments for the exact nature and extent of tectonic controls and their resulting signatures in the stratigraphic record (Olsen and others, 1989; Schlische and Olsen, 1990; Smoot, 1991; Hubert and others, 1992; Schlische, 1993) are sketchy since information concerning the character of facies changes across the basin at different stratigraphic levels is lacking. Thermal maturity indices suggest that the original size of the sedimentary basin was much larger (Smoot, 1991) and a thick sequence of upper deposits have been eroded away (1500 m according to Pratt and others, 1988a). However, changes in hydrologic drainage (open vs. closed) appear to be related to tectonic activity associated with rifting in the Hartford Basin (Table 1; Hubert and others, 1992).

SUMMARY AND CONCLUSIONS

The Jurassic East Berlin Formation of the Hartford Basin contains eight facies: trough cross-bedded sandstones (St), horizontally stratified sandstones (Sh), interbedded sandstones and mudrocks (SF), ripple cross-laminated siltstones (Fr), black shales (Fl_1), stratified mudrocks (Fl_2), disrupted shales (Fl_3), and disrupted mudstones (Fm). These facies are interpreted as a continental depositional system and are divided into two facies assemblages. The sandflat/alluvial plain facies assemblage (St, Sh, SF, Fr, Fm) is composed of sheetflood deposits from a sandflat/alluvial plain complex with the muddier sediments altered into vertisolic paleosols. The lacustrine facies assemblage (Fl_1, Fl_2, Fl_3, Fm) represents a saline lake-playa system also affected by vertisol development.

Many lines of evidence point toward elevated paleosalinities for the East Berlin lakes. Ten lake cycles are identified in the EBF in central Connecticut. These repetitive cycles, alternating between playa lake and perennial saline lake conditions, are attributed to water level variations in surface and ground waters, as influenced by climatic changes. Larger scale changes in drainage patterns in the Hartford Basin are related to tectonic activity.

ACKNOWLEDGEMENTS

Heartfelt thanks are due to Nicholas G. McDonald, Peter M. LeTourneau, Patrick J. Carey, Marie E. Grace, and the Earth and Environmental Sciences Department at Wesleyan University for support in the field. We also thank Joseph P. Smoot, Nicholas G. McDonald, R. Damian Nance, V. Paul Wright, Blas L. Valero Garcés, James L. Aronson, Eleanora I. Robbins, Michael R. Rosen, Michael R. Talbot, Robin W. Renaut, and Phillip Huber for discussion. Robert V. Demicco, John S. Bridge, James T. Teller, and John F. Hubert contributed greatly to the quality of the manuscript. Financial support was provided by the Natural Science and Engineering Research Council of Canada, Institut für Paläontologie—Freie Universität Berlin, Case Western Reserve University, Ohio University, and ACS-PRF Grant No. 14000-G2 to R. V. Demicco.

DEDICATION

This work is dedicated to the memory of Brian R. Rust, who died during the final preparation of the first draft of this manuscript. As a careful and dedicated geologist and a fine human being, he will be deeply missed.

REFERENCES

ACENOLAZA, F. G., AND BUATOIS, L. A., 1993, Nonmarine perigondwanic trace fossils from the late Paleozoic of Argentina: Ichnos, v. 2, p. 183–201.

ALLEN, J. R. L., 1987, Desiccation of mud in the temperate intertidal zone: studies from the Severn Estuary and eastern England: Philosophical Transactions of the Royal Society of London, Series B, v. 315, p. 127–156.

ALLEN, J. R. L., AND WRIGHT, V. P., 1989, Paleosols in siliciclastic sequences, *in* Postgraduate Research Institute of Sedimentology Short Course Notes, no. 1: Reading, Reading University, 97 p.

ANADON, P., ROSELL, L., AND TALBOT, M. R., 1992, Carbonate replacement of lacustrine gypsum deposits in two Neogene continental basins, eastern Spain: Sedimentary Geology, v. 78, p. 201–216.

ANDREASON, M. W., 1992, Coastal siliciclastic sabkas and related evaporative environments of the Permian Yates Formation, North Ward-Estes Field, Ward County, Texas: American Association of Petroleum Geologists Bulletin, v. 76, p. 1735–1759.

APRIL, R. H., 1980, Regularly interstratified chlorite/vermiculite in contact metamorphosed red beds, Newark Group, Connecticut Valley: Clays and Clay Minerals, v. 28, p. 1–11.

APRIL, R. H., 1981a, Clay petrology of the Upper Triassic/Lower Jurassic terrestrial strata of the Newark Supergroup, Connecticut Valley, U. S. A.: Sedimentary Geology, v. 29, p. 283–307.

APRIL, R. H., 1981b, Trioctahedral smectite and interstratified chlorite/smectite in Jurassic strata of the Connecticut Valley: Clays and Clay Minerals, v. 29, p. 31–39.

ARRIBAS, M. E., 1986, Pedogenetic facies in a lacustrine sedimentation model: Paleogene of the Tertiary Tajo Basin, *in* Rodríguez Clemente, R., and Tardy, Y., eds., Geochemistry and Mineral Formation at the Earth's Surface: Barcelona, Consejo Superior de Investigaciónes Científicas, Proceedings of International Meeting on the Geochemistry of the Earth Surface and Processes of Mineral Formation, Granada, p. 339–350.

BAULD, J., 1981, Occurrence of benthic microbial mats in saline lakes: Hydrobiologia, v. 81, p. 87–111.

BAULD, J., 1986, Benthic microbial communities of Australian saline lakes, *in* De Deckker, P., and Williams, W. D., eds., Limnology in Australia: Dordrecht, Junk Publishers, p. 95–111.

BLOKHUIS, W. A., KOOISTRA, M. J., AND WILDING, L. P., 1990, Micromorphology of cracking clayey soils (vertisols), *in* Douglas, L. A., ed., Developments in Soil Science, v. 19: Amsterdam, Elsevier, p. 123–188.

BOWLER, J. M., 1986, Spatial variability and hydrologic evolution of Australian lake basins: analogue for Pleistocene hydrologic change and evaporite formation: Palaeogeography, Palaeoclimatology, Palaeoecology, v. 54, p. 21–41.

BOWLER, J. M., AND TELLER, J. T., 1986, Quaternary evaporites and hydrologic change, Lake Tyrrell, northwest Victoria: Australian Journal of Earth Sciences, v. 33, p. 43–63.

BOWN, T. M., AND KRAUS, M. J., 1987, Intergration of channel and floodplain suites, I. Developmental sequences and lateral relations of alluvial paleosols: Journal of Sedimentary Petrology, v. 57, p. 587–601.

BREWER, R., 1976, Fabric and Mineral Analysis of Soils: New York, Robert E. Krieger Publishing Co., 482 p.

BUATOIS, L. A., AND MANGANO, M. G., 1993, Ecospace utilization, paleoenvironmental trends, and the evolution of early nonmarine biotas: Geology, v. 21, p. 595–598.

BUOL, S. W., HOLE, F. D., AND MCCRACKEN, R. J., 1989, Soil Genesis and Classification: Ames, Iowa State University Press, 446 p.

CHEADLE, B. A., 1986, Alluvial-playa sedimentation in the lower Keweenawan Sibley Group, Thunder Bay District, Ontario: Canadian Journal of Earth Sciences, v. 23, p. 527–542.

COLLINSON, J. D., 1978, Vertical sequence and sand body shape in alluvial sequences, *in* Miall, A., ed., Fluvial Sedimentology: Calgary, Canadian Society of Petroleum Geologists Memoir 5, p. 577–586.

CORNET, B., AND TRAVERSE, A., 1975, Palynological contributions to the chronology and stratigraphy of the Hartford Basin in Connecticut and Massachusetts: Geoscience and Man, v. 11, p. 1–33.

CORNEE, A., DICKMAN, M., AND BUSSON, G., 1992, Laminated cyanobacterial mats in sediments of solar salt works: some sedimentological implications: Sedimentology, v. 39, p. 599–612.

CURRAY, J. R., 1956, The analysis of two-dimensional orientation data: Journal of Geology, v. 64, p. 117–131.

DEAN, W. E., 1988, Inorganic geochemistry of fine-grained sedimentary rocks and hornfels in the Newark Supergroup: United States Geological Survey Bulletin 1776, p. 79–86.

DE DECKKER, P., 1988, Biological and sedimentary facies of Australian salt lakes: Palaeogeography, Palaeoclimatology, Palaeoecology, v. 62, p. 237–270.

DEMICCO, R. V., AND GIERLOWSKI-KORDESCH, E., 1986, Facies sequences of a semi-arid closed basin: the Lower Jurassic East Berlin Formation of the Hartford Basin, New England, U.S.A.: Sedimentology, v. 33, p. 107–118.

FITZPATRICK, E. A., 1983, Soils—Their Formation, Classification, and Distribution: Hong Kong, Longman, 353 p.

GIERLOWSKI-KORDESCH, E., 1985, Sedimentology and trace fossil paleoecology of the Lower Jurassic East Berlin Formation, Hartford Basin, Connecticut and Massachusetts: Unpublished Ph.D. dissertation, Case Western Reserve University, Cleveland, 228 p.

GIERLOWSKI-KORDESCH, E., 1991, Ichnology of an ephemeral lacustrine/alluvial plain system: Jurassic East Berlin Formation, Hartford Basin, U.S.A.: Ichnos, v. 1, p. 221–232.

GIERLOWSKI-KORDESCH, E., GOMEZ FERNANDEZ, J. C., AND MELENDEZ, N., 1991, Carbonate and coal deposition in an alluvial-lacustrine setting: Lower Cretaceous (Weald) in the Iberian Range (east-central Spain), *in* Anadón, P., Cabrera, Ll., and Kelts, K., eds., Lacustrine Facies Analysis: Oxford, International Association of Sedimentologists Special Publication, v. 13, p. 109–125.

GLENN, C. R., AND KELTS, K., 1991, Sedimentary rhythms in lake deposits, *in* Einsele, G., Ricken, W., and Seilacher, A. , eds., Cycles and Events in Stratigraphy: Berlin, Springer-Verlag, p. 188–221.

GOTTFRIED, R. M., AND KOTRA, R. K., 1988, Comparative mineralogy of clay-rich strata in selected early Mesozoic basins of the eastern United States: United States Geological Survey Bulletin 1776, p. 99–102.

GOUDIE, A. S., 1983, Calcrete, *in* Goudie, A. S. and Pye, K., eds., Chemical Sediments and Geomorphology: London, Academic Press, p. 93–131.

HAMMER, U. T., 1986, Saline Lake Ecosystems of the World: Dordrecht, Junk Publishers, 616 p.

HANDFORD, C. R., 1982, Sedimentology and evaporite genesis in a Holocene continental-sabkha playa basin- Bristol Dry Lake, California: Sedimentology, v. 29, p. 239–253.

HARDIE, L. A., SMOOT, J. P., AND EUGSTER, H. P., 1978, Saline lakes and their deposits: a sedimentological approach, *in* Matter, A. and Tucker, M. E., eds., Modern and Ancient Lake Sediments: Oxford, International Association of Sedimentologists Special Publication 2, p. 7–41.

HUBBLE, G. D., 1984, The cracking clay soils: definition, distribution, nature, genesis, and use: Reviews in Rural Science, v. 5, p. 3–11.

HUBERT, J. F., REED, A. A., AND CAREY, P. J., 1976, Paleogeography of the East Berlin Formation, Newark Group, Connecticut Valley: American Journal of Science, v. 276, p. 1183–1207.

HUBERT, J. F., GILCHRIST, J. M., AND REED, A. A., 1982, Jurassic redbeds of the Connecticut Valley: (1) Brownstones of the Portland Formation; and (2) Playa-Playa Lake-Oligomictic Lake model for parts of the East Berlin Formation, Shuttle Meadow, and Portland Formations: Hartford, State Geology and Natural History Survey of Connecticut Guidebook 5, p. 103–141.

HUBERT, J. F., FESHBACH-MERINEY, P. E., AND SMITH, M. A., 1992, The Triassic-Jurassic Hartford rift basin, Connecticut and Massachusetts: evolution, sandstone diagenesis, and hydrocarbon history: American Association of Petroleum Geologists Bulletin, v. 76, p. 1710–1734.

IGCP 219 SPANISH GROUP ON TERTIARY BASINS, 1990, Tertiary lacustrine systems in Spain (I): Tectonosedimentary and paleoclimatic constraint (abs.): Nottingham, Abstracts (Papers) 13th International Sedimentological Congress, p. 235–236.

JACKSON, M. P. A., AND GALLOWAY, W. E., 1984, Structural and depositional styles of Gulf Coast Tertiary Continental Margins: Tulsa, AAPG Continuing Education Course Notes, v. 25, p. 37–45.

KARCZ, I., 1972, Sedimentary structures formed by flash floods in southern Israel: Sedimentary Geology, v. 7, p. 161–182.

KOTRA, R. K., GOTTFRIED, R. M., SPIKER, E. C., ROMANKIW, L. A., AND HATCHER, P. G., 1988, Chemical composition and thermal maturity of kerogen and phytoclasts of the Newark Supergroup in the Hartford Basin: United States Geological Survey Bulletin 1176, p. 68–74.

KRAUS, M. J., 1987, Integration of channel and floodplain suites, II. Vertical relations of alluvial paleosols: Journal of Sedimentary Petrology, v. 57, p. 602–612.

KRAUS, M. J., AND ASLAN, A., 1993, Eocene hydromorphic paleosols: significance for interpreting ancient floodplain processes: Journal of Sedimentary Petrology, v. 63, p. 453–463.

KRUGE, M. A., HUBERT, J. F., AKES, R. J., AND MERINEY, P. E., 1990, Biological markers in Lower Jurassic synrift lacustrine black shales, Hartford Basin, Connecticut, U.S.A.: Organic Geochemistry, v. 15, p. 281–289.

KRUGE, M. A., HUBERT, J. F., BENSLEY, D. F., CRELLING, J. C., AKES, R. J., AND MERINEY, P. E., 1991, Organic geochemistry of a synrift lacustrine sequence of Early Jurassic age, Hartford Basin, Connecticut, U.S.A.: Organic Geochemistry, v. 16, p. 689–701.

KRYNINE, P. D., 1950, Petrology, stratigraphy, and origin of the Triassic sedimentary rocks of Connecticut: Hartford, Connecticut State Geology and Natural History Survey Bulletin 73, 247 p.

LAST, W. M., 1989, Sedimentology of a saline playa in the northern Great Plains, Canada: Sedimentology, v. 36, p. 109–123.

LAST, W. M., 1990, Paleochemistry and paleohydrology of Ceylon Lake, a salt-dominated playa basin in the northern Great Plains, Canada: Journal of Paleolimnology, v. 4, p. 219–238.

LAST, W. M., 1993, Geolimnology of Freefight Lake: an unusual hypersaline lake in the northern Great Plains of western Canada: Sedimentology, v. 40, p. 431–448.

LAST, W. M., AND DE DECKKER, P., 1990, Modern and Holocene carbonate sedimentology of two saline volcanic maar lakes, southern Australia: Sedimentology, v. 37, p. 967–981.

LAST, W. M., AND DE DECKKER, P., 1992, Paleohydrology and paleochemistry of Lake Beeac, a saline playa in southern Australia, *in* Robarts, R. D., and Bothwell, M. L., eds., Aquatic Ecosystems in Semi-Arid Regions: Implications for Resource Management: Saskatoon, Environment Canada, National Hydrologic Research Institute, Symposium Series No. 7, p. 63–74.

LEHMAN, E. P., 1959, The bedrock geology of the Middletown Quadrangle with map: Hartford, State Geology and Natural History Survey of Connecticut Quadrangle Report 8, 40 p.

LETOURNEAU, P. M., 1985, Alluvial fan development in the Lower Jurassic Portland Formation, central Connecticut—implications for tectonics and climate: Reston, United States Geological Circular 947, p. 17–26.

LETOURNEAU, P. M., AND MCDONALD, N. G., 1985, The sedimentology, stratigraphy, and paleontology of the Lower Jurassic Portland Formation, Hartford Basin, central Connecticut: Hartford, State Geology and Natural History Survey of Connecticut Guidebook 6, p. 353–391.

LORENZ, J. C., 1987, Triassic-Jurassic Rift Basin Sedimentology—History and Methods: New York, Van Nostrand Rheinhold, 315 p.

MAZZULLO, J., MALISCE, A., AND SIEGEL, J., 1991. Facies and depositional environments of the Shattuck Sandstone on the northwest shelf of the Permian Basin: Journal of Sedimentary Petrology, v. 61, p. 940–958.

MARTINS, U. P., AND PFEFFERKORN, H. W., 1988, Genetic interpretation of a Lower Triassic paleosol complex based on soil micromorphology: Palaeogeography, Palaeoclimatology, Palaeoecology, v. 64, p. 1–14.

MCDONALD, N. G., 1975, Fossil fishes from the Newark Group of the Connecticut Valley: Unpublished M.S. Thesis, Wesleyan University, Middletown, 230 p.

MCDONALD, N. G., 1982, Paleontology of the Mesozoic rocks of the Connecticut Valley: Hartford, State Geology and Natural History Survey of Connecticut Guidebook 5, p. 143–172.

MCDONALD, N. G., 1992, Paleontology of the Early Mesozoic (Newark Supergroup) rocks of the Connecticut Valley: Northeastern Geology, v. 14, p. 185–200.

MCDONALD, N. G., AND LETOURNEAU, P. M., 1988, Paleoenvironmental reconstruction of a fluvial-deltaic lacustrine sequence, Lower Jurassic Portland Formation, Suffield, Connecticut: United States Geological Survey Bulletin 1176, p. 24–30.

MCDONALD, N. G., AND LETOURNEAU, P. M., 1990, Revised paleogeographic model for Early Jurassic deposits, Connecticut Valley: regional easterly paleoslopes and internal drainage in an asymmetrical extensional basin (abs.): Geological Society of America Abstracts with Programs, v. 22, p. 54.

McKee, E. D., Crosby, E. J., and Berryhill, H. L., Jr., 1967, Flood deposits, Bijou Creek, Colorado, June 1965: Journal of Sedimentary Petrology, v. 37, p. 829–851.

Miller, M. F., 1984, Distribution of biogenic structures in Paleozoic nonmarine and marine-margin sequences: an actualistic model: Journal of Paleontology, v. 58, p. 550–570.

Nanson, G. C., Rust, B. R., and Taylor, G., 1986, Coexistent mud braids and anastomosing channels in an arid-zone river: Cooper Creek, central Australia: Geology, v. 14, p. 175–178.

Nettleton, W. D., and Sleeman, J. R., 1985, Micromorphology of vertisols: Madison, Soil Science Society of America Publication 15, p. 165–196.

Olsen, P. E., 1986, A 40-million year lake record of Early Mesozoic orbital climatic forcing: Science, v. 234, p. 842–848.

Olsen, P. E., 1988, Continuity of strata in the Newark and Hartford Basins: United States Geological Survey Bulletin 1776, p. 6–18.

Olsen, P. E., 1991, Tectonic, climatic, and biotic modulation of lacustrine ecosystems—examples from Newark Supergroup of eastern North America, *in* Katz, B. J., ed., Lacustrine Basin Exploration: Tulsa, American Association of Petroleum Geologists Memoir 50, p. 209–224.

Olsen, P. E., Gore, P. J. W., and Schlische, R. W., 1989, Tectonic, depositional, and paleoecological history of Early Mesozoic rift basins, eastern North America: Washington, D.C., 28th International Geological Congress, Field Trip Guidebook T-351, 174 p.

Phillips, J. D., 1988, A geophysical study of the northern Hartford Basin and vicinity, Massachusetts: United States Geological Survey Bulletin 1176, p. 235–247.

Platt, N. H, and Wright, V. P., 1991, Lacustrine carbonates: facies models, facies distributions and hydrocarbon aspects, *in* Anadón, P., Cabrera, L., and Kelts, K., eds., Lacustrine Facies Analysis: Oxford, International Association of Sedimentologists Special Publication 13, p. 57–74.

Pollard, J. E., 1981, A comparison between Triassic trace fossils of Cheshire and south Germany: Palaeontology, v. 24, p. 555–588.

Pratt, L. M., Vuletich, A. K., and Daws, T. A., 1985, Geochemical and isotopic characterization of organic matter in rocks of the Newark Supergroup: United States Geological Survey Circular 946, p. 74–78.

Pratt, L. M., Vuletich, A. K., and Burruss, R. C., 1986a, Petroleum generation and migration in Lower Jurassic lacustrine sequences, Hartford Basin, Connecticut and Massachusetts: United States Geological Survey Circular 974, p. 57–58.

Pratt, L. M., Vuletich, A. K., and Shaw, C. A., 1986b, Preliminary results of organic geochemical and stable isotope analyses of Newark Supergroup rocks in the Hartford and Newark Basins, eastern U.S.: United States Geological Survey Open-File Report 86–284, p. 1–29.

Pratt, L. M., and Burruss, R. C., 1988, Evidence for petroleum generation and migration in the Hartford and Newark Basins: United States Geological Survey Bulletin 1776, p. 74–79.

Pratt, L. M., Shaw, C. A., and Burruss, R. C., 1988, Thermal histories of the Hartford and Newark Basins inferred from maturation indices of organic matter: United States Geological Survey Bulletin 1776, p. 58–62.

Reeves, C. C., Jr., 1976, Caliche: Origin, Classification, Morphology, and Uses: Lubbock, Estacado Books, 233 p.

Renaut, R. W., 1990, Recent carbonate sedimentation and brine evolution in the saline lake basins of the Cariboo Plateau, British Columbia, Canada: Hydrobiologia, v. 197, p. 67–81.

Renaut, R. W., 1993, Morphology, distribution, and preservation potential of microbial mats in the hydromagnesite-magnesite playas of the Cariboo Plateau, British Columbia, Canada: Hydrobiologia, v. 267, p. 75–98.

Renaut, R. W., and Long, P. R., 1989, Sedimentology of the saline lakes of the Cariboo Plateau, Interior British Columbia: Sedimentary Geology, v. 64, p. 239–264.

Renaut, R. W., and Stead, D., 1991, Recent magnesite-hydromagnesite sedimentation in playa basins of the Cariboo Plateau, British Columbia: British Columbia Geological Survey Branch Paper 1991–1, p. 279–288.

Retallack, G. J., 1988, Field recognition of paleosols: Boulder, Geological Society of America Special Paper 216, p. 1–20.

Robinson, G. R., Jr. and Woodruff, L. G., 1988, Characteristics of base-metal and barite vein deposits associated with rift basins, with examples from some early Mesozoic basins of eastern North America: United States Geological Survey Bulletin 1176, p. 377–390.

Rust, B. R. and Nanson, G. C., 1989, Bedload transport of mud as pedogenic aggregates in modern and ancient rivers: Sedimentology, v. 36, p. 291–306.

Rust, B. R., and Nanson, G. C., 1991, Bedload transport of mud as pedogenic aggregates in modern and ancient rivers. Reply to discussion: Sedimentology, v. 38, p. 157–160.

Sanders, J. E., 1968, Stratigraphy and primary sedimentary structures of fine-grained well-bedded strata, inferred lake deposits, Upper Triassic, central and southern Connecticut: Boulder, Geological Society of America Special Paper 106, p. 263–306.

Schlische, R. W., 1993, Anatomy and evolution of the Triassic-Jurassic continental rift system, eastern North America: Tectonics, v. 12, p. 1026–1042.

Schlische, R. W., and Olsen, P. E., 1990, Quantitative filling model for continental extensional basins with applications to Early Mesozoic rifts of eastern North America: Journal of Geology, v. 98, p. 135–155.

Seidemann, D. E., Masterson, W. D., Dowling, M. P., and Turekian, K. K., 1984, K-Ar dates and $^{40}Ar/^{39}Ar$ age spectra for Mesozoic basalt flows of the Hartford Basin, Connecticut, and the Newark Basin, New Jersey: Geological Society of America Bulletin, v. 95, p. 594–598.

Smith, R. M. H., 1990, Alluvial paleosols and pedofacies sequences in the Permian Lower Beaufort of the southwestern Karoo Basin, South Africa: Journal of Sedimentary Petrology, v. 60, p. 258–276.

Smoot, J. P., 1981, Subaerial exposure criteria in modern playa and mud cracks (abs.): American Association of Petroleum Geologists Bulletin, v. 65, p. 994.

Smoot, J. P., 1985, The closed-basin hypothesis and its use in facies analysis of the Newark Supergroup: United States Geological Survey Bulletin 1176, p. 4–10.

Smoot, J. P., 1991, Sedimentary facies and depositional environments of early Mesozoic Newark Supergroup basins, eastern North America: Palaeogeography, Palaeoclimatology, Palaeoecology, v. 84, p. 369–423.

Smoot, J. P., and Robinson, G. R, Jr., 1988, Sedimentology of stratabound base-metal occurrences in the Newark Supergroup: United States Geological Survey Bulletin 1776, p. 356–376.

Smoot, J. P., and Lowenstein, T. K., 1991, Depositional environments of non-marine evaporites, *in* Melvin, J. L., ed., Evaporites, Petroleum, and Mineral Resources: Amsterdam, Elsevier, p.189–347.

Sneh, A., 1983, Desert stream sequences in the Sinai Peninsula: Journal of Sedimentary Petrology, v. 53, p. 1271–1279.

Southgate, P. N., Lambert, I. B., Donnelly, T. H., Henry, R., Etminan, H., and Weste, G., 1989, Depositional environments and diagenesis in Lake Parakeelya: a Cambrian alkaline playa from the Officer Basin, South Australia: Sedimentology, v. 36, p. 1091–1112.

Spiker, E. C., 1985, Stable isotope characterization of organic matter in the early Mesozoic basins of the eastern United States: United States Geological Circular 946, p. 70–73.

Spiker, E. C., Kotra, R. K., Hatcher, P. G., Gottfried, R. M., Horan, M. F., and Olsen, P. E., 1988, Source of kerogen in black shales from the Hartford and Newark Basins, eastern United States: United States Geological Survey Bulletin 1776, p. 63–68.

Suchecki, R. K., Hubert, J .F., and Birney De Wet, C. C., 1988, Isotopic imprint of climate and hydrogeochemistry on terrestrial strata of the Triassic-Jurassic Hartford and Fundy rift basins: Journal of Sedimentary Petrology, v. 58, p. 801–811.

Talbot, M. R., 1990, A review of the palaeohydrological interpretation of carbon and oxygen isotopic ratios in primary lacustrine carbonates: Chemical Geology (Isotope Geoscience Section), v. 80, p. 261–279.

Talbot, M. R., and Kelts, K., 1991, Paleolimnological signatures from carbon and oxygen isotopic ratios in carbonates from organic carbon-rich lacustrine sediments, *in* Katz, B. J., ed., Lacustrine Basin Exploration: Tulsa, American Association of Petroleum Geologists Memoir 50, p. 99–135.

TELLER, J. T., BOWLER, J. M, AND MACUMBER, P. G., 1982, Modern sedimentation in Lake Tyrrell, Victoria, Australia: Journal of the Geological Society of Australia, v. 29, p. 159–175.

TELLER, J. T. AND LAST, W. M., 1990, Paleohydrological indicators in playas and salt lakes, with examples from Canada, Australia, and Africa: Palaeogeography, Palaeoclimatology, Palaeoecology, v. 76. p. 215–240.

THOMAS, R. G., SMITH, D. G., WOOD, J. M., VISSER, J., CALVERLEY-RANGE, E. A., AND KOSTER, E. H., 1987, Inclined heterolithic stratification.—terminology, description, interpretation, and significance: Sedimentary Geology, v. 53, p. 123–179.

THORPE, M. R., 1929, A new Triassic fossil field: American Journal of Science, v. 18, p. 277–300.

TUNBRIDGE, I. P., 1984, Facies model for a sandy ephemeral stream and clay playa complex: The Middle Devonian Trentishoe Formation of North Devon, U.K.: Sedimentology, v. 31, p. 697–715.

VAN DIJK, D. E., HOBDAY, D. K., AND TANKARD, A. J., 1978, Permo-Triassic lacustrine deposits in the eastern Karoo Basin, Natal, South Africa, *in* Matter, A., and Tucker, M. E., eds., Modern and Ancient Lake Sediments: Oxford, International Association of Sedimentologists Special Publication 2, p. 225–239.

VAN HOUTEN, F. B., 1962, Cyclic sedimentation and the origin of analcime-rich Upper Triassic Lockatong Formation, west-central New Jersey and adjacent Pennsylvania: American Journal of Science, v. 260, p. 561–576.

WHEELER, W. H. AND TEXTORIS, D. A., 1978, Triassic limestone and chert of playa origin in North Carolina: Journal of Sedimentary Petrology, v. 48, p. 765–776.

WILDING, L. P. AND TESSIER, D., 1988, Genesis of vertisols: shrink-swell phenomena: College Station, Soil Management Support Services Technical Monograph 18, p. 55–81.

WILLIAMS, G. E., 1971, Flood deposits of the sand-bed ephemeral streams of central Australia: Sedimentology, v. 17, p. 1–40.

WRIGHT, V. P., AND ROBINSON, D., 1988, Early Carboniferous floodplain deposits from South Wales: a case study of the controls on palaeosol development: Journal of the Geological Society of London, v. 145, p. 847–857.

YAALON, D. H., AND KALMAR, D., 1978, Dynamics of cracking and swelling of clay soils: displacement of skeletal grains, optimum depth of slickensides, and rate of intra-pedonic turbation: Earth Surface Processes and Landforms, v. 3, p. 31–42.

ZIEGLER, D. G., 1983, Hydrocarbon potential of the Newark rift system: eastern North America: Northeastern Geology, v. 5, p. 200–208.

oriented NE-SW (Lucas, 1985). As a result, the predominant compressive tectonic conditions passed gradually to an extensional setting during Middle and Upper Permian time, and volcanism showed alkaline affinities (Bixel, 1984). The basin was composed of two subbasins (Oza-Baralet and Campo de Troya-Midi d'Ossau), both asymmetrical, with preferential subsidence on their eastern sides (Lucas, 1985; Valero Garcés, 1991).

Several stratigraphic formations have been defined within the Aragón-Béarn Basin (see Lucas, 1985, and Valero Garés, 1991, for extensive references); however, the units defined by Gisbert (1987) in the Cadi Basin of the Eastern Pyrenees have been applied here. The compressive cycle comprises a thick volcanic and volcaniclastic complex dated as Upper Stephanian age (Unidad Gris, UG, and the lower member of the Unidad de Tránsito, UT), a gray carbonate and clastic series (the upper member of the UT) and a red clastic series (the Unidad Roja Inferior, URI). The extensional cycle is represented by the Unidad Roja Superior (URS), mainly composed of clastic rocks arranged in fining-upward megasequences and interpreted as alluvial fan deposits. Carbonate rocks occur within these 'red beds'. Volcanic activity is responsible for the intercalated lava flows, sills and intrusive bodies. No paleontological dating is available for the URS, but this unit is considered of Middle to Upper Permian age over the entire Pyrenean domain (Lucas, 1985).

URS Carbonate Units

The stratigraphy of the URS in the Aragón-Béarn Basin is summarized in Figure 2. The thickness of the Unidad Roja Superior in the western subbasin (>1,500 m) is than double that in the eastern one (~700 m). Nevertheless, in both subbasins, the same number of fining-upward megasequences (three) with siliciclastic and carbonate rocks have been recognized. Numerous thin calcareous pedogenic horizons are present in the URS, but we only consider the decimeter-thick units. We group these seven interbedded URS carbonate units into three categories, based on the main lithology: URS limestone, URS dolostone, and URS gypsum units.

The limestone units are the most widespread, and crop out in the western subbasin at Baralet-Arlet, the top of the 1st megasequence in the northern area (thickness about 60 m); at Bois de Sansanet and Candanchú, both located at the bottom of the 2nd megasequence and about 40 m thick; and at Aguas Tuertas (thickness about 30 m), at the top of the 2nd megasequence in the southern area. The URS gypsum unit is about 50 m thick and composed of gypsum beds associated with limestones. It appears at the top of the 1st megasequence in the northern area of the western subbasin (Acué section). Finally, two dolostone units are present in the southeast of the western subbasin (Escalé section, 25 m thick) and at the top of the 2nd megasequence in the eastern subbasin (Fuentes de Izas section, 15 m thick). Figures 2 and 3 provide stratigraphic and geographic locations, respectively, for these URS carbonate units.

METHODOLOGY

Eight outcrop sections were measured and described in detail. Carbonate lithofacies were studied petrographically

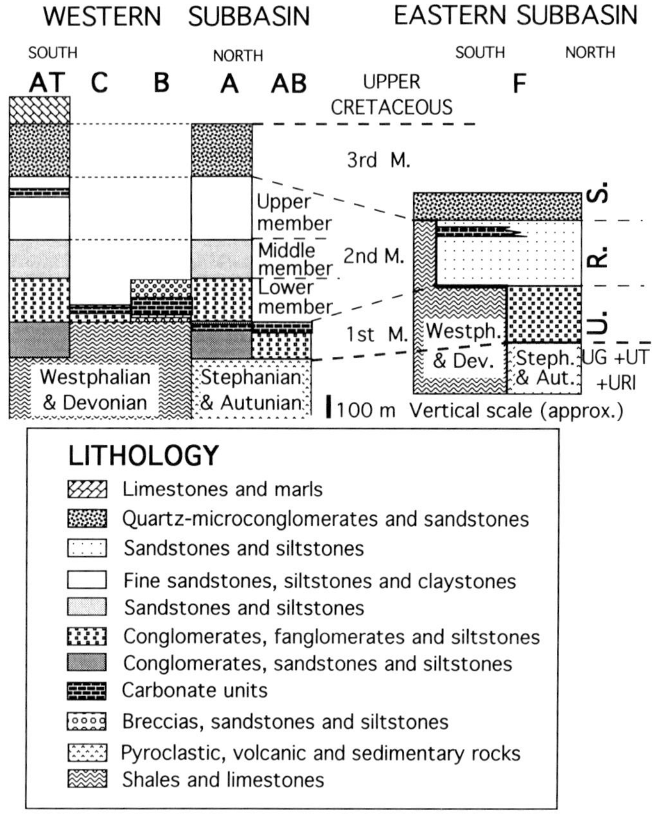

FIG. 2.—Simplified sketch showing the stratigraphy of the Permian Unidad Roja Superior (URS) in several areas of Aragón-Béarn Basin, and the stratigraphic location of the main URS carbonate units. Horizontal distances are not to scale.

in about 150 thin sections (polarized light and cathodoluminescence) and one hundred polished surfaces (binocular microscope). Common staining techniques, XRD and cathodoluminescence (CL) observations allowed the determination of carbonate mineralogy. The cathodoluminescence device was a cold cathode Technosyn 8200 MK 11. Clay mineralogy was investigated by XRD methods. A number of carbonate samples were analyzed using a JEOL model JSM-840 SEM with an energy-dispersive X-ray (EDAX) microanalyser. Total contents of organic matter and carbon analyses were calculated in two samples of black siltstones using a low temperature incinerator (LTA 302, model LFE) and chemical analysis in the Instituto del Carbón (Zaragoza), and by Pyrolysis Rock-Eval (Espitalié and others, 1977). Reflectance was measured in vitrinite particles in one organic-rich siltstone sample.

Elemental concentrations of dolostones and limestones were determined using a Perkin Elmer 40 SP inductively-coupled plasma (ICP) spectrometer. The carbonate fraction was analyzed after leaching with dilute 0.5 N HCl. Insoluble residue was about 10–15 percent (weight) and composed of quartz, clay (illite with subsidiary chlorite) and minor amounts of feldspar and hematite.

A total of 18 carbon and oxygen stable isotope analyses were performed on selected carbonates from the URS. Sampling was carried out on rock slices used for preparing

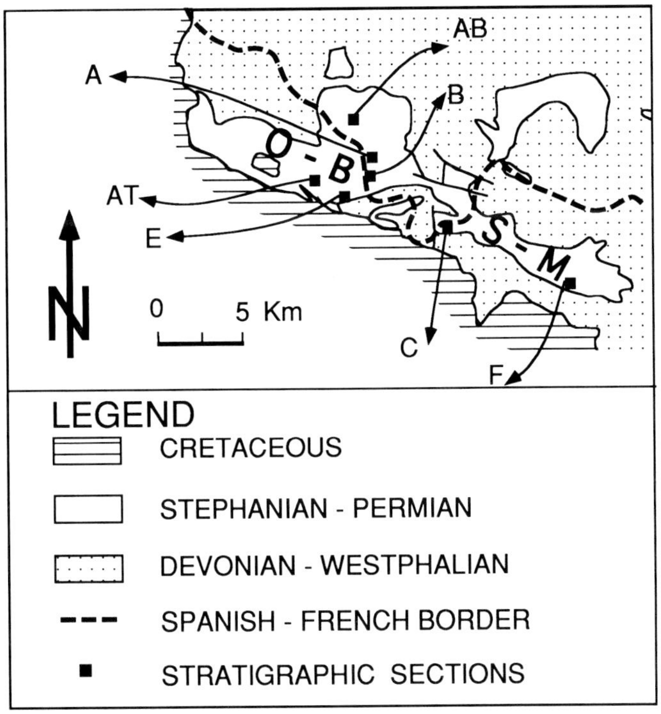

FIG. 3.—Geologic sketch of the Aragón-Béarn Basin showing the two sub-basins and the geographic location of the studied Permian Unidad Roja Superior (URS) carbonate units. Key: O-B = western subbasin (Oza—Baralet), S-M = eastern subbasin (Sallent de Gállego-Midi d'Ossau); URS Limestone units: AB = Arlet-Baralet sections, B = Bois de Sansanet section, AT = Aguas Tuertas sections, C = Candanchú section; URS gypsum unit: A = Acué section; URS dolostone units: E = Escalé section, and F = Fuentes de Izas sections.

the corresponding thin sections. Using a plain light microscope, cathodoluminescence and staining, small homogeneous areas of micrite were selected. Sample powders (about 50 mg) were drilled from the polished slides under a binocular lens using an electrical fine engraving tool. A fraction of the powder was analysed by XRD to check for dolomite. Isotopic data were obtained by standard techniques (McCrea, 1950) and the values are expressed in parts per mil with reference to the PDB.

URS LIMESTONE UNITS

Facies in the URS carbonate units have been described in detail and interpreted by Valero Garcés (1991). Here, we deal exclusively with those facies which contain evidence for saline mineral formation (pseudomorphs after gypsum and other evaporite replacement textures) or are associated with them. The studied sections are Baralet-Arlet, Bois de Sansanet, Candachú and Aguas Tuertas. The main features of these facies are summarized in Table 1.

Description

Facies.—

Gray, nodular, mottled, and brecciated limestones and red siltstones with calcite nodules are the main lithofacies in the four units (Fig. 4). The gray limestones (Facies L1) are micrites with variable quantities of bioclasts (ostracodes, molluscs, algal remains) and intraclasts. Nodular limestones (Facies L2) are defined by the presence of centimeter-sized nodules, which can represent about 80 percent by volume, within a marly or micritic matrix. In many cases, mottling is superimposed on nodular and brecciated textures (Facies L3). Brecciation and desiccation cracks (Figs. 5A, 6A), pseudomicrokarst cavities (Fig. 6B), and root structures are common features in brecciated limestones (Facies L5).

Red siltstones (Facies L4) are composed of illite, chlorite, quartz, hematite, and minor amounts of calcite and feldspar. Illite/chlorite ratio is about 6:1, and, the hematite content does not reach 5 percent. Calcite nodule sizes range from some mm to 2 cm, although in the northern outcrops, some horizons have decimeter-sized nodules and rhizocretions. Nodules usually range from 30 to 50 percent of the volume (field estimation). Some of the nodules are micritic, but most display complex fabrics (crystic, gradational, crystallaria, ferruginous, plasmic separations, internal cracks).

Non-luminescent, arrow and lens-shaped pseudomorphs after gypsum from about 100 μm up to 1 mm long, occur isolated in cm-sized clusters or lineaments (Fig. 5B). Some intraclasts also bear pseudomorphs, and these are characteristically smaller and with bright orange cathodoluminescence colors. No relicts of original gypsum or other evaporites have been detected. Lenticular crystals are now calcite, which implies replacement by dissolution and precipitation. In many cases, this early dissolution did not preserve the crystal shapes, resulting in irregular evaporite replacement textures.

Dolomite is absent in these rocks, except in the Baralet-Arlet limestone unit where yellow dolomitic mottled marls (Facies D8, Table 2) intercalate with the red siltstones.

Geochemistry.—

Calcium content in Permian limestones ranges from 35 to 39 weight percent. Magnesium and iron contents in Permian lacustrine carbonate fractions are similar and range from 2,000 to 5,000 ppm. Hundreds of ppm of strontium and manganese, and tens of ppms of barium and zinc complete the composition. Exceptionally high Ba^{2+} and Sr^{2+} values (1,000—7,000 ppm) are due to the presence of celestite-barite; EDAX determination characterizes this mineral as a barium-poor celestite. Composition values are listed in Table 3.

Interpretation

The gray limestones are interpreted as the deposits of a shallow littoral carbonate lake, based on their fossil content and sedimentary features. Energy conditions may have been locally higher (reworked limestones). The nodular and mottled limestones represent 'palustrine' lithofacies (Freytet and Plaziat, 1982; Wells, 1983; Platt, 1989). Subaerial exposure lasted long enough for pedogenic and vadose phenomena to modify primary textures. Desiccation of the unlithified sediment promoted intense cracking and brecciation, leading to brecciated limestones. Finally, red siltstones are

TABLE 1.—SALINE FACIES IN THE URS LIMESTONE UNITS (BARALET-ARLET, BOIS DE SANSANET, CANDANCHÚ AND AGUAS TUERTAS SECTIONS)

Lithofacies	Microfacies	Fossil Content	Pedogenic & Subaerial Exposure (Intensity and Features)	Early Diagenetic Cementing Phases
L1: Gray, massive to bedded limestones	Micrites & intramicrites	Scarce (ostracodes, bivalves, algae)	MEDIUM: Cavities, planar voids (mainly joint planes). Pseudomicrokarst	VS → NL → Z → BL → DL
L2: Nodular limestones	Nodular micrites & intramicrites	Very scarce	HIGH: Glaebules, crystic fabrics, mottling channels, skew, craze and curve planar voids, pseudomicrokarst pseudomorphs after gysum	VS → NL → BL
L3: Mottled limestones	Mottled micrites & intramicrites	Variable	HIGH: Mottling, clay cutans and ferrans pseudomorphs after gysum, cavities and joint planes	NL → BL
L4: Nodule–bearing red siltstones	Massive, rarely laminated siltstones chlorite <10%	None except root structures	HIGH: Glaebules, mottling, neoferrans & neocalcitans, pedotubules, sepic fabrics, skew, craze and curved planes	NL → BL
L5: Brecciated limestones	Micrites & intramicrites with evaporite replacement textures	Very scarce (ostracodes)	HIGH: Cavities, planar voids, pseudomorphs after gypsum	NL →BL → (Q & C)

Early diagenetic cement phases legend: VS, vadose silt; NL, non–luminescent calcite; Z, zoned calcite; BL, bright luminescent calcite; DL, dull luminescent calcite; Q, quartz; C, celestite-barite. Parentheses indicate rare occurrence and '→' a time—sequence.

interpreted as the deposits in the alluvial plain surrounding the lake. The rare occurrence of yellow dolomitic mottled marls in the Baralet-Arlet unit (Facies D8) is interpreted as mixed fine detrital and ash-fall deposition in the alluvial plain and subsequent early dolomitization (Valero Garcés, 1991). These lithofacies are mainly arranged in regressive sequences (Fig. 4A), but transgressive sequences (Fig. 4B) are also present.

In some places, hypersaline conditions were reached in the littoral subenvironment as result of progressive evaporation, fluctuation of the water table, and changes in groundwater salinity. Thus, gypsum was precipitated within the sediments as lenticular crystals. The mechanism involved seems to have been evaporative pumping (Hsü and Siegenthaler, 1969) because complete desiccation is not needed. Water-table fluctuations produced gypsum dissolution and calcite replacement. Lake-level changes also induced repetition of these processes and reworking of the sediments. Two or more phases can be inferred from the occurrence of pseudomorph-bearing intraclasts and pseudomorphs in matrix, or from the presence of several generations of pseudomorphs defined by different sizes and different cathodoluminescence patterns (i.e., non-luminiscent and bright luminiscent).

Some compositional differences can be drawn from this initial study between the two main carbonate lithofacies groups, palustrine and lacustrine limestones. We group within the palustrine category the carbonate samples with significant pedogenic and/or early diagenetic textures, and within the lacustrine group, those samples that lack them. Strontium, iron, manganese and magnesium contents show differential behavior in each lithofacies group (Fig. 7). Iron and magnesium contents are discriminating factors with smaller values in lacustrine than in palustrine lithofacies.

The lower Sr^{2+} content in palustrine facies, when compared to lacustrine ones, is similar to that described in marine carbonates affected by "significant" meteoric diagenetic alteration (Kinsmann, 1969; Veizer and Demovic, 1973, 1974; Veizer, 1977) and modelled by Pingitore (1978), Brand and Veizer (1980), and Pingitore and Eastman (1986). In these cases, decreasing contents of Sr^{2+} are explained by changes in mineralogy (high magnesium calcite or aragonite to low magnesium calcite) and processes of dissolution and reprecipitation of calcite in meteoric waters (low Sr/Ca ratio) and in open diagenetic systems. Neither theoretical models are completely accepted (see Tucker and Wright, 1990, and Worse and Bender, 1990, for discussion), nor are diagenetically-altered examples always poor in strontium (Brand and Veizer, 1983). In the littoral lacustrine environments, wetting and drying, alternating reducing and oxidizing conditions, and pH changes are common processes that can lead to repeated dissolution and reprecipitation of calcite. Permian meteoric waters (with a low Sr/Ca ratio) and a relatively open system may explain the loss of strontium in facies affected by repeated vadose and pedogenic processes. Strontium would be leached towards the lake where the Sr/Ca ratio would increase, leading to precipitation as celestite in lacustrine facies.

Pedogenic iron and clay plasmic separations (ferrans and argillans respectively) and early diagenetic mottling were responsible for iron enrichment in palustrine lithofacies. A magnesium increase would be related to relatively high-magnesium calcite forming during pedogenesis and early diagenesis. Manganese depletion is not so distinctive as strontium because its chemical behavior is much more influenced by redox conditions. Although manganese may be fluxed from 'palustrine' environments during reducing periods, it may precipitate during oxidizing conditions.

Using the four-fold lake classification of Wright (Tucker and Wright, 1990), these closed Permian carbonate lakes and ponds can be referred to as low gradient margins and low energy, shallow lakes. They are characterized by the dominance of 'palustrine' carbonate facies with numerous features related to lake level changes, subaerial exposure, and pedogenesis.

URS DOLOSTONE UNITS

Although thin, grey dolostone lenses may occur among thick claystone and siltstone beds, meter-thick dolomitic

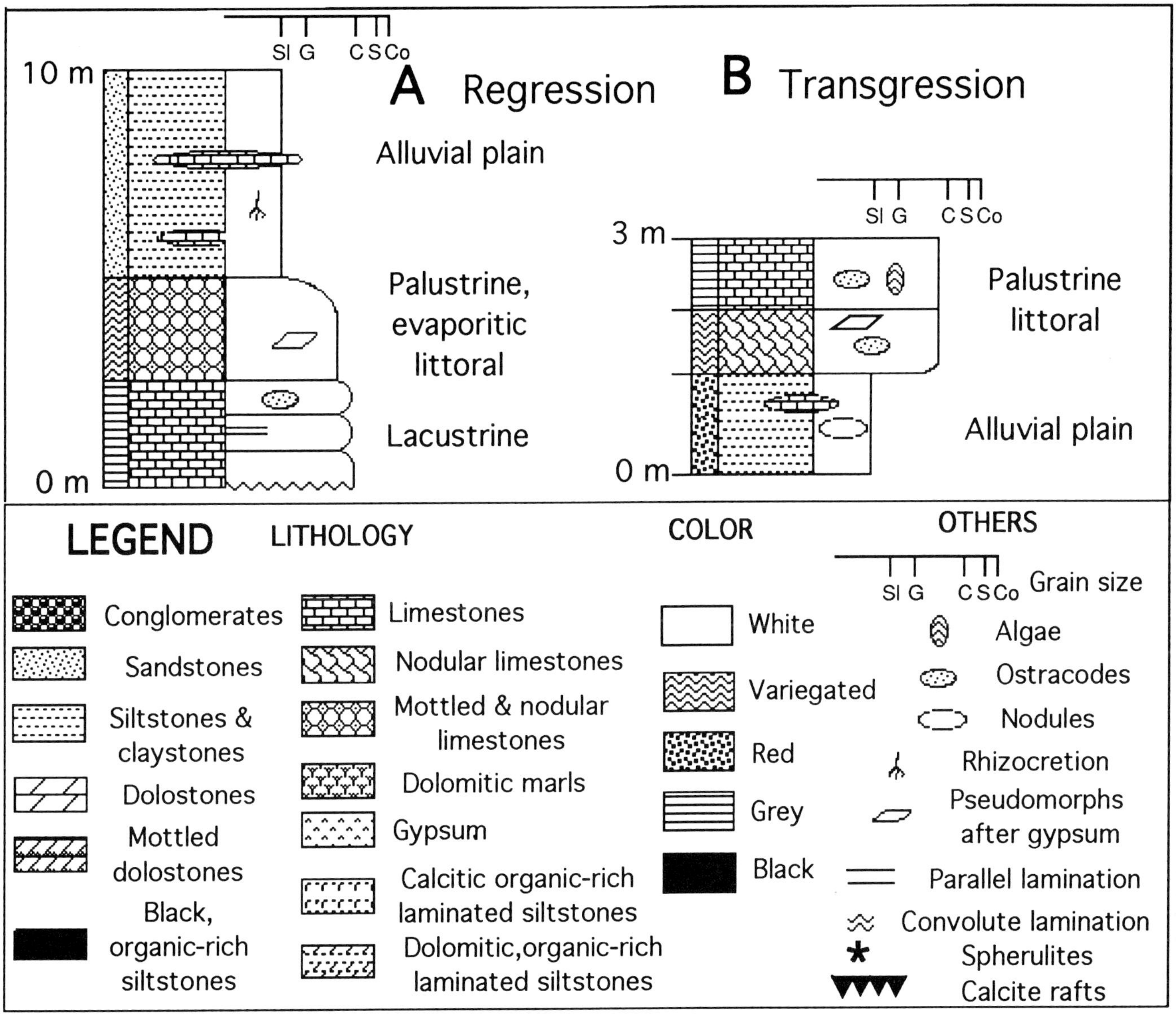

FIG. 4.—Palustrine facies associations in the URS limestone units. Regressive (A) and transgressive (B) sequences record saline conditions as pseudomorphs after gypsum in palustrine limestones. Grain size legend: Sl, siltstones, marls; G, gypsum; C, carbonates; Sd, sandstones; Co, conglomerates.

lithofacies only outcrop in the southeastern margin of the western subbasin (Escalé dolostone unit) and at the top of the 2nd megasequence in the eastern subbasin (Fuentes de Izas dolostone unit).

Escalé Dolostone Unit

Description.—

Facies.—The Escalé section is composed of gray dolostones (Facies D1), mottled dolostones (Facies D2), and red dolomitic siltstones (Facies D3) (Fig. 8). Facies description and interpretation are summarized in Table 2. In outcrop, subtabular, gray dolostone layers, 50 cm to 1 m thick, display massive, intraclastic, or diffused laminated structures. Subvertical, regularly-spaced joints may give rise to prismatic and columnar fabrics. Microfacies include dolomicrites with intraclasts, intradolomicrites, and intramicrudites. The dolomite crystals of the matrix range from several microns to several tens of microns long, but are always less than 40 μm. The grey or brown intraclasts are from 1 mm to 1 cm in size and most are micritic (Fig. 9A). Larger clasts are angular and contain mm-sized peloidal structures, intraclastic, and clotted or nodular textures. Siliciclastic components are mainly angular quartz grains, about 200-μm long. Calcite monocrystals (probably of microbial origin), several hundred μm in size, and fragmented ostracode valves (calcite) are the only biologic remains. Bioturbation textures (pedotubules, mixing fabrics) are not unusual, although biotic remains (ostracodes) are very scarce, especially when compared to the calcite-precipitating lakes. No evaporite relicts, pseudomorphs or replacement textures are present in any of the dolomitic lithofacies.

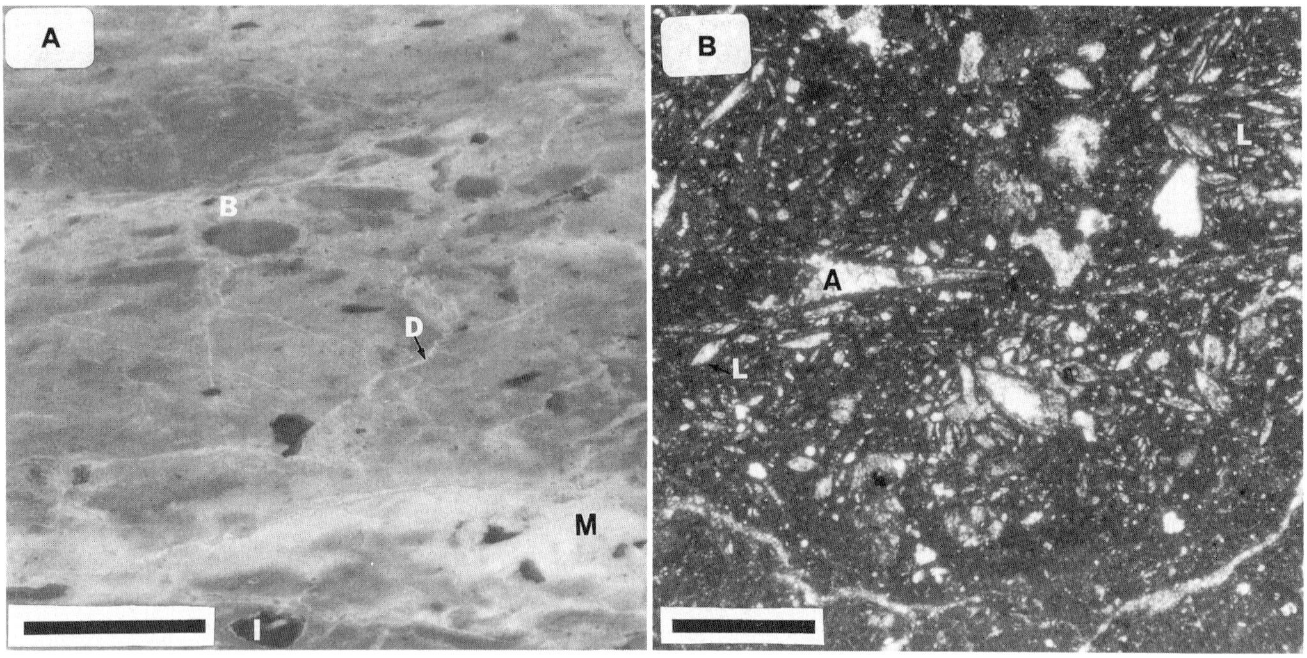

FIG. 5.—Palustrine and evaporitic limestones from the URS limestone units. (A) Polished surface of intramicrite (Facies L5). Note the network of interlinked planar voids (D), brecciation textures (B) and large spar areas (M), and the presence of both sedimentary (I) and desiccation-brecciation (B) clasts. Scale bar=7 cm. (B) Lenticular (L) and arrow-shaped (A) pseudomorphs after gypsum in nodular limestones (Facies L2). Scale bar = 0.5 mm.

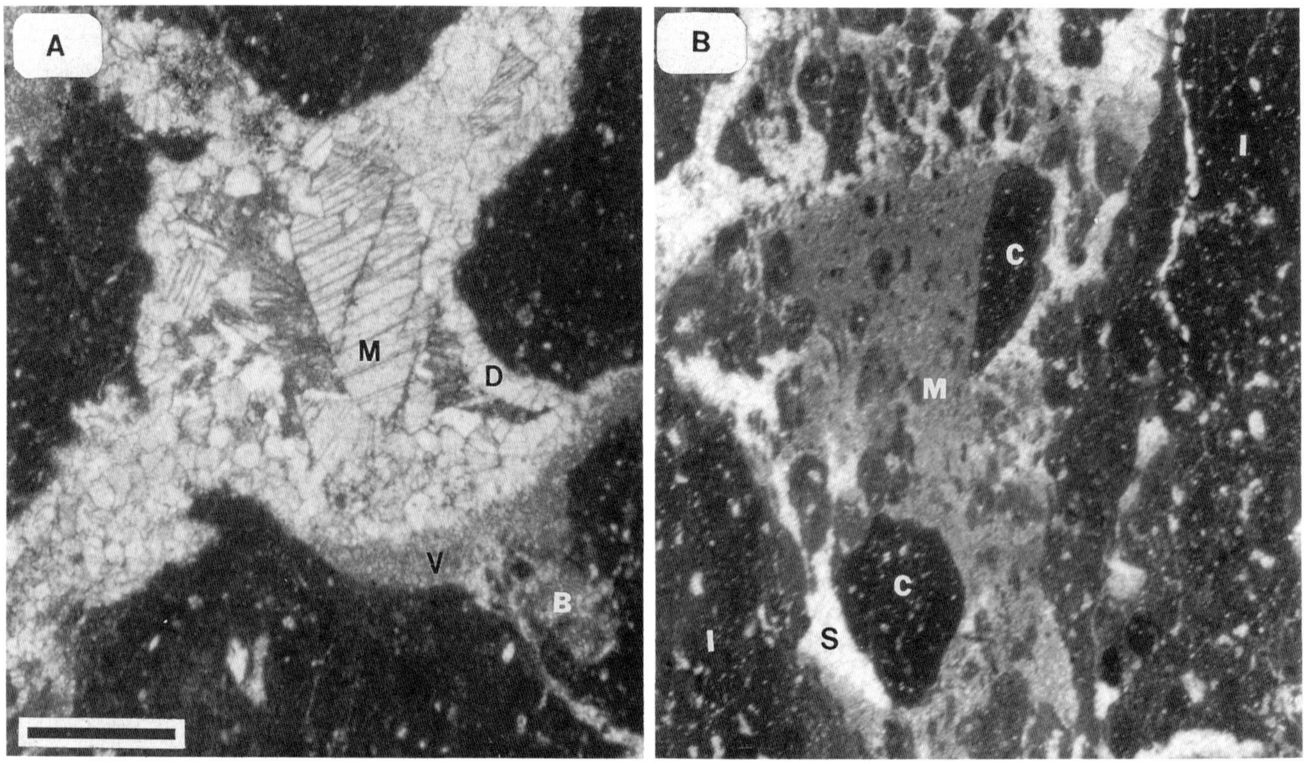

FIG. 6.—Photomicrographs of palustrine carbonates (Facies L2 and L5) from the URS limestone units showing the high secondary porosity. In (A), the cavity was filled with coarse clasts (B) and vadose crystal silt (V), and two phases of calcite cement, drusy (D) and macrosparitic (M). The smooth walls suggest dissolution played a significant role in the origin of the void. Scale bar=0.5 mm. In (B), wall brecciation originated angular clasts (C) and porosity was occluded by spar (S) and microspar (M) calcite cement. Scale bar = 0.5 mm.

TABLE 2.—FACIES IN THE URS DOLOSTONE UNITS

Lithofacies	Microfacies	Fossil Content	Pedogenesis & Subaerial Exposure (Intensity and Features)	Early Diagenetic Cementing Phases
ESCALÉ.				
D1: Grey dolostones	Dolointramicrites & Dolomicrites	Very scarce (algae remains, ostracodes).	MEDIUM: Columnar fabric planar voids, channels and calcite nodules	BL & DL
D2: Variegated (mottled) dolostones	Dolointramicrites	Very scarce (algae & ostracodes). Intense bioturbation.	HIGH: Mottling, pseudogley. Columnar fabric curved and planar voids; joints, cavities & channels. Argillans & calcitans. Pedotubules	BL & DL
D3: Red siltstones	Structureless Clotted, nodular & diffusely laminated zones	None	LOW: Horizontal joint planes	BL
FUENTES DE IZAS.				
D4: Grey marly limestones & dolostones	Marly intramicrites & marly laminated dolointramicrites	Scarce	MEDIUM in the limestones (planar voids & cavities) NONE in the dolostones.	BL & DL
D5: Black dolomitic laminated siltstones	Parallel & convolute lamination. Normal grading	None (TOC = 0.2%) (% O.M. = 1.5%).	NONE	BL
D6: Greenish—grey calcitic laminated siltstones	Parallel lamination intense hydroplastic deformation	Low: ostracods, algae, bivalves.	VERY LOW: sepic fabrics and calcite nodules	BL
D7: Massive red siltstones	Structureless Diffuse lamination	None	LOW: calcite nodules, rhizocretions, and mottling	BL & DL
BARALET—ARLET				
D8: Yellow Dolomitic mottled marls	Marly intradolomicrites with pyroclastic components.	None except root structures	HIGH: Glaebules, mottling, columnar & platty structure, pedotubules, roots & rhizocretions, chlorite sepic fabrics	(NL) → BL Silicification Chlorite authigenesis

Early diagenetic cement phases legend as in Table 1.

TABLE 3.—GEOCHEMICAL DATA OF THE CARBONATE FRACTION FROM THE URS LIMESTONE UNITS

Sample	Unit	Facies	Ca	Mg	Fe	Mn	Ba	Sr	Zn
AC-3	Acué	L	38.99	0.10	2,088	543	130	1,292	7
BR-3-A	Aguas Tuertas	P	37.44	0.36	4,661	887	115	559	17
BR-3-B	Aguas Tuertas	P	35.39	0.59	9,030	556	58	621	45
BR-4	Aguas Tuertas	P	36.77	0.26	1,725	632	6,983	6,371	16
BR-5	Aguas Tuertas	P	37.31	0.21	2,504	490	74	807	39
BR-7-A	Aguas Tuertas	P	36.82	0.22	2,181	1,043	26	560	26
BR-7-B	Aguas Tuertas	P	36.55	0.36	2,688	425	25	940	74
BR-9	Aguas Tuertas	P	36.21	0.33	5,199	402	1,692	549	30
BR-13	Aguas Tuertas	P	35.58	0.35	4,471	563	41	684	41
BR-14	Aguas Tuertas	L	36.65	0.26	2,930	446	43	618	31
BR-18	Aguas Tuertas	L	36.98	0.38	7,752	956	46	715	46
BR-19	Aguas Tuertas	L	37.06	0.21	3,332	974	68	1,038	55
BR-22	Aguas Tuertas	L	35.60	0.31	4,837	1,040	464	4,631	52
BR-23	Aguas Tuertas	L	36.27	0.26	5,270	752	1,243	638	34
BR-25	Aguas Tuertas	P	37.11	0.19	4,279	543	21	460	34
BR-26	Aguas Tuertas	L	37.56	0.18	3,498	514	22	715	22
BR-28	Aguas Tuertas	P	37.21	0.19	5,074	437	35	597	47
BS-3	Bois Sansanet	P	38.24	0.24	9,520	454	18	966	40
BS-6-A	Bois Sansanet	L	38.80	0.11	2,819	347	11	518	17
BS-6-B	Bois Sansanet	L	37.94	0.15	2,508	404	8	648	30
BS-8	Bois Sansanet	P	36.29	0.20	3,491	677	14	323	19
LA-8	Arlet-Baralet	P	37.22	0.38	3,365	478	30	394	25
LA-9-A	Arlet-Baralet	P	39.05	0.25	2,596	438	19	354	11
LA-9-B	Arlet-Baralet	P	37.65	0.37	3,179	453	47	387	19
LA-11	Arlet-Baralet	P	38.01	0.37	1,423	392	41	369	23
LA-12	Arlet-Baralet	P	37.19	0.28	1,978	297	27	336	16
E-FIZ-3	Fuentes de Izas	L	34.03	3.01	3,991	1,353	114	2,107	13
E-FIZ-5	Fuentes de Izas	L	37.58	0.43	1,583	1,945	60	1,927	6
E-FIZ-6	Fuentes de Izas	L	37.62	0.63	2,344	1,814	42	1,642	8
E-FIZ-9	Fuentes de Izas	L	31.45	0.97	13,147	3,330	154	2,117	48
E-VN-3	Fuentes de Izas	L	36.95	0.39	2,241	1,391	44	1,673	9

Ca and Mg values reported in percent and Fe, Mn, Ba, Sr, and Zn in parts per million. Samples are grouped into Palustrine (P) and Lacustrine (L) facies.

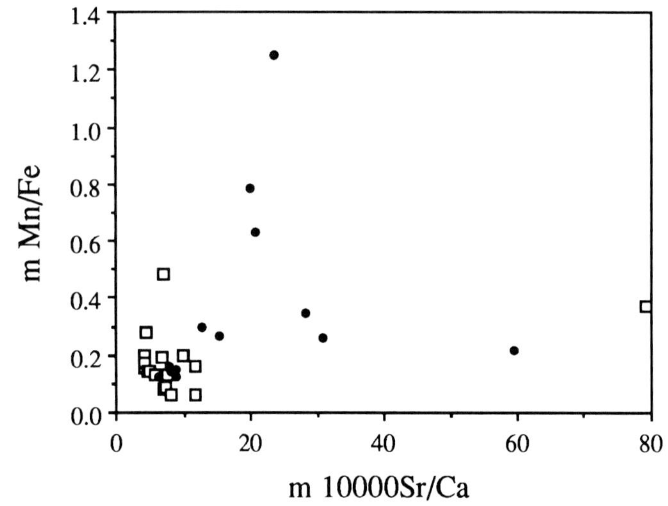

FIG. 7.—Molar Sr/Ca ratio plotted against molar Mn/Fe ratio for limestones from the URS limestone units. Palustrine samples (open squares) are depleted in Sr and Mn when compared to lacustrine ones (full circles).

The mineralogy of these dolostones (XRD, staining techniques, CL) is mainly ferroan dolomite (60 percent) with non-ferroan and ferroan calcite, quartz, feldspar and chlorites. The calcite/dolomite ratio varies from 0.1 to 0.5. Cementing phases in voids are always non-ferroan calcite; the matrix is almost exclusively composed of ferroan dolomite, but ferroan calcite is also present. Intraclasts are all ferroan dolomite.

The mottled dolostones (facies D2) occur as beds, 1 to 3 m thick, characterized by irregularly-colored patterns (purple patches a few cm across are surrounded by green and greyish-green concentric zones which gradually become red outward). In other cases, colored patterns are related to cracks (pseudogley fabrics, Retallack, 1990) or to the primary depositional intraclastic texture (grey-green haloes surrounding gray intraclasts). The clasts are of three types: (i) micritic, subrounded, 50–500 μm-sized, ferruginous or with haloes; (ii) microsparitic, up to 1 mm in size commonly non-ferruginous; and (iii) composed of subeuhedral quartz crystals, chlorites, calcite monocrystals and dolomicrite matrix. Ferruginous lutitic clasts are locally abundant. Small, angular quartz crystals and rare fragmented ostracodes are also found. Porosity is high and occurs as: (i) curved or circumgranular cracks, filled with non-ferroan calcite as irregular granular spar cement (Fig. 9B); (ii) joint planes (*sensu* Freytet and Plaziat, 1982) filled with fibrous calcite cement and oriented clay minerals (Fig. 9D); (iii) regular vertical joint planes, filled with fibrous non-ferroan calcite cement; (iv) cavities filled with dolomitic wall matrix fragments and cemented with calcite spar; and (v) channels subparallel to stratification with complex fills of dolomicrite wall fragments, calcite spar, clay minerals, and fibrous calcite. Pedotubules (300–500 μm in diameter) are filled with detrital quartz grains, calcite monocrystals and dolomicrite. Root structures are recognized by intense ferruginization, oriented clay minerals, and fibrous calcite. Mm- to cm-sized calcite nodules, displaying microsparitic and gradational granular spar textures are common (Fig.

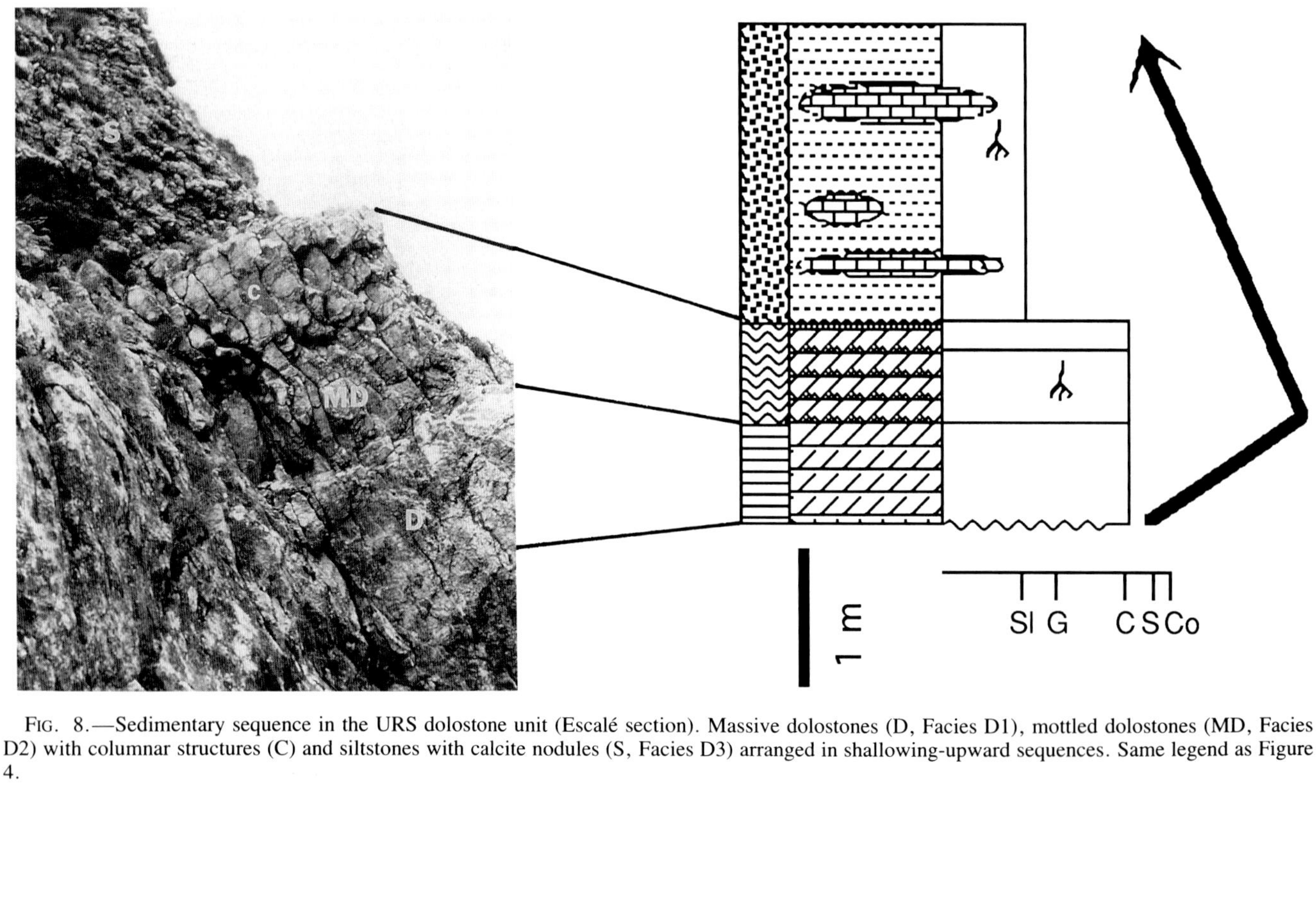

FIG. 8.—Sedimentary sequence in the URS dolostone unit (Escalé section). Massive dolostones (D, Facies D1), mottled dolostones (MD, Facies D2) with columnar structures (C) and siltstones with calcite nodules (S, Facies D3) arranged in shallowing-upward sequences. Same legend as Figure 4.

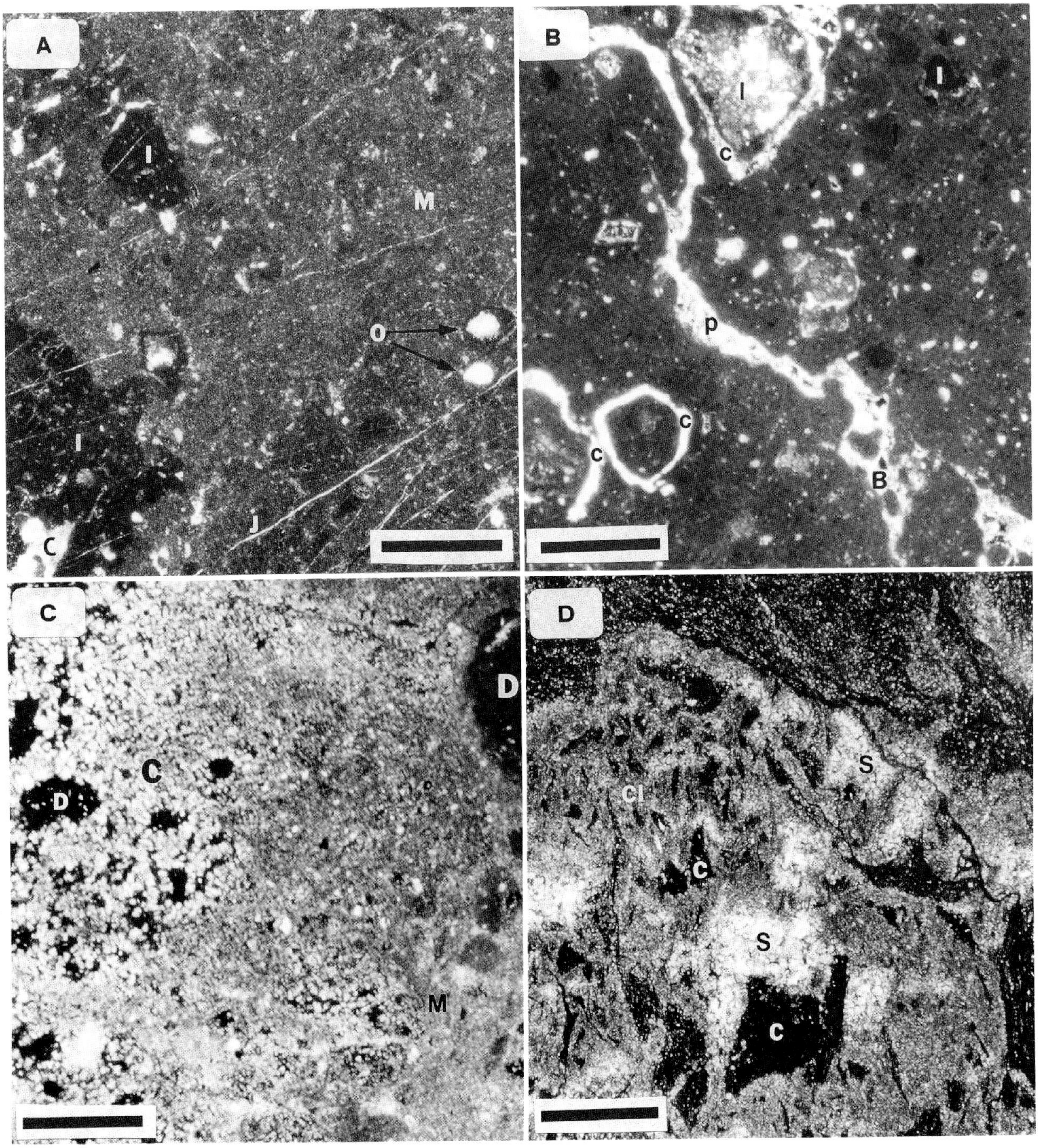

FIG. 9.—Photomicrographs of dolomitic lithofacies in the URS dolostone unit (Escalé section). (A) In the grey dolostones (Facies D1), both intraclasts (I) and matrix (M) are dolomitic in composition. Note the presence of ostracodes (O). Vadose cavities, inner voids of fossils and small cracks are filled with non-ferroan calcite (C). Fine straight cracks are tectonic in origin (J). (B) Partial and complete development of calcite spar-filled circumgranular cracks (C), planar voids (P) and wall brecciation (B) in Facies D2. C and D.- Mottled and nodular dolostones (Facies D2). In (C), a section along dolomicrite matrix (D) and calcite nodule (C) shows the gradational crystic fabrics and dolomicrite relicts within the nodule. Brecciation textures in (D) include angularly brecciated dolomicrite clasts (C), granular calcite spar mosaics (S) and slightly oriented clay minerals (Cl). Scale bars = 0.5 mm.

9C). Cathodoluminescence reveals that luminiscent (bright and dull) calcite is the only carbonate mineral in these cements. Quartz cements fill some late fine cracks and some irregular cavities and are probably tectonic in origin.

The red siltstones (Facies D3) occur as units up to 1 m thick. These are virtually structureless, both in outcrop and in thin section, although in places clotted, nodular or diffuse laminated zones are present. The mineralogy is non-ferroan calcite, ferroan dolomite, clay minerals (chlorites and illites), quartz, hematite and zircon. The calcite/dolomite ratio varies between 0.5 and 1. Clasts composed of angular quartz grains, chlorites, and calcite microspar also occur. Compared to the dolostones, the marls lack well-defined cracks. Horizontal joint planes are well developed and are filled with calcite and clay minerals.

Geochemistry.—Results are shown in Table 4. Great differences in the Ca^{2+} and Mg^{2+} contents are clear when compared with URS limestone units, but other significant features characterize these dolostones. The Mn^{2+} and Zr^{2+} values are similar to the limestone units (about 600 ppm and 300 ppm respectively), but Fe^{2+}, Ba^{2+} and Sr^{2+} values are higher than those of the limestones (Figs. 10A, 10C). These Fe^{2+} concentrations (0.7–1.2 percent) are also higher than those of ancient marine, early diagenetic dolostones (Shukla, 1988), recent lacustrine dolomites from South Australia (about 1000 ppm, see Botz and von der Borch, 1984 for references) and even slightly higher than iron-rich Jurassic lagoon dolostone values (average 5,750 ppm, Andrews and others, 1987). Staining, cathodoluminescence, and a positive association between Fe^{2+} values and Mg/Ca ratios (Fig 10B), all suggest that Fe^{2+} is preferably incorporated in the dolomite lattice. The average barium content in dolomites (about 300 ppm) is ten times higher than in the limestones, and we also consider that this results from Ba^{2+} incorporation into dolomite lattices, since barite-celestite minerals have not been detected (XRD, microscope) in the dolomitic lithofacies. The strontium content of these Permian dolostones is also higher (about 2,000 ppm) than content in the limestones. The Mn^{2+} values are lower than early diagenetic dolostones (Andrews and others, 1987; Shukla, 1988) and comparable with limestone samples from the URS limestone units.

Interpretation.—

The intraclastic gray dolostones (Facies D1) are interpreted to result from sediment reworking in a littoral lacustrine environment. However, the absence of any kind of lamination, and an intraclast microfacies similar to its matrix suggest that energy conditions were not very high. This implies only short distances of sediment transport.

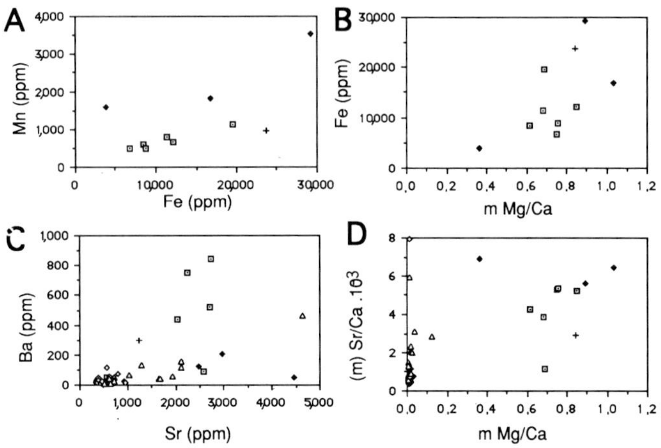

FIG. 10.—Chemical composition of dolostones. (A) Mn content plotted against Fe content. There is a positive relationship between increasing Fe and Mn content suggesting that dolomites are enriched in both cations. (B) Molar Mg/Ca ratio plotted against Fe content. There is a positive relationship between increasing Fe and Mg contents (i.e., dolomite content) which agrees with the ferroan composition of these dolomites. (C) Ba content plotted against Sr content for the dolostones and some Permian limestones. Dolostones are enriched in both cations when compared with limestones. Anomalous very high—Sr limestones are due to contamination by celestite. (D) Molar Mg/Ca ratio plotted against molar Sr/Ca ratio for dolostones and limestones. There is a positive relationship between increasing Sr and Mg content, due to the fact that dolomites are more Sr-rich. Legend: squares—dolostones from Escalé section; open diamonds—dolostones from Fuentes de Izas section; full diamonds -'palustrine' limestones from URS limestone units; open triangles—lacustrine limestones from URS limestone units; cross—a dolostone sample from Baralet-Arlet section.

Mottled dolostones (Facies D2) display many pedogenic features (Brewer, 1964; Freytet and Plaziat, 1982) including root structures, fragmented pedotubules, argilan fibrous calcitans, channel neocalcitans, and sepic and bisepic argilan fabrics. Although mottling may be an early diagenetic groundwater process (Wright and others, 1992), in these sediments it is closely related to pedogenic processes. These lithofacies are interpreted to be the result of pedogenic modification of the gray dolomite facies during the frequent

TABLE 4.—GEOCHEMICAL DATA OF THE CARBONATE FRACTION FROM THE URS DOLOSTONE UNITS

Sample	Ca	Mg	Fe	Mn	Ba	Sr	Zn	Facies
ES-1	24.29	9.93	11,461	790	435	2,040	21	D2
ES-3	23.15	10.49	8,822	497	524	2,713	27	D1
ES-6	22.18	9.14	19,612	1,110	53	574	42	D2
ES-7	24.19	8.96	8,484	603	756	2,240	26	D1
ES-8	23.58	10.63	6,754	490	840	2,739	17	D1
ES-10	22.45	11.44	12,133	647	92	2,579	24	D2
FIZ-6	37.62	.63	2,344	1,814	42	1,642	8	D4
FIZ-7	20.06	10.73	29,199	3,547	125	2,466	125	D5
FIZ-9	31.45	.97	13,147	3,330	154	2,117	45	D6
VN-2	20.85	12.96	16,844	1,813	210	2,955	28	D4
VN-3	36.95	.39	2,241	1,391	44	1,673	9	D4
FIZ-3	34.03	3.01	3,991	1,353	114	2,107	13	D4
FIZ-4	29.71	6.53	3,903	1,574	53	4,470	28	D4
FIZ-5	37.58	.43	1,583	1,945	60	1,927	6	D4
LA-6	19.34	9.76	23,761	954	298	1,225	118	D8

Ca and Mg values reported in percent and Fe, Mn, Ba, Sr and Zn in parts per million. The URS dolostone units are Escalé (ES samples), and Fuentes de Izas (FIZ and VN samples), and also is included a dolomitic sample from the Baralet—Arlet limestone unit (LA sample).

periods of subaerial exposure. They represent the 'palustrine' subenvironments of the lake margin.

The red siltstones (Facies D3) are interpreted as distal alluvial sediments that were deposited from suspension in alluvial plain environments. Some horizons can be considered poorly developed paleosols from their pedogenic features. The presence of zircons, extensive authigenic chlorite, and some clasts that could represent pyroclasts or lapilli, suggest the presence of volcanic ash falls.

These three lithofacies (D1, D2, and D3) occur together in regressive sequences about 4 m thick. The basal unit is composed of gray dolostones (0.5–1 m) which grade upward into mottled dolostones (0.5–2 m). Finally, red siltstones with units 1–3 m thick cap the sequence (Fig. 8). These regressive sequences are interpreted to represent cyclic processes of lake transgression (gray, reworked dolostones) and regression (subaerial exposure and pedogenic 'palustrine' modification) which finally lead to progradation of alluvial plain sediments over the lake margins. Repetition of these sequences suggests that the depositional setting was stable enough to maintain a lacustrine environment with conditions favorable for dolomite formation for periods longer than those of occasional small ephemeral ponds located in alluvial plains.

The dominance of dolomite is the distinctive feature of the Escalé section, being characterized mainly by high Fe^{2+} and Sr^{2+} contents. Reducing conditions seem to be necessary to explain the Fe^{2+} incorporation into the dolomite lattice, but there is no sedimentologic or petrologic evidence for anoxic conditions. "Palustrine" (mottled) dolostones (Facies D2) show higher Fe^{2+} (>1 percent) and Mn^{2+} (650–1,000 ppm) concentrations and slightly lower Mg^{2+} and Sr^{2+} concentrations than the gray dolostones (Facies D1) (Table 4); these Fe^{2+} and Sr^{2+} concentrations are similar to those documented in the palustrine limestones and are interpreted to result from pedogenesis.

The strontium content of continental carbonates is generally higher than that of marine carbonates—thousands of ppm, compared with hundreds of ppm, respectively (Shukla, 1988). In recent marine dolostones, however, values as high as 1,000 ppm have been documented and, exceptionally, more than 3,000 ppm (Land, 1980). In the Coorong lakes of South Australia, Sr^{2+} contents range from 0.3 to 1.85 percent and high concentrations are not always associated with aragonitic sediments, but also with high magnesium calcite and dolomite muds (Botz and von der Borch, 1984). Although this Sr^{2+} may have been related to small quantities of celestite precipitated directly from lake waters, Botz and von der Borch (1984) did not find celestite in most of their samples and proposed diagenetic replacement of aragonite as the major source. The significant relationship between Sr/Ca and Mg/Ca ratios (Fig. 10D) suggests that Sr^{2+} has been incorporated into the dolomite lattice. High Sr^{2+} contents and a significant relationship between Sr/Ca and Mg/Ca ratios are unusual because Sr^{2+} substitutes mostly for Ca^{2+} both in calcite and dolomite lattices, and thus is usually depleted in dolomite relative to calcite (Kretz, 1982). As far as it is known in seawater conditions, these relationships also argue against an aragonite precursor because dissolution of aragonite releases more Sr^{2+} and less Mg^{2+} than calcite and leads to anthipathetic correlation between Sr/Ca and Mg/Ca in the precipitated dolomite (Saas and Katz, 1982; Andrews and others, 1987).

Although there is much discussion about the distribution coefficient for Sr^{2+} dolomite, it is clear that enrichment in Sr^{2+} can only be achieved if dolomite precipitates from fluids with a high Sr/Ca ratio (Tucker and Wright, 1990). A similar Sr^{2+} enrichment has been documented in early diagenetic dolostones from a low salinity Jurassic lagoon by Andrews and others (1987) and in limestones from Holocene (Gasse and others, 1987) and Tertiary (Rothe and others, 1974) lacustrine closed basins. Sr-rich water inflows and an increase in the Mg/Ca and Sr/Ca ratios due to evaporation during dry periods have been invoked to explain these Sr-rich dolomites.

An ephemeral, hydrologically closed, saline lake is envisaged as model for this URS unit. The dominance of dolomite and its chemical composition support a special hydrochemistry (high Mg/Ca and Sr/Ca ratio) for the lake waters, and the sedimentologic and petrographic evidence favor primary precipitation as a very likely process of dolomite formation (see Discussion below).

Fuentes de Izas Dolostone Unit

Description—

Facies.—The 15-m thick carbonate unit is composed of grey limestones and dolostones (Facies D4), black dolomitic laminated siltstones (Facies D5), green-gray laminated siltstones (Facies D6) and massive red siltstones (Facies D7) developed over coarse conglomerates and sandstones (Fig. 11). The only gray carbonate bed varies from 3 m to 1.7 m thick in the two measured sections (Fuentes de Izas),

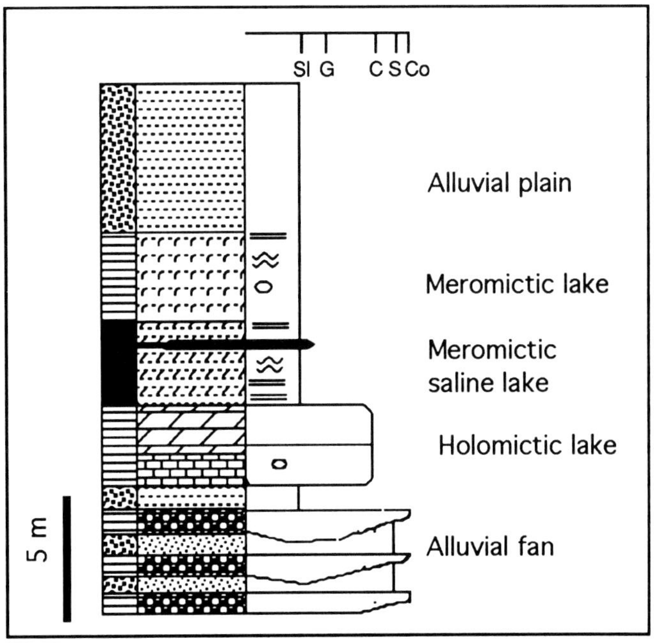

FIG. 11.—Lithofacies and sedimentological interpretation of the Fuentes de Izas dolostone unit (eastern subbasin). Same legend as Figure 4.

which are separated by some hundreds of meters. The four lithofacies are described in Table 2.

In outcrop, diffuse lamination can be recognized in the gray carbonate bed (Facies D4), but massive and intraclastic facies are also common. The basal part is a gray massive micritic limestone with a high content of terrigenous material (mainly quartz). Fenestral porosity (interconnected subhorizontal planar cavities) is common. The upper half of the bed is a dark gray, laminated, dolomitic marl. The microfacies are made up of a marly matrix (80 percent dolomicrite, 20 percent clay minerals) and minor quantities of small (less than 100 μm) angular quartz grains and lutitic, mm-sized angular extraclasts. The mineralogy is mainly ferroan dolomite (>50 percent) quartz (about 35 percent) and feldspar, ferroan calcite and clay minerals (illite and cholorite).

This gray carbonate layer is overlain by a bed about 1 m thick of black dolomitic laminated siltstones (Facies D5). In the basal part, some angular and irregular, 1 cm-sized clasts made of dolomitic marls and claystones occur. Most of the claystones clasts display oriented clay minerals. In the middle part, where the deposit is silty, the lamination consists of alternating gray and black layers, some of which are lenticular. Closer inspection reveals that most of these laminae are irregularly graded units, from 0.5–1 mm thick, composed of a basal quartz layer that grades abruptly or gradually into a fine-grained, clayey, upper part. Thicker laminae, some up to a few mm thick, incorporate dolomicrite intraclasts. Pyrite crystals occur as 1-mm sized clusters or as stringers about 6 mm long. The upper part is mainly composed of highly disrupted, laminated, fine-grained material with a few small clasts (mm sized). No fossils are present in the dolomitic siltstones. Neither porosity (cracks, cavitites) nor pedogenic or vadose cements have been detected. The mineralogy of these dolomitic siltstones comprises: quartz and chlorites, (>50 percent), ferroan dolomite (>30 percent), calcite, feldspars and pyrite. Calcite increases in abundance toward the top of the bed, while dolomite disappears.

These dolomitic laminated siltstones (Facies D5), especially the fine-grained laminae, are relatively rich in organic matter, and a carbonaceous layer, a few cm thick, intercalated at the top of the bed. Organic matter measurements have been performed in one sample from this layer (FIZ-8) resulting in 1.16 percent carbon (low temperature incinerator). Elemental chemical analysis characterizes this sample as 1.85 percent carbon, 0.40 percent hydrogen, and below detection limits on sulfur. These figures are relatively high when compared to organic-carbon content in another sample from gray-black siltstones intercalated in the URS siliciclastic red–bed deposits (0.92 percent). Pyrolysis Rock-Eval for the Facies D5 sample yielded a total organic content of 0.19 percent, too poor in organic carbon to be considered a hydrocarbon source. Due to this low organic carbon content, the kerogen type could not be determined. Maceral content is also poor, and so only four reflectance measurements were performed on some vitrinite particles. They ranged from 2.24 to 3.45 (average 2.90), indicating a post-mature stage of organic matter evolution (anthracite rank).

Greenish-gray laminated calcitic siltstones and claystones (Facies D6) overlie the dolomitic siltstones. In some places, oriented clay minerals are present. Some fossils (algal remains, fragmented ostracodes and bivalves) occur. Lamination is similar to dolomitic beds and hydroplastic deformation is intense. The rare cavities, planar voids, and incipient brecciation cracks that are present are filled with luminiscent calcite spar. Mineralogic transition between lithofacies D5 and D6 is gradual. Dolomite is virtually absent in Facies D6, which is mainly made up of clay minerals (chlorite, illite), quartz and calcite.

The whole laminated unit is a fining-upward sequence which starts with cm-scale laminae containing platy-shaped clasts and quartz grains, and ends with mm-thick, fine-grained laminae. The Fuentes de Izas section ends with a unit of red massive siltstones and claystones (Facies D6) several tens of meters thick.

Geochemistry.—The chemical composition of these dolostones is characterized by higher Mn^{2+} concentrations than in western subbasin samples (1,500 ppm and 200–1,100 ppm respectively, Table 3 and 4; Fig. 10). The associated limestones are also distinguished by their extremely high Mn^{2+} contents (1,300–2,000 ppm versus some hundreds ppm in the western samples), significantly higher Sr^{2+} content (1,000—2,000 ppm versus some hundreds in the western samples), and slightly higher Mg^{2+}content (Table 3 and Fig. 7). Surprisingly, Mn^{2+}, Sr^{2+} and Ba^{2+} values remain unchanged in the laminated layers, independent of the main carbonate lithology. Only Fe^{2+} and Zn^{2+} are closely related to dolomite.

Interpretation.—

The intramicritic limestones (Facies D4) are interpreted as the littoral deposits of a shallow lake with relatively high gradient margins. The presence of intraclasts, lamination, and terrigenous detritus points to significant alluvial and wave influence. Pedogenic features are scarce, but incipient desiccation cracks and vadose dissolutional cavities demonstrate fluctuations of lake level and the shallow nature of these deposits. The absence of the typical palustrine facies, so common in the URS limestone units, indicates a relatively high gradient along the lake margins where lake level fluctuations did not produce extensive subaerial mudflats. Carbonate precipitation was greatly influenced by progressive evaporative conditions. Mg-calcite may have been the primary precipitate. The chemical composition of these limestones (especially the higher Mg^{2+} and Sr^{2+} concentrations than in the western subbasin limestones) points to high Mg/Ca and Sr/Ca ratios in the water from which they were precipitated. The gradation from intramicritic limestones to clast-bearing laminated siltstones represents a period of high energy and alluvial influence, which can be related to lake expansion due to alluvial inflow.

The preservation of lamination, dark color, relatively high organic-matter content and pyrite authigenesis all favor deposition in anoxic, non-agitated waters for the dolomitic laminated siltstones. The thin organic layers in the Green River Formation have been attributed to settling of planktonic algal blooms or organic oozes (Eugster and Hardie, 1975; Desborough, 1978) or thin cohesive algal mats (Smoot,

1983). Both origins are possible in the URS, although the carbonaceous layers resemble more an algal mat developed in the littoral zone of the lake. The decreasing-energy events recorded by some laminae may be related to small-scale, alluvial- or lacustrine-generated currents (Platt, 1989). Some clay fabrics in the uppermost zone of this laminated unit indicate incipient pedogenic illuviation.

Soft sediment deformation of the laminae from Facies D5 and D6 is especially prevalent in some levels, but almost absent in others. Several hypotheses have been proposed to explain similar features in lacustrine sediments. Some cannot be applied to Permian lithofacies. For example, there is no evidence for syneresis and typical bench or slope facies. Lateral extent and thickness indicate that the lake was a small pond, although the marginal slope seems to be higher than in calcite-precipitating lakes from the western subbasin. Thus, chemical dewatering, rapid deposition of sediments on an unstable substratum, and slumping can be eliminated. Alternatives are seismicity (Hesselbo and Trewin, 1984), subaerial exposure (Assereto and Kendall, 1971), and water movements through laminated deposits (Boyer, 1981). Volcanism was intense during the 1st and, especially, the 2nd megasequence of the URS in the eastern subbasin (Lucas, 1985; Valero Garcés, 1991). Lava flows, sills and two major intrusive bodies demonstrate major tectonic instability that might have been recorded in the only lacustrine episode of this subbasin. Although pedogenic illuviation may have contributed to clay mineral orientation, these processes are inadequate to explain intense deformation. Water movements through laminated deposits may be responsible for this deformation. The mechanism could be similar to that proposed by Boyer (1981) for Tertiary laminated limestones where osmotic movement of fresh lake water through organic or clay laminae (acting as semipermeable membranes) into more saline connate waters lead laminated sediment to swell and wrinkle, and finally rupture, allowing more or less abrupt dewatering.

The combination of organic-matter enrichment, fine lamination and dolomite mineralogy has been documented in the Wilkins Peak Member of the Green River Formation (Eocene), Wyoming. These deposits have long been recognized as the sediments of a perennial lake (Eugster and Hardie, 1975, Smoot, 1983). Dolomite was considered as detrital in origin, having been washed into the lake (Eugster and Hardie, 1975; Smoot, 1983). In the Fuentes de Izas lacustrine episode, some of the data indicate that dolomite was formed in the lake. Within the laminated unit, the dolomite content decreases gradually until it disappears toward the top and is substituted by calcite. The total carbonate content remains constant throughout the sediment body. This change in mineralogy is difficult to interpret as a product of physical deposition, because the composition and layering style are similar, but it is closely related to organic matter content. Although dolomitic clasts are associated with quartz grains of similar size, dolomicrite is present throughout the deposits, not only in the fine silt- and clay-sized siliciclastic layers but also in coarse layers. In the Fuentes de Izas section, dolomite is restricted to the laminated facies and some carbonate layers, but it is absent in associated clastic material. Thus, the presence of detrital dolomitic particles is restricted to the lacustrine environment where dolomite formed, and this implies only smalscale local reworking. Chemical similarities between limestones and dolostones also support an authigenic origin for both carbonates. These data suggest that the chemistry of the lake waters was constant enough to allow precipitation of Mn-, Sr-, and Ba-rich calcites and dolomites, although conditions finally evolved to favor calcite. Decreasing alkalinity and Mg^{2+} depletion may have been the limiting factors for dolomite formation.

This lacustrine episode finished with the progradation of extensive fine-grained alluvial plain sediments over the lake. No significant subaerial exposure features have been detected in the laminated calcitic siltstones and the transition to red siltstones is gradual. This last stage of the lake would have been very shallow because the lake was almost filled with the earlier sediments. Rapid progradation of the alluvial plain prevented long subaerial exposure.

URS GYPSUM UNIT

Facies

At the base of the exposed section, white, alabastrine gypsum layers (Facies G2), several-meters thick occur and, in the upper portion, gray limestones (Facies G1) and red purple siltstones (Facies G3) alternate with thinner gypsum layers (Fig. 12). Facies are outlined in Table 5. The gypsum layers (Facies G2) are pure alabastrine, and lack siliciclastic and carbonate materials. They are massive, although locally an incipient decimeter-sized nodular fabric is present. In thin section, the rock is microcrystalline, and composed of subhedral crystals with non-uniform extinction crystals from 100 μm to 1 mm in size.

The thickness of the gray limestone layers (Facies G1) ranges from several decimeters (interbedded with gypsum) to about 1 m (in the upper zone). The lateral extent and variations are unknown because of poor exposure. The most common textures are brecciated and nodular, both of which are very porous. A network of subhorizontal and subvertical cracks filled with clay minerals (chlorites) and carbonates gives a brecciated appearance. Fenestral porosity is composed of interlinked subhorizontal voids about 1 cm long by 1 mm wide, and more irregular, angular voids. Where this porosity is well developed, the result is an open-mesh fabric. Micrites with intraclasts and ostracodes are common microfacies. Intraclasts are also micritic, black (1–3 mm sized) or light grey (4–5 mm), and ostracodes shells are unbroken. Siliciclastic material, except small quartz grains, is rare. Euhedral, mm-sized pyrite crystals also occur. The variability of microfacies is related to the presence and development of some fabrics and features including clotted, alveolar, vuggy and crystic fabrics, calcite rafts, spherulites, and cryptalgal structures.

Irregular, flat spar–crystal layered structures are some of the most conspicuous features of Facies G1 (Fig. 13A). The simplest structure is a layer of laterally-linked euhedral 50 μm-sized calcite crystals. In thin section, the layer has one flat side and one irregular or dentate side (Fig. 13B). In some of these structure, it is possible to recognize a thin micrite zone on the flat side. More complex fabrics can also

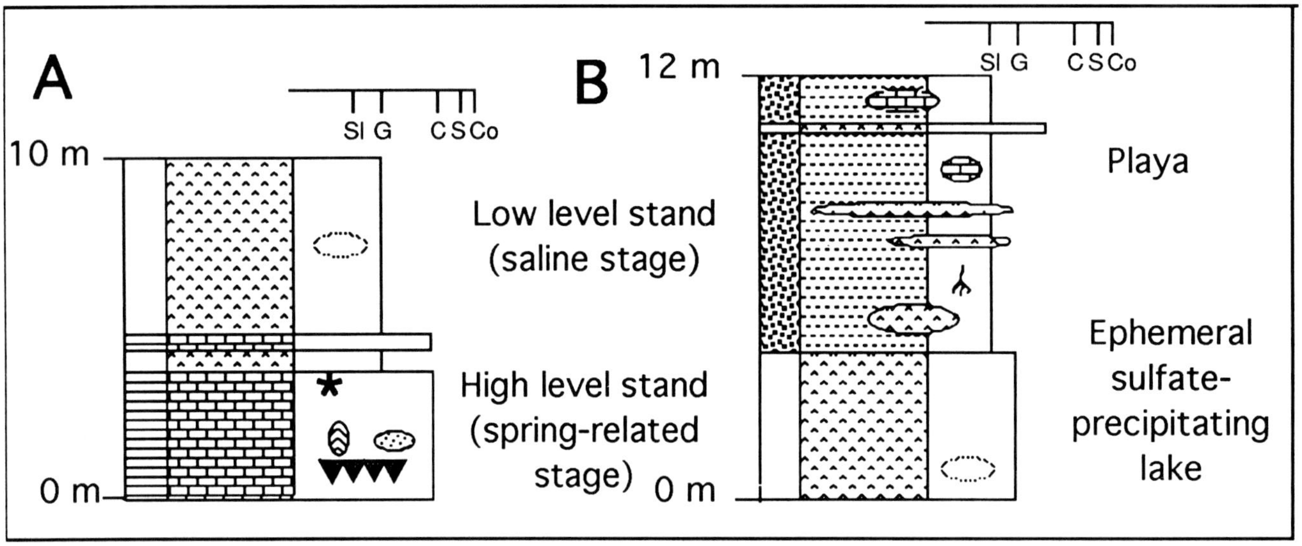

FIG. 12.—Sedimentary sequences in the URS gypsum unit (Acué section). Gypsum layers interbedded with limestones in the littoral lacustrine association (A) and with red siltstones in the playa-lake facies association (B). Same legend as Figure 4.

TABLE 5.—FACIES IN THE URS GYPSUM UNIT (ACUÉ SECTION)

Lithofacies	Microfacies	Fossil Content	Pedogenic & Subaerial Exposure (Intensity and Features)	Early Diagenetic Cementing Phases
G1: Gray limestones	Micrites, intramicrites, travertine fabrics, very low siliciclastic	High (algal remains & ostracodes)	LOW: high porosity, planar voids, cavities	BL, Q, C
G2: Massive and nodular gypsum	Microcrystalline (100 μ—1 mm sized)	None	NONE	Secondary gypsum BL
G3: Nodule-bearing red and purple siltstones	Massive		HIGH: Glaebules, mottling, neoferrans & neocalcitans, pedotubules, sepic fabrics, curved planes	NL → BL, Secondary gypsum

Early diagenetic cement phases legend as in Table 1.

occur as: (i) two joined spar layers with or without a micrite nucleus (Fig. 13C), and (ii) symmetrical and asymmetrical double-sided layers, composed of a thin, micritic, planar core surrounded by spar crystals (Figs. 13D, E). Structures with more than two sub-layers are rare. These structures are not arranged with any regular lamination.

Alveolar or vuggy fabrics are composed of isolated and linked, irregular and fenestral voids filled with spar and separated by thin linear micrite zones. In some places, casts of lenticular-shaped crystals are present. Spar mosaic textures are similar to those named 'pseudomicrospar', 'neomorphic spar' or 'clotted' in carbonate literature, but we prefer to use the term 'crystic fabric', taken from soil micromorphology (Brewer, 1964; Bal, 1975) to describe these mosaic textures composed of more or less equant calcite crystals arranged close together or separated by fine micritic or clayish material. Crystic fabrics are common in calcic soils and pedogenically modified palustrine limestones (Freytet and Plaziat, 1982; Platt, 1989; Valero Garcés, 1991). Fibrous, 40–50 μm long, calcite crystals arranged in radial fabrics form spherulites about 0.4 mm in diameter. Spherulites commonly appear grouped and related to crystic fabrics (Fig. 14A). Some display a micritic, microsparitic or isometric spar inner zone and a fibrous outer zone (Fig. 14C). Although outer spherulites boundaries are clear and sharp, the periphery is commonly composed of coarse crystic fabrics which grade into fine ones.

The micrite in the matrix of Facies G1 displays a bright orange-yellow cathodoluminescence while the spar that either fills cracks, cavities or inner pores of ostracodes, or forms large areas as crystic fabrics, displays a dull brown-orange cathodoluminescence. Layered spar structures also show bright orange-yellow CL, contrasting with the dull brown luminiscence of spherulites. Some of these show an outer zone that is even less luminiscent. Cracks and voids are filled with oriented clays, fibrous spar and/or granular mosaic spar calcite. Thin cracks and irregular areas are cemented in places by non-luminiscent quartz and celestite (blue luminescent). Some euhedral crystals with a micritic—microsparitic calcite nucleus and limpid quartz periphery appear in coarse crystic fabrics, micritic intraclasts, and micritic matrix areas (Fig. 14B). Non-luminiscent, isolated, prismatic crystals of celestite-barite, up to 1 mm long, also occur.

Red-purple siltstones (Facies G3) form massive layers a few meters thick characterized by incipient mottling, green-gray zones, and irregular calcite glaebules a few mm to a few cm in width. At some levels, slickensides and a net-

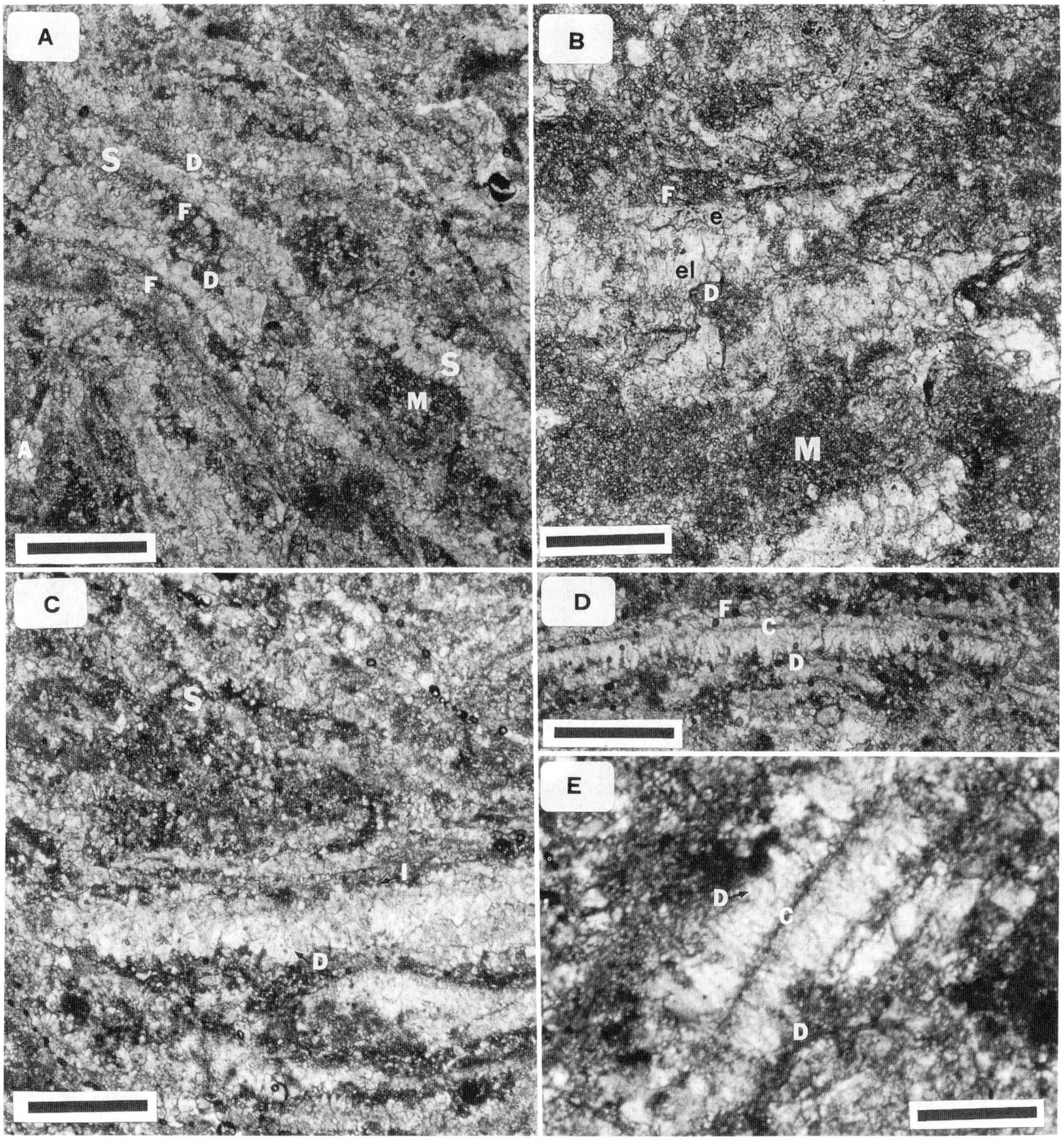

FIG. 13.—Photomicrographs of cryptalgal and travertine textures in limestones from the URS gypsum unit (Acué section, Facies G1). (A) Reworked spar–layered structures (S). Most are single with a flat side (F) and a dentate one (D); others are composed of two or more layers of spar crystals. Note the presence of granular aggregates (A) and the micrite matrix (M). (B) Single spar-layered structure. Two generations of calcite spar crystals, the first one more equidimensional (E) and the 2nd one more elongated (EL). Note the very flat (F), although slightly undulating upper side, the dentate base and the micrite matrix (M). (C) Irregular spar-layered structure. It is composed of several layers of equidimensional crystals, each with a dentate base side (D) and a straight, although irregular upper surface (F). The upper part of the photomicrograph shows smaller, irregular and reworked spar-layered structures (S). (D) Asymmetric composite micrite-spar layered structure. A thin micrite nucleus (C) is surrounded by spar calcite layers at both sides. On the basal surface, spar crystals are elongate, resulting in a dentate appearance (D), and on the upper side spar crystals are more equidimensional leading to a flat surface (F). (E) Symmetric composite micrite-spar layered structure. Two elongated spar crystals layers (D) are rooted in a thin micrite core (C). Scale bars = 0.5 mm.

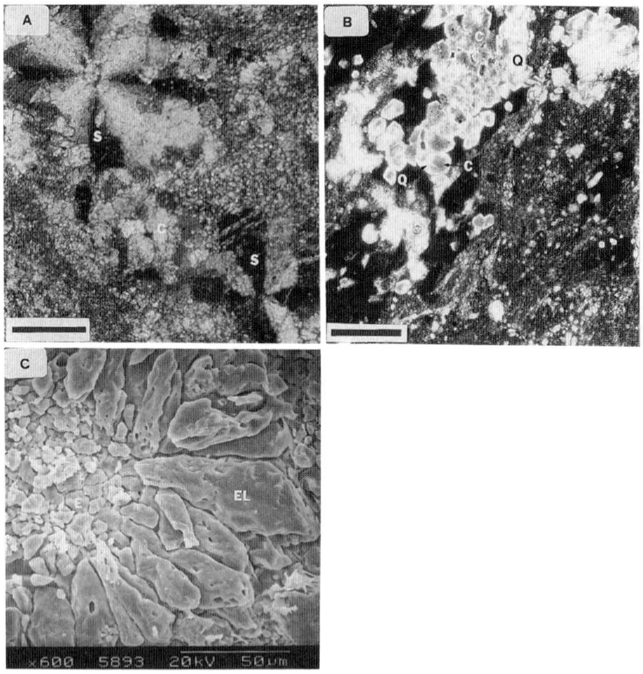

FIG. 14.—Microfacies of limestones from the URS gypsum unit (Acué section, Facies G1). (A) Granular spar mosaics (coarse crystic fabrics) (C) and fibrous spar aggregates (spherulites) (S). (B) Intraclast composed micrite and euhedral quartz crystals, displaying a calcite micrite core (C) and a limpid quartz margin (Q). (C) Scanning electron photomicrograph showing a spherulite. Note elongated calcite crystals (EL) in the outer zone and microspar core (E). Scale bars = 0.5 mm.

work composed of fine oblique cracks are filled with calcite and gypsum. The clay mineralogy is composed of chlorite and illite. In thin section, small volcanic-rock fragments, extremely angular-shaped quartz, authigenic chlorites and detrital micas, and calcite monocrystals (calcitized feldspar) are common.

Interpretation

The interbedded gypsum and limestones demonstrate significant changes in water chemistry. The gypsum layers are interpreted as precipitates from lake waters when hypersaline conditions were achieved; the limestones formed in less concentrated stages when a reduced biota could exist. The absence of primary depositional textures hinders understanding of the gypsum genesis. An incipient nodular fabric is present in the thick basal layer and, in places, a very diffuse lamination appears, but the greatest part of this lithofacies is massive and microcrystalline. Boundaries with the interbedded limestone layers are sharp whereas those with red-purple or gray-greenish siltstones are more diffuse. The thickness, purity, structureless fabric and presence of interbedded limestones seem to favor a subaqueous formation of gypsum, rather than intrasediment precipitation.

Spherulites and layered spar structures are conspicuous features that are restricted to the limestones of the Acué lacustrine episode. Layered spar structures are thought to be cryptalgal fabrics and some could be considered as "calcite rafts." These features are unusual in limestones. They have been documented mainly from modern and ancient cave deposits (Folk and others, 1985; Jones, 1989), but they have also been described in lacustrine algal bioherms (Riding, 1979) and travertine and spring carbonates (Chafetz and Butler, 1980; Chambers and others, 1982; Folk and others, 1985; Chafetz and others, 1991). In all cases, aragonite or calcite rafts are thought to form on the surface of pools where there is a maximum loss of CO_2 through degassing without agitation of the waters. The crystals grow downward into the water column until they are disturbed, usually by agitation, and then sink to the bottom. Although some rafts may be exclusively inorganic (Jones, 1989), some may be organically induced. Nevertheless, the role played by bacteria or algae in their formation is poorly understood (Folk and others, 1985). In thin section, travertine and spring deposits display many uncommon crystal morphologies (Folk and others, 1985; Chafetz and others, 1991). Spherulites from Facies G1 limestones show fibrous radial fabrics similar to those described in spring-travertine systems, but are different in size and mineralogy. Nevertheless we favor a spring-travertine origin for these spherulitic fabrics because of several structures (cryptalgal fragments, calcite rafts) that characterize these environments.

Evidence for contemporary ashfall is recognized from petrographic composition of red siltstones and includes volcanic fragments, extremely angular-shaped quartz, high chlorite and mica contents, and calcitized feldspar. Some levels can be considered as mixed ash-flow and fine-grained alluvial plain deposits,which were later reworked and pedogenically modified.

Since there is no evidences for biogenic silica, weathering of silicates seems to be the main source of dissolved silica in these waters. In saline lakes, silicification may be achieved inorganically under high pH conditions or where extensive wetting and drying occur (Eugster and Hardie, 1978). In this case, these conditions may have been achieved during the repeated cycles of lake evolution. Formation of celestite probably took place during early diagenesis, but in several phases, as the different habits (isolated crystals and crack filling) and chemistry (inferred by CL) suggest.

The association between limestones (Facies G1) and gypsum (Facies G2) (Fig. 12A) is interpreted as deposition in an ephemeral sulfate lake. The raft-, spherulitic- and algal structure-bearing limestones represent travertine deposits related to lake marginal springs. The inferred variations in water salinity indicate strong seasonal climate and/or groundwater discharge. Extensive formation of sulfates only occurs in the Aragón-Béarn basin during the Acué lacustrine episode, although intrasediment precipitation of isolated crystals is common, as described previously. The sulfate source had to be sufficient for massive gypsum precipitation. Because there are no well-developed sequences that indicate prolonged evaporative conditions, the sulfate content in the inflow water was probably high. High sulfate contents are not unusual in springs and groundwaters in volcanic areas or where the bedrock is dominated by shale or basic rocks (Eugster and Hardie, 1978). As stated, volcanism at the end of the 1st megasequence was intense. Thus, sulfate-rich inflow waters could have evaporated in the lake, leading to massive gypsum precipitation. New inflows would have diluted the waters sufficiently to support freshwater life and to become carbonate-precipitating. Near springs, travertine facies were developed and basinward micrites were the most common deposits. In shore zones, sulfates precipitated within muddy sediments of alluvial plain environments. Sequences are composed of pedogenically-modified muds with intrasedimentary gypsum (nodules, crack filling) and layers of gypsum. This alternation is interpreted as deposition in a saline mudflat, likely influenced by ash falls (Fig. 12B).

STABLE ISOTOPE ANALYSIS

Carbon and oxygen stable isotopic values (Table 6, Fig. 15) lie in the negative $\delta^{13}C$ and negative $\delta^{18}O$ quadrant,

TABLE 6.—CARBON AND OXYGEN STABLE ISOTOPE DATA FROM URS LIMESTONES

Sample	$\delta^{13}C$	$\delta^{18}O$	Lithofacies & Units
LA-15	−7.6	−9.9	L4 (glaebule)
Z-7	−6.3	−10.7	L4 (rhizocretion)
BS-6	−5.0	−11.3	L1 (Bois de Sansanet)
BS-3	−5.8	−11.3	L1 (Bois de Sansanet)
BR-23	−6.6	−9.6	L1 (Aguas Tuertas)
AC-5	−5.3	−10.2	G1 (Acué)
AC-4	−4.9	−10.1	G1 (Acué)
AC-9	−5.3	−10.4	G1 (Acué)
AC-6	−4.1	−10.3	G1 (Acué)
AC-3-B	−4.8	−9.6	G1 (Acué) (matrix)
AC-3-A	−5.0	−9.9	G1 (Acué) (intraclast)
TX-5	−5.4	−8.0	L5 (Candanchú)
FIZ-5	−4.7	−7.4	L3 (Fuentes de Izas)
LA-8	−5.8	−8.4	L2 (Baralet-Arlet)
LA-11-B	−6.2	−8.5	L2 (Baralet-Arlet)
I-54-B	−5.5	−9.5	S (microspar layer)
AA-1	−3.8	−7.0	S (micrite layer)
A-1	−3.4	−6.9	S (micrite layer)

Values are reported relative to the PDB standard.

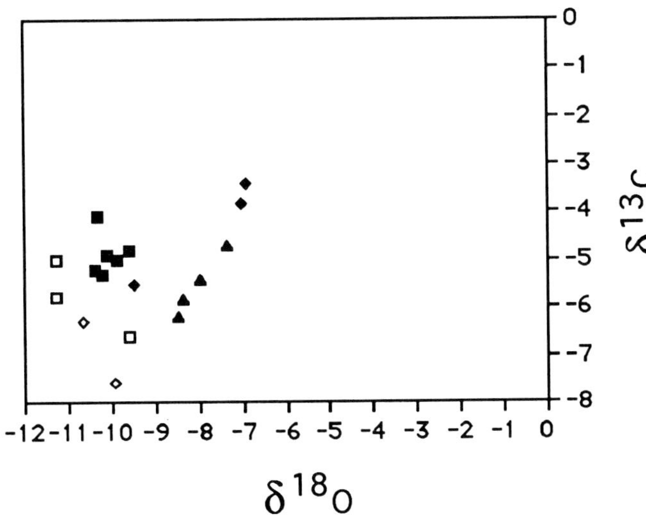

FIG. 15.—Plot of carbon and oxygen stable isotope data from the Permian URS limestones. Legend: full diamonds—Speleothem deposits; full squares—limestones from Acué (URS gypsum unit); open diamonds—calcretes; full triangles—'palustrine' limestones from URS limestone units; open squares—'lacustrine' limestones from limestone units.

with values ranging from $\delta^{13}C$ from −3 to −7‰ and for $\delta^{18}O$ from −7 to −11‰. The heaviest isotopic compositions were recorded in calcite cement samples from Permian speleothem deposits (Facies S in Table 6) located at the unconformity between the Carboniferous marine limestones and the URS continental facies, near the Escalé section. Decimeter-sized cavities are filled with red fine clastic material and several calcite cement phases. The lightest carbon isotopic compositions correspond to pedogenic glaebules (Facies L4), and the lightest oxygen isotopic composition to samples from the Bois de Sansanet limestone unit (Facies L1). Compositional values are grouped more for their stratigraphic situation than for their lithofacies. The values obtained are in agreement for average freshwater limestones (Hoefs, 1987) and pedogenic carbonates (Talma and Netterberg, 1983; Goudie, 1983), although they are strongly negative, especially for oxygen. This tendency is usually explained by surface meteoric water influx (Janaway and Parnell, 1989), although 'recrystallization' related to carbonate pedogenesis may produce up to 10‰ depletion in $\delta^{18}O$ (Cerling, 1991). Late diagenetic processes are not considered as important factors in this negative isotope shift since primary features (sedimentary and pedogenic—early diagenetic) are exceptionally well preserved in these rocks.

Lake water composition is the main source of large variation in isotopic composition in the lacustrine carbonates (e.g., Kelts and Talbot, 1990). Freshwaters are highly variable in isotopic composition and can be easily altered by a number of factors: evaporative processes (Oberhänsli and Allen; 1987; Hoefs, 1987), the contribution of soil derived organic CO_2, meteoric or groundwaters which change the hydrological balance (Gasse and others, 1987; Platt, 1989; Kelts and Talbot, 1990), and variations in the organic productivity of lake waters and stratification processes (McKenzie, 1985; Oberhänsli and Allen, 1987; Cerling and others, 1988).

The oxygen isotopic composition of the URS samples is very light, although it is similar to lacustrine Devonian calcite laminae reported by Janaway and Parnell (1989) and to dark shelly and sapropelic limestones in the Tertiary Ebro Basin (Oberhänsli and Allen, 1987). Precipitated calcites from progressively evaporated waters exhibit increasing $\delta^{18}O$ values; thus, carbonates produced during low lake stands when lake waters were more evolved record a heavier composition. This trend is recognizable in our data where carbonates from small shallow ponds in alluvial plain environments (FIZ-5, Facies L3), displaying extensive evaporite replacement textures (TX-5, Facies L4) or intense palustrine modification (LA-8, LA-11, Facies L2) are heavier in composition than those from littoral environments (BS-3, BS-6, Facies L1). The oxygen isotope composition of the meteoric input into these lakes during the Permian must, therefore, have been of a similar or even lighter isotopic composition. The isotopic composition of Permian seawater is, of course, unknown but well-preserved Permian brachiopods have near present day $\delta^{18}O$ values of about −1‰ PDB (Veizer and others, 1986). This is in contrast with Devonian and older specimens, whose compositions are of $\delta^{18}O \leq -4$‰. In our case, the lightest values correspond to carbonates precipitated during high lake stands and it may be explained by an increased inflow of upland-sourced groundwaters, as Janaway and Parnell (1989) proposed for the Orcadian Basin, or increased precipitation rates as Oberhänsli and Allen (1987) interpreted in the Tertiary Ebro Basin. Taking into account the tropical paleolatitude of the Aragón-Béarn Basin during the Permian (between 10 and 15°N; Scotese and others, 1979) and the interpreted paleogeographic setting, consistent with inland intramontane basins far from the sea (Lucas, 1985), it seems likely that both the altitude and the continentality effects (Siegenthaler and Eicher, 1986) contributed to the acquisition of light $\delta^{18}O$ values for the Permian meteoric waters.

The carbon isotopic values are heavier than those documented by Platt (1989) in Cretaceous lacustrine carbonates, and lighter than Stephanian lacustrine carbonates from the same Aragón-Béarn Basin (Valero Garcés, 1991) and the Devonian lacustrine carbonates from the Orcadian Basin (Janaway and Parrel, 1989). They are similar to spring-related Lower Permian carbonates in the Slawkow graben, southern Poland (Szulc and Cwizewicz, 1989). The lightest compositions are exhibited by pedogenic glaebules (LA-15, Facies L4), rhizocretions (Z-7, Facies L4), and thin limestone lenses interbedded with gray-green siltstones (sample BR-23, Facies L1). In the cases of the Facies L4 samples, an influx of isotopically-light soil-derived CO_2 would explain the light $\delta^{13}C$ values. The compositions of $\delta^{13}C$ of these samples are also within the range of modern soil carbonates (Talma and Netterberg, 1983; Goudie, 1983; Magaritz and others, 1980) and ancient calcretes (Purvis and Wright, 1991). Although at present, the isotopic composition of soil carbonate is mostly determined by the fraction of C_4 plants and the exchange with atmospheric CO_2, in Permian times both factors may have acted differently because of the absence of C_4 plants, the probable thin nature of soils, and the likely low levels of $P(CO_2)$ in the paleoatmosphere (Cerling, 1991; Berner, 1991). The relative decrease of $\delta^{13}C$ in the sample BR-23 (Facies L1) can be explained assuming: (i) decay of ^{12}C-rich organic matter, (ii) bacterial sulfate reduction, and (iii) the absence of photosynthetic activity and of exchange reactions with atmospheric CO_2. All three processes are possible in these relatively organic-rich sediments where anaerobic conditions favor bacterial activity.

Carbonates from the URS gypsum unit (Facies G1) show great compositional homogeneity. The values range from -4.8 to $-5.3‰$ in $\delta^{13}C$ and -9.6 to $-10.4‰$ in $\delta^{18}O$. These values are within the group of relatively heavy carbonates with regard to $\delta^{13}C$ in the sampled population of URS carbonates. These higher values of $\delta^{13}C$ can be related to biologic activity because periods of high productivity owing to ^{12}C-selective photosynthesis lead to ^{13}C enrichment in the surrounding environments (McKenzie, 1985; Oberhänsli and Allen, 1987; Szulc and Cwizewicz, 1989). On the other hand, vadose or meteoric waters usually tend to produce isotopically lighter carbonates than those from the calcite-sulfate interval. Evaporitic fractionation is not recorded in the isotopic composition of Facies G1, since $\delta^{18}O$ values are lighter than palustrine (Facies L2 and L3) and calcrete samples (Facies L4). These data are in agreement with the proposed spring-water influx for this lake.

Finally, the three speleothem deposit samples (Facies S, in Table 6) represent karstification that developed during Permian times over marine Carboniferous limestones (Valero Garcés, 1991). This was probably synchronous with the end of the 1st megasequence and the beginning of the 2nd when the basin enlarged and magmatic and tectonic activities were more intense. The $\delta^{18}O$ compositions are wider in range and slightly heavier than carbonates from the URS gypsum unit, but similar $\delta^{13}C$ compositions lead us to consider the possibility of the same groundwater recharge in both geologic settings. The relatively heavy composition of speleothem deposits seems to support long-circulating groundwater of meteoric origin rather than short-circulating, karstic-spring waters or hydrothermal waters of magmatic provenance (characteristically more negative) as in the Permian spring-related carbonates studied by Szulc and Cwizewicz (1989). These authors explain the heavy groundwater compositions as a result of the isotopic equilibrium reached between dissolved bicarbonate and the host rock (marine limestones with relatively heavy carbon isotopic compositions). The same may apply in the Aragón-Béarn Basin.

DISCUSSION

Fossil remains, sedimentologic, and petrographic evidences support a lacustrine origin for the seven URS carbonate units described in this paper. The presence of seven lacustrine episodes in the same formation makes an interesting case study for the examination of the factors controlling lake development and evolution in continental basins. Figure 16 summarizes the facies associations, intepretations and models proposed for the seven lacustrine episodes in the Aragón-Béarn Basin.

URS limestone units represent calcite-precipitating lakes developed in the distal areas of alluvial fan systems. Several subenvironments characterized by specific facies associations have been interpreted from these lake deposits (Valero Garcés, 1991), but the facies studied here correspond to the evaporitic and palustrine littoral assemblages. The low gradient margins of the lakes were responsible for subaerial exposure of large areas during low stands, and the sediments underwent strong vadose and pedogenic modification. Brief phases of more intense desiccation led to the establishment of hypersaline conditions, with intrasediment gypsum precipitation. Repeated lake level fluctuations resulted in the dissolution of evaporites and their replacement by calcite, preserving, in some cases, the lenticular shapes of the precursor crystals (Fig. 17).

Conditions for extensive and subaequous precipitation of sulfates were restricted to the Acué lacustrine episode at the top of the 1st megasequence. Evaporative precipitation was associated with incipient diagenetic silicification, but not with dolomitization. This is not the case of other lacustrine series (e.g., Cretaceous Rupelo Formation, Cameros Basin, Spain, Platt, 1989) where these processes are linked. The absence of any evaporite remains or textures can be interpreted as result of dissolution of pre-existing evaporites in the URS dolostone units. Other factors, apart from desiccation and evaporative conditions, appear to govern the

SUB BASIN	STRATIGRAPHY	TECTONIC	LACUSTRINE EPISODES	FACIES, INTERPRETATIONS & SEQUENCES	SALINE FACIES MODEL
EASTERN	2nd MEGAS. T=300 m; TOP	Stable tectonics Slow and high subsidence Small alluvial fans	**FUENTES DE IZAS** Meromixis Dolomite formation Thickness = 15 m	D4: Shallow, relatively high gradient, rarely exposed littoral D5: Oxygen depleted and saline hypolimnion D6: Oxygen depleted hypolimnion D7: Alluvial plain **D4 -> D5 -> D6 -> D7**	**MEROMICTIC DOLOMITE-PRODUCING LAKE**
WESTERN	2nd MEGASEQUENCE Thickness = 900 m; TOP	Stable tectonics Slow and high subsidence Large alluvial fans	**AGUAS TUERTAS** Well-developed palustrine facies Thickness = 30 m	L1: lacustrine littoral, frequently exposed L2: lacustrine littoral, with frequent emersion and soil development L3: lacustrine littoral with frequent water level changes L4: Alluvial plain fringed lake, with intense pedogenesis L5: Evaporitic littoral **L1 -> L2, L3 -> L4** **L4 -> L2, L3 -> L1** **L5 -> L4**	**CALCITE-PRECIPITATING, PALUSTRINE LAKE**
	BASE	Tectonic pulsations Basin extension Smaller alluvial fans Rapid and high subsidence	**CANDANCHÚ** Evaporative pumping Thickness = 40 m		
			BOIS SANSANET Alluvial influx Incipient meromixis? Thickness = 40 m		
	1st MEGASEQUENCE; NORTH (TOP) Thickness = 300 m	Stable tectonics volcanic influence (ash fall)	**ACUÉ** Travertine & springs Gypsum precipitation Groundwaters/springs Thickness = 60 m	G1: Springs & saline lake littoral (fresher stage) G2: Saline lake littoral (hypersaline stage) G3: Alluvial plain with ash-fall influence **G1 -> G2** **G2 -> G3**	**SULFATE-PRECIPITATING LAKE**
		Large alluvial fans	**BARALET ARLET** Well-developed palustrine facies Rare dolomite Thickness = 60 m	D8: Alluvial plain with ash-fall influence, hydromorphism and dolomitization **L1 -> L2, L3 -> L4, D8**	**CALCITE-PRECIPITATING, PALUSTRINE LAKE**
	SOUTH	Tectonic pulsation volcanic influence (ash fall)	**ESCALÉ** Primary dolomite formation Intense pedogenesis Groundwaters/ springs Thickness = 25 m	D1: Ephemeral dolomitic littoral frequently exposed D2: Ephemeral dolomitic littoral with intense pedogenesis D3: Alluvial plain with pedogenesis and ash-fall influence **D1 -> D2 -> D3**	**HOLOMICTIC DOLOMITE-PRODUCING LAKE**

FIG. 16.—Summary of the sedimentologic features of the lacustrine episodes developed in the Permian Unidad Roia Superior in the Aragón-Béarn Basin (western Pyrenees).

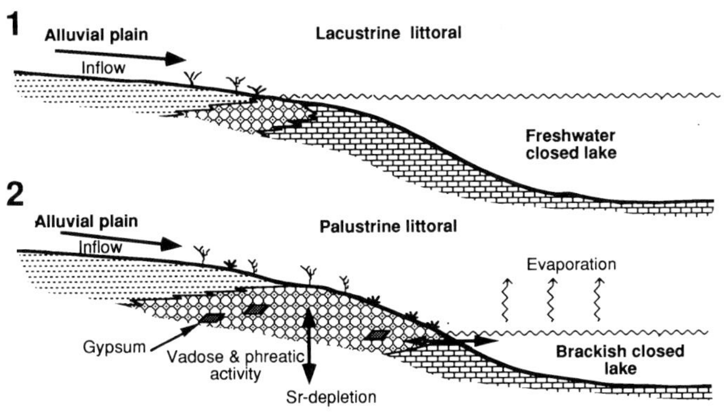

FIG. 17.—Summary diagram to illustrate the main features of the evolution of the Permian shallow, calcite-precipitating lakes (Baralet-Arlet, Bois de Sansanet, Candanchú and Aguas Tuertas limestone units). See text for discussion.

precipitation of gypsum and dolomite. The uniqueness of this saline episode is stressed by the fact that this is the only gypsum occurrence in all the Pyrenean Permian basins. Travertine fabrics (cryptalgal textures, calcite-rafts, spherulites), rather unusual in lacustrine limestones, suggest an important role for groundwaters in the hydrology of this lake. The relatively heavy $\delta^{13}C$ values (about −5‰) indicate a higher organic productivity than in the calcite-precipitating lakes, which is in accordance with an oasis-like sedimentary setting as proposed by Szulc and Cwizewicz (1989) in the Permian Slawkow Graben (Poland). The values of $\delta^{18}O$ (about −10‰ PDB), which are lighter than those from limestones with pseudomorphs after gypsum, are congruent with the absence of well-developed evaporitic sequences and the deduced strong variation in water chemistry. A playa lake system whose hydrology was mainly controlled by sulfate-rich groundwater recharge from marginal springs is envisaged as the sedimentary setting for these sulfate-precipitating lakes.

Dolomite mineralogy is the most conspicuous feature of the Escalé and Fuentes de Izas lacustrine episodes. In both, small lateral extent and thickness suggest small-sized lakes (ponds). In the Escalé lacustrine episode, the dolostone texture is homogeneous, micritic, and aphanitic. There is general agreement that the fine homogeneous texture is a result of primary precipitation or early replacement. "Protodolomite" with a similar fine texture is found in modern lakes (Müller and others, 1972) and in supratidal environments. In the marine environment, the texture is fine- grained where

lakes (Facies L2, L3 and L4), meter-thick gypsum layers (Facies G2) only occur in the Acué section. Some specific features characterize the limestones (Facies G1) of this saline episode and include travertine fabrics (cryptalgal textures, calcite-rafts, spherulites), the relatively heavy $\delta^{13}C$ values (about −5‰), explained as the result of high organic productivity during high lake stands, and the values of $\delta^{18}O$ (about −10‰ PDB), which are lighter than those from limestones with pseudomorphs after gypsum in calcite lakes (Facies L2, L3 and L4). These features, together with the absence of well-developed evaporitic sequences and the deduced strong variation in water chemistry, are in agreement with a lake system whose hydrology was mainly controlled by sulfate-rich groundwater recharge from marginal springs.

4. Ephemeral, meromictic, carbonate-producing lake is the only lacustrine episode developed in the eastern sub-basin, located at the top of the 2nd megasequence. The association between carbonates (limestones and dolostones, Facies D4) and organic-rich laminated dolomitic siltstones (Facies D5) is thought to represent a chemically-induced water stratification event (ectogenic meromixis).

The differences among lacustrine episodes developed in the same basin illustrate the fundamental roles of depositional subenvironment and paleohydrology (sources of waters, paleochemistry) in the genesis and evolution of small, low gradient carbonate lakes in arid closed basins. Although global scale factors like tectonics and climate are fundamental, local small-scale factors such as sedimentary subenvironment, groundwater/runoff ratio or particular water chemistry, must be taken into account for a more accurate interpretation of saline lakes.

ACKNOWLEDGMENTS

This paper is mainly based on data from the Ph. D. dissertation of B. Valero Garcés submitted at the Departamento de Geologia, Universidad de Zaragoza (Spain) and mainly funded by a 'Formacion del Personal Investigador' fellowship of the Spanish Science and Education Ministry. Many colleagues from this department and friends are thanked for their cooperation during dissertation work and, consequently, for their contribution to this paper. Thin sections were produced by Manolo Tricás. Help with photography came from Juan Mandado who also advised and assisted with carbonate chemistry. ICP analyses were made at the Universidad de Zaragoza. Stable isotope analyses were carried out at the Universidad de Barcelona, with helpful assistance from Juan José Pueyo, Rosa Utrilla, Anna Travé and Conchita Taberner. SEM facilities were also provided by the Universidad de Barcelona. The staff at the Instituto de Carboquímica de Zaragoza, especially Ana María Mastral, helped with organic geochemistry. Vitrinite reflectance was measured by J. G. Prado of the Instituto Nacional del Carbón in Oviedo. Pyrolysis Rock-Eval was carried out by REPSOL.

Thanks also go to Eliseo Asso for his company and constant encouragement during the Ph. D. research and to Carmen Valero for her help in translation. Elizabeth Gierlowski-Kordesch and Kerry Kelts gave valuable discussion and advice. The authors are very grateful to two anonymous reviewers for their helpful suggestions and constructive comments that have greatly improved the manuscript.

REFERENCES

ANDREWS, J. E. , HAMILTON, P. J., AND FALLICK, A. E., 1987, The geochemistry of early diagenetic dolostones from a low-salinity Jurassic lagoon: Journal of the Geological Society of London, v.144, p. 687–698.

ASSERETO, R. L., AND KENDALL, C. G. ST. C., 1971, Megapolygons in Ladinian limestones of the Triassic of the Southern Alps: Evidence of deformation by penecontemporaneous desiccation and cementation: Journal of Sedimentary Petrology. v. 41, p. 715–723.

BAL, L., 1975, Carbonate in soils: a theoretical consideration, and proposal for its fabric analysis. 1. Crystic, calcic and fibrous plasmic fabric: Netherland Journal of Agricultural Sciences, v. 23. p. 18–35.

BERNER, R., 1991, A model for atmospheric CO_2 over Phanerozoic time: American Journal of Science, v. 291, p. 339–376.

BIXEL, F., 1984, Le volcanisme stephano-permien des Pyrénées: Unpublished Ph. D. Dissertation, Université Paul Sabatier, Toulouse, p. 415.

BIXEL, F., AND LUCAS, C., 1983, Magmatisme tectonique et sedimentation dans les fosses stephano-permiens des Pyrénées occidentales: Revue de Géologie Dynamique et Géographie Physique, v. 24, p. 329–342.

BOTZ, R. W., AND VON DER BORCH, C., 1984, Stable isotope study of carbonate sediments from the Coorong Area, South Australia: Sedimentology, v.31, p. 837–849.

BOYER, B. W., 1981, Tertiary lacustrine sediment from Sentinel Butte, North Dakota and the sedimentary record of ectogenic meromixic: Journal of Sedimentary Petrology, v. 51, p. 429–440.

BRAND, V., AND VEIZER, J., 1980, Chemical diagenesis of a multicomponent carbonate system. 1: trace elements: Journal of Sedimentary Petrology, v. 50, p. 1219–1236.

BRAND, U., AND VEIZER, J., 1983, Origin of coated grains: Trace element constraints, *in* Peryt, T. M., ed., Coated Grains: Berlin, Springer-Verlag, p. 9–26.

BREWER, R., 1964, Fabric and Mineral Analysis of soils: New York, Wiley, 470 p.

CERLING, T., 1991, Carbon dioxide in the atmosphere: evidence from Cenozoic and Mesozoic paleosols: American Journal of Science, v. 291, p. 377–400.

CERLING, T., BOWMAN, J., AND O'NEILL, J. R., 1988, An isotopic study of a fluvial lacustrine sequence: the Plio-Pleistocene Koobi Fora sequence, East Africa: Palaeogeography, Palaeoclimatology, Palaeoecology, v. 63, p. 335–356.

CHAFETZ, H. S., AND BUTLER, J. C., 1980. Petrology of caliche pisolites, spherulites and speleothem deposits from central Texas: Sedimentology, v. 27, p. 497–570.

CHAFETZ, H. S., RUSH, P. F., AND UTECH, N. M., 1991, Microenvironmental controls on mineralogy and habit of $CaCO_3$ precipitates: an example from an active travertine system: Sedimentology, v. 38, p. 107–126.

CHAMBERS, L. A., BURNE, R. V., AND FERGUSON, J., 1982, Lithification of peritidal carbonates by continental brines at Fisherman Bay, South Australia, from a megapolygon/spelean limestone association: Journal of Sedimentary Petrology, v. 52, p. 1127–1145.

DALLONI, M. G., 1950, Revisión de la feuille d'Urdos au 80.000. Terrains primaires: Bulletin de Service de la Carte Géologique de France, v. 48, p. 209–213.

DESBOROUGH, G. A, 1978, A biogenic-chemical stratified lake model for the origin of oil shale of the Green River Formation: An alternative to the playa-lake model: Geological Society of American Bulletin, v. 89, p. 961–971.

ESPITALIE, J., LAPORTE, J. L., MADEC, M., MARQUIS, F., LEPLAT, P., PAULET, J., AND BOUTEFEU, A., 1977, Méthode rapide de caracterisation des roches meres, de leur potential petrolier et de leur degree d'évolution: Revue Institut Français du Petrole, v. 32, p. 23–42.

EUGSTER, H. P., AND HARDIE, L. A., 1975, Sedimentation in an ancient playa lake complex: Wilkins Peak Member of The Green River Formation, Wyoming: Geological Society of American Bulletin, v. 86, p. 319–334.

EUGSTER, H. P., AND HARDIE, L. A., 1978, Saline lakes *in* Lerman, A., ed., Lakes: Chemistry, Geology, Physics: Berlin, Springer-Verlag, p. 237–293.

FOLK, R. L., CHAFETZ, H. S., AND TIEZZI, P. A., 1985, Bizarre forms of depositional and diagenetic calcite in hot-spring travertines, central Italy, *in* Schneidermann, N., and Harris, P. M., eds., Carbonate Cements: Tulsa, Society of Economic Paleontologist and Mineralogist Special Publication 36, p. 349–369.

FREYTET, P., AND PLAZIAT, J. C., 1982, Continental Carbonate Sedimentation and Pedogenesis Late Cretaceous and Tertiary of Southern France: Stuttgart, E. Schweizerbart'sche Verlagsbuchhandlung (Nägele u. Obermiller), Contribution to Sedimentology 12, 273 p.

GASSE, F., FONTES, J. C., PLAZIAT, J. C., CARBONEL, P., KACZAMARSKA, I., DE DECHER, P., SOULIE-MARSCHE, I., CALLOT, Y., AND DUPEUBLE, P. A., 1987, Biological remains, geochemistry and stable isbtopes for the reconstruction of environmental and hydrological changes in the Holocene lakes from North Sahara: Palaeogeography, Palaeoclimatology, Palaeoecology. v. 60, p. 1–46.

GISBERT, J., 1981, Estudio geológico-petrológico del Estefaniense-Pérmico de la Sierra del Cadí. Diagénesis y Sedimentologia: Unpublished Ph.D. Dissertation, University of Zaragoza, Zaragoza, 314 p.

GOUDIE, A. S., 1983, Calcrete, *in* Goudie, A. S., and Pye, K., eds., Chemical Sediments and Geomorphology: London, Academic Press, p. 93–131.

HARDIE, L. A., 1987, Dolomitization: a critical view of some current views: Journal of Sedimentary Petrology, v. 57, p. 166–183.

HESSELBO, S. P., AND TREWIN, N. H., 1984, Deposition, diagenesis and structures of the Cheese Bay Shrimp Bed, Lower Carboniferous, East Lothian: Scottish Journal of Geology, v. 20, p. 281–296.

HOEFS, J., 1987, Stable Isotope Geochemistry: Berlin, Springer-Verlag, 241 p.

HSÜ, K. J., AND SIEGENTHALER, C., 1969, Preliminary experiments on hydrodynamic movement induced by evaporation and their bearing on the dolomite problem: Sedimentology, v.12, p. 11–25.

JANAWAY, T. M., AND PARNELL, J., 1989, Carbonate production within the Orcadian basin, northern Scotland: a petrographic and geochemical study: Palaeogeography, Palaeoclimatology, Palaeoecology, v. 70, p. 89–106.

JONES, B., 1989, Calcite rafts, peloids and micrite in cave deposits from Cayman Brac, British West Indies: Canadian Journal of Earth Sciences, v. 26, p. 654–664.

KASTNER, M., 1984, Control of dolomite formation: Nature, v. 311, p. 410–411.

KELTS, K., AND TALBOT, M., 1990, Lacustrine carbonates as geochemical archives of environmental change and biotic/abiotic interactions, *in* Tilzer, M. M., and Serruya, C., eds., Large Lakes: Ecological Structure and Function: Berlin, Springer-Verlag, p. 288–315.

KINSMAN, D. J. J., 1969, Interpretation of Sr^{2+} concentrations in carbonate minerals and rocks: Journal of Sedimentary Petrology, v. 39, p. 486–508.

KRETZ, R., 1982, A model the distribution of trace elements between calcite and dolomite: Geochimica et Cosmochimica Acta, v. 46, p. 1979-1981.

LAND, L. S., 198O, The isotopic and trace element geochemistry of dolomite: the state of the art, *in* Zenger, R. D., Dunham, J. B., and Ethington, R. L., eds., Concepts and Models of Dolomitization: Tulsa, Society of Economic Paleontologist and Mineralogist Special Publication 28, p. 87–110.

LAST, W. M., 1990. Lacustrine dolomite-an overview of modern Holocene and Pleistocene occurrences: Earth-Science Reviews, v. 27, p. 221–263.

LUCAS, C., 1985, Le grés rouge du versant nord des Pyrénées. Essai sur la Géodynamique de depots continentaux du Permien et du Trias: Unpublished Ph.D. Dissertation, University of Toulouse 111, Toulouse, 269 p.

MARGARITZ, M., KAUFMAN, A., AND YAALON, D. H., 1980, Calcium carbonate nodules in soils: $^{18}O/^{16}C$ and $^{13}C/^{12}C$ ratios and ^{14}C contents: Geoderma, v. 25, p. 157–172.

MCCREA, J. M., 1950, On the isotopic chemistry of carbonates and a paleotemperature scale: Journal of Chemical Physics, v. 18, p. 849–857.

MCKENZIE, J. A., 1985, Carbon isotopes and productivity in the lacustrine and marine environment, *in* Stumm W., ed., Chemical Processes in Lakes: New York, Wiley, p. 99–118.

MIROUSE, R., 1966, Recherches Géologiques dans la Partie Occidentales de la Zone Primaire Axiale des Pyrénées: Paris, Memoires Carte Géologique de France, 451 p.

MUIR, M., LOCK, D., AND VON DER BORCH, C., 1980, The Coorong model for penecontemporaneous dolomite formation in The Middle Proterozoic McArthur Group, Northern Territory, Australia, *in* Zenger, R. D., Dunham, J. B., and Ethington, R. L., eds., Concepts and Models of Dolomitization: Tulsa, Society of Economic Paleontologist and Mineralogist Special Publication 28, p. 51–67.

MÜLLER, G., IRION, G., AND FORSTNER, V., 1972, Formation and diagenesis of inorganic Ca-Mg carbonates in the lacustrine environments: Naturwissenschaften, v. 59, p. 158-164.

OBERHÄNSLI, H., AND ALLEN, P. A., 1987, Stable isotopic signatures of Tertiary lake carbonates, Eastern Ebro Basin, Spain: Palaeogeography, Palaeoclimatology, Palaeoecology, v. 60, p. 59–75.

PIERRE, C., ORTLIEB, L., AND PERSON, A., 1984, Supratidal evaporitic dolomite at Ojo de Liebre lagoon: mineralogical and isotopic arguments for primary crystallization: Journal of Sedimentary Petrology, v. 54, p. 1049–1061.

PINGITORE, N. E., 1978, The behaviour of Zn^{2+} and Mn^{2+} during carbonate diagenesis: theory and applications: Journal of Sedimentary Petrology, v. 48, p. 799–814.

PINGITORE, N. E., AND EASTMAN, M., 1986, The coprecipitation of Sr^{2+} with calcite at 25°C and 1 atm: Geochimica et Cosmochimica Acta, v. 50, p. 2195–2203.

PLATT, N. H., 1989, Lacustrine carbonates and pedogenesis: sedimentology and origin of palustrine deposits from The Early Cretaceous Rupelo Formation, W. Cameros Basin, N. Spain: Sedimentology, v. 36, p. 665–684.

PURVIS, K., AND WRIGHT. V. P., 1991, Calcretes related to phreatophitic vegetation from the Middle Triassic Otter Sandstones of South West England: Sedimentology, v. 38, p. 539–551.

RETALLACK, G. J., 1990. Soils of the Past, an Introduction to Paleopedology: London, Unwin Hyman, 520 p.

RIDING, R., 1979, Origin and diagenesis of lacustrine algal bioherms at the margin of Ries Crater, Upper Miocene, S. Germany: Sedimentology, v. 26, p. 645–680.

ROSEN, M. R., MISER, D. E., AND WARREN, J. K., 1988, Sedimentology, mineralogy and isotopic analysis of Pellet Lake, Coorong Region, South Australia: Sedimentology, v. 35, p. 105–122.

ROSEN, M. R., MISER, D. E., STARCHER, M. A., AND WARREN, J. K., 1989, Formation of dolomite in the Coorong region, South Australia: Geochimica et Cosmochimica Acta, v. 53, p. 661–669.

ROTHE, P., HOEFS, J., AND SONNE, V., 1974, The isotopic composition of Tertiary carbonates from the Mainz Basin: an example of isotopic fractionation in 'closed basins': Sedimentology, v. 21, p. 373–395.

SAAS, E., AND KATZ, A., 1982, The origin of platform dolomites: new evidence: American Journal of Science, v. 282, p. 1184-1213.

SEUNES, J., 1894, Vallée d'Ossau et vallée d'Aspe: Bulletin de Service de la Carte Géologique de France, v. 6, p. 97–102.

SHUKLA, V., 1988, Sedimentology and Geochemistry of a regional dolostone: correlation of trace elements with dolomite fabrics, *in* Shukla, V., and Baker, P., eds., Sedimentology and Geochemistry of Dolostones: Tulsa, Society of Economic Paleontologist and Mineralogist Special Publication 43, p. 145–159.

SIEGENTHALER, U., AND EICHER, U., 1986, Stable oxygen and carbon isotopic analysis, *in* Berlung, B., ed., Handbook of Palaeoecology and Palaeohydrology: New York, J. Wiley and Sons, p. 407–422.

SMOOT, J. P., 1983. Depositional subenvironments in an arid closed basin: The Wilkins Peak Member of the Great River Formation (Eocene), Wyoming, U.S.A.: Sedimentology, v. 30, p. 801–827.

SZULC, J., AND CWIZEWICZ, M., 1989, The Lower Permian freshwater carbonates of the Slawkow Graben, southern Poland: sedimentary facies context and stable isotope study: Palaeogeography, Palaeoclimatology, Palaeoecology, v. 70, p. 107–120.

TALBOT, M. R., AND KELTS, K., 1986, Primary and diagenetic carbonates in the anoxic sediments of Lake Bosumtwi Ghana: Geology, v. 14, p. 912–916.

TALMA, A. S., AND NETTERBERG, F., 1983, Stable isotope abundances in calcretes, *in* Wilson, R. L. L., ed., Residual Deposits: Surface Related Weathering Processes and Materials: London, Blackwell, p. 221–233.

TUCKER, M. E., AND WRIGHT, V. P., 1990, Carbonate Sedimentology: London, Blackwell, 482 p.

VALERO GARCÉS, B. L., 1991, Los sistemas lacustres carbonatados del Stephaniense y Pérmico en el Pirineo Central y Occidental: Unpublished Ph.D. Dissertation, University of Zaragoza, Zaragoza, 413 p.

VEIZER, J., 1977, Diagenesis of pre-Quaternary carbonates as indicated by tracer studies: Journal of Sedimentary Petrology, v. 47, p. 565–581.

VEIZER, J., AND DEMOVIC, R., 1973, Environmental and climatic controlled fractionation of elements in the Mesozoic carbonate sequences of the Western Carpathians: Journal of Sedimentary Petrology, v. 43, p. 258–277.

VEIZER, J., AND DEMOVIC, R., 1974, Strontium as a tool in facies analysis: Journal of Sedimentary Petrology, v. 44, p. 93–115.

VEIZER, J., FRITZ, P., AND JONES, B., 1986, Geochemistry of brachiopods: oxygen and carbon isotopic records of Paleozoic ocean: Geochimica et Cosmochimica Acta, v. 50, p. 1679–1696.

WELLS, N. A., 1983, Carbonate deposition, physical limnology and environmentally controlled chert formation in Paleocene-Eocene Lake Flagstaff, central Utah: Sedimentary Geology, v. 35, p. 263–296.

WORSE, J. W., AND BENDER, M., 1990, Partition coefficients in calcite: Examination of factors influencing the validity of experimental results and their application to natural systems: Chemical Geology, v. 82, p. 265–277.

WRIGHT, V. P., SLOAN, R. J., VALERO GARCÉS, B. L, AND GARVIE, L. A. J., 1992, Groundwater ferricretes from the Silurian of Ireland and Permian of the Spanish Pyrenees: Sedimentary Geology, v. 77, p. 37–49.

PART IV
ANCIENT SALT LAKE/MARINE SYSTEMS

Sedimentology and Geochemistry of Modern and Ancient Saline Lakes, SEPM Special Publication No. 50

THE EVOLUTION OF AN INLAND SEA OF MARINE ORIGIN TO A NON-MARINE SALINE LAKE: THE PENNSYLVANIAN PARADOX SALT

SHERILYN WILLIAMS-STROUD
United States Geological Survey, Branch of Sedimentary Processes, Denver, Colorado 80225

ABSTRACT: The Paradox Formation has long been identified as having a marine origin. However, the magnesium sulfate-poor potash assemblage in the salt cannot be explained by simple equilibrium evaporation of seawater. Dolomitization of associated carbonate rocks may be the key to the initial depletion of seawater sulfate in the form of additional gypsum and anhydrite deposition, which leads to non-deposition of magnesium sulfate minerals and the precipitation of sylvite and carnallite. The economic-grade potash ore in several of the evaporite cycles occurs as sylvite with only minor carnallite. The deficiency of magnesium chloride minerals indicates that the water chemistry was not merely the result of cation exchange of Ca^{2+} for Mg^{2+} but that a considerable amount of brine mixing with waters of other compositions beside calcium-enriched seawater took place. Replacement fabrics and depositional textures as well as sedimentary facies relationships indicate probable shallow water deposition of saline minerals in the basin, and subaerial exposure on the basin margins. The chemistry of the evaporite minerals can be explained by an altered seawater brine mixing with meteoric water in the form of stream runoff and flash floods in an arid basin.

INTRODUCTION

The episodic incursions of seawater into the rapidly subsiding Paradox Basin alternated with evaporitic periods which resulted in the deposition of a cyclic succession of salt beds separated by clastic interbeds (Raup and Hite, 1992). Hite (1960) identified 29 evaporite cycles in the salt, consisting of beds of halite, separated by interbeds of anhydrite, silty dolomite and black shale. Recent drilling data indicate that there are possibly 33 salt-bearing cycles in the sequence. Eighteen of the salt cycles contain potash, mostly in the form of sylvite (KCl) and some carnallite ($KCl \cdot MgCl_2 \cdot 6H_2O$) (Hite, 1961, 1985), and almost no magnesium sulfate minerals. Previous workers have identified the Paradox Formation as a marine evaporite deposit based on the proximity of the evaporites to carbonate rocks of marine origin (Sloss, 1953; Hite and Buckner, 1981) and on the bromine geochemistry of the halite (Raup and others, 1970; Raup and Hite, 1978; Hite, 1983). However, the chemical paragenesis of the saline minerals present is inconsistent with a seawater origin (Harvie and others, 1980; Hardie, 1984). The paucity of magnesium sulfate minerals in potash deposits has been explained by various processes, including incongruent dissolution of carnallite (Garrett, 1970; Braitsch, 1971), dolomitization (Shearman, 1966; Hite, 1983), bacterial reduction of sulfate (Borchert, and Muir, 1964), and alteration of seawater by mixing with hydrothermal brines rich in $CaCl_2$ (Hite, 1975; Hardie, 1990).

The recognition of primary, or syndepositional, textures in evaporites is critical because saline minerals are readily altered upon even shallow burial (Casas and Lowenstein, 1989; Lowenstein, 1982). Excellent preservation in the Paradox salt of primary textures, such as cumulate halite textures and selenite gypsum (pseudomorphously replaced by halite and anhydrite), provides clues to deciphering the sedimentary environment. The textural and depositional relationships of the carnallite and sylvite zones of the salt probably resulted from syndepositional precipitation from original brines, rather than secondary dissolution and recrystallization. Geochemical models suggest that the majority of solutes entered the basin from two sources. The first, as seawater, which flowed into the basin during sealevel highstand or percolated through surrounding carbonates, and the second, as meteoric water from streams, storms and flash-flood runoff.

Lithologic and petrologic evidence indicate that the sources of the two major water types were from opposite sides of the basin. A considerable amount of detrital carbonate was brought into the basin from the western and southern margins during the marine incursions. The periods of high sea-stands more or less correlate with pluvial climatic phases when evaporite deposition ceased. The deeper water shales show an increased clastic input toward the southeast basin margin (Fetzner, 1960), and the apparent source area was the San Luis uplift (Hite and others, 1984). During most of the history of the salt depositing phase of the basin, the drops in sea level below the marginal topographic highs caused complete disconnection of the ocean from the brine in the basin, after which brine salinities increased by evaporative drawdown, in a scenario similar to that proposed for the Silurian Michigan basin (Cercone, 1988). The environment of deposition of the evaporites can be characterized as a perennial saline lake.

LOCATION AND GENERAL GEOLOGY

The Paradox Basin is located in southeastern Utah and southwestern Colorado and contains extensive evaporite deposits of Middle Pennsylvanian age. The boundaries of the evaporite basin are usually determined by the extent of Paradox salt (Fig. 1). The basin is in the Colorado Plateau province, bounded on the northeast by the Uncompahgre uplift, on the east by the San Luis uplift, on the southwest by the Monument upwarp and Defiance Uplift, and on the northwest by the San Rafael Swell. The salt basin itself is an elongate asymmetrical trough, with the deepest part adjacent to the Uncompahgre fault line on the northeast basin margin. Wengerd (1958) identified the uplifts and swells to the southwest and west as the basin shelf. The Uncompahgre uplift was a high to moderately high positive feature. Subsurface data along the Uncompahgre front indicate that the Uncompahgre fault was a Pennsylvanian-Permian low-angle thrust fault (Frahme and Vaughn, 1983). The asymmetrical deepening of the basin near the northeast border is interpreted to be the result of a basin foredeep related to the thrusting (Johnson and others, 1991).

Sedimentology and Geochemistry of Modern and Ancient Saline Lakes, SEPM Special Publication No. 50

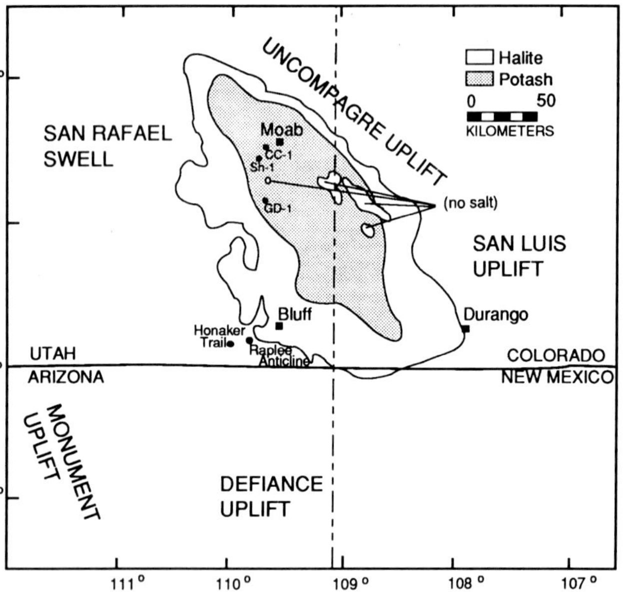

FIG. 1.—Location map of Paradox basin showing extent of potash and locations of wells discussed in text. (CC-1) Delhi-Taylor Cane Creek No. 1, (Sh-1) Delhi-Taylor Shafer No. 1, (GD-1) D.O.E. Gibson Dome No. 1.

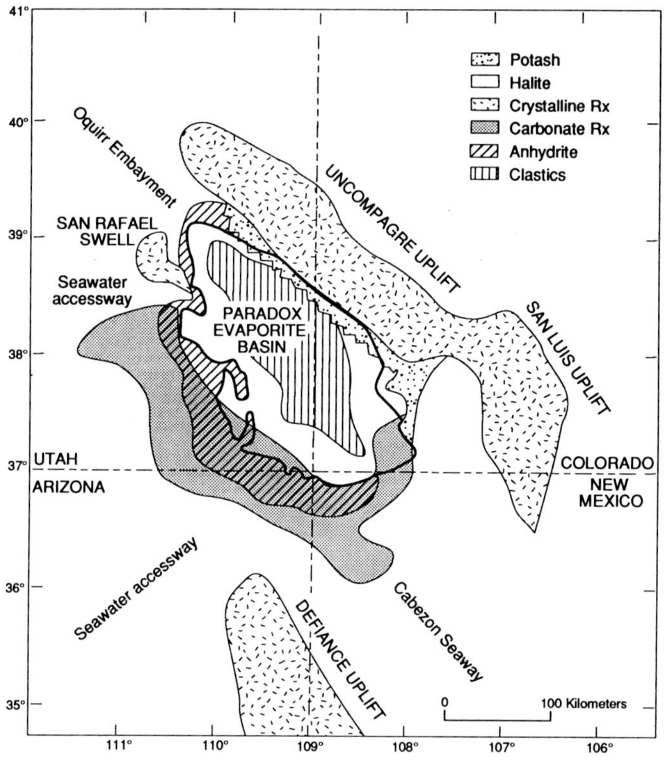

FIG. 2.—Facies map of Paradox Formation and related rocks showing Pennsylvanian structural elements and facies (after Peterson and Hite, 1969).

The distribution of Pennsylvanian sedimentary facies and structural elements in the Paradox Basin is shown in Figure 2. The Uncomphaghre Uplift is postulated to have been a structure of moderately high to high relief (Mallory, 1975; Johnson and others, 1991). Seawater access into the deeper basin is inferred to have been through the moderate to low relief platforms to the southeast, west and northwest (Hite, 1970).

The Paradox Formation overlies the Pinkerton Trail Formation, and is in turn overlain by the Honaker Trail Formation. The three formations constitute the Hermosa Group (Fig. 3). The Honaker Trail Formation is similar in lithology to the Pinkerton Trail Formation, consisting of gray to reddish-gray fine-grained to coarse-grained limestone with black and red chert, and reddish-gray to buff-gray carbonaceous sandy siltstones. More detailed descriptions of the lithology and stratigraphy of the carbonate rocks can be found in Clair (1952), Wengerd and Strickland (1954), and Wengerd (1958).

LITHOLOGIES AND SEDIMENTARY FACIES OF THE PARADOX FORMATION

Basinal Evaporite Facies

Wengerd and Strickland (1954) and Wengerd (1958) divided the Paradox Formation into three members. The lower member is a complex of arkosic granulite, sandy and gypsiferous siltstones in the southeastern part of the basin. To the southwest, it consists of dark gray shales, gray porous dolomite, and gray cherty limestones that grade into interbedded black shale, siltstone, gypsum and dolomite toward the central part of the basin. The upper member is similar in lithology to the lower member with penesaline deposits in the central basin and biostromal-biohermal dolomite and limestone, gray shale, and silty dolomite on the southwest shelf area.

The middle, or "salt member" (Wengerd, 1958) contains interbedded silty dolomite, anhydrite, black shale and halite, with or without potash. The limestone facies on the shelf is considered to be contemporaneous with the salt (Wengerd and Strickland, 1954; Hite, 1970; Hite and Buckner, 1981). The original maximum depositional thickness of the salt facies is estimated to be ~2,000 m, pinching out to zero on the margins of the basin.

Textures in the halite beds of the evaporite facies range from euhedral cumulates with grain sizes of a few tenths of a millimeter, bottom growth halite crusts with crystal sizes up to one half centimeter in diameter, draped by fine grained anhydrite, to mosaic fabrics found in the tops of the potash-bearing halite where the rock is more than 50 percent sylvite. Potash is present in the salt over a large extent of the salt deposit (Fig. 1). The sylvite zones occur in the tops of the halite beds, and are commonly red due to hematite inclusions (Raup and Hite, 1991a, 1991b). The thicknesses of the individual salt beds range from 7 to 270 m in the center of the basin to zero on the basin flanks. Because of flowage of salt in the basin center, salt in the anticlines has increased in total thickness to as much as 4,700 m.

The base of each evaporite cycle is interpreted to represent the beginning of a period of influx of seawater, or a freshening of the brine in the basin due to sea-level rise and climatic changes that resulted in increased precipitation

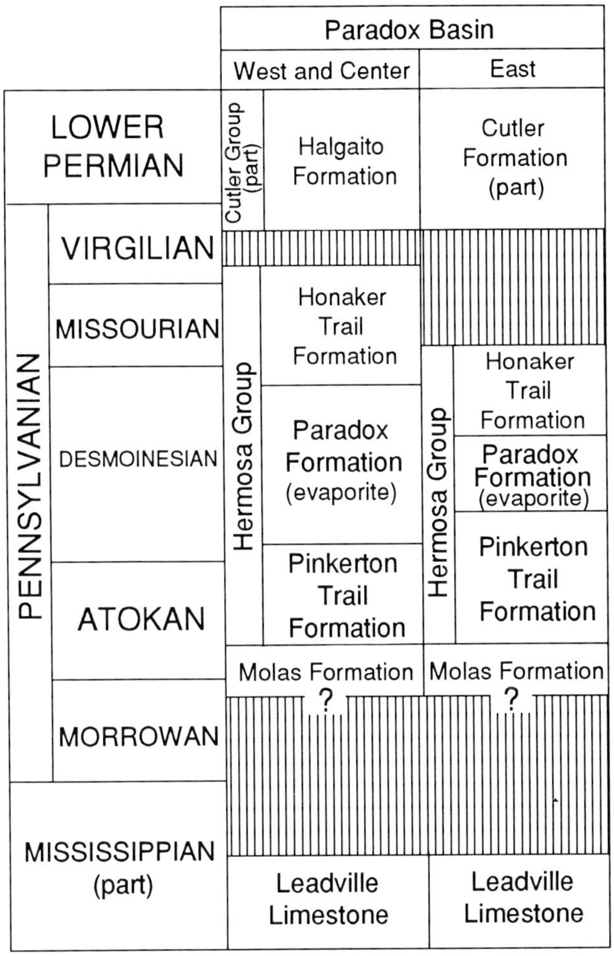

FIG. 3.—Stratigraphic column of upper Paleozoic and Lower Permian rocks in the Paradox basin (from Peterson and Hite, 1969).

(Hite and Buckner, 1981). After sea level fell again, evaporative concentration resulted in precipitation of evaporite minerals. The base of an 'idealized' cycle (Peterson and Hite, 1969) is an erosional or dissolution contact between anhydrite and an underlying halite bed. The anhydrite grades upward into silty dolomite, shale (which ranges from black to tan or gray in various cycles), dolomite, anhydrite then halite.

Anhydrite occurs in beds that are fine-grained and laminated, nodular, or pseudomorphous after selenitic gypsum. Pseudomorphs of selenitic gypsum are also replaced by halite. Anhydrite is also present in beds with millimeter-scale laminations defined by carbonate mud, or in beds with wavy to lenticular laminations containing large amounts of silty material.

The fine-grained dolomite usually shows little to no visible texture, though some beds are faintly laminated, are interlaminated with shaley or anhydritic layers, or show laminations disrupted by bioturbation. Some silica fossil detritus, such as diatoms and sponge spicules are found in the black shales, but except for palynomorphs which are well preseved in the salt and the black shale (B. F. Rueger, pers. commun., 1991), the other lithologies in the evaporite facies are unfossiliferous.

The presence of silt in the dolomite is attributed to reworking of sediments from the basin margins and the transportation of fine grained detritus via low-density inflow over the high salinity brines in the basin (Hite, 1970). Hite (1983) suggested that the dolomite, which is mostly microcrystalline but also occurs as rhombs that appear to be detrital, was a primary precipitate. Erosional surfaces and evidence of subaerial exposure found in the marginal carbonates are found in outcrops of the marginal carbonates along the Honaker Trail, and in cores through the carbonate mound facies of the basin (Peterson and Hite, 1969, Goldhammer and others, 1991). I suggest here that the silty dolomite beds are a clastic deposit which is a gradational end member to terriginous clastics in the southeastern part of the basin.

The average thickness of the shale beds is about 10 m. Most of the interbed sequences in the 33 cycles contain some black shale beds, but they cannot always correlated. There are 6 exceptionally thick black shale units which are present throughout the basin that have high organic carbon content, nearly 13 percent in some beds (Hite and others, 1984). They have a distinctive signature on gamma-ray logs and have been given specific names by the petroleum industry (Fig. 4). The shales, together with the lithologic sequence of the interbeds of each cycle, serve as marker beds for identification of strata within the sequence.

Shelf Carbonate Facies

Where they are not separated by the evaporite facies, the Pinkerton Trail and Honaker Trail Formations are not distinguished from each other and form the Hermosa Formation. On the low shelves to the southwest, the Hermosa Formation consists of interbedded black shale, dark-gray

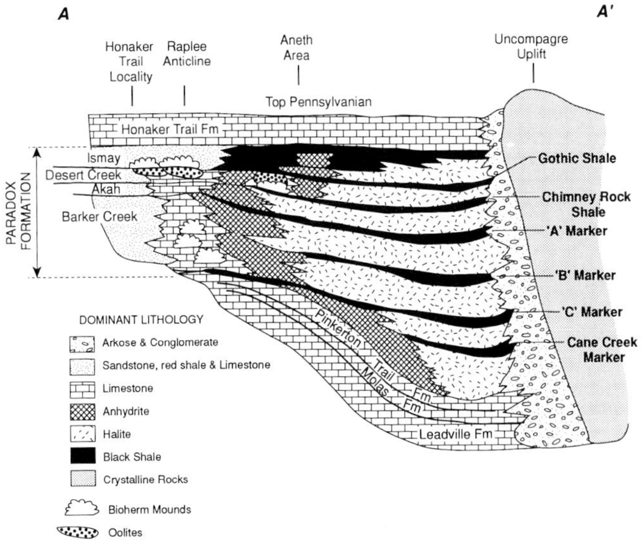

FIG. 4.—General cross section through the Paradox Basin, modified from Peterson and Hite (1969), and Hite and Buckner (1981). Location of cross section shown in Figure 1.

siltstone, anhydrite, dolomite and limestone. Higher on the shelf are dark gray shales, gray porous dolomite and gray cherty limestone. Along the edge of the San Luis uplift to the east, the lower member is a penesaline complex of buff arkose and greenish-gray sandy siltstone overlain by gypsiferous siltstones.

On the southwest shelf of the Paradox basin, the carbonate facies contains mound-like buildups of biogenic carbonates. A narrow belt of mound-bearing carbonates is also found along the northeast flank of the Paradox basin between the arkosic clastic facies and the evaporites (Peterson and Hite, 1969). The principle mound-building organisms have been identified as the calcareous phylloid algae *Ivanovia* (Murray, 1960). The mound facies is disconformably overlain by the pelletal-foraminiferal limestone 'cap facies' which represents the beginning of the next depositional cycle. Other rocks present in the carbonate shelf include the porous algal facies, marked by the Horn Point limestone marker bed at the Honaker Trail locality near Mexican Hat, Utah (Pray and Wray, 1963). Remains of other organisms found in the carbonate rocks include crinoid columnals, bryozoan debris, brachiopods, chaetetid coral heads, conodonts, and fusulinids. Also found on the shelf are fine-grained cherty limestone beds that are interpreted to be a deeper-water deposit, equivalent to the black shale beds in the evaporite facies (Hite and Buckner, 1981). Most of the marginal carbonate rocks exposed in the Honaker Trail section of the southwest shelf are limestone, though dolomite occurs locally.

Marginal Clastic Facies

The northeastern margin of the evaporite basin consists mainly of a thick wedge of arkosic sediments derived from the Uncompahgre Uplift. Although the foredeepening of the basin along the Uncompahgre front trapped most of the sediments in the trough adjacent to the northwest margin (Johnson and others, 1991), arkosic debris was carried westward almost completely across the basin at the end of Pennsylvanian time and during Permian time (Peterson and Hite, 1969).

The clastic facies of the Paradox Formation on the southeast basin margin was interpreted by Fetzner (1960) to be part of a large fan-delta complex, which he named the Silverton fan delta. Spoelhof (1976) defined 19 depositional cycles in the complex, and suggested that their cyclical nature was related to eustatic sea-level changes. The source area for the Silverton fan delta was probably the San Luis uplift, which, as indicated by their thickening toward the southeast, is also the source area for the black shale facies.

MINERALOGY AND PETROGRAPHY OF THE EVAPORITE FACIES

Petrographic examination of the rocks from drill cores in the evaporite facies of the Paradox Formation reveals a wide variety of textures. Cores drilled in Utah through Cane Creek Anticline (S25, T26S, R20E), Shafer Dome (S15 , T27S, R20E) and Gibson Dome (S21, T30S, R.21E) were studied, and the clastic interbed relationships in cycles 4 and 5 were compared. The locations of the drill cores are shown in Figure 1.

Halite

The most distinctive texture found in the halite beds is the halite/anhydrite couplets composed of layers of euhedral halite crystals draped by fined grained anhydrite (Fig. 5A). The anhydrite drapes also contain clays and carbonate. The largest halite crystals are about 0.5 cm across, with slight vertical elongation indicating their formation as bottom growth crusts (Fig. 5B). In some places there is a sharp boundary between the top of a layer of halite crust fabric containing anhydrite drapes and the bottom of a layer of nearly pure halite. This indicates that some dissolution preceded precipitation of the halite cubes of the next crust layer. The euhedral texture of halite cubes is not preserved where there is no anhydrite drape over the halite grains.

In the potash-bearing salt of cycle 5, sylvinite layers alternate with layers of fine grained anhydrite plus clay and carbonate (Fig. 6). In cycle 13 sylvinite, halite crystals are found as poikilitic inclusions floating in larger anhedral sylvite crystals (Figs. 7A, 7B). The texture of the sylvinite in potash ore from cycle 5 is an anhedral mosaic (Fig. 7C). The layers of halite sandwiched between the sylvite zones and the anhydrite laminae consist of small (0.1 mm), fluid inclusion-rich halite crystals (Fig. 8).

In the halite of cycle 10, the crystal sizes are much larger than the crystals in the other cycles, on the order of 1 to 2 cm (Fig. 9A). Fluid inclusion banding of chevron halite is found in some places, but the texture appears to be largely recrystallized (Figs. 9B, 9C). Patches of clear halite with black muddy inclusions are found adjacent to cloudy halite.

Anhydrite

Anhydrite and halite pseudomorphs after swallow-tail selenite gypsum crystals are common in anhydrite layers di-

FIG. 5.—(A) Core photo of halite crust texture from Dehli-Taylor Cane Creek No. 1 core, showing pure halite layer and euhedral halite (h) with anhydrite drapes (a); and (B) Photomicrograph of halite crust showing muddy anhydrite drapes (a) over elongated halite crystals. Scale bar = 1 mm.

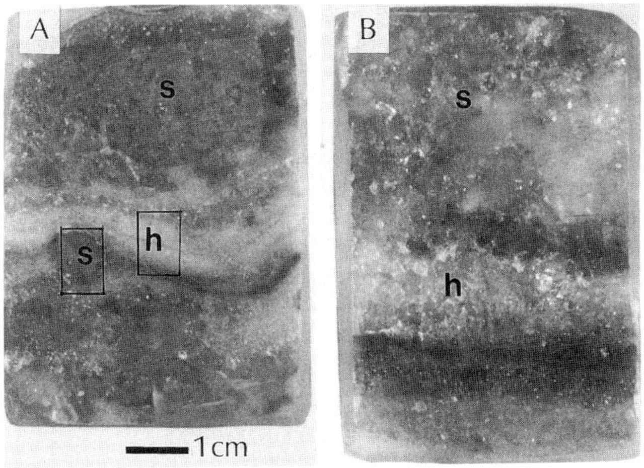

FIG. 6.—(A) Photo of sample from sylvite zone in cycle 13 from Shafer No. 1 core. The small box labeled "h" is area of photomicrograph in Figure 8, and box labeled "s" is area of photomicrograph in Figure 7B. (B) Photo of sylvite ore taken from potash mine near Moab, Utah. H = halite, S = sylvite.

rectly below halite (Fig. 10A). The crystals are commonly oriented vertically or as radial splays extending from bedding planes. The crystals are 3 to 4 cm long near the tops of the anhydrite beds and become progressively smaller near the basal contact with silty dolomite.

The basal anhydrite bed of each cycle is in erosional contact with the underlying halite of the previous cycle. Laminations are 2 to 3 mm thick, defined by fine grained anhydrite alternating with clayey dolomite partings (Fig. 10B). In thin section, the tabular anhydrite crystals of the laminated anhydrite show a horizontal preferred orientation. In some beds, the fine-grained, laminated anhydrite grades upward into nodular anhydrite, then to silty dolomite or black shale.

Figure 11A shows an example from the Shafer No. 1 core of selenite replaced by fine-grained anhydrite in which the gypsum crystal morphology is well-preserved. Nodular anhydrite commonly occurs in the silty dolomite of each cycle (Fig. 11B). Nodules can form discrete layers comprised of felted tufts of anhydrite crystals with small amounts of included silty or dolomitic material.

The nodular anhydrite that occurs in shaley beds (Fig. 11C) usually shows a more developed "chicken wire" texture, with some horizontal relect bedding traces. More detailed descriptions of the stratigraphy and lithology of the Cane Creek and Shafer Dome cores can be found in Raup and Hite (1991a, b, 1992).

Dolomite

The dolomite is fine-grained and faintly laminated in some places, but generally is structureless. There are some casts of what appear to be recrystallized fossil debris. Bioturbation and possible microbial laminae are present in some silty dolomite beds. The interbed sequence of cycle 4, of the D.O.E. Gibson Dome No. 1 core contains a silty, anhydritic dolomite bed with laminations showing evidence of extensive bioturbation (Fig. 12). The color in the silty, anhydritic dolomite ranges from gray near the top to brown and dark gray at the bottom, where it grades into a black shale. Grains of quartz silt are dispersed throughout. The dolomite is usually gradational to laminated anhydrite and/or black shale.

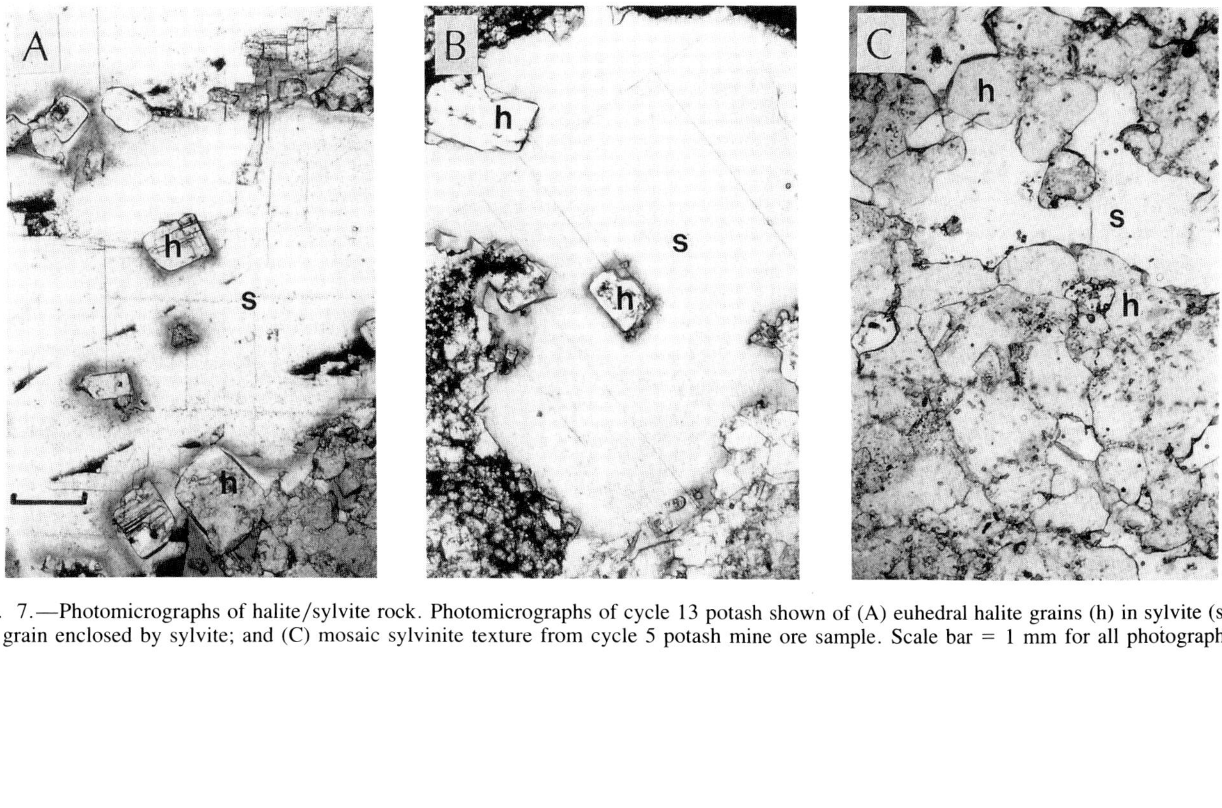

FIG. 7.—Photomicrographs of halite/sylvite rock. Photomicrographs of cycle 13 potash shown of (A) euhedral halite grains (h) in sylvite (s); (B) halite grain enclosed by sylvite; and (C) mosaic sylvinite texture from cycle 5 potash mine ore sample. Scale bar = 1 mm for all photographs.

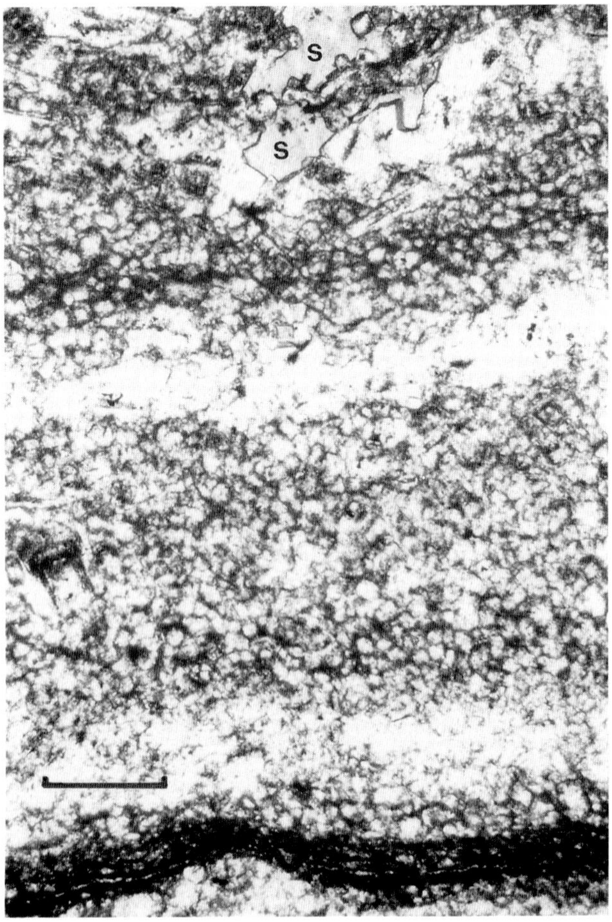

FIG. 8.—Fine grained euhedral to subhedral halite in anhydrite layers of cycle 5 sylvinite from area in box labeled "h," shown in Figure 6. (s) denotes sylvite. Scale bar = 1 mm.

Shale

The shale beds that serve as marker beds are brownish-black to black, slightly pyritic, calcareous, and highly organic. In places the black shale shows millimeter-scale lamination. The sparse fossils found in the beds include brachiopods, carbonaceous fossil fragments, conodonts, and a few fish remains. Palynomorphs are found in the black shale, as well as the halite and silty dolomites. The lower part of each black shale unit has a substantial amount of terriginous quartz and silt, and contains increasing amounts of dolomite toward the top where, in most cycles, it grades into a silty dolomite (Hite and Buckner, 1981; Raup and Hite, 1992).

Some shale beds are above or below nodular anhydrite with a black shale matrix. Figure 13 shows a portion of core from cycle 21 in the GD-1 hole. The shale bed shown is just above the nodular anhydrite in Figure 11C. The layering is disturbed, with possible rip-up clasts (Fig. 13A), and mud cracks now filled with halite (Fig. 13B.)

Exposures of gypsum and black shale on the southeast shelf of the Paradox Basin at Hermosa Mountain near Durango, Colorado, show alternating centimeter-scale gypsiferous and silty layers (Fig. 14). The gypsum occurs in beds which are nodular to fine-grained massive to laminated. The exact cycle in which this particular gypsum occurs cannot be determined from the outcrop, but it probably correlates to the upper middle part of the Paradox Formation (K. J. Franczyk, pers. commun., 1991). The contacts between the different lithologies are usually abrupt. The texture of the gypsum is recrystallized and nodular, and there are no selenite crystals or pseudomorphs. Relect lamination can be seen in the nodular beds. The gypsum here has probably rehydrated from secondary anhydrite after exposure to meteoric water.

ENVIRONMENTS OF DEPOSITION

The origin of the Paradox Formation has been attributed to evaporite deposition in a restricted marine environment (Hite, 1970, 1983) and to playa salt deposition in a deep desiccated basin (Kendall, 1987, 1988). There is evidence for the existence of both of these depositional end members in the Paradox basin (Fig. 15). The high content of clastics in the dolomite or of carbonates in the shales and siltstones suggests deposition in a marine environment, but where the meteoric water that flowed into the basin as runoff from the major uplifts greatly influenced the character and composition of the basin waters. Wengerd and Strickland (1954) first suggested that the bioherms which are visible in the San Juan Canyon at the Honaker Trail locality acted as depositional barriers that caused the Paradox Basin to be cut off cyclically from normal seawater. Hardie and others (1981) define a perennial saline lake as a surface body of brine that persists for many years (tens, hundreds or thousands) without drying up. In the Paradox Basin, the halite beds alone represent a depositional period on the order of thousands of years, therefore, the majority of evaporite deposition probably occurred in an environment that is most accurately described as a perennial saline lake.

The sedimentary structures observed in the Paradox basin cores are both primary and diagenetic. The halite cumulate textures, subvertical selenite crystals on a horizontal substrate, and fine laminated anhydrite and shale are all indicators of subaqueous deposition. Absolute brine depth cannot be determined from the textures alone, but many of the textures, such as the bottom growth halite crust fabrics, and the swallow-tail selenite gypsum, in the evaporite beds indicate that the environment of deposition was probably relatively shallow water.

The position of the laminated anhydrite at the erosional contact with the underlying halite beds indicates a sudden lowering of salinity of the basin brines, due to marine transgression and perhaps also the beginning of a pluvial climatic phase (Hite and Buckner, 1981). Only a small fraction of the halite is dissolved, because in the beginning of the rise of sea level, the seawater supplied to the basin comes in as intermittent overflows, possibly from storm surges over the barrier. A relatively thin layer of seawater covered the brine after each influx, which then evaporated and precipitated a layer of gypsum. The first one or two centimeters of fine laminated anhydrite that mark the sharp

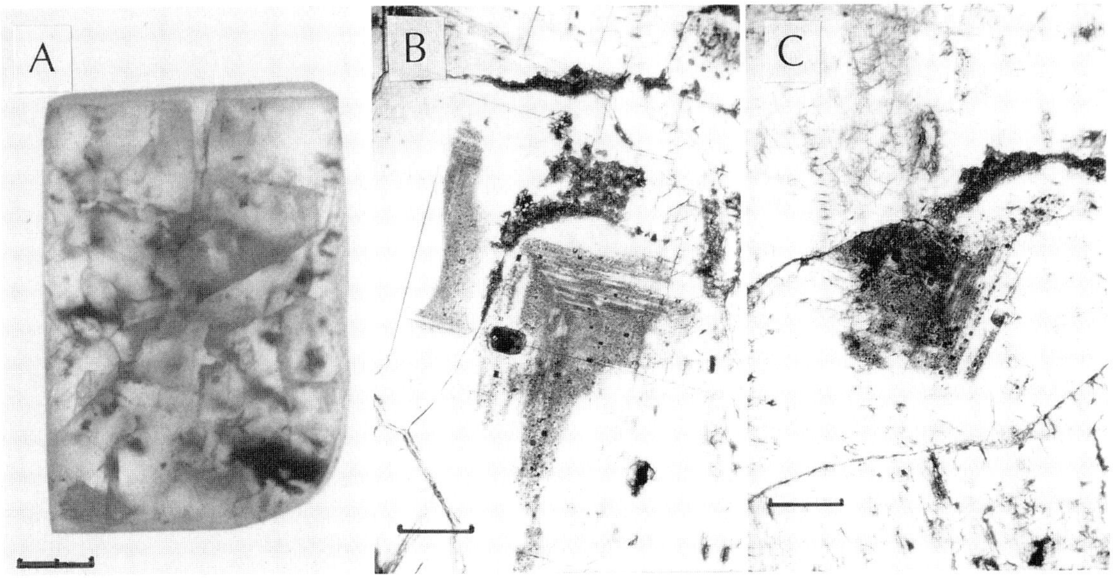

FIG. 9.—(A) Photo of core slab from halite in cycle 10. Scale bar = 1 cm; (B) Photomicrograph showing large halite grains and chevron halite fluid-inclusion bands; and (C) Halite cornet from the same specimen. Note concentration of anhydrite on tops of rounded grain boundaries. Scale bar on photomicrographs = 1 mm.

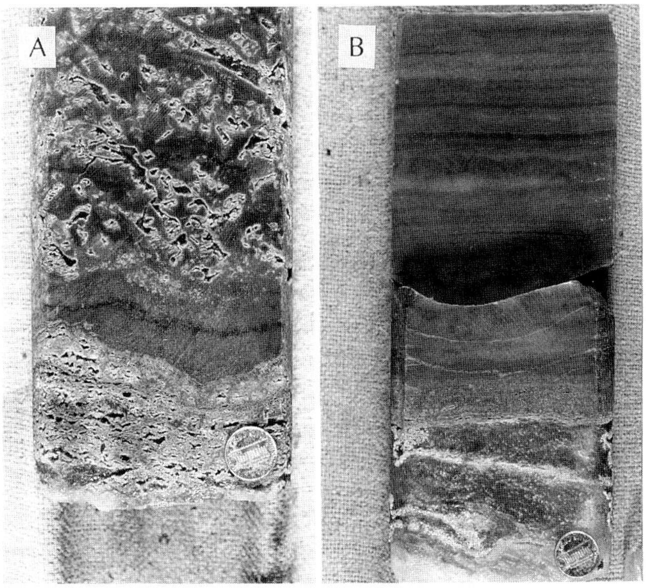

FIG. 10.—Core photographs of anhydrite and dolomite in interbed sequences, D.O.E. Gibson Dome No. 1 core, cycle 4 interbed. (A) Selenite replaced by halite and anhydrite, 895 m; and (B) fine grained, laminated basal anhydrite, 905 m. Diameter of coin = 19 mm.

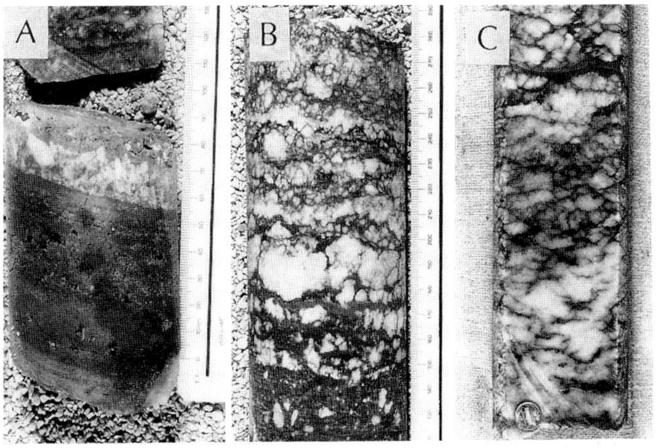

FIG. 11.—(A) Bedded selenite crystals above silty dolomite, below halite bed, now pseudomorphically replaced by anhydrite, cycle 4 interbed, Delhi-Taylor Shafer No. 1 core, 800 m; (B) nodular anhydrite in silty dolomite, Delhi-Taylor Cane Creek No. 1, 806 m; and (C) Nodular anhydrite in black shale, D.O.E. Gibson Dome No. 1 core, interbed cycle 21–24. Diameter of coin = 19 mm.

contact at the base of a cycle could be a residue formed by dissolution of halite in the halite/anhydrite couplets, as was suggested for the Palo Duro Basin evaporite cycles (Fracasso and Hovorka, 1986). The residues formed when the brine became undersaturated with respect to halite after repeated seawater influxes, followed by dissolution of halite until the bottom brine became saturated and prevented further halite dissolution. R. J. Hite (pers. commun., 1991) estimates that as much as 1 m of halite dissolves at the top of each bed.

Gypsum and anhydrite structures in the Paradox basin provide the best record of environmental indicators of all the rock types present. The nodular anhydrite texture found in some cores usually occurs above and/or within a silty dolomite. The silty dolomites, which were probably brought into the basin by periodic floods of seawater from the west and southwest, were deposited on top of gypsum crystals growing on the bottom of the water body. Discrete layers

FIG. 12.—Bioturbated silty dolomite, D.O.E. Gibson Dome No. 1 core, cycle 4, 901 m. Diameter of coin = 19 mm.

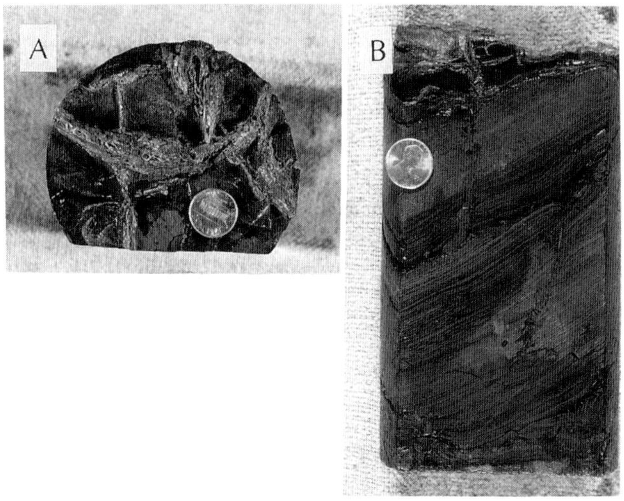

FIG. 13.—Anhydrite and black shale from D.O.E. Gibson Dome No. 1 core, interbed cycle 21–24. (A) Black shale with halite-filled cracks, and (B) disturbed bedding indicating possible mudflat deposit. Diameter of coin = 19 mm.

FIG. 14.—Outcrop photograph of gypsum and shale at Hermosa Creek near Durango, Colorado, showing sharp contacts between beds. G = massive to nodular gypsum, GL = laminated shaley gypsum, and S = gypsiferous shale.

of nodules could have formed from dehydration of bottom-growth selenite crystals when lake salinities increased periodically to the anhydrite stability field. The very well preserved pseudomorphic textures of bottom growth selenite in silty dolomite in the Gibson Dome 1 core is a possible precursor texture for the layered nodular anhydrite (Fig. 12B). This one-centimeter layer of anhydrite after gypsum is only a few centimeters below a halite bed, and it is possible that in this example, the brine did not stay in the stability field of anhydrite long enough to precipitate additional calcium sulfate and the intrasediment growth texture of anhydrite nodules did not develop. Dehydration of selenite to anhydrite probably occured after lithification of the silty dolomite, thereby accounting for the preservation of the selenite crystal morphology.

In the selenite beds that are gradational to halite, replacement of selenite gypsum is by both anhydrite and halite. Selenite occurs as layers of millimeter to centimeter-sized vertically oriented gypsum crystals (now replaced by anhydrite and halite) growing from a common substrate of fine-grained anhydrite in the upper anhydrite bed in the cycle 4 interbed of the Shafer No. 1 core. The grain size of the selenite pseudomorphs gradually increases to centimeter size toward the top of the anhydrite bed. Primary bedded selenite of this type is found throughout the Upper Miocene Solfifera in Sicily (Hardie and Eugster, 1971; Schreiber, 1988), and in the Holocene gypsum salina deposits in south Australia (Warren and Kendall, 1985). The environment of deposition was probably shallow, such that wind-induced mixing could contribute to continued bottom growth of the selenite with progressive evaporation. Selenite bottom growth is observed in shallow water environments such as Lake Assale in the Danakil Depression, Djibouti (Schreiber, 1988), in the shallow lagoon deposits of Marion Lake in South Australia (Dickinson and King, 1951; Crawford, 1965), and in the modern Persian Gulf (Illing and others, 1965). The replacement of the larger selenite crystals near the tops of the gypsum beds by halite was probably syndepositional to

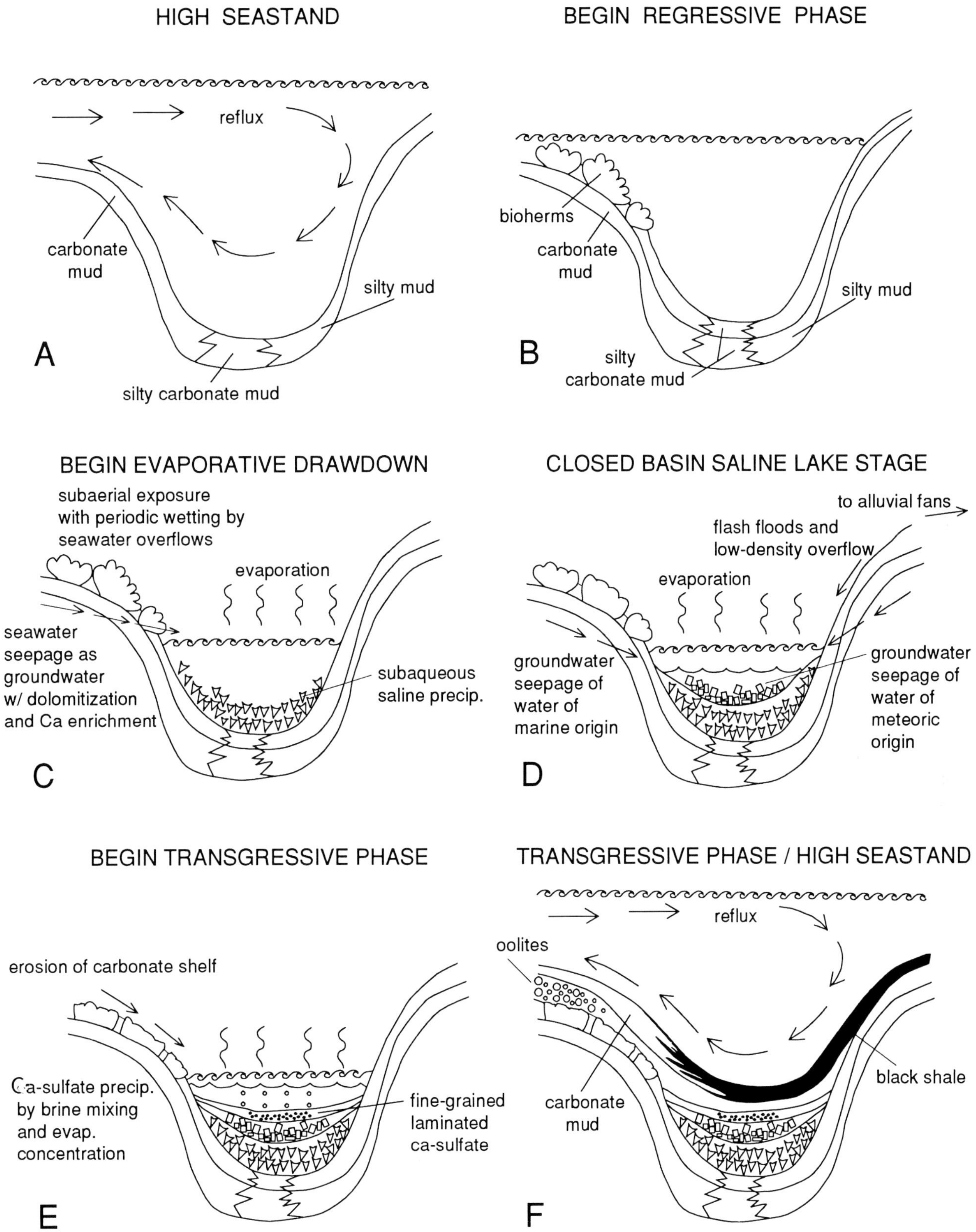

FIG. 15.—Stages in the depositional environment of the Paradox basin, starting with high sea-stand and open communication with the ocean (A), through regressive phase (B), and subsequent evaporative drawdown (C) in the basin to closed-basin, saline lake stage (D). Marine transgression during saline lake stage gradually fills basin with lower salinity water (E) until open communication with ocean is again established with the next high sea-stand (F).

very early diagenetic, and occurred when halite saturated brines that were undersaturated with respect to calcium sulfate reacted with the previously precipitated gypsum.

The halite/anhydrite couplets in the halite beds vary in thickness throughout the halite bed, with thicknesses between anhydrite partings ranging from 20 cm to 1 or 2 cm. The fine-grained anhydrite drapes appear to contribute to the excellent preservation of the bottom-growth crust fabric of vertically elongated halite crystals. Halite crusts have been observed in many other ancient evaporite deposits (e.g., Lowenstein, 1982; Hardie and others, 1983; Lowenstein and Spencer, 1990). The crust texture does not imply any maximum or minimum depth, but their position in the sequence above the selenite bottom growth textures, and the fact that a mixed brine column is required in order to promote bottom growth, suggests that they too were deposited in a shallow brine.

In individual halite beds, the texture commonly grades upward from the primary sedimentary texture of the cumulates to a mosaic texture of anhedral millimeter-sized sylvite and halite crystals. In the potash ore zone of cycle 5 in the Paradox Basin, the sylvite is sandwiched between two halite layers, which in turn are sandwiched by anhydrite laminae. In the Oligocene potash deposits in the Rhine Graben, primary, centimeter-sized chevron halite is found underlying millimeter-sized crystals of sylvite with an equigranular mosaic texture (Lowenstein and Spencer, 1990). Millimeter-sized cubes of halite with primary fluid-inclusion bands occur in the sylvite layers, indicating that this texture is either primary, or that the sylvite was precipitated as a very early cement.

In cycle 13 in the Shafer No. 1 core, a fine grained halite cumulate layer that overlies the anhydrite lamination (Figs. 6A, 8) thins over a large sylvite crystal (Fig. 7B). The cumulates were formed when the evaporation of a perennial brine body resulted in the precipitation of small cubes of halite at the air-brine interface. As the crystals grew and the surface tension of the brine could no longer support them, they would founder and fall to the bottom of the water body. The gradation of textures from mostly halite to sylvite, with individual halite grains floating poikilitically in sylvite suggests an original precipitate of halite crust where void spaces were filled with sylvite cement. Precipitation of halite on top of the sylvite layer resulted from reduction in brine salinity at the surface of the brine body with maintenance of a sylvite-saturated brine at the brine-sediment interface. This is consistent with subaqueous precipitation of salts within a periodically stratified perennial brine body, similar to the interpretation given by Lowenstein and Spencer (1990) for the Rhine Graben. The lack of a well-defined crystal cummulate texture in some of the sylvite layers near the tops of the cycles in the Paradox Basin may have resulted from dissolution and reworking near the beginning of the next marine transgression.

Figure 15 shows the sequence of events that could have lead to cyclic deposition of evaporites in the Paradox basin. After a sea-level drop, the water in the basin no longer has open circulation with the ocean (Fig. 15B). Meteoric water, which was also supplied to the basin during high sea level, would have a more important influence on the brine chemistry during the closed-basin phases. Evaporative drawdown lowers the water level and reduces the areal extent of the brine (Fig. 15D). Precipitation of halite from the air-brine interface results in the accumulation of halite cubes on the sediment-brine interface. Intermittent influxes of seawater into the basin increase in volume and frequency during the transgressive phase, gradually raising the basin brine level until it is the same as sea level, and open communication with the ocean is re-established (Figs. 15E, F).

PERENNIAL SALINE LAKE ENVIRONMENT

Halite and anhydrite are important minerals in both marine and non-marine evaporite deposits, and their presence does not preclude a non-marine origin (see Hardie, 1984, p. 195). Likewise, the thickness of a halite deposit is not necessarily an indicator of marine origin, as shown, for example, by the several thousand meters of halite and anhydrite found in the Dead Sea (Zak and Bentor, 1972). The Paradox basin was probably open to seawater for part of its depositional history. However, it may not be possible to establish a horizontal salinity gradient in an open basin into which seawater flows over a shallow sill, which would result in concentration to halite saturation at the distal end of the basin. It is more likely that complete disconnection from the ocean may be required to reach not only halite, but perhaps also gypsum saturation (Lucia, 1972; Shaw, 1977).

In the closed basin phase of deposition in the Paradox Basin, surface inflow of seawater is cut off, but solutes would be supplied to the basin via meteoric precipitation, river flow and floods, and groundwater. A large portion of the saline brines in the basin could have been marine derived, but seawater percolating through permeable carbonate rocks would have reacted with the rock until it entered the basin as groundwater with a composition which was no longer that of seawater. After sea level dropped below the level of the carbonate shelf forming marginal highs, the rate of inflow into the basin was less than the rate of evaporation and evaporative drawdown resulted. The resulting isolation of the basin from the ocean created a perennial saline lake which persisted until the next rise in sea level. If evaporite deposition occurred during a closed-basin stage accompanied by exposure of the marginal carbonates, then the halite and the carbonate deposition is no longer contemporaneous. Figure 16 is a reinterpretation of the cycle boundaries shown in Hite and Buckner (1981), which accounts for non-deposition of carbonates while evaporites were being precipitated in the basin.

The black shales, which probably represent the extent of the Paradox Sea at the highest level and lowest salinities, grade into red and greenish-black arkosic siltstones and coarser grained clastics toward the major uplift areas (Wengerd and Strickland, 1954; Peterson and Hite, 1969). The presence of pyrite and the high-organic content of the shales suggest sulfate reduction by micro-organisms, which resulted in the development of an anoxic environment in the clays within the evaporite facies. Though the presence of anoxic bottom waters is usually attributed to deep-water environments, black sulfurous muds are found in many modern examples of shallow perennial saline lakes, such as the

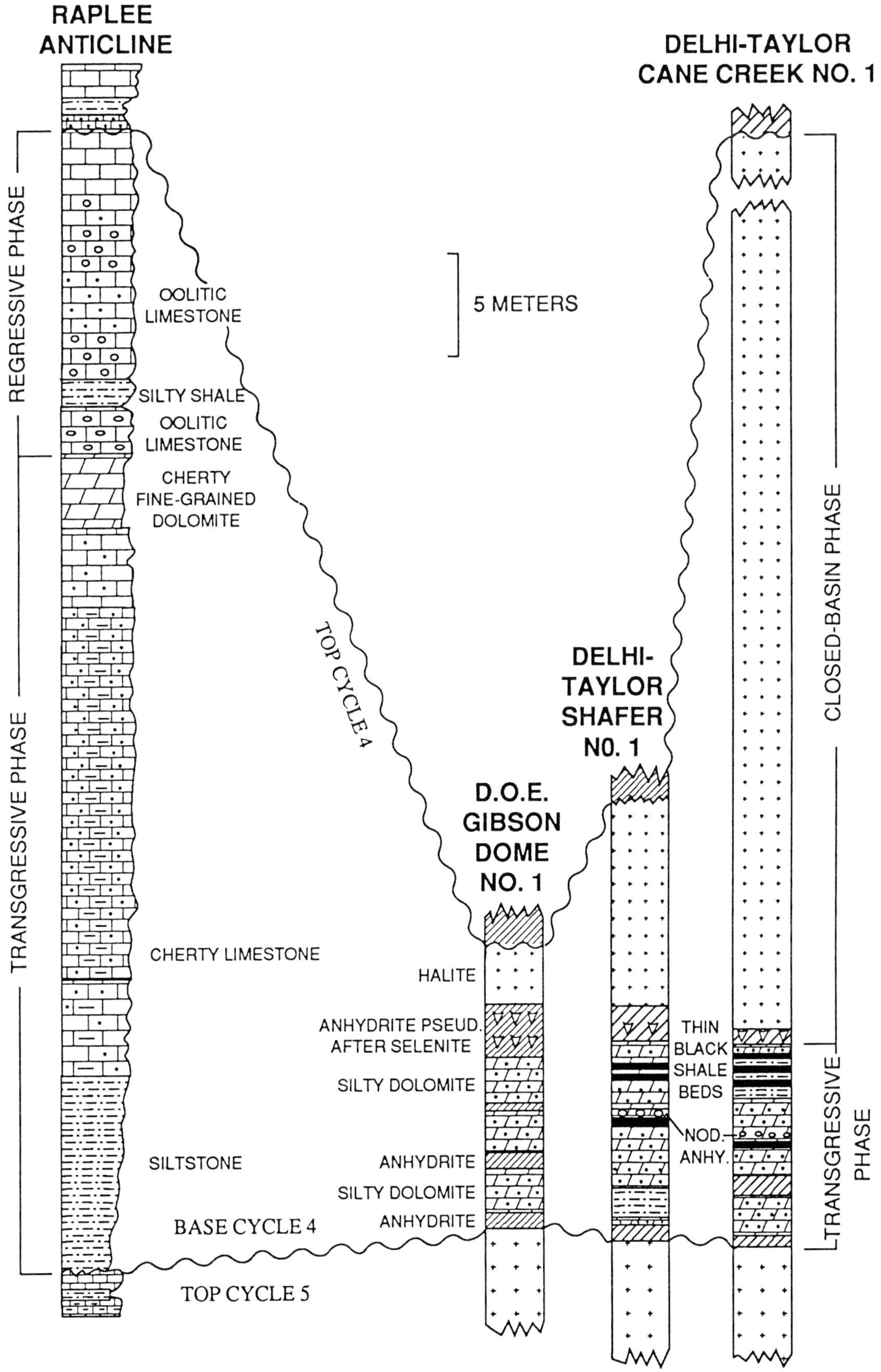

FIG. 16.—Correlation of measured section from Hite and Buckner (1981) and interbed sequences in cycle 4 of Delhi-Taylor Cane Creek No. 1, Shafer No.1 and D.O.E. Gibson Dome No. 1 cores.

Great Salt Lake in Utah (Eardley, 1938; Hardie and others, 1981), indicating that deep water is not required for their development.

The coarser sediments along the northeast margin of the Paradox basin are alluvial fan deposits from the Uncompahgre uplift, a region that represents a major source of meteoric water inflow to the basin. Hite and Buckner (1981) described cores in Salt Valley (northeastern Paradox Basin) containing sandstones within the evaporite facies and suggest that their source could be storms or floods from the Uncompahgre Uplift. The sandstones have graded bedding, poor sorting, small-scale cross laminations and sole marks, with cross laminations accentuated by concentrations of clay. Also found in the sandstones are macerated remains of vascular plants and thin clasts of gray shale. Hite and Buckner (1981) interpreted these sandstones as turbidites. However, studies of modern saline lakes and their subenvironments (Hardie and others, 1981) describe clastic deposition from flash floods on the dry mudflat subenvironment of saline lake basins with many of the same features found in the Salt Valley core.

The effect of the meteoric water input on the brine chemistry was more pronounced during the drawdown phases when salts were precipitated. Estimates for rates of deposition of various lithologies within the evaporite facies (Hite and Buckner, 1981) suggest that the salt precipitating phase occupied the shortest interval of geologic time, on the order of thousands of years for each halite bed. Limiting the analogy to the Paradox Formation only, the depositional environments found in the Great Salt Lake correlate very closely with the Paradox depositional environments. The Wasatch Mountains in Utah serve as the catchment for a plentiful supply of meteoric water into the Great Salt Lake via the Bear, Jordan and Weber Rivers. When the lake level is low, precipitation of halite occurs. Thrombolitic carbonate bioherms are found on the east side of the Great Salt Lake, and surrounding the lake are dry mudflats that originated as lake bottom sediments when the lake level was high. When the lake level is low, the carbonate bioherms and oolitic shoals become subaerially exposed. Periodic flooding into the Great Salt Lake basin deposits clastics and mud layers, and contributes more solutes to the lake. The lake itself is shallow, (maximum depth ~10 m), perennial and stratified. The Paradox Basin was bounded on the northeast by the high relief of the Uncompaghre highlands, and its catchment supplied meteoric water into the basin via the river that produced the Silverton fan delta. Carbonate bioherms formed on the shelf margins when the water level was high, and after evaporative drawdown, the bioherms were exposed and dry mudflats formed from former lake bottom sediments. Silty dolomites indicate a high clastic input into the Paradox Basin, and the depositional features in the halite and gypsum indicate shallow water.

Although the Great Salt Lake contains many environmental features that make it a good process analog for the depositional environment of the Paradox Formation, the Great Salt Lake is more than 1000 km from the sea and the solute composition of its waters would lead to precipitation of minerals that are generally recognized as characteristic of continental evaporite deposits, such as the sodium sulfates, mirabilite ($Na_2SO_4 \cdot 10H_2O$) and thenardite (Na_2SO_4). The potassium-rich composition of the Paradox Formation, the presence of carnallite, and a small amount of magnesium sulfate minerals in at least one cycle, indicate that the Paradox Basin perennial saline lake was marine-influenced.

The sediments of the Danakil Depression in Ethiopia are a modern example of an evaporite deposit that originated from seawater, but which now precipitate minerals which suggest a very strong influence from hydrothermal and meteoric waters (Holwerda and Hutchinson, 1968). Potash-bearing evaporites are found in a depression more than 100 m below sea level, formed by an extensional triple junction between tectonic plates. The depression is separated from the Red Sea by carbonate highlands formed by fault-block tilting from tectonic extension. Seawater percolating through the coastal carbonates to the east and meteoric water runoff from the Balakia Mountains to the west are, at present, the major sources of solutes for the sylvite-precipitating brines found in the basin. Of the three potash-bearing members of the Houston Formation in the Danakil depression, only the intermediate member contains a near normal seawater sequence that includes magnesium sulfate minerals. The sylvite content increases toward the top of the potash interval, and the characteristic marine evaporite minerals are no longer present, indicating an evolution of salt-precipitating brines away from seawater composition. The composition of the hot brine pools in the basin could not have resulted from simple evaporation of seawater, and is probably the result of increasing influence of both meteoric and hydrothermal water inflow over time.

The salt pan of the Danakil Depression only is 45 km from the coast of the Red Sea, and it is very likely that seawater percolating through the rocks as groundwater has a strong influence on the chemistry of the brine pools. The fact that sylvite precipitates from the brines indicates the presence of a magnesium sink, such as is required for the Paradox Basin. In both the Paradox Basin and the Danakil Depression, the magnesium sink was probably carbonate rocks reacting with groundwater flow of seawater origin. In addition to the magnesium sink provided by dolomitization in the Danakil Depression, the inflow from hydrothermal brines strongly affects the brine composition. Such brines are high in Na^+, K^+, Ca^{2+} and Cl^-, but low in Mg^{2+} and SO_4^{2-}, and the influence of these brines leads to the precipitation of magnesium sulfate-poor potash evaporites. There is no indication of hydrothermal activity in the Paradox Basin during Pennsylvanian time, but mass balance calculations indicate that the volume of carbonate rocks that could have been dolomitized as seawater percolated into the basin as groundwater is more than enough to change the Mg/Ca ratio to the point where magnesium sulfate minerals would not precipitate. The predominance of sylvite over carnallite suggests an additional magnesium sink, probably related to refluxing of dense brines that reacted with clays interbedded with the evaporites.

During the perennial saline lake stage of the Paradox Basin, the edge of the brine, or the maximum extent of salt, was a minimum of 150 km from the ocean. The carbonate shelves to the southwest contain numerous exposure horizons in the form of caliche crusts, geopetal sediments, and moldic, solution-enlarged porosity (Goldhammer and others, 1991). The exposure horizons represent periods of non-deposition of carbonates that correlate with the accumulation of salts in the basin. The Pennsylvanian aged sediments in the basins to the north of the Paradox Basin that correlate to the sea-level lowstands during which salt was deposited in the Paradox Basin are primarily eolian sandstone and alluvial plain deposits. The distance of the salt-precipitating brine body in the Paradox Basin from direct seawater inflow in all directions indicates that the basin was probably more strongly influenced by the input of non-marine inflow waters. Using inflow volumes based on rates of river flow in the Great Salt Lake and the volume of halite in the Paradox Basin, estimates were made of the relative volumes of seawater, seawater-derived brines, and meteoric water that entered the Paradox Basin during the perennial saline lake stage. Calculations indicate that inflow of nearly twice as much meteoric water as seawater is required to produce the mineral assemblage found in the Paradox Formation. In addition, the presence of rare-earth element minerals in the Paradox Formation indicate the presence of artesian springs that can provide relatively high salinity input (relative to meteoric water) that is not seawater-derived (Raup, 1972).

CONCLUSIONS

Comparison of two possible modern analogs to the Paradox, the Great Salt Lake in Utah, and the Danakil Depression in Ethiopia, suggests that the Paradox salt was deposited in an environment that contained elements present in both. Chemical and hydrological constraints in the Paradox basin were similar to those found in the Danakil Depression. At present the Danakil depression is disconnected from the ocean, but a substantial sea-level rise would flood the basin again, and a sequence of marine to non-marine evaporite deposition would recommence. Geomorphic similarities between the Great Salt Lake basin and the Paradox Basin are indicated by comparing depositional environments to the facies present in the Paradox Formation.

The cyclic evaporite deposition in the Paradox probably resulted both from climatic cycles associated with sea-level fluctuations and pluvial phases, and from tectonic pulses related to thrusting on the Uncompahgre fault (Hite and others, 1984; Johnson and others, 1991). Between the time when the basin water level was high, evaporative draw-down resulted in reduction of the brine body, and eventual precipitation of evaporite minerals. Though the duration of salt deposition probably represents the smallest amount of time, the rapid rates of precipitation of halite resulted in that lithology having the largest proportional thickness. After the basin was isolated, the influence of altered seawater, which percolated through carbonate barriers, and meteoric water, which entered the basin via rivers and springs, became more important than the seawater that initially filled the basin. It was not necessary for the basin to be deep, as continued syndepositional subsidence of the basin foredeep provided enough space to accommodate thick halite beds. Although there were periods when the Paradox Basin was best described as a normal marine environment, the deposition of the evaporite minerals, beginning possibly with calcium sulfate, occurred during a time when the water body was isolated, shallow, and had brine characteristics which were predominantly non-marine, with the sedimentation controlled by morphologic features similar to those found in the Great Salt Lake and the Danakil Depression today. The Paradox Formation is the result of deposition in a closed evaporite basin where the volume of continental-derived inflow waters exceeds marine-derived inflow waters by nearly two to one. The record is one of a marine-influenced, penecontinental perennial saline lake which existed for thousands of years at a time.

ACKNOWLEDGEMENTS

I would like to thank Bob Hite and Omer Raup of the U.S.G.S. for many helpful discussions on the Paradox Basin. I am grateful to Curt Huffman and Omer Raup for helpful criticisms that greatly improved the original manuscript. The author, however, takes full responsibility for the ideas presented in this paper.

REFERENCES

BRAITSCH, O., 1971, Salt Deposits. Their Origin and Composition: Springer, New York, 297 p.

BORCHERT, H., AND MUIR, R. O., 1964, Salt Deposits: D.Van Nostrand, London, 338 p.

CASAS, E., AND LOWENSTEIN, T. K., 1989, Diagenesis of saline pan halite: comparison of petrographic features of Modern, Quaternary and Permian Halites: Journal of Sedimentary Petrology, v. 59, p. 724–739.

CERCONE, K. R., 1988, Evaporative sea-level drawdown in the Silurian Michigan Basin: Geology, v. 16, p. 387–390.

CLAIR, J. R., 1952, Paleozoic Rocks of the Southern Paradox Basin: Geological Symposium of the Four Corners Region: Durango, Four Corners Geological Society, p. 36–39.

CRAWFORD, A. R., 1965, The geology of Yorke Peninsula: South Australia Geological Survey Bulletin, v. 39, p. 51–52.

DICKINSON, S. B. AND KING, D., 1951, The Stenhouse Bay gypsum deposits: South Australia Mining Review, v. 91, p. 95–113.

EARDLEY, A. J., 1938, Sediments of Great Salt Lake, Utah: American Association of Petroleum Geologists Bulletin, v. 22, p. 1305–1411.

FETZNER, R. W., 1960, Pennsylvanian Paleotectonics of the Colorado Plateau: American Association of Petroleum Geologists Bulletin, v. 44, p. 1371–1413.

FRACASSO, M. A., AND HOVORKA, S. D., 1986, Cyclicity in the middle Permian San Andres Formation, Palo Duro Basin, Texas Panhandle: Austin, Texas Bureau of Economic Geology Report of Investigations 156, 48 p.

FRAHME, C. W., AND VAUGHN, E. B., 1983, Paleozoic geology and seismic stratigraphy of the Northern Uncompahgre Front, Grand County, Utah, *in* Lowell, J. D. and Gries, R., eds., Rocky Mountain Foreland Basins and Uplifts: Denver, Rocky Mountain Association of Geologists, p. 201–211.

GARRETT, D. E., 1970, The chemistry and origin of potash deposits, *in* Rau, J. L, and Dellwig, L. F., eds., Third International Symposium on Salt: Cleveland, Northern Ohio Geological Society, v.1, p. 211–222.

GOLDHAMMER, R. K., OSWALD, E. J., AND DUNN, P. A., 1991, Hierarchy of stratigraphic forcing: Example from Middle Pennsylvanian shelf carbonates of the Paradox Basin, *in* Fransee, E. K., Watney, W. L., Kendall, C. G., and Ross, W., eds., Sedimentary Modeling: Computer Simulations and Methods for Improved Parameter Definition: Lawrence, Kansas Geological Survey Bulletin 233, p. 361–413.

HARDIE, L. A., 1984, Evaporites: Marine or non-marine?: American Journal of Science, v. 284, p. 193–240.

HARDIE, L. A., 1990, The roles of rifting and hydrothermal $CaCl_2$ brines in the origin of potash evaporites: an hypothesis: American Journal of Science, v. 290, p. 43–106.

HARDIE, L. A., AND EUGSTER, H. P., 1971, The depositional environment of marine evaporites, a case for shallow, clastic accumulation: Sedimentology, v. 16, p. 187–220.

HARDIE, L. A., LOWENSTEIN, T. K., AND SPENCER, R. J., 1983, The Problem of distinguishing between primary and secondary features in evaporites, *in* Schreiber, B. C., and Warner, H. L., eds., Sixth International Symposium on Salt: Alexandria, Salt Institute, v. 1, p. 11–39.

HARDIE, L. A., SMOOT, J. P., AND EUGSTER, H. P., 1981, Saline lakes and their deposits: a sedimentological approach, *in* Matter, A., and Tucker, M. E., eds., Modern and Ancient Lake Sediments: Oxford, International Association of Sedimentologists Special Publication 2, Blackwell Scientific Publications, p. 7–41.

HARVIE, C. E., WEARE, J. H., HARDIE, L. A., AND EUGSTER, H. P., 1980, Evaporation of seawater: calculated mineral sequences: Science, v. 208, p. 498–500.

HITE, R. J., 1960, Stratigraphy of the saline facies of the Paradox member of the Hermosa Formation of southeastern Utah and southwestern Colorado, *in* Smith, K. G. ed., Geology of the Paradox Basin Fold and Fault Belt: Durango, Four Corners Geological Society 3rd Field Conference Guidebook, p. 86–89.

HITE, R. J., 1961, Potash-bearing evaporite cycles in the salt anticlines of the Paradox Basin, Colorado and Utah: Washington, D. C., United States Geological Survey Professional Paper 424-D, p. 135–138.

HITE, R. J., 1970, Shelf carbonate sedimentation controlled by salinity in the Paradox Basin, southeast Utah, *in* Third International Symposium on Salt: Cleveland, Northern Ohio Geological Society, p. 48–66.

HITE, R. J., 1975, A suggested origin for the tachyhydrite-producing brines of the Sergipe Basin, Brazil, *in* 27th Congresso Brasileiro de Geologia: Aracaju, v. 2, p. 55–56.

HITE, R. J., 1983, The sulfate problem in marine evaporites, *in* Sixth International Symposium on Salt: Alexandria, Salt Institute, v. 1, p. 217–230.

HITE, R. J., 1985, Preliminary mineralogical and geochemical data from D.O.E. Gibson Dome Corehole no. 1, San Juan County, Utah: Denver, United States Geological Survey Open File Report 83–780, 57 p.

HITE, R. J., ANDERS, D. E., AND GING, T. G., 1984, Organic-rich source rocks of Pennsylvanian age in the Paradox Basin of Utah and Colorado, *in* Woodward, J., Meissner, F. F., and Clayton, J. L., eds., Hydrocarbon Source Rocks of the Greater Rocky Mountain Region: Denver, Rocky Mountain Association of Geologists, 1984 Field Conference, p. 255–274.

HITE, R. J., AND BUCKNER, D. H., 1981, Stratigraphic correlations, facies concepts, and cyclicity in Pennsylvanian rocks of the Paradox Basin, *in* Wiegand, D. L., ed., Geology of the Paradox Basin: Denver, Rocky Mountain Association of Geologists 1981 Field Conference, p. 147–159.

HOLWERDA, J. G., AND HUTCHINSON, R. W., 1968, Potash-bearing evaporites in the Danakil area, Ethiopia: Economic Geology, v. 63, p. 124–150.

ILLING, L. V., WELLS, A. J., AND TAYLOR, J. C. M., 1965, Penecontemporaneous dolomite in the Persian Gulf, *in* Murray, R. C., and Pray, L. C., eds., Dolomitization and Limestone Diagenesis: Tulsa, Society of Economic Paleontologists and Mineralogists Special Publication 13, p. 89–111.

JOHNSON, S. Y., CHAN M. A., AND KONOPKA, E. H., 1991, Pennsylvanian and Early Permian Paleogeography of the Uinta-Piceance Basin region, Northwest Colorado and Northeast Utah: United States Geological Survey Bulletin 17857-CC, 35 p.

KENDALL, A. C., 1987, Depositional model for carbonate-evaporite cyclicity: Middle Pennsylvanian of Paradox Basin (abs.): American Association of Petroleum Geologists Bulletin, v. 71, p. 576.

KENDALL, A. C., 1988, Aspects of evaporite basin stratigraphy, *in* Schreiber, B. C., ed., Evaporites and Hydrocarbons: New York, Columbia University Press, p. 11–65.

LOWENSTEIN, T. J., 1982, Primary features in a potash evaporite deposit, the Permian Salado Formation of west Texas and New Mexico, *in* Handford, C. R., Loucks, R. G., and Davies, G. R., eds., Depositional and Diagenetic Spectra of Evaporites-A Core Workshop: Tulsa, Society of Economic Paleontologists and Mineralogists Core Workshop 3, p. 276–304.

LOWENSTEIN, T. J., AND SPENCER, R. J., 1990, Syndepositional origin of potash evaporites: petrographic and fluid inclusion evidence: American Journal of Science, v. 290, p. 1–42.

LUCIA, F. J., 1972, Recognition of evaporite-carbonate sedimentation, *in* Rigby J. K., and Hamblin, W. K., eds., Recognition of Ancient Sedimentary Environments: Tulsa, Society of Economic Paleontologists and Mineralogists Special Publication 16, p. 160–191.

MALLORY, W. W., 1975, Middle and southern Rocky Mountains, Northern Colorado Plateau, and Eastern Great Basin Region, *in* McKee, E. D., and Crosby, E. J., eds., Paleotectonic Investigations of the Pennsylvanian System in the United States: Washington, D. C., United States Geological Survey Professional Paper 853-N, p. 265–278.

MURRAY, R. C., 1960, Origin of porosity in carbonate rocks: Journal of Sedimentary Petrology, v. 30, p. 59–84.

PETERSON, J. A., AND HITE, R. J., 1969, Pennsylvanian evaporite-carbonate cycles and their relation to petroleum occurrence, southern Rocky Mountains: American Association of Petroleum Geologists Bulletin, v. 53, p. 884–908.

PRAY, L. C., AND WRAY, J. L., 1963, Porous algal facies (Pennsylvanian) Honaker Trail, San Juan Canyon, Utah: Durango, Four Corners Geological Society, 4th Field Conference, p. 204–234.

RAUP, O. B., 1972, Origin of rare-earths in a marine evaporite mineral, *in* Richter-Bernberg, G., ed., Geology of Saline Deposits: UNESCO, p. 103–110.

RAUP, O. B., AND HITE, R. J., 1978, Bromine distribution in marine halite rocks, *in* Dean, W. E., and Schreiber, B. C., eds., Marine Evaporites: Tulsa, Society of Economic Paleontologists and Mineralogists Short Course 4, p. 106–124.

RAUP, O. B., AND HITE, R. J., 1991a, Preliminary stratigraphic and lithologic data from the Delhi-Taylor Oil company, Shafer No. 1 corehole, San Juan County, Utah: Denver, United States Geological Survey Open File Report 91–373, 34 p.

RAUP, O. B., AND HITE, R. J., 1991b, Prelimary lithologic and mineralogical data from the Delhi-Taylor Oil Company, Cane Creek No. 1 corehole, Grand County, Utah: Denver, United States Geological Survey Open File Report 91–324, 21 p.

RAUP, O. B., AND HITE, R. J., 1992, Lithology of evaporite cycles and cycle boundaries in the upper part of the Paradox Formation of the Hermosa Group of Pennsylvanian age in the Paradox Basin, Utah and Colorado: United States Geological Survey Bulletin 2000-B, 37 p.

RAUP, O. B., HITE, R. J., AND GROVES, H. L., 1970, Bromine distribution and paleosalinities from well cuttings, Paradox Basin, Utah and Colorado, *in* Rau, J. L., and Dellwig, L. F., eds., Third Symposium on Salt: Cleveland, Northern Ohio Geological Society, p. 40–47.

SCHREIBER, B. C., 1988, Subaqueous evaporite deposition, *in* Schreiber, B.C., ed., Evaporites and Hydrocarbons: New York, Columbia University Press, p. 182–255.

SHAW, A. B., 1977, A review of some aspects of evaporite deposition: Mountain Geologist, v. 14, p. 1–16.

SHEARMAN, D. J., 1966, Origin of marine evaporites by diagenesis: Institute of Mining and Metallurgy Transactions, v. 379, p. 155–162.

SLOSS, L. L., 1953, The significance of evaporites: Journal of Sedimentary Petrology, v. 23, p. 143–161.

SPOELHOF, R. W., 1976, Pennsylvanian stratigraphy and paleotectonics of the western San Juan Mountains, Colorado, *in* Epis, R., and Weimer, R. J., eds., Studies in Colorado Field geology: Golden, Professional Contributions of Colorado School of Mines 8, p. 159–179.

WARREN, J. K, AND KENDALL, C. St. C., 1985, Comparison of sequences formed in marine sabkha (subaerial) and salina (subaqueous) settings—modern and ancient: American Association of Petroleum Geologists Bulletin, v. 69, p. 1013–1023.

WENGERD, S. A., 1958, Pennsylvanian stratigraphy, southwest Shelf, Paradox Basin, *in* Sanborn, A. F., ed., Guidebook to the Geology of the Paradox Basin: Salt Lake City, Intermountain Association of Petroleum Geologists 9th Annual Field Conference, Intermountain Assosiaiton of Petroleum Geologists, p. 109–134.

WENGERD, S. A., AND STRICKLAND, J. W., 1954, Pennsylvanian Stratigraphy of Paradox Salt Basin, Colorado and Utah: American Association of Petroleum Geologists, v. 38, p. 2157–2199.

ZAK, I., AND BENTOR, Y. K., 1972, Some new data on the salt deposits of the Dead Sea area, Israel, *in* Richter-Bernberg, G. ed., Geology of Saline Deposits: Paris, UNESCO, p. 137–146.

THE SEDIMENTOLOGICAL SIGNIFICANCE OF CHEVRON HALITE: A REAPPRAISAL

DONALD CHIPLEY AND T. KURTIS KYSER

Department of Geological Sciences, University of Saskatchewan Saskatoon, Saskatchewan S7N 0W0, Canada

ABSTRACT: Growth of chevron halites during evaporation of water in saline pans, tidal flats, and in laboratory experiments has been cited as evidence that chevron halite in ancient evaporite deposits also formed from evaporating brines at the earth's surface. Permian halites in the Palo Duro Basin (Texas) contain fluid inclusions that have stable isotopic compositions that may be comparable to evaporated seawater. In contrast, waters from chevron halite fluid inclusions in potash ore of the Devonian Prairie Evaporite Formation (Saskatchewan) have δD values near −100 and $\delta^{18}O$ values near −12‰, which are substantially different from values expected from evaporated seawater. These values are comparable to basinal brines that probably affected evaporites in the subsurface.

The thermometric properties of the fluid inclusions in chevron halites from the central Western Canada Sedimentary Basin indicate that the fluids are Na-K-Ca-Mg-Cl brines, whereas inclusions from the eastern portion of the basin have identical isotopic compositions but thermometric properties of Na-K-Cl brines. These Na-K-Cl brines do not represent the expected composition for simple evaporation of seawater to potash facies.

Chemical and isotopic compositions from fluid inclusions in chevron halites from the Prairie Evaporite Formation indicate that, as well as forming at the surface, chevron halite may form in the subsurface or, less likely, recrystallize without destruction of its delicate texture. Laboratory experiments show that brines will precipitate chevron halites simply by cooling in closed systems. Thus, fluids trapped in evaporites may represent warm basinal waters that dissolved the evaporite and precipitated halite during cooling. Therefore, from microthermometric and stable isotope compositions of fluid inclusion waters in chevron halites, the assumption that chevron texture in halite is unique to surface processes in both non-marine and marine evaporite deposits is not valid in all instances.

INTRODUCTION

Chemical and thermometric analyses of fluid inclusion waters in halite from ancient evaporite deposits have commonly been used to determine the chemical compositions of the ancient oceans and atmosphere by assuming that when chevron halite precipitated, the original brine was trapped in its fluid inclusions (Das and others, 1989; Horita, 1990; Horita and others, 1990; Lowenstein and Spencer, 1990). However, several studies have indicated that evaporite minerals can recrystallize in the subsurface and that chevron halites may not form exclusively by precipitation from evaporating brine at the earth's surface, as commonly thought (Brookins and others, 1980; Chipley and Kyser, 1989). These and other studies clearly indicate that, in contrast to the opinion of some workers (e.g., Hardie, 1991), it is possible to determine the sources of fluids trapped as fluid inclusions in evaporite minerals, either during crystallization or recrystallization, by using the hydrogen and oxygen isotopic compositions of those inclusion fluids.

Precipitation of halite during evaporation of open brine ponds (both salinas and continental saline pans) results in three basic textures: (a) chevron, (b) hopper, and (c) clear halite (Dellwig, 1955; Wardlaw and Schwerdtner, 1966). Chevron halites usually grow upward from the bottom of brine ponds and are recognized by alternating fluid inclusion-rich and -poor bands parallel to the [100] crystallographic planes that produce the upward pointing chevron-texture (Roedder, 1984; Fig. 1). Chevron-texture is attributed to variable growth rates (Kendall, 1984). Similar fluid inclusion textures are found in hopper crystals that form at the surface of evaporating brine (Roedder, 1984). Floating hopper crystals form rafts that eventually sink to the sediment-water interface, forming randomly oriented aggregates (Arthurton, 1973; Kendall, 1984). Clear halite cubes form as floating crystals and grow upward from the bottom of the brine pond (Arthurton, 1973).

FIG. 1.—Photomicrograph of chevron halite from Potash Corporation of Saskatchewan (PCS) Lanigan mine in Saskatchewan. Growth zones parallel to [100] are outlined by fluid inclusion-rich and -poor bands, resulting in the chevron-texture. Scale bar = 100 μm.

PREVIOUS STUDIES

The presence of chevron halite in modern and ancient evaporite deposits has prompted experiments to determine

Sedimentology and Geochemistry of Modern and Ancient Saline Lakes, SEPM Special Publication No. 50

the origin and mechanism of formation of the chevron texture. Artificial evaporation of seawater and synthetic brines produced chevron textures and led to the assumption that chevron halites could form only by evaporation of seawater (Dellwig, 1955; Wardlaw and Schwerdtner, 1966; Arthurton, 1973). Stratified brine experiments, in which concentrated brine was overlain by a dilute brine, also precipitated chevron halite, but only if the lower brine was exposed and evaporated (Sloss, 1969).

Although these experiments showed that chevron halite could form by subaerial evaporation of brine, they did not preclude formation in other environments or by other mechanisms. Raup (1970), for example, showed that mixing $MgCl_2$ and NaCl solutions decreased the solubility of NaCl and resulted in precipitation of chevron halite crystals. The chevron halite crystals formed at the mixing interface between the two solutions, demonstrating that chevron texture could develop by mixing brines in diverse geologic environments, thereby eliminating the necessity for surface evaporation.

The possibility of using salt deposits for storage and disposal of radioactive waste prompted a search for evidence of fluid migration within and through evaporite horizons, and for dissolution and recrystallization of evaporite minerals by these fluids. Hydrostatic measurements in the Palo Duro Basin of Texas showed the potential for fluid migration through the evaporite horizon (Bair and O'Donnell, 1985). This was supported by chemical and stable isotopic analyses of fluids in surrounding aquifers that suggested some fluid migration through the evaporite had probably occurred (Kreitler and others, 1985). Fluids migrate through evaporites along crystal boundaries and fractures (Roedder, 1984; Kreitler and others, 1985). Such fluids in the subsurface environment can dissolve existing chevron halite and recrystallize to form new halite with a texture identical to chevron halite, as discussed below.

Both recent and ancient evaporites provide many examples of chevron and hopper halites that were not formed in an environment where seawater was undergoing evaporation. Large, isolated, halite hopper crystals in muds near the Dead Sea formed either by downward diffusion of continental brine into the mud or by upward diffusion of brine into the soil profile after exposure (Gornitz and Schreiber, 1981). Dunn (1982) observed that chevron halites grew displacively in lime muds and as post-depositional pore-fillings in fossiliferous limestones of the Middle Devonian Dawson Bay Formation of Saskatchewan, Canada. Fluids actively leaking into potash mines at 1-km depth in the Devonian Prairie Evaporite Formation in Saskatchewan commonly form halite stalactites. Microscopic examination reveals a texture identical to typical chevron halite (Fig. 2).

Fluid access to the Prairie Evaporite Formation in the geologic past is shown by leach and collapse structures in the potash ore horizons. Leaches are areas where sylvite (KCl) and carnallite ($KCl \cdot MgCl_2 \cdot 6H_2O$) have been partially or completely removed and replaced by halite (McIntosh and Wardlaw, 1968). Complete dissolution has resulted in collapse of the overlying rocks into the void, forming a breccia with a halite matrix (DeMille and others, 1964). Halite with chevron-texture has not been found in collapse structures, probably due to slow crystallization rates of the halite, as demonstrated by large crystals, some exceeding 30 cm. The vertical extent of collapse structures into the rocks overlying the Prairie Evaporite Formation, indicates that dissolution of the evaporite minerals has occurred repeatedly since the Late Devonian (McTavish and Vigrass, 1987).

FIG. 2.—Photomicrograph of halite from a stalactite from PCS Rocanville mine showing chevron-texture in the halite. Scale bar = 100 μm.

This evidence is sufficient to question the generally accepted theory that chevron halite forms exclusively from evaporating brine at the earth's surface. In this paper, we propose that chevron halite can form in a variety of environments and show that the environment can be identified by measurement of the chemical and hydrogen and oxygen isotopic compositions of the inclusion fluids.

HYDROGEN AND OXYGEN STABLE ISOTOPIC COMPOSITIONS OF SURFICIAL FLUIDS

Numerous waters at and near the earth's surface can be distinguished by their hydrogen and oxygen isotopic ratios (Fig. 3). Seawater (SMOW) is the standard relative to which the hydrogen and oxygen isotopic compositions of waters are measured and by definition has δD and $\delta^{18}O$ values of 0‰. According to laboratory experiments, the isotopic composition of the residual water during evaporation of seawater follows a trajectory that depends on atmospheric conditions such as humidity (Lloyd, 1966; Fig. 3). Precipitation that falls on the earth as rain and snow has isotopic compositions that plot along the meteoric water line (MWL). Near-equatorial precipitation in coastal regions is similar in

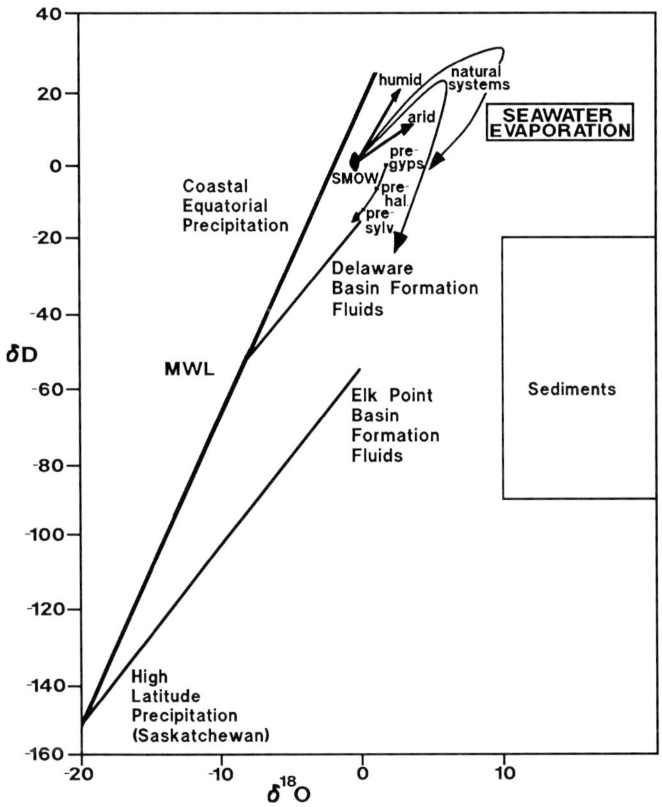

FIG. 3.—δD and $\delta^{18}O$ values of common sediments (Sheppard, 1986) and various surface and near-surface waters that may be associated with the formation and alteration of ancient evaporite deposits. Shown are the stable isotopic compositions of seawater (SMOW), basinal fluids from the Elk Point Basin (Wittrup and Kyser, 1990), and Delaware Basin (O'Neil and others, 1986), the meteoric water line (MWL), and evaporation paths of seawater from natural systems and experimental studies (from Knauth and Beeunas, 1986). The isotopic compositions of evaporating seawater measured in this study from salt crystallizing ponds on Bonaire Island are indicated by pre-gypsum, pre-halite and pre-sylvite brines.

isotopic composition to seawater, whereas precipitation at higher latitudes, higher elevations, and farther inland has lower δD and $\delta^{18}O$ values (Craig, 1961; Friedman and others, 1964). Fluids in many sedimentary basins have variable stable isotopic compositions; near-surface aquifers contain fluids similar in isotopic composition to local modern meteoric water and become more D and ^{18}O-enriched with depth. The isotopic compositions of these fluids commonly define a linear array (Fig. 3). The isotopic compositions of fluids in basins are enigmatic and may result from mixing meteoric water with older, evolved evaporite brines, or seawater, or may result from reaction between meteoric water and minerals (e.g., Clayton and others, 1966; Hitchon and Friedman, 1969). These waters, which may also recrystallize the evaporites to obtain their high salinities, are distinct in their isotopic compositions and, therefore, can be recognized as components of fluids trapped as inclusions during recrystallization of the evaporite minerals.

ANALYTICAL METHODS AND RESULTS

Potash ore samples containing halite were obtained from several potash mines in Saskatchewan. The potash mines are about 1 km below the land surface and located in the Middle Devonian Prairie Evaporite Formation (Wardlaw, 1968). Halite crystals were removed from the hand samples and cleaved to yield thin (<1 mm) slabs for microscopic examination. Groups of fluid inclusions were distinguished by textural differences, chemical compositions, and daughter mineral dissolution temperatures, which were observed using a U.S.G.S.-type fluid inclusion stage. Determination of the chemical compositions of the fluids was by identification of daughter minerals present and melting characteristics of the frozen fluid. Daughter mineral dissolution temperatures were determined by heating the fluid inclusion until sylvite or carnallite daughter minerals dissolved. Extraction of the fluids and analysis of hydrogen and oxygen isotopic compositions was undertaken as outlined by Koehler and others (1991).

Temperatures and Chemical Compositions of Fluids

The potash ore zones of the Prairie Evaporite Formation consist predominantly of halite, sylvite, and carnallite with minor amounts of clay minerals, carbonate, and anhydrite (Wardlaw, 1968). Fluids interacting with chloride minerals in the potash ore zone should reach chemical equilibrium quickly through dissolution (Wittrup and Kyser, 1990), and the resultant fluids trapped in the fluid inclusions should be saturated with respect to the surrounding chloride minerals. Consequently, the temperatures at which sylvite and carnallite daughter minerals in fluid inclusions dissolve should indicate the minimum temperatures of the fluids at the time of trapping.

Daughter mineral dissolution temperatures of fluids determined from fluid inclusions in chevron halites from the Prairie Evaporite Formation are between 16 and 40°C. Although these temperatures are comparable to evaporating seawater, the stable isotopic compositions of the fluid extracted from the fluid inclusions do not resemble evaporated seawater, as will be discussed.

Qualitative chemical compositions of the fluids, determined by freezing point depression, indicate that chevron halites from different localities formed from chemically distinct fluids. Inclusion fluids in chevron halite from potash mines in central Saskatchewan melt between −60 and −45°C, indicative of Na-K-Ca-Mg-Cl brine (Davis and others, 1990) that cannot be distinguished from evaporated seawater. Inclusion fluids in chevron halites from eastern Saskatchewan potash mines melt between −24 and −23°C and are characteristic of Na-K-Cl brines (Davis and others, 1990). The chemical compositions of fluids from the eastern potash mines are very different from those of fluids in chevron halites from central Saskatchewan and do not resemble the chemical composition of seawater evaporated to sylvite facies.

Stable Isotopic Compositions of Inclusion Fluids in Chevron Halite

Fluids extracted from inclusions in chevron halite from the Permian evaporites of the Palo Duro Basin have stable isotopic compositions that may be consistent with formation of the minerals from evaporating seawater (Knauth and

crystals also have chevron-texture cores surrounded by clear halite, and indicate that chevron halite can form on an existing halite substrate when growth rates are relatively fast. Thus, a warm brine may dissolve downward into evaporites, possibly along fractures, creating a depression in which the warm brine can cool in isolation from the flow (Fig. 7). Chevron halite could precipitate due to reaction of the brine with Mg-rich evaporite minerals during dissolution, or during relatively rapid cooling of the isolated brine. Progressively slower cooling of the brine, or cooling after reaction of the brine with Mg-rich minerals, would cause precipitation of clear halite resulting in chevron halite surrounded by clear halite.

CONCLUSIONS

Numerous studies have shown that chevron halites can form in environments other that evaporating seawater or continental meteoric water at the earth's surface. The chemical and stable isotopic compositions of fluids extracted from chevron halite fluid inclusions from ancient marine evaporites do not always resemble evaporated seawater and, therefore, indicate that chevron halite can form in the subsurface environment. This process has been verified by experiments whereby simple cooling of brine and artificial seawater results in the formation of halite crystals with chevron halite cores and clear halite rims—textures identical to those observed in many ancient evaporite deposits. The stable isotopic compositions of inclusion fluids in chevron halites from the Devonian Prairie Evaporite Formation in Saskatchewan can only be explained by subsurface recrystallization during the interaction of formation fluids or modern meteoric waters with the evaporite deposit. Chevron halites may form in many environments, including evaporation of continental or marine brines at the earth's surface and through subsurface recrystallization by meteoric waters or formation fluids long after evaporite deposition. The possibility of subsurface genesis of chevron halite should be considered when interpreting ancient non-marine and marine evaporite sequences.

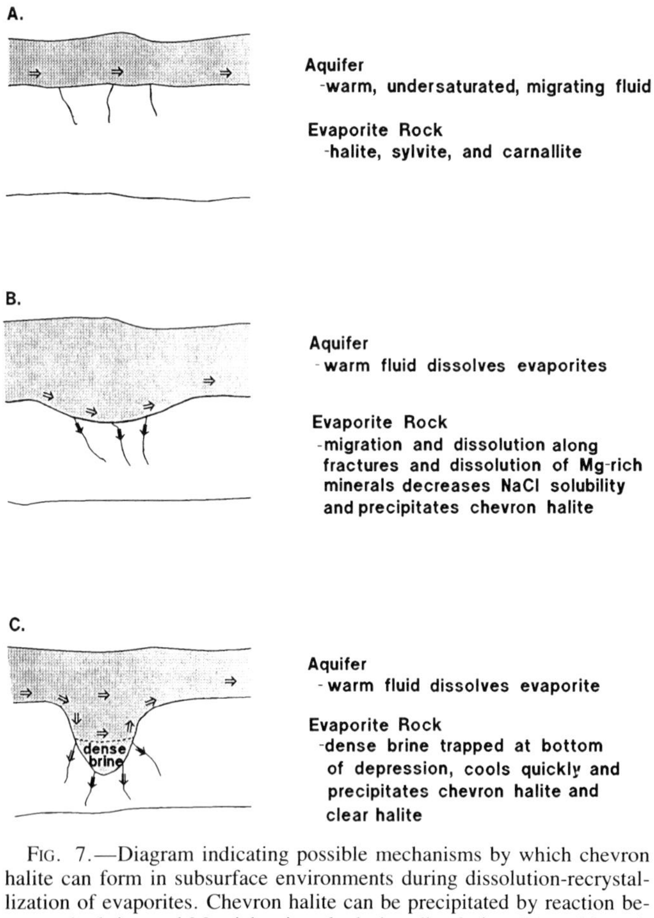

FIG. 7.—Diagram indicating possible mechanisms by which chevron halite can form in subsurface environments during dissolution-recrystallization of evaporites. Chevron halite can be precipitated by reaction between the brine and Mg-rich minerals during dissolution (B) and by relatively rapid cooling of dense isolated brine (C).

ACKNOWLEDGMENTS

The authors would like to thank the member companies of the Saskatchewan Potash Producers Association for continued funding of research on evaporites in Saskatchewan, and NSERC for Equipment, Infrastructure, and Operating grants. Special thanks are extended to Terry Danyluk of the Potash Corporation of Saskatchewan for many helpful discussions and sampling tours, and Robin Renaut for suggesting our participation in the Saline Lakes conference. H. E. Dunsmore of Energy Mines and Resources, Canada, graciously supplied the Bonaire Island brine samples. The authors would also like to thank L. S. Land and E. Roedder for their constructive comments and suggestions.

REFERENCES

ARTHURTON, R. S., 1973, Experimentally produced halite compared with Triassic layered halite-rock from Cheshire, England: Sedimentology, v. 20, p. 145–160.

BAIR, E. S. AND O'DONNELL, T. P., 1985, Hydrodynamics of aquifers and aquitards at proposed high-level nuclear-waste repository sites, Palo Duro basin, Texas, U.S.A.: Washington, D. C., International Association of Hydrogeologists Memoir, v. 17, p. 596–611.

BROOKINS, D. G., REGISTER, J. K., AND KRUEGER, H. W., 1980, Potassium-argon dating of polyhalite in southeastern New Mexico: Geochimica et Cosmochimica Acta, v. 44, p. 635–637.

CHIPLEY, D. AND KYSER, T. K., 1989, Fluid inclusion evidence for the deposition and diagenesis of the Patience Lake Member of the Devonian Prairie Evaporite Formation, Saskatchewan, Canada: Sedimentary Geology, v. 64, p. 265–271.

CLAYTON, R. N., FRIEDMAN, I., GRAF, D. L., MAYEDA, W. F., AND SHIMP, N. F., 1966, The origin of saline formation waters, 1. Isotopic composition: Journal of Geophysical Research, v. 71, p. 3869–3882.

CRAIG, H., 1961, Isotopic variations in meteoric waters: Science, v. 133, p. 1702–1703.

DAS, N., HORITA, J., AND HOLLAND, H. D., 1989, Chemistry of fluid inclusions in halite from the Salina Group of the Michigan Basin: Implications for Late Silurian seawater and the origin of sedimentary brines: Geochimica et Cosmochimica Acta, v. 54, p. 319–327.

DAVIS, D. W., LOWENSTEIN, T. K., AND SPENCER, R. J., 1990, Melting behaviour of fluid inclusions in laboratory-grown halite crystals in the systems $NaCl-H_2O$, $NaCl-KCl-H_2O$, $NaCl-MgCl_2-H_2O$, $NaCl-CaCl_2-H_2O$: Geochimica et Cosmochimica Acta, v. 54, p. 591–601.

DELLWIG, L. F., 1955, Origin of the Salina salt of Michigan: Journal of Sedimentary Petrology, v. 25, p. 83–110.

DEMILLE, G., SCHOULDICE, J. R., AND NELSON, H. W., 1964, Collapse structures related to the evaporites of the Prairie Evaporite Formation, Saskatchewan: Geological Society of America Bulletin, v. 75, p. 307–316.

DUNN, C. E., 1982, Geology of the Middle Devonian Dawson Bay Formation in the Saskatchewan potash mining district: Regina, Saskatchewan Energy and Mines Report 194, 117 p.

FRIEDMAN, I., REDFIELD, A. C., SCHOEN, B., AND HARRIS, J., 1964, The variation of the deuterium content of natural waters in the hydrologic cycle: Reviews of Geophysics, v. 2, p. 177–224.

GORNITZ, V. M. AND SCHREIBER, B. C., 1981, Displacive halite hoppers from the Dead Sea: Some implication for ancient evaporite deposits: Journal of Sedimentary Petrology, v. 51, p. 787–794.

HARDIE, L. A., 1991, On the significance of evaporites: Annual Review of Earth and Planetary Sciences, v. 19, p. 131–168.

HITCHON, B. AND FRIEDMAN, I., 1969, Geochemistry and origin of formation waters in the Western Canada Sedimentary Basin: Stable isotopes of hydrogen and oxygen: Geochimica et Cosmochimica Acta, v. 33, p. 1321–1349.

HORITA, J., 1990, Stable isotope palæoclimatology of brine inclusions in halite: Modelling and application to Searles Lake, California: Geochimica et Cosmochimica Acta, v. 54, p. 2059–2073.

HORITA, J., FRIEDMAN, T. J., LAZAR, B., AND HOLLAND, H. D., 1990, The composition of Permian seawater: Geochimica et Cosmochimica Acta, v. 55, p. 417–432.

KENDALL, A. C., 1984, Evaporites, *in* Walker, R. G., ed., Facies Models (2nd ed.): Toronto, Geoscience Canada Reprint Series 2, p. 259–296.

KNAUTH, L. P. AND BEEUNAS, M. A., 1986, Isotope geochemistry of fluid inclusions in Permian halite with implications for the isotopic history of ocean water and the origin of saline formation waters: Geochimica et Cosmochimica Acta, v. 50, p. 419–433.

KOEHLER, G. D., CHIPLEY, D., AND KYSER, T. K., 1991, Measurement of the hydrogen and oxygen isotopic compositions of concentrated chloride brines and brines from fluid inclusions in halite: Chemical Geology, v. 94, p. 45–54.

KREITLER, C. W., FISHER, R. S., SENGER, R. K., HOVORKA, S. D., AND DUTTON, A. R., 1985, Hydrogeology of an evaporite aquitard: Permian evaporite strata, Palo Duro Basin, Texas, in Hydrogeology of Rocks of Low Permeability: Washington, D. C., International Association of Hydrogeologists Memoir, v. 17, p. 150–168.

LLOYD, R. M., 1966, Oxygen isotope enrichment of seawater by evaporation: Geochimica et Cosmochimica Acta, v. 30, p. 801–814.

LOWENSTEIN, T. K. AND SPENCER, R. J., 1990, Syndepositional origin of potash evaporites: Petrographic and fluid inclusion evidence: American Journal of Science, v. 290, p. 1–42.

MCTAVISH, G. J. AND VIGRASS, L. W., 1987, Salt dissolution and tectonics, south-central Saskatchewan, *in* Carlson, C. G. and Christopher, J. E., eds., Fifth International Williston Basin Symposium: Grand Forks, North Dakota Geological Survey, p. 157–168.

MCINTOSH, R. A. AND WARDLAW, N. C., 1968, Barren halite bodies in the sylvinite mining zone at Esterhazy, Saskatchewan: Canadian Journal of Earth Sciences, v. 5. p. 1221–1240.

O'NEIL, J. R., JOHNSON, C. M., WHITE, L. D., AND ROEDDER, E., 1986, The origin of fluids in the salt beds of the Delaware Basin, New Mexico and Texas: Applied Geochemistry, v. 1, p. 265–271.

RAUP, O. B., 1970, Brine mixing: an additional mechanism for formation of basin evaporites: American Association of Petroleum Geologists Bulletin, v. 54, p. 2246–2259.

ROEDDER, E., 1984, Fluid Inclusions: Washington, D.C., Mineralogical Society of America, Reviews in Mineralogy, v. 12, 644 p.

SHEPPARD, S. M. F., 1986, Characterization and isotopic variations in natural waters, *in* Valley, J. W., Taylor, H. P., and O'Neil, J. R., eds., Stable Isotopes in High Temperature Processes: Washington, D.C., Mineralogical Society of America, Reviews in Mineralogy, v. 16, p. 165–183.

SLOSS, L. L., 1969, Evaporite deposition from layered solutions: American Association of Petroleum Geologists Bulletin, v. 53, p. 776–789.

WARDLAW, N. C., 1968, Carnallite-sylvite relationships in the Middle Devonian Prairie Evaporite Formation, Saskatchewan: Geological Society of America Bulletin, v. 79, p. 1273–1294.

WARDLAW, N. C. AND SCHWERDTNER, W. M., 1966, Halite-anhydrite seasonal layering in the Middle Devonian Evaporite Formation, Saskatchewan, Canada: Geological Society of America Bulletin, v. 77, p. 331–342.

WITTRUP, M. B. AND KYSER, T. K., 1990, The petrogenesis of brines in the Devonian potash deposits of western Canada: Chemical Geology, v. 82, p. 103–128.

RELATIONSHIP OF ORGANIC GEOCHEMISTRY TO SEDIMENTATION UNDER HIGHLY VARIABLE ENVIRONMENTS, LORCA BASIN, SPAIN: PRELIMINARY RESULTS

SAID BENALIOULHAJ[1, 2] B. CHARLOTTE SCHREIBER[1]

[1]*Department of Geology, Queens College (C.U.N.Y.), Flushing, New York 11367*

AND

R. PAUL PHILP[2]

[2]*School of Geology and Geophysics, University of Oklahoma, Norman, Oklahoma 73019*

ABSTRACT: Detailed analysis of a sedimentary sequence of the Lorca basin (Upper Miocene, southeast Spain), employing stratigraphy, sedimentary petrology, paleontology and organic geochemistry, permitted the establishment of sediment-biomarker relationships of deposits formed under highly variable conditions. The conditions of sedimentation in this sequence range from open marine to strongly hypersaline (both marine and non-marine), as well as a number of marked variations in circulation, water-body chemistry and salinity. Most of the section formed under marine conditions, but the upper part of the sequence is the product of increasingly non-marine waters, suggesting that the basin sporadically may have become a non-marine, hypersaline lake. Sedimentologically, many details of water circulation and the increasing amount of non-marine input are not particularly apparent, but the distribution of *n*-alkanes and isoprenoids permit these changes to be evaluated in a more convincing fashion.

INTRODUCTION

Immature organic-rich sequences in evaporitic systems contain biomarkers that have fingerprints characteristic of the original biomass and/or the environmental conditions prevailing at the time of deposition. In many cases, interpretations based on the hydrocarbon distributions may be difficult because of the alterations occurring during diagenesis of the organic matter. Because geologists employ the paradigm that the present is the key to the past, modern evaporitic environments have been used to obtain a better understanding of the processes leading to the accumulation and preservation of organic matter in evaporitic systems (Waples and others 1974; Philp and others, 1978; Philp, 1980; Albaiges and others, 1984; Rowland and others, 1985; Jiamo and others, 1986; Volkman, and others, 1986; Kenig and others, 1990; Barbe and others, 1990; Wakeham and others, 1990). Despite the efforts of the last ten years toward the identification of organic compounds present in recent evaporitic environments, there is still relatively little data which establishes direct links between the micro-organisms and the biomarkers encountered in the different facies of evaporites. On the contrary, geochemical techniques traditionally applied to the study of the organic geochemistry of petroleum and ancient sediments, are more frequently employed in studying the microbiology and biogeochemistry of saline lakes with high biological activity (Javor, 1989).

The work described here is an attempt to further our understanding of the association between organic matter and evaporites. The regulating factors of this association are not always well established, and the alteration of the organic matter in the depositional environment and during its sedimentary evolution only complicates the situation. The case of pristane (Pr) and phytane (Ph) is important from this point of view. A wide range of examples discussing the origin of these compounds and the validity of the use of the relative Pr/Ph ratio as an indicator of depositional conditions in terms of Eh and salinity exists in the literature (Brooks and others, 1969; Powell and McKirdy 1973; Didyk and others, 1978; Risatti and others, 1984; Goosens and others, 1984; Volkman and Maxwell 1986; ten Haven and others, 1985, 1987, 1988; Moldowan and others, 1986; Freeman and others, 1989, 1990; de Leeuw and Sinninghe Damste, 1990).

The study of biomarkers in the Lorca Basin sediments further illustrates the complexity surrounding the pristane and phytane distribution. It has been shown that four organo-facies can be distinguished in the stratigraphic section of the basin based on lithological and sedimentological characteristics and saturated hydrocarbon distributions (notably C_{22}, Pr and Ph). In this paper, we discuss these organo-sedimentary variations in order to reconstruct the depositional history of the variations of the Lorca Basin.

GEOLOGICAL SETTING

The earliest Tertiary deposition within the area of the Lorca basin began toward the upper Lower Miocene and continued sporadically until the Serravallian (Middle Miocene). Most of this section is either covered or has been removed by erosion, except for a few patches around the periphery of the basin (Fig. 1). After a phase of regional tectonism due to major changes in plate motion (Dewey and others, 1989), deposition during the lower part of the Upper Miocene began with shallow-water marine to estuarine deposits on top of a tilted, truncated Middle Miocene (Serravalian) section, and passes upward into the largely open-marine, Upper Miocene section (Tortonian-Messinian) (Fig. 2). Most of this Upper Miocene sequence is composed of fossiliferous, calcareous to siliciclastic turbidites interlayered with calcareous hemipelagic marine marls. In this section, however, there is evidence of sporadic periods of restricted circulation within the basin. These restricted phases are represented by thin diatomites apparently formed during stratification of the water column. Stratification produced a depositional sequence composed of finely laminated diatomite-coccolith oozes (marine species with a low diversity) that originated from high productivity in the photic zone within a basin that had a poorly oxygenated to anoxic bottom. These biogenic sediments are largely laminar, unburrowed, with no benthic faunal remains, and have little or no organic content (on outcrop). Many diatomites within the lower part of the section are silicified to chert. In most

Sedimentology and Geochemistry of Modern and Ancient Saline Lakes, SEPM Special Publication No. 50
 ISBN 1-56576-014-X

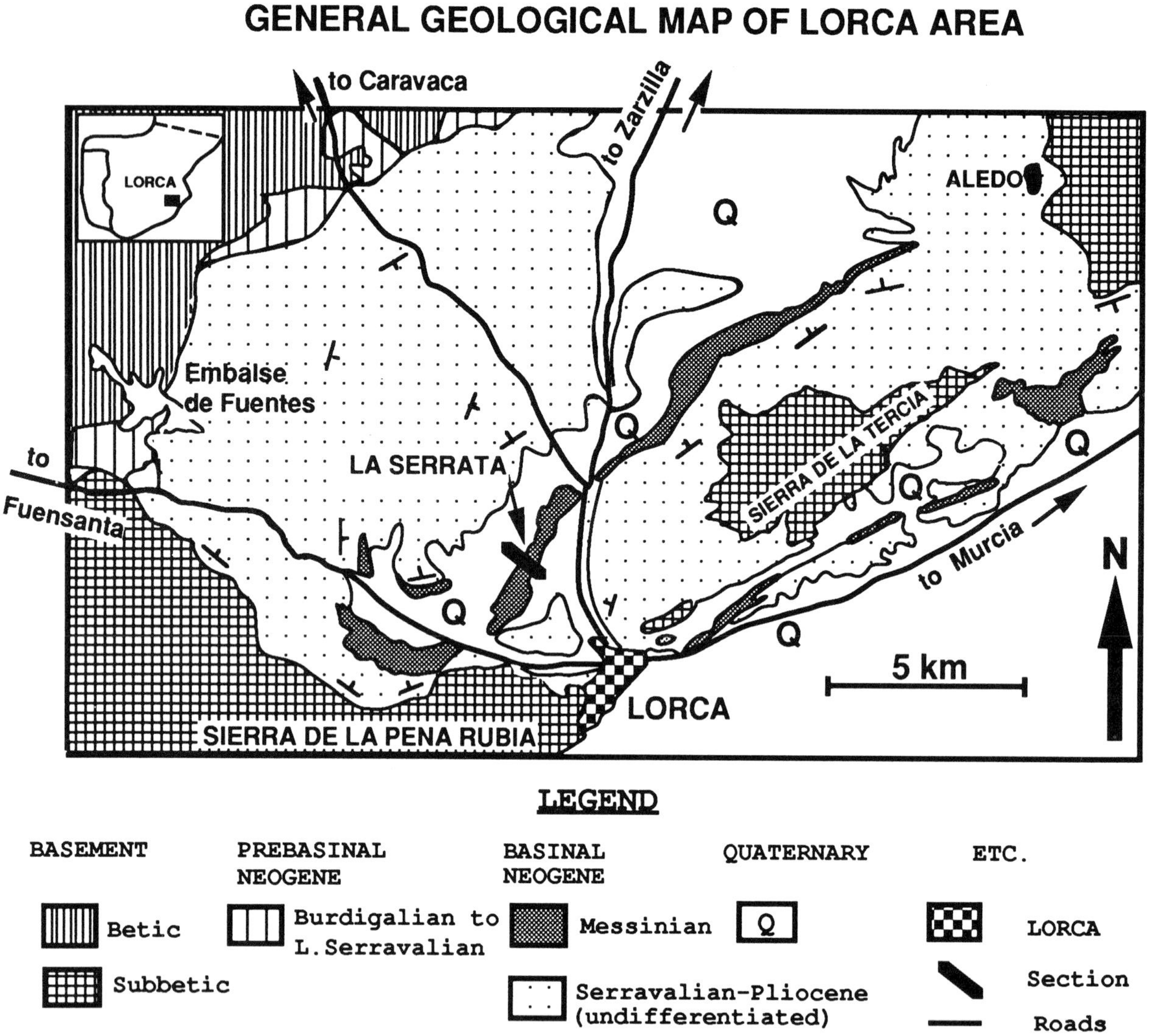

FIG. 1.—Geological map of the region near Lorca, southeastern Spain. Inset shows location map. Adapted from Geel (1976/1978).

examples, the diatomites throughout the column are directly overlain by fossiliferous marls that indicate a return to normal, open conditions with good bottom circulation (i.e., a diverse benthic foraminiferal population).

In several parts of the sequence, the layers of diatomite are much thicker, relatively unaltered, and some are overlain by exceptionally organic-rich, laminar, calcareous, pelitic shales (TOC >25 percent). A few of the diatomite-shale pairs are overlain by somewhat sandy, porous, unfossiliferous limestones with numerous lenticular to nodular vugs, in a few places filled with sulfur, that have a lower organic content than the underlying shales (TOC <2 percent). In thin section, some of the pore-filling cement in the limestone is gypsum together with a coarse calcite mosaic, and the voids are relics of lenticular, rosette and nodular displacive gypsum (formed from evaporative waters, within soft sediments). This specific change in the style of sedimentation, with the grouping of diatomites, organic-rich shales and gypsiferous limestone, was probably produced by varying degrees of basinal restriction and stratification that took place sporadically during the deposition of the sequence. The most restricted component, the gypsiferous limestone, formed only five times in the entire sequence. After each restriction, the basin returned to open marine conditions, but circulation became increasingly limited upsection.

The uppermost part of the section at Lorca is the product of a period of marked restriction in circulation, during which the sporadic influx of continental (meteoric) water became increasingly important. The tangible effects of continental water input toward the top of the section are expressed by the accumulation of several massive beds of unfossiliferous sandstone (tidal to fluvial) containing several thin intercalations of dark-colored, clayey mudstone with well-preserved land plant remains (TOC = 0.31 percent) (Fig. 2, indicated by *). The establishment of evaporative conditions with massive deposition of gypsum and halite only took place during Latest Miocene (Upper Messinian) time,

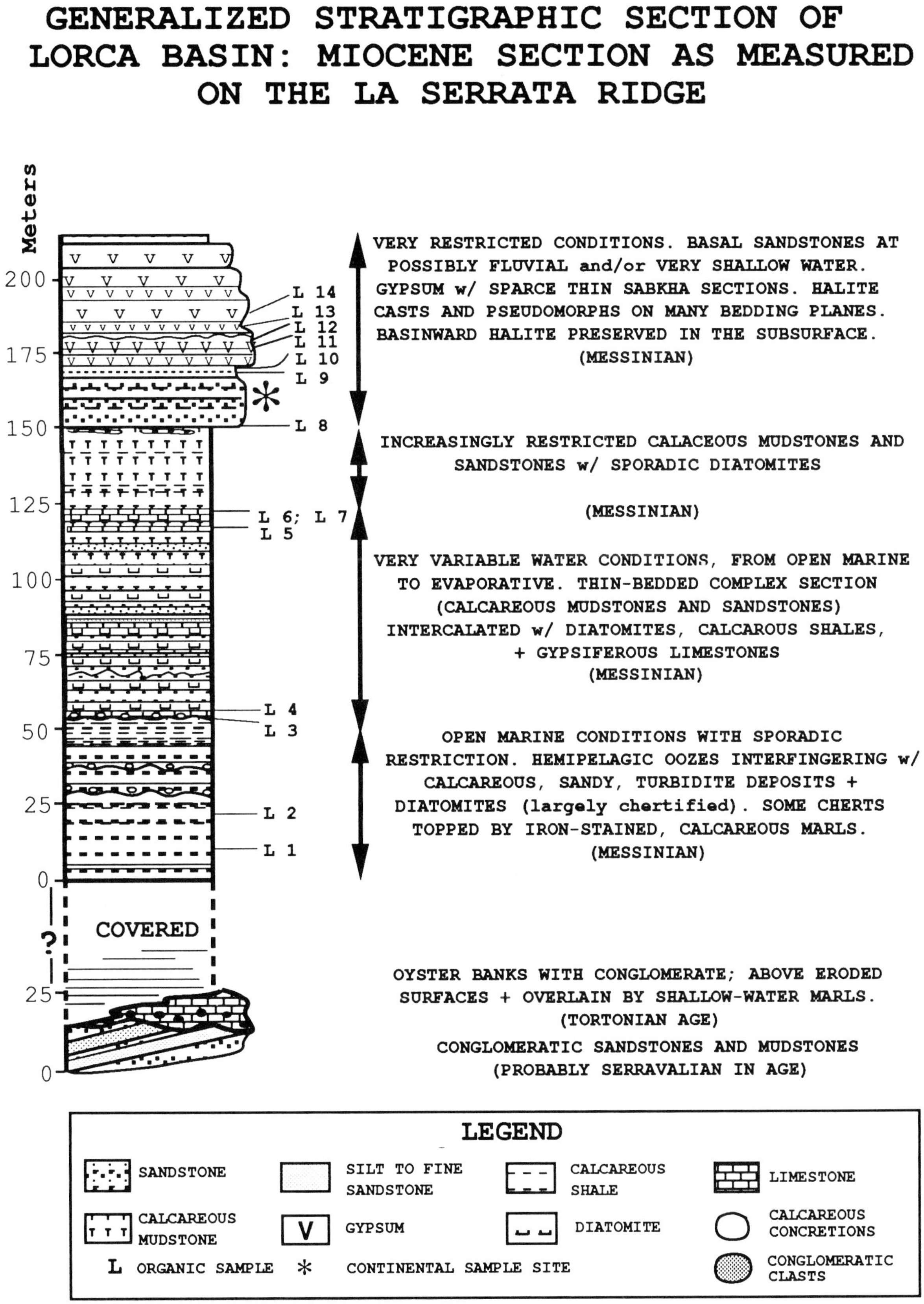

FIG. 2.—Stratigraphic section of the "La Serrata" section (outcrop), Lorca basin, southeastern Spain. Note that lower part of the column is taken from a site 1 km from the Serrata and is incomplete because the section is partially covered by Pleistocene-Recent sands and soils. Age designation for the "Serravalian" (Middle Miocene) sandstones and mudstones is poorly constrained because of a paucity of fossils, and may be Lower Tortonian.

as seen in the upper part of the exposed sections and in boreholes (Ortí, 1990). Based on the morphology of the gypsum and halite, deposition took place within very shallow water under these extreme evaporative conditions, and preservation of organic matter was reduced. The evaporites themselves contain little organic residue (TOC <0.46 percent). Stable isotope studies of this gypsum ($\delta^{34}S$ and $\delta^{18}O$) by Pierre (1982) and Utrilla (1989) strongly suggest that the gypsum section formed from mixed waters (highly evaporated marine to continental). The bromine profile from the basin-center halite section (boreholes) also demonstrates an increasing proportion of continental water (Veigas and others, 1990), and suggests that the basin had become a saline lake toward the top of the section.

BIOMARKERS

Evaporite environments are sites of elevated biological activity, but although conditions are very favorable for the formation, accumulation and the conservation of organic matter, these sediments have rarely been the object of organic geochemical studies to determine the signatures of their biological precursors and diagenetic evolution (cf. Busson, 1979; Rouchy, 1982; Eugster, 1985, Warren, 1986; Evans and Kirkland, 1988; Hite and Anders, 1991). Organic compounds or lipids can be extracted from these sediments through treatment by organic solvents such as dichloromethane. Biomarkers are compounds, generally hydrocarbons, present in geological samples whose structures can be related to functionalized precursor molecules occurring in living systems (Philp and Lewis, 1987). Identification of the biomarkers permits the establishment of the origin or evolution of organic matter from which they originated, together with inferences concerning the depositional environments and extent of the diagenesis of this organic matter. Several studies concerned with biomarker distributions in carbonate and evaporite environments have been reported in the literature (ten Haven and others, 1985, 1988; Connan and others, 1986; de Leeuw and Sinninghe Damste, 1990). To understand the origin of various biomarkers in diverse sediments, we have undertaken a study of the hydrocarbons contained in the lipid extract of the organic matter associated with the diverse sedimentary facies found in the Lorca Basin. We then discuss the facies based on the analyses of their *n*-alkanes and isoprenoid distributions.

Experimental Procedure

Sampling.—

The samples examined in this study were collected from fourteen levels in the La Serrata section that were chosen as representative of diverse lithologies of the sedimentary facies present in the exposed Messinian section of the Lorca basin (Figs. 1 and 2 for sample location, depths, sedimentology and general mineralogy).

Analytical Techniques.—

Total organic carbon (TOC) determinations of the samples were made using a LECO carbon analyzer CR-12. The samples were extracted with a mixture of dichloromethane and methanol (9:1) in a Soxhlet apparatus. After the removal of asphaltene from the extract by *n*-pentane precipitation, the extract was fractionated into saturate, aromatic and polar fractions with thin layer chromatography (TLC: silica gel plates - 0.25 mm thickness). The saturate fraction was analyzed by gas chromatography using a Varian 3300GC and G/C conditions include: spitless injection system; injection temperature of 300°C; oven temperature programmed from 40°C to 130°C at 15°C/min, to 300°C at 4°C/min, and held isothermally for 25 minutes. The instrument was equipped with DB-5 fused-silica capillary column (30 m × 0.25 mm) and helium was used as the carrier gas at a flow rate of 1 ml/min.

RESULTS AND DISCUSSION

Quantitative Evaluation of the Different Fractions from Lipid Extracts

The relative proportions of extractable organic matter (EOM) relative to the total organic carbon and the amount of sample extracted are given in Table 1. These values vary from 5.45 to 28.57 percent and from 3.92 to 9.02 percent, respectively. The values for TOC and EOM were quite variable, possibly due to the diversity of the original carbon input into the Lorca basin, and/or the variability of the physio-chemical depositional and diagenetic conditions, but there is a good correlation between the amount of EOM and TOC values. Such a correlation is characteristic of the presence of immature organic matter in the sediments (Tissot and Welte, 1984). The proportions of the different fractions relative to the quantity of the lipid extract (Table 1) emphasize the relatively high concentrations of polar compounds (64–92 percent) in all samples, which is also characteristic of immature organic matter (Tissot and Welte, 1984). The relative proportions of aromatic hydrocarbons vary from 2.6 to 13.3 percent and predominate over those of the saturate hydrocarbons, which vary from 1.4 to 31.4 percent, characteristic of samples having a marine origin (Tissot and Welte, 1984). Samples L-8, L-9 and L-10 differ from the other samples examined on the basis of their relatively high concentration of non-aromatic hydrocarbons: 26.3, 31.4 and 14.1 percent, respectively. The *n*-alkane distributions of these samples are also dominated by hydrocarbons of continental origin, whereas all the other samples examined have an *n*-alkane distribution characteristic of organic matter derived mainly from marine material with a minor continental influence.

Saturate Hydrocarbons.—

The distribution of the *n*-alkanes and the ratios of Pr/Ph and $R_{22} = 2\ C_{22}/(C_{21} + C_{23})$ permit the samples examined to be divided into distinct types that correspond with four lithologic groupings, and reflect the differences in the origin of the organic matter and the conditions of its deposition (Figs. 3A, B).

Class I.—This type is represented by the organic-rich, calcareous, pelitic shale of samples L-5 and L-6, whose *n*-alkanes are characterized by a bimodal distribution ranging from *n*-C_{15} to *n*-C_{35}, with *n*-C_{22} and *n*-C_{31} as maxima, and with phytane the most abundant component in the chro-

TABLE 1.—SUMMARY OF QUANTITATIVE DATA OBTAINED FROM THE 14 EXTRACTS FROM THE LORCA BASIN

Sample	% Organic	Lipid Extract		Saturates % Total Extract	Aromatics % Total Extract	Polars % Total Extract	Pr/Ph	Pr/nC_{17}	Ph/nC_{18}	$R_{22} = \frac{2C_{22}}{C_{21} + C_{23}}$
		% Rock	% Organic							
L-1	1.88	0.15	7.98	3.4	4.4	92.2	0.64	0.52	0.90	1.85
L-2	<0.10	—	—	—	—	—	—	—	—	—
L-3	0.55	0.03	5.45	5.3	4.5	90.2	0.72	0.41	1.96	5.70
L-4	1.20	0.09	7.50	2.2	10.9	86.9	0.54	0.58	2.62	2.70
L-5	25.02	2.56	10.23	1.4	13.3	85.2	0.15	0.62	7.94	4.78
L-6	27.75	3.92	14.13	1.5	10.2	88.3	0.12	0.68	16.66	6.50
L-7	0.35	0.05	14.28	4.4	7.2	71.1	0.73	0.44	0.54	3.03
L-8	0.31	0.04	12.90	26.3	2.6	64.0	1.50	0.73	0.84	1.45
L-9	0.21	0.06	28.57	31.3	4.6	82.0	—		—	—
L-10	0.32	0.03	9.68	14.1	3.9	99.0	1.10	0.49	0.60	3.00
L-11	0.30	0.02	8.33	2.3	5.7	88.3	0.66	0.59	2.40	1.91
L-12	0.46	0.04	8.26	7.2	4.4	—	0.71	0.79	4.00	2.50
L-13	<0.10	—	—	—	—	—	—	—	—	—
L-14	<0.10	—	—	—	—	—	—	—	—	—

— Represents values too low to be measured.

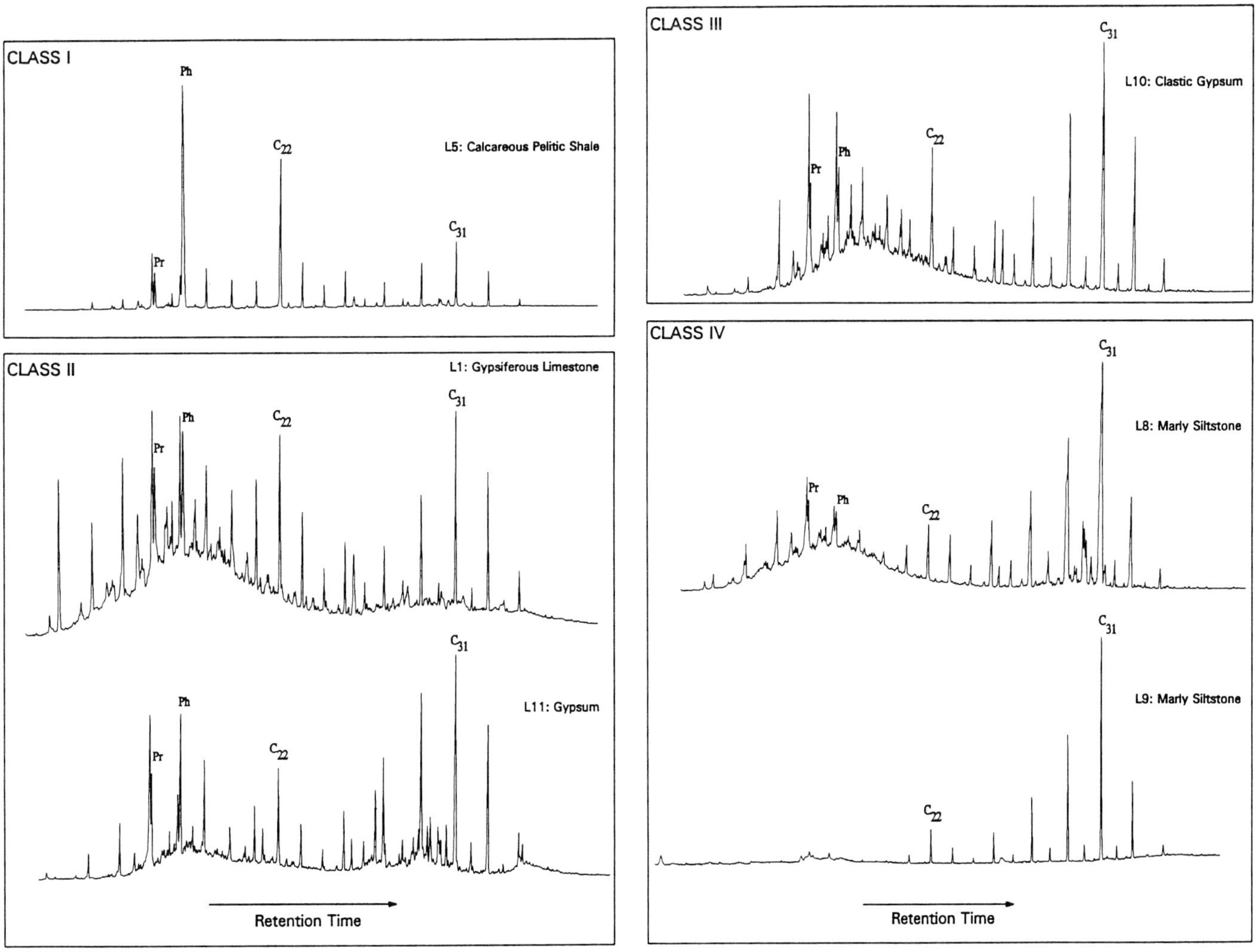

FIG. 3.—Gas chromatograms of saturate hydrocarbons from representative samples taken from the Lorca Basin: (A) Class I (L5) and Class II (L1 and L11) and (B) Class III (L10) and Class IV (L8 and L9).

matograms (Fig. 3A). The strong odd/even predominance of the components, greater than 23, indicates the immaturity of the organic matter as well as the higher plant input into the organic matter (Table 1). The pristane/phytane ratio (Pr/Ph) is very low (less than 0.15) and, based on the traditional interpretation for this ratio, would suggest reducing conditions of deposition (Powell and McKirdy, 1973; Didyk and others, 1978). This is also indicated by the very high amount of organic carbon (greater than 25 percent). However, care needs to be taken in interpreting Pr/Ph ratios in extreme environments because it has been shown that phytane has a number of other origins, such as archæobacteria containing bisphytanyl ethers rather than chlorophyll (ten Haven and others, 1988; de Leeuw and Sinninghe Damste, 1990). Freeman (1989) has also demonstrated different sources for Pr and Ph based on isotopic measurements. Low Pr/Ph ratios have also been found in organic matter from other evaporative sequences (ten Haven and others, 1985, 1988; de Leeuw and Sinninghe Damste, 1990; Hite and Anders, 1991).

The ratios of Pr/n-C_{17}, and particularly the high Ph/n-C_{18} values of 7.94 and 16.66 (Table 1) for the L-5 and L-6 samples respectively, can be considered a consequence of the high pristane and phytane content for these two samples, plus their low levels of maturity. The predominance of n-C_{22} has been interpreted as a marker for hypersaline environments (ten Haven and others, 1988). Therefore, the high values of the $R_{22} = 2\,C_{22}/(C_{21}+C_{23})$ index (Tables 1, 2) also suggest a hypersaline environment for these two samples (4.78 for L-5 and 6.50 for L-6), although care must be taken in the case of samples with low TOC and EOM because n-C_{22} has also been found as a contaminant in plastic containers.

Class II.—Samples L-1, L-3, L-4, L-7, L-11 and L-12 represent the second sample class, composed of gypsiferous limestones and gypsum. They are distinguished from those of Class I mainly by a higher Pr/Ph ratio, ranging between 0.54 and 0.73, and a lower R_{22}, ranging between 1.85 and 5.70 (Tables 1, 2).

The ratios Pr/n-C_{17} and Ph/n-C_{18} range between 0.41 and 0.59, and 0.54 and 4.00, respectively, and are lower than those of Class I (L-5 and L-6, Tables 1, 2 and Fig. 3A). This cannot be interpreted as a difference due to maturity alone, because the geothermal gradient was and is modest, based both on lack of thermally controlled clay diagenesis within the section and on the maturity of the kerogen present in the sediment. The odd/even predominance of the components greater than C_{23}, and the evaluation of the different fractions from the lipid extract, also indicate a low grade of maturity for all the samples. These differences in organic composition may be connected to the conditions of deposition and diagenesis, as well as the possibility of changes in source material, leading to a lower input of archæobacteria thought to be responsible for higher phytane contents in Class I samples.

Class III.—The third type of sample is represented by L-10, a layer of reworked clastic gypsum, and is characterized by a ratio Pr/Ph = 1.1, reflecting oxidizing conditions in the environment (Tables 1, 2; Fig. 3B). However, the index of R_{22}= 3 still reflects hypersaline conditions. Compared to the samples from the two preceding types, L-10 has a distribution of n-alkanes that shows a greater contribution of organic matter from terrestrial origins, which also correlates with the relatively high proportion of saturates, 14.1 per cent, in this sample.

Class IV.—Two samples, of marly siltstones between terrestrial sandstones (L-8 and L-9), represent this class. Sample L-8 is characterized by a high predominance of terrestrial organic matter; the n-alkane distribution of the sample L-9 (Fig. 3B) is unimodal (n-C_{21} - n-C_{33}) with a strong odd-even predominance. Such a distribution indicates an exclusively continental character for the organic matter and its low maturity. The alkanes and isoprenoids in the n-C_{17}—n-C_{20} range are not detectable in sample L-9. Because much of any discussion concerning origin, deposition or diagenesis is based on the presence of pristane, phytanes or n-alkanes, little further interpretation of this sample can be made. L-8 shows a high Pr/Ph ratio of 1.5 and a relatively low R_{22} ratio of 1.45 (Tables 1, 2).

INTERPRETATION

The variations in the origin of organic matter, as well as the conditions of its deposition, are reflected in the great variability of the lithologies present in the stratigraphic section of the Lorca basin (Fig. 2). The organochemical fluctuations observed in the n-alkanes and isoprenoids illustrate environmental groupings that originated under diverse depositional conditions, much in the same way as the sedimentological observations.

Calcareous pelitic shales represented by samples from Class I are characterized by elevated amounts of organic matter (TOC greater than 25 percent). Their sedimentological characteristics indicate deposition in a stratified body of water similar to the models described by Por (1972) and Busson (1978). Stratification within a such a water body resulted in restricted and highly saline bottom waters for the basin where the calcareous, pelitic shales were deposited. This explains their exceptional richness in organic matter and the duality of having the greatest reducing and the high-

TABLE 2.—THE GEOCHEMICAL CHARACTERISTICS FOR THE FOUR CLASSES OF SAMPLES DISCUSSED IN THIS PAPER.

Sediment Type	% Organic Carbon	Pr/Ph	Pr/n-C_{17}	Ph/n-C_{18}	$R_{22} = \frac{2C_{22}}{C_{11} + C_{23}}$	Organic Grouping
calcareous pelitic shale	25–27.75	.12–.15	.62–.68	7.94–16.66	4.78–7.94	CLASS I
gypsiferous limestone and gypsum	<.1–.55	.54–.73	.41–.59	.54–4.0	1.85–5.70	CLASS II
clastic gypsum	.32	1.1	.49	.60	3	CLASS III
marly siltstone	.31	1.5	.73	.84	1.45	CLASS IV

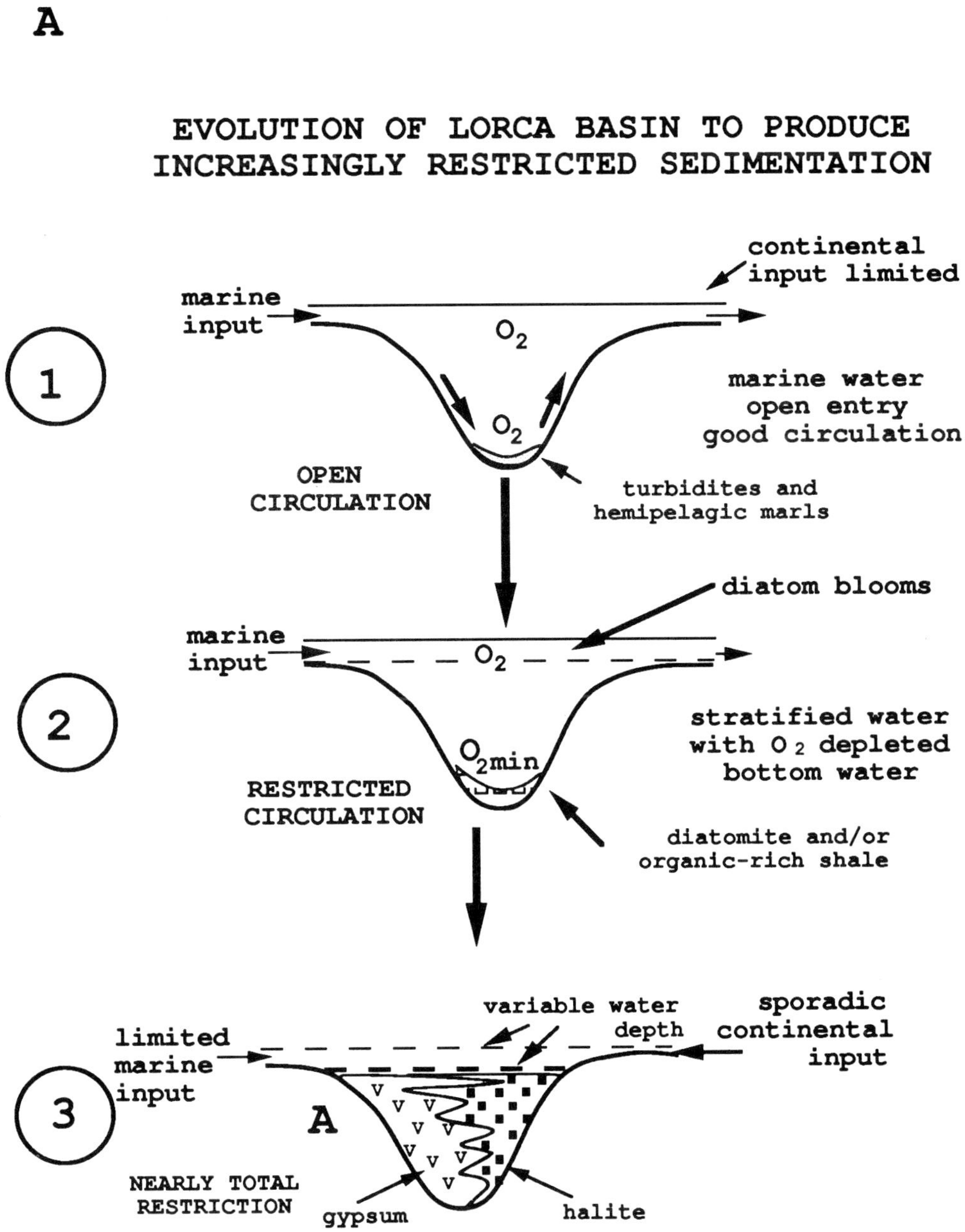

FIG. 4.—Diagrammatic representation of the evolution of the water and related sedimentation in the Lorca basin. (A) General evolution going from open marine to evaporative; (B) specific basinal conditions related to restrictive phases, forming diatomites-organic rich shales-evaporative limestones, that then revert to open-marine conditions.

est salinity conditions in this class of organic facies. This interpretation also is supported by a study of the distribution of the biomarkers. The gypsiferous limestones of Class II correspond to the final phases of restriction, leading to the formation of each of the repetitive organic-rich sequences that contain the calcareous pelitic shale. The physio-chemical conditions are indicated by the presence of gypsum. Formation of gypsum requires neutral or oxidizing conditions that also allow for the efficient preservation of calcareous material (Busson, 1979). This is also supported by the higher Pr/Ph ratio and the lower R_{22} ratio. The waters in which the limestones formed were relatively more oxidizing than those of the underlying organic-rich shale.

The samples of Class IV (L-8 and 9), which come from the marly siltstone facies (Fig. 2), were formed under conditions associated with a high input of terrigenous material. This was associated the development of strongly oxidizing conditions and the accumulation of organic matter of continental origin. The clastic gypsum (Class III) that lies just above this continental sequence, was deposited from reworked and slightly earlier basin-marginal gypsums. This episode is represented by samples from the upper section

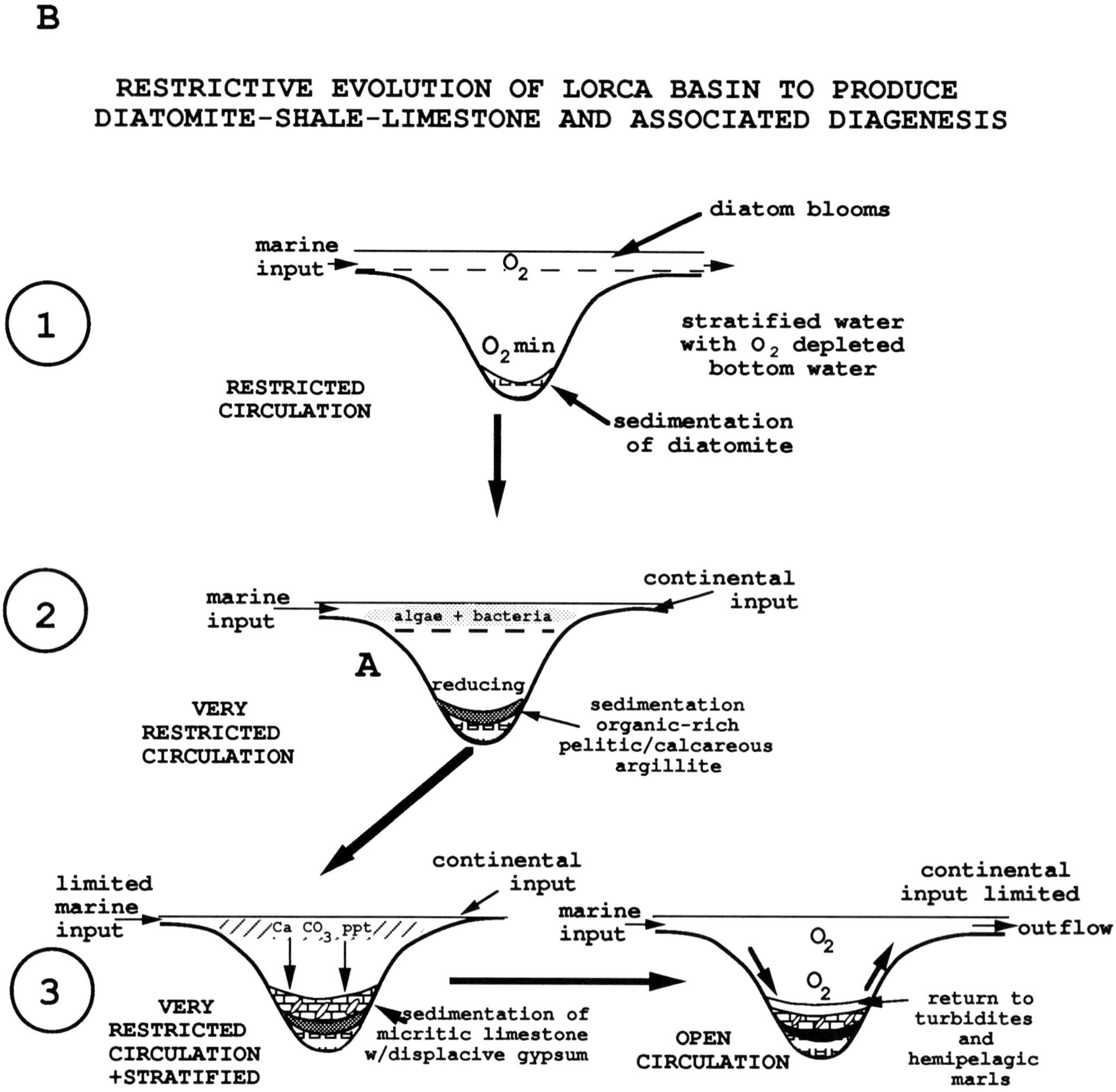

FIG. 4.—Continued

belonging to Class II and was deposited under intermediate Eh conditions between that of Class I, on the one hand, and Classes III and IV on the other hand. Based on the presence of terrestrial organic matter brought in during the deposition of the gypsum, continental influx was lower, but still effective. A reduction in both the marine and continental influx permitted increasing concentration of the brines and led to formation of the evaporitic deposits that cap the section. In the study areas, there was continuing but limited continental influx, mixed with restricted amounts of marine water within the gypsum section that reflects the aridity of the basin and of the surrounding hinterlands. Based on sedimentological features, the gypsum deposits formed in fairly shallow waters with sporadic desiccation. At the basin center, in somewhat deeper water, halite was deposited (Veigas and others, 1990).

The gypsum layers have almost no organic content, suggesting that rapid sedimentation under oxidizing to intermediate conditions allowed the formation of gypsum, and were inhibiting factors for the incorporation and preservation of the organic matter. Comparable modern environments commonly have high organic productivity, but much of the organic matter is not preserved within the gypsum itself.

CONCLUSIONS

After a period of open marine circulation, characterized by the presence of turbidites and hemipelagic marls and the virtual absence of organic matter, periods of episodic restrictive conditions resulted in the stratification of the water body, and led to the formation of repetitive organic-rich sequences of diatomite-calcareous, pelitic shale, and gypsiferous limestone. Near the top of the section, the development of extreme restriction, associated with a strong continental input, resulted in the deposition of several continental beds capped by reworked gypsum. The persistence of marine restriction, associated with a depletion in the continental influx, then led to the concentration of the brines and formation of gypsum within the main basin. The deposition of halite, both within inter-gypsum layers and at the basin center, occurred periodically due to nearly complete desiccation of the brine. This resulted from the aridity of the hinterland as well as a nearly total decoupling from marine influx.

A reconstruction summarizing the different stages of the evolution of the Lorca basin water body, together with the sedimentation and the environmental interrelationships, is shown in Figures 4A and 4B. Based on our analysis, it is evident that during the Upper Miocene a gradual but episodic development of extreme basinal restriction governed sedimentation in the Lorca Basin. The nature of the organic matter and the overprints caused by the actual conditions of deposition, as demonstrated by the saturated hydrocarbons, have allowed us to follow evolution of the basin in parallel to changes in the sediments.

ACKNOWLEDGMENTS

The authors wish to acknowledge the Department of Energy to the Research Foundation of the City University of New York (Grant # DE FG 02–88-ER13961) for support that has permitted this study.

REFERENCES

ALBAIGES, J., ALGABA, J., AND GRIMALT, J., 1984, Extractable and sound neutral lipids in some lacustrine sediments: Organic Geochemistry, v. 6, p. 223–236.

ALBAIGES, J., AND TORRADIAS, J. M., 1974, Significance of the even carbon *n*-paraffin preference of a Spanish crude oil: Nature, v. 250, p. 567–568.

BARBE, A., GRIMALT, J. O., PUEYO, J. Y., AND ALBAIGES, J., 1990, Characterization of model evaporitic environments through the study of lipid components, *in* Advances in Organic Geochemistry, 1989: Organic Geochemistry, v. 16, p. 815–828.

BROOKS, J. O., GOULD, K., AND SMITH, J. W., 1969, Isoprenoid hydrocarbons in coal and petroleum: Nature, v. 222, p. 257–259.

BUSSON, G., 1979, Couches laminées riches en matière organique et précedant les roches salines: les enseignements d'un enchaînement de faciès. Documents des Laboratoires de Géologie Lyon, no. 75, p. 5–18.

CONNAN, J., BOUROULLEC, J., DESSORT, D., AND ALBRECHT, P., 1986, The microbial input into carbonate, anhydrite facies of a sabkha paleoenvironmental from Guatemala: A molecular approach, *in* Advances in Organic Geochemistry, 1985: Organic Geochemistry, v. 10, p. 29–50.

DE LEEUW, J. W. AND SINNINGHE DAMSTE, J. S., 1990, Organic sulfur compounds and other biomarkers as indicators of paleosalinity, *in* Wilson, L. O. and White, K. M., eds., Geochemistry of Sulfur in Fossil Fuels: Washington, D. C., American Chemical Society, p. 417–443.

DEWEY, J. F., HELMAN, M. L., TURCO, E., HUTTON, D. W. H., AND KNOTT, S. D., 1989, Kinematics of the western Mediterranean, *in* Coward, M. P., Dietrich, D., and Park, R. G., eds., Alpine Tectonics: Oxford, Blackwell, Osney Mead, Geological Society Special Publication 45, p. 265–283.

DIDYK, B. M., SIMONEIT, B. R. T., BRASSELL, S. C., AND EGLINTON, G., 1978, Organic geochemical indicators of paleoenvironmental conditions of sedimentation: Nature, v. 272, p. 216–222.

EUGSTER, H. P., 1985, Oil shales, evaporites, and ore deposits: Geochimica et Cosmochimica Acta, v. 49, p. 619–635.

EVANS, R., AND KIRKLAND, D. W., 1988, Evaporitic environments as a source for petroleum, *in* Schreiber, B. C., ed., Evaporites and Hydrocarbons: New York, Columbia University Press, p. 256–299.

FREEMAN, K. H., 1989, Distinct origins of pristane and phytane: isotopic evidence: Geological Society of America Abstracts with Programs, v. 21, p. A10.

FREEMAN, K. H., HAYES, J. M., TRENDEL, J. M,. AND ALBRECHT, P., 1990, Evidence from carbon isotope measurements for diverse origins of sedimentary hydrocarbons: Nature, v. 343, p. 254–256.

GEEL, T., 1976/1978, Messinian gypsiferous deposits of the Lorca basin (province of Murcia, SE Spain): Memorie Società Geologica Italiana, v. 16, p. 369–385.

GOSSENS, H., DE LEEUW, J. W., SCHENCK, P. A., AND BRASSELL, S. C., 1984, Tocopherols as likely precursors of pristane in ancient sediments and crude oils: Nature, v. 312, p. 440–442.

HITE, R. J., AND ANDERS, D. E., 1991, Petroleum and evaporites, *in* Melvin, J. L., ed., Evaporites, Petroleum and Mineral Resources: Amsterdam, Elsevier, p. 349–411.

JAVOR, B., 1989, Hypersaline Environments: Microbiology and Biogeochemistry: Berlin, Springer, 328 p.

JIAMO, F., GUOYING, S., PINGAN, P., BRASSELL, S. C., EGLINTON, G., AND JIGANG, J., 1986, Peculiarities of salt lake sediments as potential source rocks in China, *in* Advances in Organic Geochemistry, 1985: Organic Geochemistry, v. 10, p. 119–126.

KENIG, F., HUC, A. Y., PURSER, B., AND OUDIN, J. -L., 1990, Sedimentation, distribution and diagenesis of organic matter in a recent carbonate environment, Abu Dhabi, U.A.E, *in* Advances in Organic Geochemistry, 1989: Organic Geochemistry, v. 16, p. 735–747.

MOLDOWAN, J. M., SUNDARAMAN, P., AND SCHOELL, M., 1986, Sensitivity of biomarker properties to depositional environment and/or source input in the Lower Toarcian of SW Germany, *in* Advances in Organic Geochemistry, 1985: Organic Geochemistry, v. 10, p. 915–926.

ORTI, F., 1990, Yesos di Lorca (Messinienese), *in* Ortí Cabo, F., and Salvany Duran, J. M., eds., Formaciones Evaporíticas de la Cuenca del Ebro y Cadenas Periféricas, y de la Zona de Levante: Barcelona, ENRESA, Departamente de Geoquimica Petrologia i Prospecció Geologia, Universidad Barcelona, p. 297–298.

PHILP, R. P., 1980, Studies of recent algal mats, coal shales and Precambrian sediments, *in* Prashnoushy, A., ed., Alfred Treibs Symposium: Wurzburg, B. J. M. Universitat, p. 159–169.

PHILP, R. P., BROWN, S., CALVIN, M., BRASSEL, S., AND EGLINTON, G., 1978, Hydrocarbons and fatty acid distributions in recently deposited algal mats at Lajuna Guerrero; Baja California, *in* Krumbein, W. E., ed., Environmental Biogeochemistry and Geomicrobiology: Ann Arbor, University of Michigan, p. 255–270.

PHILP, R. P., AND LEWIS, C. A., 1987, Organic geochemistry of biomarkers: Annual Reviews of the Earth and Planetary Sciences, v. 15, p. 363–395.

PIERRE, C., 1982, Teneurs en isotopes stables ($\delta^{18}0$, $\delta^{2}H$, $\delta^{34}S$) et conditions de genèse des évaporites marines: application à quelques milieux actuels et au Messinien de la Méditerranée: Thése de doctorat d'Etat, Université Paris-Sud, Orsay, 222 p.

POR, F. D., 1972, Hydrological notes on the high-salinity waters of the Sinai Peninsula: Marine Biology, v. 14, p. 11–119.

POWELL, T. G., AND MCKIRDY, D. M., 1973, Relationship between ratio of pristane to phytane, crude oil composition and geological environment in Australia: Nature, v. 243, p. 37–39.

RISATTI, J. B., ROWLAND, S. J., YON, D. A., AND MAXWELL, J. R., 1984, Stereochemical studies of acylic isoprenoids—XII: Lipids of methanogenic bacteria and possible distributions to sediments, *in* Advances in Organic Geochemistry, 1983: Organic Geochemistry, v. 6, p. 93–104.

ROUCHY, J. M., 1982, La genèse des évaporites messiniennes de Méditerranée: Memoire Museum Natural Histoire Paris (C), v.2, 267 p.

ROWLAND, S. J., YON, D. A., LEWIS, C. A., AND MAXWELL, J. R., 1985, Occurrence of 2,6,10-trimethyl-7-(B-methylbutyl)-dodecane and related hydrocarbons in the green alga *Enteromorpha prolifera* and sediments: Organic Geochemistry, v. 8, 207–213.

TEN HAVEN, H. L., DE LEEUW, J. W., AND SCHENK, P. A, 1985, Organic geochemical studies of a Messinian evaporite basin, Northern Apennines (Italy). 1. Hydrocarbon biological markers for a hypersaline environment: Geochimica et Cosmochimica Acta, v. 49, p. 2181–2191.

TEN HAVEN, H. L., DE LEEUW, J. W., RULLKOTTER, J., AND SINNINGHE DAMSTE, J. S., 1987, Restricted utility of the pristane/phytane ratio as a palæoenvironmental indicator: Nature, v. 330, p. 641–643.

TEN HAVEN, H. L., DE LEEUW, J. W., SINNINGHE DAMSTE, J. S., SCHENK, P. A., PALMER, S. E., AND ZUMBERGE, J. E., 1988, Application of biological markers in the recognition of paleohypersaline environments, *in* Fleet, A. J. U., Kelts, K., and Talbot, M. R., eds., Lacustrine Petroleum Source Rocks: Oxford, Blackwell, Geological Society Publication 40, p. 123–130.

TISSOT, J. B., AND WELTE, D. H., 1984, Petroleum Formation and Occurrence (2nd edition): Berlin, Springer Verlag, 538 p.

UTRILLA, R., 1989, Les composicions isotòpiques ($\delta^{18}O$, $\delta^{34}S$) del sulfat, com a indicadores de l'origen de les evaporites del Mesozoic i del Cenozoic de la Peninsula Ibèrica i les Illes Balears: Unpublished Ph. D. Dissertation. Universidad de Barcelona, Barcelona, 276 p.

VEIGAS, J. G., ROSELL, L., UTRILLA, R., AND ORTI, F., 1990, Aportaciones geoquímicas al conocimiento de las evaporitas messinienes: cuencas de Palma (Mallorca), San Miguel de Sallnas (Alicante) y Lorca (Murcia), *in* Ortí Cabo, F., and Salvany Duran, J. M., eds., Formaciones Evaporíticas de la Cuenca del Ebro y Cadenas Periféricas, e de Zona de Levante: Barcelona, ENRESA, Departamente de Geoquimica Petrologia i Prospecció Geologia, Universidad Barcelona, p. 257–266.

VOLKMAN, J. K., ALLEN, D. I., PHILIP, L. S., AND BURTON, H. R., 1986, Bacterial and algal hydrocarbons in sediments from a saline Antarctic lake, Ace Lake, *in* Advances in Organic Geochemistry, 1985: Organic Geochemistry, v. 10, p. 671–681.

VOLKMAN, J. K., AND MAXWELL, J. R., 1986, Acyclic isoprenoids as biological markers, *in* Johns, R. B., ed., Biological Markers in the Sedimentary Record: Amsterdam, Elsevier, p. 1–42.

WAKEHAM, S. G., AND CANVEL, E. A., 1990, Fatty acids and sterols of particulate matter in brackish and seasonally anoxic coastal salt pond, in Advances in Organic Geochemistry, 1989: Organic Geochemistry, v. 16, p. 703–713.

WAPLES, D. W., HAUG, P., AND WELTE, D. H., 1974, Occurrences of a regular C_{25} isoprenoid hydrocarbon in Tertiary sediments representing a lagoonal, saline environment: Geochimica et Cosmochimica Acta, v. 38, p. 381–387.

WARREN, J. K., 1986, Shallow-water evaporitic environments and their source-rock potential: Journal of Sedimentary Petrology, v. 56, p. 442—454.

SUBJECT INDEX

SUBJECT INDEX

Sedimentology and Geochemistry of Modern and Ancient Saline Lakes, SEPM Special Publication No. 50